Tight Junctions

Tight Junctions

Edited by Gwen Johnston

SYRAWOOD
PUBLISHING HOUSE
New York

Published by Syrawood Publishing House,
750 Third Avenue, 9th Floor,
New York, NY 10017, USA
www.syrawoodpublishinghouse.com

Tight Junctions
Edited by Gwen Johnston

International Standard Book Number: 978-1-64740-365-2 (Hardback)

Cataloging-in-publication Data

Tight junctions / edited by Gwen Johnston.
p. cm.
Includes bibliographical references and index.
ISBN 978-1-64740-365-2
1. Tight junctions (Cell biology). 2. Cell junctions. 3. Cytology. I. Johnston, Gwen.
QH603.C4 T54 2023
571.6--dc23

TABLE OF CONTENTS

PREFACE

Tight junctions are multiprotein junctional complexes whose canonical function includes the prevention of leakage of seals, water and solutes among the epithelial cells. These junctions can also act as leaky pathways for small cations, water or anions by establishing selective channels. Tight junctions are mostly found in vertebrates. Tight junction proteins (TJ proteins) are molecules found in the tight junctions of myelinated, epithelial, and endothelial cells. This multiprotein junctional complex has a regulating function in the passage of solutes, ions and water by the paracellular pathway. It may also help in coordinating the movement of proteins and lipids among the basolateral and apical regions of the plasma membrane. It conducts signaling molecules, which effect polarity, differentiation, and proliferation of cells. Hence, tight junctions play an important role in maintaining trans-cellular transport and osmotic balance of tissue-specific molecules. This book contains some path-breaking studies on the role of tight junctions in cell biology. It is an upcoming area of study that has undergone rapid development over the past few decades. The extensive content of this book provides the readers with a thorough understanding of tight junction.

This book is a comprehensive compilation of works of different researchers from varied parts of the world. It includes valuable experiences of the researchers with the sole objective of providing the readers (learners) with a proper knowledge of the concerned field. This book will be beneficial in evoking inspiration and enhancing the knowledge of the interested readers.

In the end, I would like to extend my heartiest thanks to the authors who worked with great determination on their chapters. I also appreciate the publisher's support in the course of the book. I would also like to deeply acknowledge my family who stood by me as a source of inspiration during the project.

Editor

1

Claudin-2: Roles beyond Permeability Functions

Shruthi Venugopal, Shaista Anwer and Katalin Szászi *

Keenan Research Centre for Biomedical Science of the St. Michael's Hospital and Department of Surgery, University of Toronto, Toronto, ON M5B 1W8, Canada; shruti.venugopal@gmail.com (S.V.); shaista.anwer@mail.utoronto.ca (S.A.)

* Correspondence: katalin.szaszi@unityhealth.to.

Abstract: Claudin-2 is expressed in the tight junctions of leaky epithelia, where it forms cation-selective and water permeable paracellular channels. Its abundance is under fine control by a complex signaling network that affects both its synthesis and turnover in response to various environmental inputs. Claudin-2 expression is dysregulated in many pathologies including cancer, inflammation, and fibrosis. Claudin-2 has a key role in energy-efficient ion and water transport in the proximal tubules of the kidneys and in the gut. Importantly, strong evidence now also supports a role for this protein as a modulator of vital cellular events relevant to diseases. Signaling pathways that are overactivated in diseases can alter claudin-2 expression, and a good correlation exists between disease stage and claudin-2 abundance. Further, loss- and gain-of-function studies showed that primary changes in claudin-2 expression impact vital cellular processes such as proliferation, migration, and cell fate determination. These effects appear to be mediated by alterations in key signaling pathways. The specific mechanisms linking claudin-2 to these changes remain poorly understood, but adapters binding to the intracellular portion of claudin-2 may play a key role. Thus, dysregulation of claudin-2 may contribute to the generation, maintenance, and/or progression of diseases through both permeability-dependent and -independent mechanisms. The aim of this review is to provide an overview of the properties, regulation, and functions of claudin-2, with a special emphasis on its signal-modulating effects and possible role in diseases.

Keywords: claudin-2; tight junctions; epithelium; cancer; inflammation; fibrosis; paracellular permeability; proliferation; migration

---

1. Introduction

Claudin-2 and the Claudin Family of Tight Junction Proteins

Epithelial cells maintain tissue homeostasis by forming a protective layer to separate the internal milieu from the outside. The epithelium mediates highly regulated directional transport and thus controls the exchange of water and solutes. Tight junctions (TJ) are found at the apical side of the junctional complexes that connect epithelial cells. Their primary role is to generate cell polarity (referred to as fence function), control paracellular transport (gate function), and provide signaling input for a wide variety of cellular events (reviewed in [1–3]). The TJs are comprised of a multitude of transmembrane and intracellular proteins. The claudin family of TJ membrane proteins consists of 26 (human) or 27 (rodents) small (20-25 kDa) tetraspan proteins expressed in a tissue-specific manner in epithelial and endothelial cells [4–6]. The number of claudins varies between species, and in some species e.g., fish there are a much higher number of claudins, probably due to a larger need for osmoregulation. In this review, we only focus on the mammalian claudin-2. Tsukita and colleagues found that the mouse and human orthologs mostly clustered together, except for claudin-13, which is absent in humans [6–8].

Claudins form the backbone of the TJ strand and determine junctional permeability. Based on permeability properties, members of the claudin family can be classified into channel forming and sealing proteins [2,4,9,10]. The composition of the TJ strand, i.e., the types and ratio of claudins expressed, determines the permeability of the TJs, which can be altered by selectively replacing specific claudins [11].

Claudin-2 was one of the first claudins identified by Tsukita and co-workers as a member of a new class of TJ transmembrane proteins that share tetraspan topology with occludin but have no homology with it [12,13]. Subsequent studies revealed that claudin-2 has a high degree of homology with claudins 1–10, 14, 15, 17 and 19, a group referred to as classic claudins [10]. The mouse and human orthologs of claudin-2 share 70% sequence homology, while their promoters possess a homology of 84% [14,15]. Claudin-2 is a cation-selective channel forming TJ protein expressed in leaky epithelia [16–19]. In this review, we will refer to paracellular channel formation by claudin-2 as its *permeability function*. Importantly, expression of claudin-2 is dynamically modulated by a variety of conditions, and compelling evidence now indicates that altered claudin-2 expression affects vital biological processes, such as proliferation, migration and cell fate decision. These effects cannot be explained by the permeability function of claudin-2, and they appear to be mediated by specific signaling pathways and transcription factors. This newly emerging, incompletely understood role of claudin-2 will be referred to as its signal-modulating function and will be the focus of this review. In fact, such a novel role is in line with the emerging non-classical functions described for several other claudins [20–22].

We will first provide an overview of the properties of claudin-2, and the key inputs regulating its expression. This will be followed by a discussion of its functions including the mounting evidence of its signal-modulating function, and its link to various diseases.

2. Properties of Claudin-2

2.1. Expression

Under physiological conditions, claudin-2 mRNA is enriched in the kidneys, where it is exclusively expressed in the proximal tubules, and in the gastrointestinal tract, where the highest expression was reported in the small intestine, liver, gall bladder, and pancreas [23–25]. It is also detectable at lower levels in the stomach and colon [24]. Interestingly, claudin-2 expression was also reported outside of epithelia, e.g., in endothelial cells [26,27] and macrophages (e.g., [28]), but it is not clear whether this represents a physiological localization. As discussed later (see Section 5), pathological conditions affect claudin-2 levels, and may also alter its expression pattern, therefore these observations might indicate pathological expression. Indeed, interleukins and Transforming Growth Factor β1 (TGFβ1) were shown to induce claudin-2 expression in macrophages [28]. Whether claudin-2 has a role in non-epithelial cells remains undetermined.

2.2. Structure and Interactions

Claudin-2 is a 230 amino acid transmembrane protein, with a calculated molecular mass of 24.5 kDa. Alternatively spliced transcript variants with different 5′ untranslated regions have been found for the claudin-2 gene, although their relevance remains unknown. Like other members of the family, claudin-2 is a tetraspan membrane protein consisting of two extracellular domains (ECL1 and ECL2), a small intracellular loop connecting the second and third transmembrane sections and short intracellular N and C terminal portions (Figure 1) ([12,13] and see [5,10] for review on claudin structure). ECL1 (amino acids 29–81), is responsible for paracellular charge selectivity and permeability [29]. The narrow fluid-filled pores of the claudin-2 channel are lined by polar side chains of ECL1 [16,29–32]. Mutagenesis studies revealed that the cation specificity of the channel is due to electrostatic interactions between the transported cations and the negatively charged carboxyl sidechain of D65 and the aromatic residue of Y67 [16,31]. These residues allow cations to permeate in a fully or partially dehydrated state.

Intramolecular cis interactions between C54 and C64, two highly conserved cysteines in ECL1 stabilize the TJ pore [33].

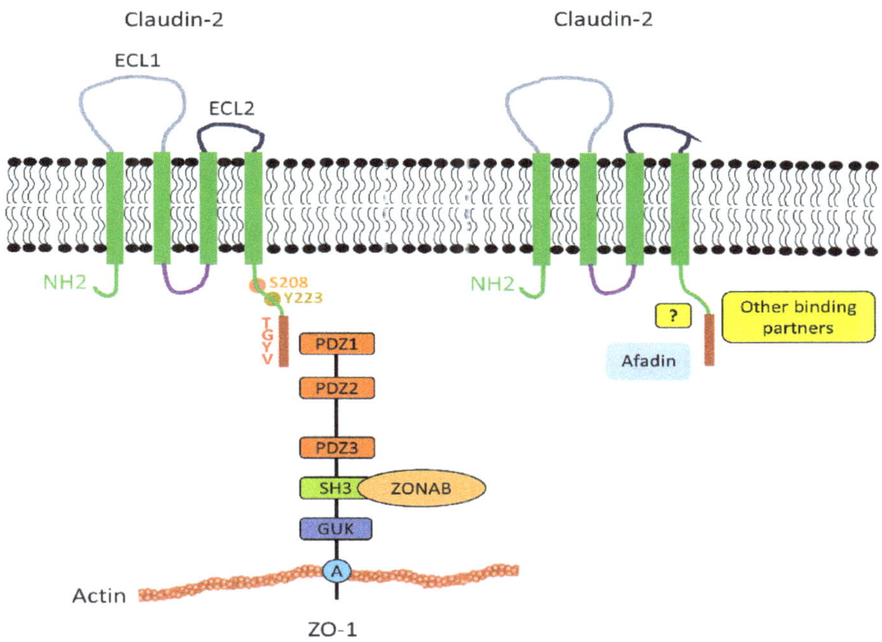

Figure 1. Claudin-2 structure and interactions with cytosolic multidomain adapters. Claudin-2 consists of two extracellular loops, the longer ECL1 (grey) and shorter ECL2 (black). Claudin-2 also contains 4 transmembrane domains (green box), a cytoplasmic loop (purple), a short N-terminal region and a longer C-terminal region (green). Claudin-2 interacts with the ZO family of TJ plaque proteins (for clarity only ZO1 is depicted) through its PDZ-binding motif - TGYV located at the end of the C-terminus (indicated by the red box). The domains of ZO1 depicted include PDZ1, PDZ2, and SH3, mediating binding of a variety of proteins to create large multiprotein complexes; GUK and A, for actin binding segment. The SH3 domain of ZO-1 was shown to bind to the transcription factor ZONAB, which may play a role in the proliferative effects of claudin-2 (Section 4.2). Afadin is a recently identified claudin-2 partner. The mode of coupling (direct biding or indirect association through adapters) remains to be established. Other newly identified candidate binding partners for claudin-2 include Scrib, Arhgap21, PDLIM2/7, and Rims-2 [34]. The claudin-2 tail contains many potential phosphorylation sites. Among these, Y223 affects the affinity of the PDZ binding domain [35], and S208 appears to be a switch for membrane retention and lysosomal or nuclear localization (see Sections 3.4 and 4.2).

Claudin-2 forms both homo- and heterotypic adhesions, via cis (i.e., molecules in the same cell) and trans (between molecules in neighboring cells) interactions. Cis homodimerization of claudin-2 requires the transmembrane domains and may have a role in organizing the TJ strands [36]. On the other hand, trans homophilic interactions are thought to promote cell-cell adhesion (e.g., [37]). Atomic force microscopy studies probing properties of the interactions between the extracellular loops of claudin-2 revealed that ECL1 was sufficient for homophilic trans interaction, and ECL2 did not mediate these [38]. This is different from what was shown for some other claudins (e.g., claudin-5) [39], where ECL2 had a role in trans interactions, and the reason for such difference remains unknown. Thus, the exact role of the claudin-2 ECL2 is not yet established, since it was not found to be involved in determining permeability or homotypic interactions. Of note, however, a peptide (DFYSP) mimicking a highly conserved region in claudin-2, induced endocytosis, and degradation of the protein [40]. This finding suggests that ECL2 might be key for maintaining claudin-2 at the membrane through a yet to be revealed mechanism.

Some heterotypic trans interactions of claudin-2 have also been described. Claudin-2 can bind to claudin-3 but not to claudin-1 on neighboring cells [41]. However, to date, there is limited information on these interactions, and the exact structural basis for the trans interactions of claudin-2 remains to be mapped. Interestingly, an antagonistic relationship was demonstrated between claudin-2 and two other claudins, and this was implicated in cytokine-induced junction reorganization. Fluorescence recovery after photobleaching (FRAP) analysis of GFP-tagged claudin-4 revealed that claudin-2 and -4 compete with one another for residency at the TJs [42]. Further, claudin-8 was shown to displace claudin-2 from the junctions, resulting in elevated transepithelial resistance (TER) [43].

The intracellular interactions of claudin-2 are pivotal for its signal-modulating functions. However, we only have a limited understanding of the protein network associated with claudin-2. The last four C-terminal amino acids of claudins correspond to a PDZ domain-binding motif, and in classic claudins, the two C-terminal amino acids (YV) are 100% conserved. Specificity for different PDZ domains is determined by variation in the neighboring amino acids close to the PDZ domain-binding motifs [44]. In claudin-2, the last 4 amino acids are T-G-Y-V (N- to C-term). This PDZ-binding motif was shown to associate with a variety of TJ plaque scaffold proteins including the membrane-associated guanylate kinase (MAGUK) family adapters ZO-1-3 [45] (Figure 1). Recently, Tabaries et al. identified several new candidate claudin-2 binding partners [34]. These will be described in more detail in Section 4.4.

The intracellular portions of claudin-2 also undergoes post-translational modifications including phosphorylation [46–48], sumoylation [49], and nitration [50] (see Section 3.4). The effects of posttranslational modifications are not fully understood.

3. Regulation of Claudin-2

3.1. Context-Dependent Regulation of Claudin-2

The TJs are continuously remodeled in response to environmental cues. In line with this, claudin-2 expression is dynamically modulated by a multitude of inputs in a context-dependent manner that affect both synthesis and turnover (Figure 2). Protein half-life and trafficking are fine-tuned by a variety of post-translational modifications and interactions with a large array of proteins. Many of the pathways that control claudin-2 are dysregulated in various diseases, resulting in disease-associated changes in claudin-2 expression (see Section 5). In general, studies to decipher the mechanisms controlling claudin-2 were performed in cell lines, and it is important to emphasize that significant context-dependent discrepancies are abundant in the literature. Specifically, some treatments were found to be stimulatory in one cell type, but inhibitory in another. Some of these differences may reflect the complex and likely organ-specific regulation of the protein. In support of this notion, the claudin-2 promoter was affected in the opposite direction by EGF in Caco2 colon cancer and in MDCK tubular epithelial cells [51]. However, some caution is warranted when evaluating data in the literature on claudin-2 regulation in cell lines, as the experiments are impacted not only by difference in origin, culture conditions and passage number of the cells, but also a multitude of other parameters. For example, our studies call attention to the importance of cell confluence and junction maturity as a factor in claudin-2 expression regulation. We found that in LLC-PK1 tubular cells, claudin-2 expression was low in subconfluent cultures and increased as confluence was established [52]. Treatment time is another important parameter that is not always taken into consideration when comparing studies. In tubular and intestinal cell lines the effect of TNFα proved to be opposite depending on the length of treatment [53]. Despite these discrepancies in findings, several mechanisms of claudin-2 regulation are now well established.

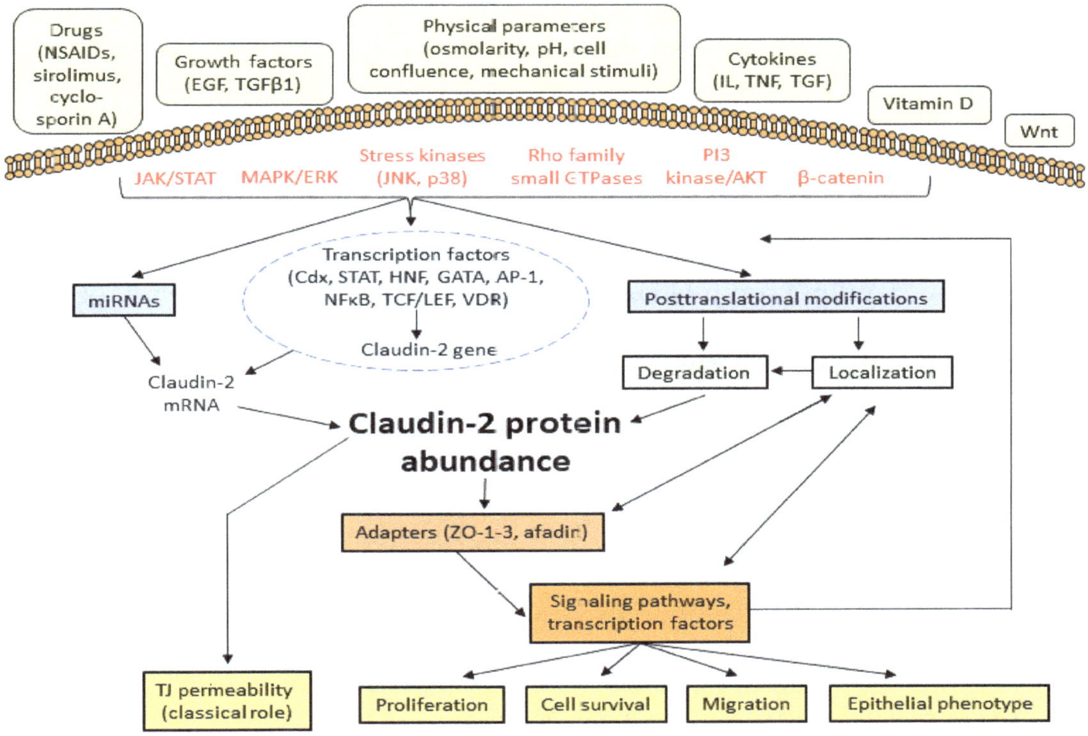

Figure 2. Schematic overview of claudin-2 regulation and its downstream effects. Various extracellular stimuli activate signaling pathways that impact claudin-2 expression by altering its synthesis, and via post-transcriptional and post-translational modifications. In addition to its permeability effects, altered claudin-2 expression also modulates various cellular processes likely acting via a number of downstream signaling pathways and transcription factors (see Table 1 for details on the signaling and effects downstream from claudin-2).

3.2. Signal Transduction Pathways Regulating Claudin-2 Expression

A multitude of signaling pathways, many of which are pro-proliferative and cancer-related, were shown to affect claudin-2 abundance. These pathways affect all aspects of claudin-2 regulation, including the control of synthesis via a set of transcription factors, as well as turn-over and localization of the protein. These latter are affected by post-translational modifications and interactions with various proteins (Figure 2). Many pathways affect multiple aspects of claudin-2 control, and the transcriptional effects are often intertwined with the control of turnover. In this section we will summarize the most studied stimuli and pathways that affect claudin-2 expression, highlighting the mode of their action. The transcription factors affecting claudin-2 will be discussed in more detail in Section 3.3.

One of the best-explored regulators of claudin-2 is the epidermal growth factor receptor (EGFR) and its effectors. Both the Ras/Raf/MEK/ERK and the phosphoinositide 3-kinase (PI3K)/Akt/nuclear factor-κB (NF-κB) pathways were found to target claudin-2 [54–56]. EGFR and ERK signaling were shown to affect claudin-2 in many tissues and could mediate effects of a variety of inputs. As mentioned above, both negative and positive effects were reported, highlighting the importance of context (e.g., [51,53,57–62]. EGFR can also mediate effects of other stimuli. For example, histone deacetylases including HDAC4 appear to affect claudin-2 via EGFR and ERK signaling [63,64]. Thus, HDAC inhibitors, that are in use or under consideration for therapy may have a major impact on claudin-2.

The effects of various cytokines on claudin-2 were extensively explored. For example, several interleukins, including IL-1β, 6, 9, 13, and 22 control claudin-2, and some were shown to act via the Jak/Stat pathway [65–67]. The effect of TNFα might be organ-specific, as it was verified to have different effects on claudin-2 in intestinal and tubular cells (e.g., [53,56]).

Kinases, downstream from various receptors may alter claudin-2 synthesis, but direct phosphorylation of the protein could also affect turnover (see Section 3.4). For most of the kinases, these details have yet to be established. Src kinases reduced claudin-2 expression [68,69], while protein kinase C (PKC) family members and the stress kinase JNK augmented it (e.g., [70–73]). Regulation of claudin-2 by Wnt/β-catenin signaling and Rho family small GTPases could also be significant both physiologically and in diseases [52,53,74]. RhoA/Rho kinase and Rac/Pak control both resting claudin-2 levels [52,53,75] and its localization [76].

Finally, several miRNAs, including miR-488, miR-16, and the miR-199a-5p/214-3p gene cluster were also implicated in claudin-2 control [54,77,78]. For example, the miR-199a-5p/214-3p gene cluster was found to be downstream from serum response factor (SRF), a pro-fibrotic transcription factor, and was implicated in high glucose-induced claudin-2 downregulation in peritoneal mesothelial cells undergoing epithelial-mesenchymal transition (EMT) [78].

3.3. *Transcription Factors controlling claudin-2 expression*

The TATA-less claudin-2 promoter has binding sites for several transcription factors [14]. Regulation by Caudal-related homeobox (Cdx) homeodomain proteins, hepatocyte nuclear factor (HNF)-1, GATA 2 and -4, AP1, Vitamin D Receptor, TCF/LEF1 and STATs was verified experimentally [14,23,55,72,74,79]. Many of these have significant roles in pathology (see Section 5). HNF-1α and GATA-4 are responsible for claudin-2 expression in specific locations of the gastrointestinal tract, but not in the kidney, highlighting a key organ-specific difference [23,80]. In contrast, GATA-2 was suggested to control claudin-2 in lung and kidney cells [70]. Cdx proteins are intestine-specific transcription factors with significant functions in gut differentiation and carcinogenesis (reviewed in [81]). The claudin-2 promoter is activated by both Cdx1 and 2, which might mediate the effects of the ERK pathway in cancer cells [14,55,82–84]. Interestingly, Cdx factors can collaborate with β-catenin-T cell factor (TCF)/lymphoid enhancer factor (LEF) [74] and HNF-1α [14]. The Cdx1 binding site also contains a functional vitamin D response element, which can directly interact with the Vitamin D Receptor transcription factor [79]. This could be significant for intestinal Ca2+ absorption. Finally, the AP1 complex can mediate the effects of JNK and ERK on the claudin-2 promoter [55,60,72,85], while STATs are effectors of interleukins [65,72].

Many stimuli were shown to decrease claudin-2 levels. This downregulation could be due to inhibition of the above signaling pathways and transcription factors. For example, hyperosmolarity was found to reduce claudin-2 promoter activity by inhibiting PKCβ-dependent GATA-2 transcriptional activity [70]. In addition, transcriptional repressors might also act directly on the claudin-2 promoter. Several EMT-inducing transcription factors are known repressors of tight junction protein genes [86,87]. Accordingly, the transcription factor Snail can induce a drop in claudin-2 [88]. Significantly, the exact mechanisms responsible for claudin-2 downregulation remain mostly unclear. For example, we do not yet have a mechanistic understanding of claudin-2 downregulation in tubular cells (Section 5.3), and the reported negative effects of Src kinases remain unexplained [68,69]. Considering the emerging consequences of reduced claudin-2 expression, a better understanding of these mechanisms will be critical.

3.4. *Claudin-2 Turnover, Trafficking, and Posttranslational Modifications*

Claudin-2 is a high turnover protein that dynamically cycles between the membrane and intracellular pools [89]. Claudins are connected to both the microtubules and the actomyosin belt through adapter proteins, and these play key roles in assembly and remodeling of the TJs [90–92]. Junctional localization of claudin-2 is not only a prerequisite for permeability functions, but also increases the protein's half-life (e.g., [52,93]). TJ insertion can be promoted or hindered via interactions with membrane and adapter proteins. As mentioned in Section 2.2. claudin-4 and 8 were shown to exclude claudin-2 from the membrane [42,43]. In contrast, ZO family adapters augment the localization of claudin-2 to the TJs. Silencing of ZO-1 and 2 reduced claudin-2 levels, likely in part by enhancing its degradation [52,67,94,95]. Interestingly, however, loss of ZO-1 also inhibited the claudin-2 promoter, suggesting the existence of feedback regulation between claudin-2 turnover and synthesis [52].

Claudin-2 is retrieved from the TJs by endocytosis. This process was found to be clathrin-dependent in lung and kidney cells [40,48,58]. Other studies have demonstrated a direct binding between claudin-2 and the endocytic scaffold protein caveolin-1 [96,97]. Cytokine-induced gut permeability increase was found to be dependent on caveolin-1 and myosin light chain-dependent endocytosis of TJ proteins [98]. In general, the actomyosin belt plays an essential role in promoting TJ protein retrieval induced by cytokines [91,99], although its specific role in claudin-2 retrieval remains less well established.

The fate of endocytosed claudin-2 is poorly understood. One study found recycling of the endocytosed protein back to the TJs [89]. Several other studies showed that it is targeted for lysosomal degradation (e.g., [40,47,71]). The small GTPases Rab14 and the atypical PKCι, an essential regulator of apico-basal polarity were implicated in the control of claudin-2 trafficking [71]. The autophagy pathway can also target claudin-2 [100,101]. TNFα-induced inhibition of autophagic degradation was suggested to elevate claudin-2 expression in intestinal cells [102]. TNFα also caused an acute ERK-dependent decrease in claudin-2 degradation in tubular cells [53]. Interestingly, similar to some other claudins, claudin-2 was also shown to translocate to the nucleus [103] (see Section 4.2).

Post-translational modifications appear to be key determinants of claudin-2 localization, and turnover, and likely control protein-protein interactions. The claudin-2 intracellular segment contains several potential phosphorylation sites. Phosphorylation of Y224 in human claudin-2 (referred to in some papers as Tyr-6) was shown to affect the affinity of the ZO-1 PDZ binding domain [35], which may impact trafficking. In many publications, the numbering of the position of the C-terminal claudin residues follows the convention that the C-terminal amino acid is referred to as the P0 residue. Subsequent residues toward the N terminus are termed P−1, P−2, etc. Thus, the tyrosine in question is often referred to as position P-6. The possible importance of this site is supported by the fact that many claudins have a tyrosine at this position [35]. S208 is another site that was shown to be phosphorylated. It was implicated in the control of membrane retention, lysosomal localization, and degradation, as well as nuclear localization [46,48,103]. Accordingly, a non-phosphorylatable mutant (S208A) was found to be more cytosolic and lysosomal, while the phospho-mimetic mutant (S208E) localized more to the plasma membrane. Interestingly, mutants, that prevent claudin-2 from insertion to the TJs were found to be poorly phosphorylated, suggesting that TJ insertion is a necessary step for the protein to become phosphorylated. Since the S208 site was implicated both in membrane retention and nuclear localization [46,48,103], it is conceivable that a coupling between claudin-2 TJ insertion and nuclear localization may exist. Accordingly, S208 phosphorylation (e.g., by PKC) could induce plasma membrane localization of claudin-2, while dephosphorylation of this site might promote nuclear translocation [46,48,103].

Claudin-2 also undergoes other post-translational modifications. Sumoylation was found to control membrane localization, ubiquitination, and degradation of claudin-2 [49], and nitration of tyrosine residues was associated with reduced expression [50]. Despite these emerging data, however, we still poorly understand how claudin-2 trafficking and fate are determined.

Importantly, many details of the complex, multilevel regulation of claudin-2 are closely linked to its pathogenic roles, and therefore an improved understanding of the molecular details will be pivotal for further insight into its dysregulation.

4. Functions of Claudin-2

4.1. Permeability Functions

Claudin-2 forms paracellular cation and water permeable channels. An elegant study by Weber et al. used a novel trans-TJ patch-clamp technique to show that the claudin-2 channel exhibits symmetrical and reversible conductance of ~90 pS [32]. They found that the channel is gated with one open and two distinct closed states. Claudin-2, on the other hand, is relatively impermeable to uncharged solutes [17,18], but permeable to cations and water [104]. Interestingly, water transport is mediated by the same pore that allows cation transport ([105] and reviewed in [106]).

These permeability properties are well verified in cell lines and knockout mouse models. Claudin-2 overexpression and silencing in various cell lines decreases and increases TER, respectively [17,18,53, 104,107]. Claudin-2 silencing also eliminates preference of the paracellular pathway towards Na^+ over Cl^- [19].

Knockout mouse models also revealed that the selective, site-specific localization and cation channel properties of claudin-2 are essential for highly specialized functions in the kidneys and the bile duct (reviewed in [108]). In the kidney, claudin-2 is pivotal for efficient Na^+ and water reabsorption in the proximal tubules. Accordingly, the S2 segment of the proximal tubules in claudin-2 KO mice showed significantly reduced net transepithelial reabsorption of Na^+, Cl^-, and water [109]. Although the mice had normal Na^+ homeostasis when fed a regular diet, administration of salt revealed Na^+ handling deficiencies in the kidneys. This suggests that under resting conditions the kidneys could compensate for the loss of the passive claudin-2-mediated Na^+ transport, but under stress, this capacity was insufficient. This conclusion was also supported by findings from Pei et al. [110]. They showed that in claudin-2 knockout mice, upregulation of active transcellular pathways took over the role of passive paracellular Na^+ transport. However, claudin-2 appears to be vital for energy efficiency of the proximal tubular Na^+ reabsorption process, and the higher energy-demand for transcellular transport resulted in medullary hypoxia and increased susceptibility to renal ischemia-reperfusion injury in the claudin-2 KO mice (see also Section 5.3).

In addition to a significant role in ion transport, studies using KO mice also revealed that claudin-2-mediated water transport contributes 23-30% of total water absorption in the proximal tubules [106]. Taken together, the permeability function of claudin-2 in the kidneys is fundamental for Na^+ and water homeostasis and blood pressure regulation [106,109,111].

In the liver and bile system, claudin-2 is highly expressed in hepatocytes and cholangiocytes and plays a central role in water transport associated with bile generation. Accordingly, in claudin-2 KO mice, the rate of bile flow was found to be reduced by half, resulting in a significant increase in bile concentration, and gallstones [112].

In the small intestine, claudin-2 plays a role in luminal Na^+ homeostasis. Surprisingly, however, it was found to be dispensable for Na^+-driven nutrient transport [113]. This contrasted with the role of claudin-15, which proved to be indispensable for both luminal Na^+ transport and Na^+-driven absorption of vital nutrients. The absence of these two claudins in double knockout mice resulted in severe malnutrition [114].

Taken together, claudin-2 plays a key role as a paracellular permeability molecule. However, as discussed in the next sections, emerging evidence suggests that its functions go beyond permeability.

4.2. Role in Proliferation

Evidence is mounting that claudin-2 is not only a paracellular channel protein, but also acts as a signal modulator and integrator (Table 1). This conclusion is derived in part from the correlation between claudin-2 expression and altered outcomes, and in part from overexpression and silencing studies that reveal the effects of altered claudin-2 expression on various outcomes. Importantly, dysregulation of claudin-2 expression appears to be an important event in several diseases (see Section 5).

As discussed in Section 3.2, many proliferative pathways affect claudin-2 expression, and these are often dysregulated in diseases, including in cancer. In fact, a strong correlation has been established by many studies between proliferation/cell viability and altered claudin-2 expression. For example, the silencing of the TJ adapter protein cingulin in kidney tubular cells simultaneously elevated claudin-2 levels and caused a RhoA-dependent increase in G1/S phase transition [75]. In A549 lung adenocarcinoma cells, ephrinA1 reduced both claudin-2 expression and proliferation, and both were mediated by Cdx-2 [115]. Recently, miR-488 was shown to suppress both claudin-2 expression and cell viability in colorectal carcinoma cells [77].

These correlations, however, do not prove a causative connection. More definitive proof is provided by gain- and loss-of-function experiments in a variety of cell lines, showing that primary claudin-2 expression changes can alter proliferation (e.g., [67,103,116]). The experimental evidence connecting claudin-2 with downstream effects is summarized in Table 1. Experimental interventions that reduce claudin-2 expression in lung cancer cells reduce proliferation (see Section 5.1) [64,117]. Thus, claudin-2 might be a permissive factor for proliferation and may act as a mediator of pro-proliferative signaling. Accordingly, silencing claudin-2 in the colon cancer cell line Caco-2 and in tubular cells prevented EGF- and TNFα-induced increase in cell proliferation [51,67].

Claudin-2 impacts cell cycle progression and expression of cell cycle regulators both in cultured cells and in a transgenic mouse model [67,103,116]. In a Villin-claudin-2 transgenic mouse model that overexpresses claudin-2 in the colon, the proliferation of colon cells was augmented by claudin-2 overexpression via the PI-3Kinase/Bcl-2 pathway [116]. In lung cancer cells claudin-2 controls the G1/S transition through cyclin D1 and E1 [103]. This effect was mediated by the Y-box transcription factor Zonula occludens 1-associated nucleic acid-binding protein (ZONAB). ZONAB is known to shuttle between the TJs and the nucleus. Interestingly, claudin-2 was also found in the nucleus where it formed a complex with ZO-1 and ZONAB [103]. Moreover, its nuclear levels increased during the G1/S transition. Interaction between claudin-2 and ZONAB was also demonstrated in colon cancer cells, where symplekin, a transcriptional regulator that cooperates with nuclear ZONAB, was shown to control claudin-2 levels [118]. The molecular underpinning of the nuclear translocation of claudin-2 remains insufficiently understood. No classic nuclear localization sequence is recognizable in claudin-2, but mutagenesis studies revealed that the C-terminal cytosolic segment (but not the PDZ domain) was essential for nuclear transport. As discussed in Section 3.4 S208 phosphorylation is a key regulator of nuclear localization, and a phospho-incompetent mutant (S208A) was found to be more nuclear.

Another cell cycle regulator affected by claudin-2 is $p27^{kip1}$ (cyclin-dependent kinase inhibitor 1B, CDKN1B). The expression of this molecule is high in quiescent cells, where it blocks cell cycle entry. We found that claudin-2 silencing elevated $p27^{kip1}$ levels, and this effect was mediated by the GEF-H1/RhoA pathway [67]. Since $p27^{kip1}$ has tumor suppressor effects, its regulation by claudin-2 might be relevant for carcinogenesis.

Altered claudin-2 expression affects several other transcription factors too. Sp1, a zinc finger transcription factor belonging to the Sp/KLF family and c-jun, part of the AP1 early response complex were found to be affected by claudin-2. Specifically, claudin-2 knockdown decreased the levels of phosho-c-Jun and reduced nuclear Sp1 [119,120]. Importantly, these transcription factors also regulate cell proliferation (e.g., [121]) suggesting another possible link.

Recently, claudin-2 was shown to affect miRNAs. This effect was implicated in the control of self-renewal in colon cancer stem cells [122]. Although the mechanisms that connect claudin-2 to miRNA regulation remain undefined, such effects might link claudin-2 to many downstream events.

Finally, yet another intriguing observation is that claudin-2 can localize along the primary cilium [123]. Claudin-2 is dispensable for ciliogenesis, and its exact role in the cilium remains unclear. However, since the primary cilium is present only in quiescent cells and its disassembly is necessary for cell proliferation [124], this intriguing observation clearly warrants further exploration.

4.3. Migration

The effects of claudin-2 on cell migration may be relevant in both cancer biology and tissue regeneration. Although many studies found that claudin-2 affects migration, significant contradictions are notable among these (Table 1). Specifically, both claudin-2 overexpression and silencing were shown to enhance migration, pointing to context-dependent factors.

In support of a role for claudin-2 in augmenting motility, in various cancer cells claudin-2 overexpression correlates with enhanced migration. In gastric cancer cells, the Helicobacter pylori virulence factor cytotoxin-associated gene (CagA) augmented both claudin-2 expression and invasiveness [82] (see Section 5.1). In line with a stimulatory role, in several cells, a decrease in claudin-2 abundance reduced migration. Non-steroidal anti-inflammatory drugs (NSAIDs), that may provide protection against cancer, reduced both migration and claudin-2 expression in colon, stomach, and lung cancer cell lines [125]. In these studies, the authors provided further evidence for a causal link, by showing that overexpression of claudin-2 stimulated migration and that claudin-2 re-expression restored migration following drug treatment.

Underlying mechanisms for a positive effect of claudin-2 on migration are yet to be unraveled, but a few possible players have emerged. Matrix metalloproteases (MMP) might be important mediators of the effect. This family of zinc-dependent proteases contains over 25 members. Several of them were shown to aid migration through the digestion of extracellular substrates [126] and are considered as potential therapeutic targets. Claudin-2 was shown to affect MMP9 activity, although surprisingly claudin-2 decrease and increase exerted a similar augmenting effect [120,127]. Inhibition of A549 cancer cell migration in a wound-healing assay by claudin-2 knockdown was attributed to reduced MMP-9 expression and activity due to Sp1 inhibition [120]. However, this might not be a universal effect, as claudin-2 overexpression-induced augmented migration of Caco-2 cells proved independent of MMPs [128].

As opposed to the above-described effects, some studies have found that claudin-2 exerts an inhibitory effect on migration. In MDCK tubular cells inducible knockdown of claudin-2 increased MMP-9 mRNA and activity, and this enhanced migration in a wound-healing assay [120]. Similarly, hyperosmotic stress-induced decrease in claudin-2 expression in the same cell line led to augmented migration, and the phenotype was rescued by re-expression of claudin-2 [70]. Of note, MDCK cells are not derived from cancer, but from normal renal tubular epithelial cells, which might explain the different effects on migration. Claudin-2 loss was also shown to augment migration downstream from the transcription factor Spi-B. This factor is normally restricted to the lymphocyte lineage, but it is frequently expressed in lung cancer. Further, it was shown to repress the transcription of claudin-2, which led to enhanced invasiveness [129]. Another study suggested that in osteosarcoma cells claudin-2 loss augmented migration via the afadin/ERK pathway [130]. Along the same lines miR-488 was shown to regulate invasion and lymph node metastasis in colon cancer cells through claudin-2-dependent control of the MAPK pathway [77].

Another mechanism whereby claudin-2 might affect migration is through effects on EMT [131,132]. This process involves a coordinated genetic reprogramming, during which cells lose their epithelial characteristics and gain mesenchymal properties, including enhanced motility. This process is key during embryonic development and wound healing, but also plays a role in cancer metastasis formation. Tight junction downregulation is a hallmark of full-blown EMT [86], but interestingly, multiple studies now suggest a reciprocal correlation too, namely active EMT-regulating roles for claudins. For example, claudin-1 overexpression was shown to promote EMT, while claudin-3 suppressed it [133,134]. While the exact role of claudin-2 in this process remains to be fully established, we recently found that claudin-2 silencing enhanced RhoA-mediated activation of Myocardin-Related Transcription Factor (MRTF), and upregulated Slug, two pro-EMT transcription factors [67]. The potential role of these in altered migration following claudin-2 loss remains unknown.

Taken together, claudin-2 was shown to have both a positive and a negative effect on migration, suggesting a need for optimal claudin-2 levels for efficient migration regulation. Clearly, further mechanistic inquiries will be required to establish the context-dependent mechanisms and importance of these effects.

Table 1. Summary of the experimental evidence for a causal link between claudin-2 expression and functional outcomes.

Cell Line or Transgenic Mouse	Change in Claudin-2 Expression	Downstream Functional Effect	Signaling Components Downstream from Claudin-2	Ref.
Villin-claudin-2 transgenic mice	Claudin-2 overexpression	Increased colonocyte proliferation	PI-3K/Bcl-2 pathway	[116]
Caco2 human colon cancer cell	Claudin-2 silencing	Reduced EGF-induced proliferation		[51]
SW480 and HCT116 colon cancer cell lines	Claudin-2 overexpression	Increased proliferation and anchorage-independent growth		
LLC-PK1 porcine kidney tubular cells	Claudin-2 silencing	Reduced proliferation	GEF-H1-mediated RhoA activation, increase in $p27^{kip1}$	[67]
		Pro-fibrotic epithelial shift	RhoA-mediated MRTF activation	
A549 human lung adenocarcinoma cells	Claudin-2 downregulation/ knockdown	Reduced G1/S transition	Cyclin D1 and E1, ZONAB	[103]
		Increased sensitivity to anti-cancer agents; increased intracellular drug accumulation and reduced efflux	Decrease in phosho-c-Jun and nuclear Sp1; reduced expression of multidrug resistance-associated protein/ABCC2	[119]
		Reduced migration	Decreased Sp1, reduced MMP-9 expression, activity	[120]
A549 human lung adenocarcinoma cells	Flavonoid- or epigenetic inhibitor-induced claudin-2 downregulation;	Reduced proliferation; partial rescue of phenotype by claudin-2 reexpression		[64,117]
HT-29 colorectal cancer cell line	Symplekin silencing-induced claudin-2 downregulation;	Reduced anchorage-dependent growth and Zonab nuclear localization; rescue of phenotype by claudin-2 reexpression		[118]
A549 human lung adenocarcinoma cells	Endocytosis and lysosomal degradation of claudin-2 induced by a peptide mimic (DFYSP) of a conserved ECL2 region	Claudin-2 accumulation in the lysosomes, cellular injury and necrotic cell death	Cathepsin B release from lysosomes	[40]
MDCK canine tubular cells	Inducible knockdown of claudin-2	Enhanced migration in a wound-healing assay	Increased MMP-9 mRNA and activity	[120]
Human colon cancer stem-like cells (patient-derived CCP1 cells)	Claudin-2 overexpression	Self-renewal of cancer stem cells; enhanced tumor initiation, progression, and metastasis	YAP and miRNAs (especially miR-222-3p)	[122]
Caco2 human colon cancer cell	Claudin-2 overexpression	Enhanced migration	Effect independent from MMP-2 and 9	[128]
AGS stomach carcinoma cells	Claudin-2 overexpression	Enhanced migration	Likely via increased MMP-1, -2 and 9 expression	[125]

Table 1. *Cont.*

Cell Line or Transgenic Mouse	Change in Claudin-2 Expression	Downstream Functional Effect	Signaling Components Downstream from Claudin-2	Ref.
T-84 colonic adenocarcinoma, AGS and KATO-III stomach carcinoma cells; A549 lung adenocarcinoma cell lines	Claudin-2 silencing Claudin-2 overexpression Non-steroidal anti-inflammatory drugs (NSAIDs)-induced claudin-2 downregulation	Reduced migration Augmented migration Reduced migration; claudin-2 reexpression counteracted the effect		[125]
U2OS osteosarcoma cell line	Claudin-2 overexpression	Reduced migration and invasion	Afadin-mediated ERK inhibition	[130]
Fetal osteoblast cell line hFOB.1.19	Claudin-2 silencing	Augmented migration and invasion	ERK activation, afadin reduction	
Claudin-2 KO mice	Claudin-2 KO	Augmented TNFα-induced colorectal inflammation	NFκB, myosin light chain kinase	[135]
Claudin-2 KO mice	Claudin-2 KO	Augmented energy demand of transport processes; increased ischemia-reperfusion kidney injury		[110]

4.4. Signal-Modulating Effects of Claudin-2: Emerging Mechanisms Underlying Roles in Biological Processes

As discussed in Sections 4.2 and 4.3, evidence is mounting that claudin-2 affects cell behavior (Table 1). However, a coherent picture of the underlying mechanisms has yet to emerge. Many studies suggest that claudin-2 may be a signal modulator and integrator protein. Nevertheless, an in-depth understanding of how claudin-2 is linked with altered signaling pathways remains elusive.

The TJ cytoplasmic plaque contains a large array of signaling and regulatory proteins. These are connected to the TJ membrane proteins by various adapters. For many of these proteins, recruitment to the junctions results in inactivation, or a spatial restriction of activity (for review see [90,136,137]). The specific role of individual claudins in the organizing of the cytoplasmic plaque is not yet resolved. As mentioned in Section 2.2, the cytoplasmic tail of claudin-2 interacts with multiple proteins, but the claudin-2 interactome has not yet been fully mapped, and context-dependent regulation of the interactions remains unknown. As mentioned earlier, a recent study from the Siegel group identified some candidate binding partners of the claudin-2 PDZ domain [34]. Using metastatic breast cancer cells with elevated claudin-2 expression, they demonstrated that the last three C-terminal amino acids of claudin-2 (comprising the PDZ-binding motif) are necessary for anchorage-independent growth. They identified several potential partners of this motif, including afadin, which is an actin filament- and Rap1 small GTPase-binding protein encoded by the MLLT4/AF-6 gene. Loss of afadin impaired colony formation of breast cancer cells in soft agar and reduced lung and liver metastasis, verifying the significance of the claudin-2/afadin complex. Importantly, some of their findings suggested the existence of afadin-independent mechanisms too. The identified potential other partners, including the polarity protein Scrib, the small GTPase regulator Arhgap21, PDZ-LIM domain-containing proteins (PDLIM) 2 and 7, and the exocytosis-controlling Rims-2 (regulating synaptic membrane exocytosis protein 2) may mediate such downstream effects. However, whether these are direct binding partners or interact via adapters, and their potential functional significance remains to be established.

The crosstalk between claudin-2 and the cytoskeleton may play a central role in mediating downstream effects. Claudin-2, as all TJ membrane proteins, is anchored to the junctional actomyosin belt and microtubules through adaptor proteins. Among these, ZO family proteins directly bind claudins, and connect to F-actin either directly, or indirectly through actin-binding proteins [138,139]. Thus, it is conceivable that altered claudin-2 expression might affect the organization of the prejunctional cytoskeleton. In support of this, we recently found that in tubular cells, claudin-2 silencing altered acto-myosin organization through RhoA [67].

The microtubules also play a role in junction regulation and might be affected by the TJ organization. Cingulin and its paralog paracingulin interact with TJ proteins, likely through ZO adapters, and thus connect to actin. Cingulin also binds the microtubules and has a role in organizing the microtubular planar apical network in mammary epithelial cells [140,141]. Further, association of cingulin with actin filaments and the microtubules was found to be regulated by phosphorylation and plays a crucial role in the control of the epithelial barrier [141]. The specific roles of individual claudins as organizers of the cytoplasmic plaque complexes and their impact on the cytoskeleton are yet to be uncovered. Nevertheless, it is conceivable that some of the signal modulating effects of claudin-2 is mediated by an impact on the acto-myosin or the MTs.

Interestingly, claudin-2 appears to have a bidirectional relationship with the cytoskeleton and some signaling pathways. Specifically, some of the pathways modulated by claudin-2 can also regulate its expression. This suggests the existence of various regulatory feedback cycles (Figure 2). Such regulatory loops also pose a challenge for defining clear "upstream" and "downstream" events. The ERK and Rho signaling pathways are prominent examples of this complexity. The role of MEK/ERK signaling in basal and stimulus-induced claudin-2 expression is well documented [53,55,60,61,85,142,143]. Besides, a recent study reported the converse effect: claudin-2 silencing was found to augment MEK/ERK1/2 signaling [130], an effect that may help to restore claudin-2 levels. A complex relationship also exists between Rho family small GTPases and claudin-2. We found that RhoA/Rho-kinase are negative regulators of claudin-2 expression in cultured tubular cells [53]. On the other hand, as mentioned above, in the same cells claudin-2 silencing enhanced RhoA activity through the exchange factor GEF-H1, suggesting that in resting cells, claudin-2 might suppress RhoA activation [67]. Thus, claudin-2 loss can augment Rho activity, which in turn may further reduce claudin-2 expression, in a possible self-augmenting cycle.

5. Claudin-2 in Diseases

A growing number of studies document altered claudin-2 expression, phosphorylation and/or localization in various pathological conditions. Initially, the described claudin expression alterations were considered an epiphenomenon, i.e., they were assumed to arise due to the underlying disease process but were thought not to be causally contributing to pathogenesis. Nevertheless, interest in pathological alterations in claudin-2 expression was boosted by the hope that it can be used as a diagnostic and/or prediction marker. As described in Section 4, in the past years, strong evidence accumulated in support of a causal link between claudin-2 dysregulation and functional alterations, the key points of which are the following. First, signaling pathways that are known to be overactivated in diseases can alter claudin-2 expression, and a good correlation exists between disease stage and claudin-2 expression. Second, loss-of-function and gain-of-function studies recapitulate some aspects of the functional changes. Thus, primary changes in claudin-2 expression can alter cell behavior. The mounting experimental evidence prompted a paradigm shift, favoring a pathogenic role for claudin-2. Although dysregulation of claudin-2 is likely not a primary cause, pathological changes in claudin-2 abundance and/or localization might play significant roles in the generation, maintenance and/or progression of diseases.

In the following sections, we will provide an overview of the evidence linking claudin-2 to cancer, and various non-malignant pathologies, such as inflammatory gastrointestinal and kidney diseases.

5.1. Claudin-2 in Cancer and Metastasis Formation

An expanding body of literature documents dysregulated claudin-2 expression in gastric, colorectal, lung, breast, and renal carcinomas and in osteosarcoma (e.g., [51,144–148]. As described in Sections 4.2 and 4.3, claudin-2 levels affect processes underlying carcinogenesis and metastasis formation, including proliferation, migration and epithelial-mesenchymal transition. The following overview however also highlights that the role of claudin-2 is likely complex, and differences in its impact might exist not only based on tissue origin but also cancer stage.

Cancers of the gastrointestinal tract—Claudin-2 is highly expressed in gastric and colorectal cancers and its expression level shows a good correlation with the development of these tumors. For example, a gradual increase was observed in claudin-2-positivity in various stages of gastric carcinogenesis from no expression in gastritis to elevated expression in dysplasia and gastric intestinal-type adenocarcinoma [149]. In fact, in this study, 73% of adenocarcinoma samples were found to be claudin-2 positive.

As described in Sections 3.2 and 3.3, claudin-2 is the target of key signaling pathways and transcription factors central to gastrointestinal cancers. One prominent example is the significant correlation between Cdx2 and claudin-2 expression in gastric dysplasia and cancer [150]. Further, a potentially important link between claudin-2 and H. pylori was also uncovered. H. pylori infection shows a strong correlation with gastric cancer [151], and the virulence factor CagA is considered a gastric oncogene. During infection, this protein is translocated into and reprograms the gastric cells. Interestingly, CagA augmented claudin-2 expression in gastric cells, and promoted invasiveness through effects on Cdx2 [82].

Hyperactivation of Wnt signaling and the consequent increase in gene transcription by TCF/ LEF is a hallmark of colon cancer (reviewed in [152]). Wnt-1 was shown to increase claudin-2 promoter activity through β-catenin/LEF-1 [74]. Increased claudin-2 and β-catenin expression were detected in active inflammatory bowel disease (IBD), adenomas, and IBD-associated dysplasia, but not in acute, self-limited colitis [153]. These data suggest that β-catenin might mediate an increase in claudin-2 during carcinogenesis.

EGFR signaling is yet another key pathway linking colon cancer and claudin-2. EGFR activation upregulated claudin-2 in colon cancer cells, and claudin-2 expression was decreased in the colon of waved-2 mice that have EGFR activation deficiency [51] (see also Section 3.2 on the effects of EGFR on claudin-2).

Taken together, these studies suggest that the initial upregulation of claudin-2 could be due to the overactivation of the above-discussed pathways, and this may promote carcinogenesis. To prove a causal link between elevated claudin-2 and the properties of cancer, exogenous claudin-2 was overexpressed in colorectal cancer cell lines. An increase in claudin-2 expression promoted colonocyte proliferation and anchorage-independent colony formation and stimulated tumor formation in colorectal cancer xenografts (e.g., [51,116]). Further, claudin-2 expression was a negative predictor for post-chemotherapy disease-free survival of colon cancer patients [51].

Another key role of claudin-2 could be its effect on colorectal cancer stem-like cells [122]. Claudin-2 promoted self-renewal of these cells via miRNAs and the hippo effector yes-associated protein (YAP). Interestingly, claudin-2 was also detected in human colorectal cancer-associated fibroblasts (CAFs). This finding is especially intriguing since claudin-2 is regarded as an epithelial molecule. Its expression in CAFs was linked to KRAS mutation status and correlated with reduced progression-free survival [154]. Of note, claudin-2 was also shown to be enriched in macrophages associated with mammary tumors [28]. Thus, an attractive hypothesis is that claudin-2 may increase proliferation/survival and migration of fibroblasts and macrophages in the tumor environment.

Lung cancer—Primary lung cancer is among the deadliest types of tumors worldwide, and adenocarcinomas are the most frequent forms of non-small cell lung cancer. Claudin-2 expression in normal bronchial epithelium is low [85]. In contrast, according to one study, two-thirds of examined lung adenocarcinoma samples overexpressed claudin-2 [145]. Surprisingly, in these cells, claudin-2 was found mostly in cytoplasmic granules. Adenocarcinoma is often associated with upregulated EGFR activity and KRAS/MEK/ERK signaling. Indeed, EGFR signaling is key in upregulating claudin-2 in the human lung adenocarcinoma cell line A549 [60]. These cells were used in a series of studies to demonstrate that claudin-2 may be a therapeutic target in lung cancer. The studies showed that rapid proliferation of A549 cells required claudin-2. Accordingly, claudin-2 knockdown impaired cell growth and migration [103,120] and elevated sensitivity for anti-cancer agents [119]. Importantly, a variety of interventions, including epigenetic inhibitors and various flavonoids reduced claudin-2 expression and decreased proliferation. These effects were counteracted by retransfection of claudin-2 [64,117]. Claudin-2 abundance can also be reduced by a peptide that mimics a sequence in ECL2. This peptide-induced endocytosis and lysosomal accumulation of claudin-2, resulting in lysosomal damage and necrotic cell death [40]. Taken together, these studies provided strong evidence that targeting claudin-2 can mitigate proliferation and raised hope that this approach will prove useful in cancer therapy (see further elaboration in Section 5.4).

Osteosarcoma and breast cancer—Interestingly, in some tumors, claudin-2 expression is decreased. Osteosarcoma cells have lower claudin-2 expression than normal osteoblasts [130], and this is associated with enhanced migration. In breast cancer, increased or decreased claudin-2 expression might be associated with different cancer types and stages [155,156]. As discussed below, in this type of tumor, claudin-2 expression appears to correlate with metastasis formation.

Metastasis—The importance of claudin-2 in metastasis is also emerging. Claudin-2 might affect migration via a variety of effects. Its possible role in migration was discussed in Section 4.3. Beyond general effects on migration, claudin-2 also appears to affect metastasis through mediating specific adhesion of the metastatic cells, thereby promoting invasion into distant organs. Interestingly, the exact role of claudin-2 in metastasis appears to vary based on the site of invasion. In some breast cancer reduction in claudin-2 was associated with lymph node metastasis and higher clinical stage [156]. In contrast, claudin-2 overexpression was shown to augment breast carcinoma metastasis to the liver. Claudin-2 expression is elevated in liver metastases compared to primary breast cancer cells [157].

Indeed, claudin-2 promoted attachment and survival of metastatic cells specifically in the liver and enhanced their interactions with hepatocytes [37]. Due to these effects, claudin-2 was found to be a negative prognostic factor that predicts liver metastasis in breast cancer [158]. Metastatic breast cancer cells showed a general increase in adhesion to the extracellular matrix, as claudin-2 promoted surface expression of $\alpha2\beta1$- and $\alpha5\beta1$-integrins [157]. However, while these effects might contribute to invasiveness in general, enhanced interaction between hepatocytes and breast cancer cells proved to be independent of integrins. Instead, the effect required the first claudin-2 extracellular loop (ECL1) and was likely attributable to homotypic trans interactions between claudin-2 molecules. Thus, claudin-2 in the metastatic breast cancer cells can bind to claudin-2 on hepatocytes, thereby promoting their adhesion to the liver [157]. Further, the PDZ-binding motif (YV) in claudin-2 was found to be necessary for anchorage-dependent growth and metastases [34]. These exciting findings raise hope that blocking peptides or neutralizing antibodies against the identified domain might be effective in reducing liver metastasis (see Section 5.4).

5.2. *Claudin-2 in Gut Inflammation*

Claudin-2 emerged as an important factor in IBD, a term describing conditions such as ulcerative colitis and Crohn's disease. These chronic diseases are hallmarked by compromised epithelial barrier and are thought to be caused by a dysregulated immune response in the gut (e.g., reviewed in [159,160]). Increased paracellular permeability is likely a critical pathogenic factor. Specifically, the disease appears to be maintained by a vicious cycle: increased epithelial permeability due to dysregulation of TJ proteins elicits an aberrant immune response, and the ensuing inflammation further disrupts TJs and aggravates the condition. The effects of inflammation on epithelial permeability have been long known. Elevated claudin-2 expression in IBD is attributed to the presence of cytokines (e.g., [56,161,162] and reviewed in [159,163]). The permeability function of claudin-2 appears to be central in IBD, although a causal link is hard to decipher. Studies in transgenic mice explored and verified the role of the permeability function of claudin-2. In contrast, little is known about the possible significance of claudin-2 as a signal modulator in IBD. Somewhat surprisingly, recent studies point to a possible protective role of claudin-2 against injury. A transgenic mouse with targeted overexpression of claudin-2 in the colon was protected against colitis-associated injury [116]. This study also highlighted the complex roles of claudin-2 in intestinal homeostasis and IBD. Claudin-2 overexpression in the colon augmented mucosal permeability but did not by itself cause inflammation. Instead, the mice had longer colons and elongated crypts, a result of accelerated colonocyte proliferation. Most notably, despite the leaky colon, the mice were significantly protected against experimental colitis, and inflammation-associated gene expression was sharply downregulated. Some of the effects could be the result of reduced apoptosis and augmented regeneration due to faster proliferation. Thus, in this respect, claudin-2 elevation may at least initially confer some protection in IBD. This notion is also supported by another study that used claudin-2 null mice to explore the effects of intraperitoneally injected TNFα and experimental colitis [135]. The data revealed augmented colorectal inflammation in the KO mice compared to WT animals. These effects were mediated by NF-κB signaling. The mice also had increased expression of IL-6 and IL-1β and higher intestinal myosin light chain kinase levels. Overall, these studies suggest that increased claudin-2 expression might suppress inflammatory signals. Importantly, although the proliferative effect observed could be beneficial for regeneration, this is a double-edged sword. IBD elevates the risk for developing colorectal cancer and augmented claudin-2 expression could contribute to this. As mentioned in Section 5.1, claudin-2 levels rise with the development of dysplasia and cancer. This raises the intriguing possibility that elevated claudin-2 expression might be a crucial connection between inflammation and cancer. Thus, while elevated claudin-2 might protect against injury and enhance regeneration, it could raise the risk of cancer.

The role of claudin-2 in intestinal diseases might go beyond IBD, as it was also implicated in diseases caused by enteropathogenic bacteria. Salmonella invasion elevated claudin-2 protein and mRNA levels both in cultured cells and in the intestine of mice. Upregulation of claudin-2, in turn, elevated permeability and promoted internalization of the bacteria [73]. Finally, recent studies raise the possibility that claudin-2 might contribute to chronic pancreatitis, a progressive inflammatory disease that is frequently associated with alcohol consumption. A genome-wide association search found that polymorphisms of the claudin-2 locus confer an increased risk of alcohol-induced pancreatitis [164]. Since claudin-2 is essential for bile formation, it will be interesting to see if claudin-2 polymorphism might also predispose for gall bladder disease and gallstones.

5.3. *Claudin-2 in Kidney Disease*

Claudin-2 mRNA levels are highest in the kidney, where it is a key mediator of cation and water transport in the proximal tubules (reviewed in [165–167]). Although a direct role of claudin-2 in kidney disease has not been definitively established, several observations suggest a possible link.

Claudin-2 is affected by inflammatory cytokines in tubular cells. Specifically, short-term TNFα treatment caused an increase in claudin-2 abundance [53]. In contrast, prolonged (>8h) TNF-α or IL-1β treatment downregulated claudin-2 expression [53,67]. Interestingly, a large variety of potentially harmful stimuli can downregulate claudin-2 expression in tubular cells. In addition to these cytokines, metabolic acidosis [168], changes in osmolarity [48,127], oxidants [169], EGF [58,170], sheer stress [171], the flavonoid quercetin [172], and drugs, such as the immunosuppressants sirolimus and cyclosporine A [173], reduce claudin-2 levels. Reduced claudin-2 expression was also recorded in various kidney disease animal models, including cisplatin-induced nephrotoxicity [174], diabetic nephropathy [50] and obstructive nephropathy-induced fibrosis [67]. These findings are of significance because claudin-2 was found to exert a protective effect against kidney injury by reducing the energy demand of transport processes [110]. Further, as mentioned above loss of claudin-2 in tubular cells induced RhoA activation, and claudin-2 knockout mice had elevated RhoA activity [67]. This led to reduced proliferation and activation of MRTF, a transcription factor that mediates profibrotic epithelial transition [67,175]. Thus, reduced claudin-2 expression in stress conditions might increase susceptibility for injury and may promote fibrosis.

Finally, the proximal tubular paracellular pathway also plays a central role in Ca2+ reabsorption. Accordingly, claudin-2 null mice were hypercalciuric, a condition that enhances kidney stone formation [109]. Thus, conditions that reduce the abundance of claudin-2 in the kidney might also lead to kidney stone disease, a condition affecting a significant portion of the population.

5.4. *Development of Therapies Targeting Claudin-2*

Several studies raise hope that targeting claudin-2 could have beneficial effects. Indeed, in a proof of principle study, a monoclonal antibody that recognizes the first extracellular loop (ECL1) of claudin-2 was shown to prevent TNFα-induced TJ disruption [176]. The same group recently generated a human-rat chimeric monoclonal antibody against claudin-2 and showed that it accumulated in claudin-2 positive sarcoma xenografts. The antibody also suppressed tumor growth [177]. Importantly, no major adverse effects were found.

Taken together, accumulating evidence supports that claudin-2 can be considered a diagnostic and prognostic marker in various diseases and is an exciting potential therapeutic target.

6. Open Questions

Claudin-2 is now established as a pathogenic factor in several diseases due to its permeability-dependent and -independent functions. The recognition that claudins have major signal modulator, permeability-independent functions reenergized claudin research, leading to an explosion in the number of studies published. Although the fact that altered claudin-2 expression affects signaling and cell behavior is firmly established, many questions remain unanswered. It is worthwhile to articulate some of these as they set the direction for future research. Importantly, our understanding of mechanisms mediating pathological dysregulation of claudin-2 is still incomplete, which precludes the design of interventions. There are also fundamental unanswered questions regarding the mechanisms that link claudin-2 to various downstream events. We do not know the role of claudin-2 localization in its signal-modulating effects. Are these effects mediated by TJ localized claudin-2? Given that claudin-2 was shown to reside in other compartments (cytosolic vesicles, nucleus), some of the non-classical effects may well be mediated by these claudin-2 sub-pools. Further, many studies revealed the importance of altered claudin-2 expression in the downstream effects, but we do not know whether these are exclusively related to the number of molecules or if further regulation (e.g., by distinct posttranslational modifications) are also involved. Extra- and intracellular binding partners of claudin-2 are likely crucial mediators of signal-modulating effects. Indeed, TJ-localized claudin-2 can interact with other membrane proteins and may act as a sensor of neighboring cells. However, it is not clear whether claudin-2 needs to be engaged for its signal-modulating effects. Further, the claudin-2 interactome is only now starting to emerge and uncovering the context-dependent regulation of interactions will be key for a better understanding. Adapters in the TJ plaque generate signaling complexes, that may change depending on the availability of claudin-2, leading to recruitment or release of signaling intermediates and alterations in the cytoskeleton organization. However, the specific role of claudin-2 in such events remains unknown. Ongoing research from many groups will likely help fill these knowledge gaps leading to a better mechanistic understanding of the signal-modulating effects of claudin-2. The emerging knowledge has already informed the design of interventions targeting claudin-2. Such future therapies have the potential to benefit many patients with a broad spectrum of diseases.

Author Contributions: Writing–Original Draft Preparation, S.V., S.A., K.S.; Review & Editing, S.V., S.A., K.S.; Visualization, S.A.; Supervision, K.S.; Funding Acquisition, K.S.

Acknowledgments: The authors would like to express their gratitude to Andras Kapus and Mirjana Jerkic for critical reading of the manuscript.

Abbreviations

CagA	CagA: cytotoxin-associated gene
Cdx	Caudal-related homeobox
EGF	epidermal growth factor
EGFR	epidermal growth factor receptor
EMT	epithelial-mesenchymal transition
HNF	hepatocyte nuclear factor
IBD	inflammatory bowel disease
IL	Interleukin
MMP	matrix metalloprotease
TGFβ1	transforming growth factor β1
TNF	tumor necrosis factor
PDZ	postsynaptic density, disc-large, ZO-1
TCF/LEF	T cell factor/lymphoid enhancer factor
ZONAB	Zonula occludens 1-associated nucleic acid-binding protein
TJ	tight junctions
ZO	zonula occludens

References

1. Zihni, C.; Mills, C.; Matter, K.; Balda, M.S. Tight junctions: From simple barriers to multifunctional molecular gates. *Nat. Rev. Mol. Cell Biol.* **2016**, *17*, 564–580. [CrossRef] [PubMed]
2. Van Itallie, C.M.; Anderson, J.M. Architecture of tight junctions and principles of molecular composition. *Semin. Cell Dev. Biol.* **2014**, *36*, 157–165. [CrossRef] [PubMed]
3. Shen, L.; Weber, C.R.; Raleigh, D.R.; Yu, D.; Turner, J.R. Tight junction pore and leak pathways: A dynamic duo. *Annu. Rev. Physiol.* **2011**, *73*, 283–309. [CrossRef] [PubMed]
4. Gunzel, D.; Fromm, M. Claudins and other tight junction proteins. *Compr. Physiol.* **2012**, *2*, 1819–1852. [PubMed]

5. Tsukita, S.; Tanaka, H.; Tamura, A. The Claudins: From Tight Junctions to Biological Systems. *Trends Biochem. Sci.* **2019**, *44*, 141–152. [CrossRef] [PubMed]
6. Mineta, K.; Yamamoto, Y.; Yamazaki, Y.; Tanaka, H.; Tada, Y.; Saito, K.; Tamura, A.; Igarashi, M.; Endo, T.; Takeuchi, K.; et al. Predicted expansion of the claudin multigene family. *FEBS Lett.* **2011**, *585*, 606–612. [CrossRef] [PubMed]
7. Lal-Nag, M.; Morin, P.J. The claudins. *Genome Biol.* **2009**, *10*, 235. [CrossRef] [PubMed]
8. Wu, J.; Helftenbein, G.; Koslowski, M.; Sahin, U.; Tureci, O. Identification of new claudin family members by a novel PSI-BLAST based approach with enhanced specificity. *Proteins* **2006**, *65*, 808–815. [CrossRef] [PubMed]
9. Furuse, M. Molecular basis of the core structure of tight junctions. *Cold Spring Harb. Perspect. Biol.* **2010**, *2*, a002907. [CrossRef] [PubMed]
10. Krause, G.; Winkler, L.; Mueller, S.L.; Haseloff, R.F.; Piontek, J.; Blasig, I.E. Structure and function of claudins. *Biochim. Biophys. Acta* **2008**, *1778*, 631–645. [CrossRef] [PubMed]
11. Capaldo, C.T.; Nusrat, A. Claudin switching: Physiological plasticity of the Tight Junction. *Semin. Cell Dev. Biol.* **2015**, *42*, 22–29. [CrossRef] [PubMed]
12. Furuse, M.; Fujita, K.; Hiiragi, T.; Fujimoto, K.; Tsukita, S. Claudin-1 and -2: Novel integral membrane proteins localizing at tight junctions with no sequence similarity to occludin. *J. Cell Biol.* **1998**, *141*, 1539–1550. [CrossRef] [PubMed]
13. Furuse, M.; Sasaki, H.; Fujimoto, K.; Tsukita, S. A single gene product, claudin-1 or -2, reconstitutes tight junction strands and recruits occludin in fibroblasts. *J. Cell Biol.* **1998**, *143*, 391–401. [CrossRef] [PubMed]
14. Sakaguchi, T.; Gu, X.; Golden, H.M.; Suh, E.; Rhoads, D.B.; Reinecker, H.C. Cloning of the human claudin-2 5′-flanking region revealed a TATA-less promoter with conserved binding sites in mouse and human for caudal-related homeodomain proteins and hepatocyte nuclear factor-1alpha. *J. Biol. Chem* **2002**, *277*, 21361–21370. [CrossRef] [PubMed]
15. Gunzel, D.; Yu, A.S. Claudins and the modulation of tight junction permeability. *Physiol. Rev.* **2013**, *93*, 525–569. [CrossRef] [PubMed]
16. Yu, A.S.; Cheng, M.H.; Angelow, S.; Gunzel, D.; Kanzawa, S.A.; Schneeberger, E.E.; Fromm, M.; Coalson, R.D. Molecular basis for cation selectivity in claudin-2-based paracellular pores: Identification of an electrostatic interaction site. *J. Gen. Physiol.* **2009**, *133*, 111–127. [CrossRef] [PubMed]
17. Furuse, M.; Furuse, K.; Sasaki, H.; Tsukita, S. Conversion of zonulae occludentes from tight to leaky strand type by introducing claudin-2 into Madin-Darby canine kidney I cells. *J. Cell Biol.* **2001**, *153*, 263–272. [CrossRef] [PubMed]
18. Amasheh, S.; Meiri, N.; Gitter, A.H.; Schoneberg, T.; Mankertz, J.; Schulzke, J.D.; Fromm, M. Claudin-2 expression induces cation-selective channels in tight junctions of epithelial cells. *J. Cell Sci.* **2002**, *115*, 4969–4976. [CrossRef] [PubMed]
19. Hou, J.; Gomes, A.S.; Paul, D.L.; Goodenough, D.A. Study of claudin function by RNA interference. *J. Biol. Chem.* **2006**, *281*, 36117–36123. [CrossRef] [PubMed]
20. Gupta, I.R.; Ryan, A.K. Claudins: Unlocking the code to tight junction function during embryogenesis and in disease. *Clin. Genet.* **2010**, *77*, 314–325. [CrossRef] [PubMed]
21. Kwon, M.J. Emerging roles of claudins in human cancer. *Int. J. Mol. Sci.* **2013**, *14*, 18148–18180. [CrossRef] [PubMed]
22. Hagen, S.J. Non-canonical functions of claudin proteins: Beyond the regulation of cell-cell adhesions. *Tissue Barriers* **2017**, *5*, e1327839. [CrossRef] [PubMed]
23. Escaffit, F.; Boudreau, F.; Beaulieu, J.F. Differential expression of claudin-2 along the human intestine: Implication of GATA-4 in the maintenance of claudin-2 in differentiating cells. *J. Cell Physiol.* **2005**, *203*, 15–26. [CrossRef] [PubMed]
24. Aung, P.P.; Mitani, Y.; Sanada, Y.; Nakayama, H.; Matsusaki, K.; Yasui, W. Differential expression of claudin-2 in normal human tissues and gastrointestinal carcinomas. *Virchows Arch.* **2006**, *448*, 428–434. [CrossRef] [PubMed]
25. Markov, A.G.; Veshnyakova, A.; Fromm, M.; Amasheh, M.; Amasheh, S. Segmental expression of claudin proteins correlates with tight junction barrier properties in rat intestine. *J. Comp. Physiol. B* **2010**, *180*, 591–598.

[CrossRef] [PubMed]

26. Lippoldt, A.; Liebner, S.; Andbjer, B.; Kalbacher, H.; Wolburg, H.; Haller, H.; Fuxe, K. Organization of choroid plexus epithelial and endothelial cell tight junctions and regulation of claudin-1, -2 and -5 expression by protein kinase C. *Neuroreport* **2000**, *11*, 1427–1431. [CrossRef] [PubMed]
27. Tan, X.; Li, D.; Wang, X.; Zeng, Y.; Yan, Y.; Yang, L. Claudin-2 downregulation by KSHV infection is involved in the regulation of endothelial barrier function. *J. Cutan Pathol.* **2014**, *41*, 630–639. [CrossRef] [PubMed]
28. Van den Bossche, J.; Laoui, D.; Morias, Y.; Movahedi, K.; Raes, G.; De Baetselier, P.; Van Ginderachter, J.A. Claudin-1, claudin-2 and claudin-11 genes differentially associate with distinct types of anti-inflammatory macrophages in vitro and with parasite- and tumour-elicited macrophages in vivo. *Scand J. Immunol.* **2012**, *75*, 588–598. [CrossRef] [PubMed]
29. Colegio, O.R.; Van Itallie, C.; Rahner, C.; Anderson, J.M. Claudin extracellular domains determine paracellular charge selectivity and resistance but not tight junction fibril architecture. *Am. J. Physiol. Cell Physiol.* **2003**, *284*, C1346–C1354. [CrossRef] [PubMed]
30. Colegio, O.R.; Van Itallie, C.M.; McCrea, H.J.; Rahner, C.; Anderson, J.M. Claudins create charge-selective channels in the paracellular pathway between epithelial cells. *Am. J. Physiol. Cell Physiol.* **2002**, *283*, C142–C147. [CrossRef] [PubMed]
31. Li, J.; Zhuo, M.; Pei, L.; Rajagopal, M.; Yu, A.S. Comprehensive cysteine-scanning mutagenesis reveals Claudin-2 pore-lining residues with different intrapore locations. *J. Biol. Chem* **2014**, *289*, 6475–6484. [CrossRef] [PubMed]
32. Weber, C.R.; Liang, G.H.; Wang, Y.; Das, S.; Shen, L.; Yu, A.S.; Nelson, D.J.; Turner, J.R. Claudin-2-dependent paracellular channels are dynamically gated. *Elife* **2015**, *4*, e09906. [CrossRef] [PubMed]
33. Li, J.; Angelow, S.; Linge, A.; Zhuo, M.; Yu, A.S. Claudin-2 pore function requires an intramolecular disulfide bond between two conserved extracellular cysteines. *Am. J. Physiol. Cell Physiol.* **2013**, *305*, C190–C196. [CrossRef] [PubMed]
34. Tabaries, S.; McNulty, A.; Ouellet, V.; Annis, M.G.; Dessureault, M.; Vinette, M.; Hachem, Y.; Lavoie, B.; Omeroglu, A.; Simon, H.G.; et al. Afadin cooperates with Claudin-2 to promote breast cancer metastasis. *Genes Dev.* **2019**, *33*, 180–193. [CrossRef] [PubMed]
35. Nomme, J.; Antanasijevic, A.; Caffrey, M.; Van Itallie, C.M.; Anderson, J.M.; Fanning, A.S.; Lavie, A. Structural Basis of a Key Factor Regulating the Affinity between the Zonula Occludens First PDZ Domain and Claudins. *J. Biol. Chem.* **2015**, *290*, 16595–16606. [CrossRef] [PubMed]
36. Van Itallie, C.M.; Mitic, L.L.; Anderson, J.M. Claudin-2 forms homodimers and is a component of a high molecular weight protein complex. *J. Biol. Chem.* **2011**, *286*, 3442–3450. [CrossRef] [PubMed]
37. Tabaries, S.; Dupuy, F.; Dong, Z.; Monast, A.; Annis, M.G.; Spicer, J.; Ferri, L.E.; Omeroglu, A.; Basik, M.; Amir, E.; et al. Claudin-2 promotes breast cancer liver metastasis by facilitating tumor cell interactions with hepatocytes. *Mol. Cell Biol.* **2012**, *32*, 2979–2991. [CrossRef] [PubMed]
38. Lim, T.S.; Vedula, S.R.; Hunziker, W.; Lim, C.T. Kinetics of adhesion mediated by extracellular loops of claudin-2 as revealed by single-molecule force spectroscopy. *J. Mol. Biol.* **2008**, *381*, 681–691. [CrossRef] [PubMed]
39. Piontek, J.; Winkler, L.; Wolburg, H.; Muller, S.L.; Zuleger, N.; Piehl, C.; Wiesner, B.; Krause, G.; Blasig, I.E. Formation of tight junction: Determinants of homophilic interaction between classic claudins. *FASEB J.* **2008**, *22*, 146–158. [CrossRef] [PubMed]
40. Ikari, A.; Taga, S.; Watanabe, R.; Sato, T.; Shimobaba, S.; Sonoki, H.; Endo, S.; Matsunaga, T.; Sakai, H.; Yamaguchi, M.; et al. Clathrin-dependent endocytosis of claudin-2 by DFYSP peptide causes lysosomal damage in lung adenocarcinoma A549 cells. *Biochim. Biophys. Acta* **2015**, *1848*, 2326–2336. [CrossRef] [PubMed]
41. Furuse, M.; Sasaki, H.; Tsukita, S. Manner of interaction of heterogeneous claudin species within and between tight junction strands. *J. Cell Biol.* **1999**, *147*, 891–903. [CrossRef] [PubMed]
42. Capaldo, C.T.; Farkas, A.E.; Hilgarth, R.S.; Krug, S.M.; Wolf, M.F.; Benedik, J.K.; Fromm, M.; Koval, M.; Parkos, C.; Nusrat, A. Proinflammatory cytokine-induced tight junction remodeling through dynamic self-assembly of claudins. *Mol. Biol. Cell* **2014**, *25*, 2710–2719. [CrossRef] [PubMed]
43. Angelow, S.; Schneeberger, E.E.; Yu, A.S. Claudin-8 expression in renal epithelial cells augments the paracellular barrier by replacing endogenous claudin-2. *J. Membr Biol.* **2007**, *215*, 147–159. [CrossRef] [PubMed]

44. Zhang, Y.; Yeh, S.; Appleton, B.A.; Held, H.A.; Kausalya, P.J.; Phua, D.C.; Wong, W.L.; Lasky, L.A.; Wiesmann, C.; Hunziker, W.; et al. Convergent and divergent ligand specificity among PDZ domains of the LAP and zonula occludens (ZO) families. *J. Biol. Chem.* **2006**, *281*, 22299–22311. [CrossRef] [PubMed]
45. Itoh, M.; Furuse, M.; Morita, K.; Kubota, K.; Saitou, M.; Tsukita, S. Direct binding of three tight junction-associated MAGUKs, ZO-1, ZO-2, and ZO-3, with the COOH termini of claudins. *J. Cell Biol.* **1999**, *147*, 1351–1363. [CrossRef] [PubMed]
46. Van Itallie, C.M.; Tietgens, A.J.; LoGrande, K.; Aponte, A.; Gucek, M.; Anderson, J.M. Phosphorylation of claudin-2 on serine 208 promotes membrane retention and reduces trafficking to lysosomes. *J. Cell Sci.* **2012**, *125*, 4902–4912. [CrossRef] [PubMed]
47. Yamauchi, K.; Rai, T.; Kobayashi, K.; Sohara, E.; Suzuki, T.; Itoh, T.; Suda, S.; Hayama, A.; Sasaki, S.; Uchida, S. Disease-causing mutant WNK4 increases paracellular chloride permeability and phosphorylates claudins. *Proc. Natl. Acad. Sci. USA* **2004**, *101*, 4690–4694. [CrossRef] [PubMed]
48. Fujii, N.; Matsuo, Y.; Matsunaga, T.; Endo, S.; Sakai, H.; Yamaguchi, M.; Yamazaki, Y.; Sugatani, J.; Ikari, A. Hypotonic Stress-Induced Down-Regulation of Claudin-1 and -2 Mediated by Dephosphorylation and Clathrin-Dependent Endocytosis in Renal Tubular Epithelial Cells. *J. Biol. Chem.* **2016**, *291*, 24787–24799. [CrossRef] [PubMed]
49. Van Itallie, C.M.; Mitic, L.L.; Anderson, J.M. SUMOylation of claudin-2. *Ann. N. Y. Acad. Sci.* **2012**, *1258*, 60–64. [CrossRef] [PubMed]
50. Molina-Jijon, E.; Rodriguez-Munoz, R.; Namorado Mdel, C.; Pedraza-Chaverri, J.; Reyes, J.L. Oxidative stress induces claudin-2 nitration in experimental type 1 diabetic nephropathy. *Free Radic. Biol. Med.* **2014**, *72*, 162–175. [CrossRef] [PubMed]
51. Dhawan, P.; Ahmad, R.; Chaturvedi, R.; Smith, J.J.; Midha, R.; Mittal, M.K.; Krishnan, M.; Chen, X.; Eschrich, S.; Yeatman, T.J.; et al. Claudin-2 expression increases tumorigenicity of colon cancer cells: Role of epidermal growth factor receptor activation. *Oncogene* **2011**, *30*, 3234–3247. [CrossRef] [PubMed]
52. Amoozadeh, Y.; Anwer, S.; Dan, Q.; Venugopal, S.; Shi, Y.; Branchard, E.; Liedtke, E.; Ailenberg, M.; Rotstein, O.D.; Kapus, A.; et al. Cell confluence regulates claudin-2 expression: Possible role for ZO-1 and Rac. *Am. J. Physiol. Cell Physiol.* **2018**, *314*, C366–C378. [CrossRef] [PubMed]
53. Amoozadeh, Y.; Dan, Q.; Xiao, J.; Waheed, F.; Szaszi, K. Tumor necrosis factor-alpha induces a biphasic change in claudin-2 expression in tubular epithelial cells: Role in barrier functions. *Am. J. Physiol. Cell Physiol.* **2015**, *309*, C38–C50. [CrossRef] [PubMed]
54. Sonoki, H.; Sato, T.; Endo, S.; Matsunaga, T.; Yamaguchi, M.; Yamazaki, Y.; Sugatani, J.; Ikari, A. Quercetin Decreases Claudin-2 Expression Mediated by Up-Regulation of microRNA miR-16 in Lung Adenocarcinoma A549 Cells. *Nutrients* **2015**, *7*, 4578–4592. [CrossRef] [PubMed]
55. Suzuki, T.; Yoshinaga, N.; Tanabe, S. Interleukin-6 (IL-6) regulates claudin-2 expression and tight junction permeability in intestinal epithelium. *J. Biol. Chem.* **2011**, *286*, 31263–31271. [CrossRef] [PubMed]
56. Mankertz, J.; Amasheh, M.; Krug, S.M.; Fromm, A.; Amasheh, S.; Hillenbrand, B.; Tavalali, S.; Fromm, M.; Schulzke, J.D. TNFalpha up-regulates claudin-2 expression in epithelial HT-29/B6 cells via phosphatidylinositol-3-kinase signaling. *Cell Tissue Res.* **2009**, *336*, 67–77. [CrossRef] [PubMed]
57. Singh, A.B.; Harris, R.C. Epidermal growth factor receptor activation differentially regulates claudin expression and enhances transepithelial resistance in Madin-Darby canine kidney cells. *J. Biol. Chem.* **2004**, *279*, 3543–3552. [CrossRef] [PubMed]
58. Ikari, A.; Takiguchi, A.; Atomi, K.; Sugatani, J. Epidermal growth factor increases clathrin-dependent endocytosis and degradation of claudin-2 protein in MDCK II cells. *J. Cell Physiol.* **2011**, *226*, 2448–2456. [CrossRef] [PubMed]
59. Kim, T.I.; Poulin, E.J.; Blask, E.; Bukhalid, R.; Whitehead, R.H.; Franklin, J.L.; Coffey, R.J. Myofibroblast keratinocyte growth factor reduces tight junctional integrity and increases claudin-2 levels in polarized Caco-2 cells. *Growth Factors* **2012**, *30*, 320–332. [CrossRef] [PubMed]
60. Ikari, A.; Sato, T.; Watanabe, R.; Yamazaki, Y.; Sugatani, J. Increase in claudin-2 expression by an EGFR/MEK/ERK/c-Fos pathway in lung adenocarcinoma A549 cells. *Biochim. Biophys. Acta* **2012**, *1823*, 1110–1118. [CrossRef] [PubMed]
61. Lipschutz, J.H.; Li, S.; Arisco, A.; Balkovetz, D.F. Extracellular signal-regulated kinases 1/2 control claudin-2 expression in Madin-Darby canine kidney strain I and II cells. *J. Biol. Chem.* **2005**, *280*, 3780–3788. [CrossRef] [PubMed]

62. Martin-Martin, N.; Ryan, G.; McMorrow, T.; Ryan, M.P. Sirolimus and cyclosporine A alter barrier function in renal proximal tubular cells through stimulation of ERK1/2 signaling and claudin-1 expression. *Am. J. Physiol. Renal Physiol.* **2010**, *298*, F672–F682. [CrossRef] [PubMed]
63. Ahmad, R.; Kumar, B.; Pan, K.; Dhawan, P.; Singh, A.B. HDAC-4 regulates claudin-2 expression in EGFR-ERK1/2 dependent manner to regulate colonic epithelial cell differentiation. *Oncotarget* **2017**, *8*, 87718–87736. [CrossRef] [PubMed]
64. Hichino, A.; Okamoto, M.; Taga, S.; Akizuki, R.; Endo, S.; Matsunaga, T.; Ikari, A. Down-regulation of Claudin-2 Expression and Proliferation by Epigenetic Inhibitors in Human Lung Adenocarcinoma A549 Cells. *J. Biol. Chem.* **2017**, *292*, 2411–2421. [CrossRef] [PubMed]
65. Wang, Y.; Mumm, J.B.; Herbst, R.; Kolbeck, R.; Wang, Y. IL-22 Increases Permeability of Intestinal Epithelial Tight Junctions by Enhancing Claudin-2 Expression. *J. Immunol.* **2017**, *199*, 3316–3325. [CrossRef] [PubMed]
66. Tian, L.; Li, Y.; Zhang, J.; Chang, R.; Li, J.; Huo, L. IL-9 promotes the pathogenesis of ulcerative colitis through STAT3/SOCS3 signaling. *Biosci. Rep.* **2018**, *38*, 9476–9487. [CrossRef] [PubMed]
67. Dan, Q.; Shi, Y.; Rabani, R.; Venugopal, S.; Xiao, J.; Anwer, S.; Ding, M.; Speight, P.; Pan, W.; Alexander, R.T.; et al. Claudin-2 suppresses GEF-H1, RHOA, and MRTF thereby impacting proliferation and profibrotic phenotype of tubular cells. *J. Biol. Chem.* **2019**, *294*, 15446–15465. [CrossRef] [PubMed]
68. Tabaries, S.; Annis, M.G.; Hsu, B.E.; Tam, C.E.; Savage, P.; Park, M.; Siegel, P.M. Lyn modulates Claudin-2 expression and is a therapeutic target for breast cancer liver metastasis. *Oncotarget* **2015**, *6*, 9476–9487. [CrossRef] [PubMed]
69. Garcia-Hernandez, V.; Flores-Maldonado, C.; Rincon-Heredia, R.; Verdejo-Torres, O.; Bonilla-Delgado, J.; Meneses-Morales, I.; Gariglio, P.; Contreras, R.G. EGF regulates claudin-2 and -4 expression through Src and STAT3 in MDCK cells. *J. Cell Physiol.* **2015**, *230*, 105–115. [CrossRef] [PubMed]
70. Ikari, A.; Fujii, N.; Hahakabe, S.; Hayashi, H.; Yamaguchi, M.; Yamazaki, Y.; Endo, S.; Matsunaga, T.; Sugatani, J. Hyperosmolarity-Induced Down-Regulation of Claudin-2 Mediated by Decrease in PKCbeta-Dependent GATA-2 in MDCK Cells. *J. Cell Physiol.* **2015**, *230*, 2776–2787. [CrossRef] [PubMed]
71. Lu, R.; Dalgalan, D.; Mandell, E.K.; Parker, S.S.; Ghosh, S.; Wilson, J.M. PKCiota interacts with Rab14 and modulates epithelial barrier function through regulation of claudin-2 levels. *Mol. Biol. Cell* **2015**, *26*, 1523–1531. [CrossRef] [PubMed]
72. Al-Sadi, R.; Ye, D.; Boivin, M.; Guo, S.; Hashimi, M.; Ereifej, L.; Ma, T.Y. Interleukin-6 modulation of intestinal epithelial tight junction permeability is mediated by JNK pathway activation of claudin-2 gene. *PLoS ONE* **2014**, *9*, e85345. [CrossRef] [PubMed]
73. Zhang, Y.G.; Wu, S.; Xia, Y.; Sun, J. Salmonella infection upregulates the leaky protein claudin-2 in intestinal epithelial cells. *PLoS ONE* **2013**, *8*, e58606. [CrossRef] [PubMed]
74. Mankertz, J.; Hillenbrand, B.; Tavalali, S.; Huber, O.; Fromm, M.; Schulzke, J.D. Functional crosstalk between Wnt signaling and Cdx-related transcriptional activation in the regulation of the claudin-2 promoter activity. *Biochem. Biophys. Res. Commun.* **2004**, *314*, 1001–1007. [CrossRef] [PubMed]
75. Guillemot, L.; Citi, S. Cingulin regulates claudin-2 expression and cell proliferation through the small GTPase RhoA. *Mol. Biol. Cell* **2006**, *17*, 3569–3577. [CrossRef] [PubMed]
76. Bruewer, M.; Hopkins, A.M.; Hobert, M.E.; Nusrat, A.; Madara, J.L. RhoA, Rac1, and Cdc42 exert distinct effects on epithelial barrier via selective structural and biochemical modulation of junctional proteins and F-actin. *Am. J. Physiol. Cell Physiol.* **2004**, *287*, C327–C335. [CrossRef] [PubMed]
77. Wang, Y.B.; Shi, Q.; Li, G.; Zheng, J.H.; Lin, J.; Qiu, W. MicroRNA-488 inhibits progression of colorectal cancer via inhibition of the mitogen-activated protein kinase pathway by targeting claudin-2. *Am. J. Physiol. Cell Physiol.* **2019**, *316*, C33–C47. [CrossRef] [PubMed]
78. Che, M.; Shi, T.; Feng, S.; Li, H.; Zhang, X.; Feng, N.; Lou, W.; Dou, J.; Tang, G.; Huang, C.; et al. The MicroRNA-199a/214 Cluster Targets E-Cadherin and Claudin-2 and Promotes High Glucose-Induced Peritoneal Fibrosis. *J. Am. Soc. Nephrol.* **2017**, *28*, 2459–2471. [CrossRef] [PubMed]
79. Zhang, Y.G.; Wu, S.; Lu, R.; Zhou, D.; Zhou, J.; Carmeliet, G.; Petrof, E.; Claud, E.C.; Sun, J. Tight junction CLDN2 gene is a direct target of the vitamin D receptor. *Sci. Rep.* **2015**, *5*, 10642. [CrossRef] [PubMed]
80. Lepage, D.; Belanger, E.; Jones, C.; Tremblay, S.; Allaire, J.M.; Bruneau, J.; Asselin, C.; Perreault, N.; Menendez, A.; Gendron, F.P.; et al. Gata4 is critical to maintain gut barrier function and mucosal integrity following epithelial injury. *Sci. Rep.* **2016**, *6*, 36776. [CrossRef] [PubMed]
81. Guo, R.J.; Suh, E.R.; Lynch, J.P. The role of Cdx proteins in intestinal development and cancer. *Cancer Biol.*

Ther. **2004**, *3*, 593–601. [CrossRef] [PubMed]

82. Song, X.; Chen, H.X.; Wang, X.Y.; Deng, X.Y.; Xi, Y.X.; He, Q.; Peng, T.L.; Chen, J.; Chen, W.; Wong, B.C.; et al. pylori-encoded CagA disrupts tight junctions and induces invasiveness of AGS gastric carcinoma cells via Cdx2-dependent targeting of Claudin-2. *Cell Immunol.* **2013**, *286*, 22–30. [CrossRef] [PubMed]
83. Sakamoto, H.; Mutoh, H.; Sugano, K. Expression of Claudin-2 in intestinal metaplastic mucosa of Cdx2-transgenic mouse stomach. *Scand J. Gastroenterol.* **2010**, *45*, 1273–1280. [CrossRef] [PubMed]
84. Wang, W.; Chen, S.W.; Zhu, J.; Zuo, S.; Ma, Y.Y.; Chen, Z.Y.; Zhang, J.L.; Chen, G.W.; Liu, Y.C.; Wang, P.Y. Intestinal alkaline phosphatase inhibits the translocation of bacteria of gut-origin in mice with peritonitis: Mechanism of action. *PLoS ONE* **2015**, *10*, e0124835. [CrossRef] [PubMed]
85. Peter, Y.; Comellas, A.; Levantini, E.; Ingenito, E.P.; Shapiro, S.D. Epidermal growth factor receptor and claudin-2 participate in A549 permeability and remodeling: Implications for non-small cell lung cancer tumor colonization. *Mol. Carcinog.* **2009**, *48*, 488–497. [CrossRef] [PubMed]
86. Wang, Y.; Shi, J.; Chai, K.; Ying, X.; Zhou, B.P. The Role of Snail in EMT and Tumorigenesis. *Curr. Cancer Drug Targets* **2013**, *13*, 963–972. [CrossRef] [PubMed]
87. Burns, W.C.; Thomas, M.C. The molecular mediators of type 2 epithelial to mesenchymal transition (EMT) and their role in renal pathophysiology. *Expert Rev. Mol. Med.* **2010**, *12*, e17. [CrossRef] [PubMed]
88. Carrozzino, F.; Soulie, P.; Huber, D.; Mensi, N.; Orci, L.; Cano, A.; Feraille, E.; Montesano, R. Inducible expression of Snail selectively increases paracellular ion permeability and differentially modulates tight junction proteins. *Am. J. Physiol. Cell Physiol.* **2005**, *289*, C1002–C1014. [CrossRef] [PubMed]
89. Dukes, J.D.; Whitley, P.; Chalmers, A.D. The PIKfyve inhibitor YM201636 blocks the continuous recycling of the tight junction proteins claudin-1 and claudin-2 in MDCK cells. *PLoS ONE* **2012**, *7*, e28659. [CrossRef] [PubMed]
90. Vasileva, E.; Citi, S. The role of microtubules in the regulation of epithelial junctions. *Tissue Barriers* **2018**, *6*, 1539596. [CrossRef] [PubMed]
91. Shen, L. Tight junctions on the move: Molecular mechanisms for epithelial barrier regulation. *Ann. N. Y. Acad. Sci.* **2012**, *1258*, 9–18. [CrossRef] [PubMed]
92. Nusrat, A.; Turner, J.R.; Madara, J.L. Molecular physiology and pathophysiology of tight junctions. IV. Regulation of tight junctions by extracellular stimuli: Nutrients, cytokines, and immune cells. *Am. J. Physiol. Gastrointest Liver Physiol.* **2000**, *279*, G851–G857. [CrossRef] [PubMed]
93. Van Itallie, C.M.; Colegio, O.R.; Anderson, J.M. The cytoplasmic tails of claudins can influence tight junction barrier properties through effects on protein stability. *J. Membr. Biol.* **2004**, *199*, 29–38. [CrossRef] [PubMed]
94. Van Itallie, C.M.; Fanning, A.S.; Bridges, A.; Anderson, J.M. ZO-1 stabilizes the tight junction solute barrier through coupling to the perijunctional cytoskeleton. *Mol. Biol. Cell* **2009**, *20*, 3930–3940. [CrossRef] [PubMed]
95. Raya-Sandino, A.; Castillo-Kauil, A.; Dominguez-Calderon, A.; Alarcon, L.; Flores-Benitez, D.; Cuellar-Perez, F.; Lopez-Bayghen, B.; Chavez-Munguia, B.; Vazquez-Prado, J.; Gonzalez-Mariscal, L. Zonula occludens-2 regulates Rho proteins activity and the development of epithelial cytoarchitecture and barrier function. *Biochim. Biophys. Acta* **2017**, *1864*, 1714–1733. [CrossRef] [PubMed]
96. Ares, G.; Buonpane, C.; Sincavage, J.; Yuan, C.; Wood, D.R.; Hunter, C.J. Caveolin 1 is Associated with Upregulated Claudin 2 in Necrotizing Enterocolitis. *Sci. Rep.* **2019**, *9*, 4982. [CrossRef] [PubMed]
97. Itallie, C.M.; Anderson, J.M. Caveolin binds independently to claudin-2 and occludin. *Ann. N. Y. Acad. Sci.* **2012**, *1257*, 103–107. [CrossRef] [PubMed]
98. Marchiando, A.M.; Shen, L.; Graham, W.V.; Weber, C.R.; Schwarz, B.T.; Austin, J.R., 2nd; Raleigh, D.R.; Guan, Y.; Watson, A.J.; Montrose, M.H.; et al. Caveolin-1-dependent occludin endocytosis is required for TNF-induced tight junction regulation in vivo. *J. Cell Biol.* **2010**, *189*, 111–126. [CrossRef] [PubMed]
99. Cunningham, K.E.; Turner, J.R. Myosin light chain kinase: Pulling the strings of epithelial tight junction function. *Ann. N. Y. Acad. Sci.* **2012**, *1258*, 34–42. [CrossRef] [PubMed]
100. Nighot, P.K.; Hu, C.A.; Ma, T.Y. Autophagy enhancement of intestinal epithelial tight junction barrier function by targeting claudin-2 degradation. *J. Biol. Chem.* **2015**, *290*, 7234–7246. [CrossRef] [PubMed]
101. Zhang, C.; Yan, J.; Xiao, Y.; Shen, Y.; Wang, J.; Ge, W.; Chen, Y. Inhibition of Autophagic Degradation Process Contributes to Claudin-2 Expression Increase and Epithelial Tight Junction Dysfunction in TNF-alpha Treated Cell Monolayers. *Int. J. Mol. Sci.* **2017**, *18*, 910–919.
102. Huang, L.; Jiang, Y.; Sun, Z.; Gao, Z.; Wang, J.; Zhang, D. Autophagy Strengthens Intestinal Mucosal Barrier by Attenuating Oxidative Stress in Severe Acute Pancreatitis. *Dig. Dis. Sci.* **2018**, *63*, 910–919. [CrossRef] [PubMed]

103. Ikari, A.; Watanabe, R.; Sato, T.; Taga, S.; Shimobaba, S.; Yamaguchi, M.; Yamazaki, Y.; Endo, S.; Matsunaga, T.; Sugatani, J. Nuclear distribution of claudin-2 increases cell proliferation in human lung adenocarcinoma cells. *Biochim. Biophys. Acta* **2014**, *1843*, 2079–2088. [CrossRef] [PubMed]
104. Rosenthal, R.; Milatz, S.; Krug, S.M.; Oelrich, B.; Schulzke, J.D.; Amasheh, S.; Gunzel, D.; Fromm, M. Claudin-2, a component of the tight junction, forms a paracellular water channel. *J. Cell Sci.* **2010**, *123*, 1913–1921. [CrossRef] [PubMed]
105. Rosenthal, R.; Gunzel, D.; Krug, S.M.; Schulzke, J.D.; Fromm, M.; Yu, A.S. Claudin-2-mediated cation and water transport share a common pore. *Acta Physiol. (Oxf)* **2017**, *219*, 521–536. [CrossRef] [PubMed]
106. Rosenthal, R.; Gunzel, D.; Theune, D.; Czichos, C.; Schulzke, J.D.; Fromm, M. Water channels and barriers formed by claudins. *Ann. N. Y. Acad. Sci.* **2017**, *1397*, 100–109. [CrossRef] [PubMed]
107. Borovac, J.; Barker, R.S.; Rievaj, J.; Rasmussen, A.; Pan, W.; Wevrick, R.; Alexander, R.T. Claudin-4 forms a paracellular barrier, revealing the interdependence of claudin expression in the loose epithelial cell culture model opossum kidney cells. *Am. J. Physiol. Cell Physiol.* **2012**, *303*, C1278–C1291. [CrossRef] [PubMed]
108. Tanaka, H.; Tamura, A.; Suzuki, K.; Tsukita, S. Site-specific distribution of claudin-based paracellular channels with roles in biological fluid flow and metabolism. *Ann. N. Y. Acad. Sci.* **2017**, *1405*, 44–52. [CrossRef] [PubMed]
109. Muto, S.; Hata, M.; Taniguchi, J.; Tsuruoka, S.; Moriwaki, K.; Saitou, M.; Furuse, K.; Sasaki, H.; Fujimura, A.; Imai, M.; et al. Claudin-2-deficient mice are defective in the leaky and cation-selective paracellular permeability properties of renal proximal tubules. *Proc. Natl. Acad. Sci. USA* **2010**, *107*, 8011–8016. [CrossRef] [PubMed]
110. Pei, L.; Solis, G.; Nguyen, M.T.; Kamat, N.; Magenheimer, L.; Zhuo, M.; Li, J.; Curry, J.; McDonough, A.A.; Fields, T.A.; et al. Paracellular epithelial sodium transport maximizes energy efficiency in the kidney. *J. Clin. Investig.* **2016**, *126*, 2509–2518. [CrossRef] [PubMed]
111. Schnermann, J.; Huang, Y.; Mizel, D. Fluid reabsorption in proximal convoluted tubules of mice with gene deletions of claudin-2 and/or aquaporin1. *Am. J. Physiol. Renal Physiol.* **2013**, *305*, F1352–F1364. [CrossRef] [PubMed]
112. Matsumoto, K.; Imasato, M.; Yamazaki, Y.; Tanaka, H.; Watanabe, M.; Eguchi, H.; Nagano, H.; Hikita, H.; Tatsumi, T.; Takehara, T.; et al. Claudin 2 deficiency reduces bile flow and increases susceptibility to cholesterol gallstone disease in mice. *Gastroenterology* **2014**, *147*, 1134–1145. [CrossRef] [PubMed]
113. Tamura, A.; Hayashi, H.; Imasato, M.; Yamazaki, Y.; Hagiwara, A.; Wada, M.; Noda, T.; Watanabe, M.; Suzuki, Y.; Tsukita, S. Loss of claudin-15, but not claudin-2, causes Na+ deficiency and glucose malabsorption in mouse small intestine. *Gastroenterology* **2011**, *140*, 913–923. [CrossRef] [PubMed]
114. Wada, M.; Tamura, A.; Takahashi, N.; Tsukita, S. Loss of claudins 2 and 15 from mice causes defects in paracellular Na+ flow and nutrient transport in gut and leads to death from malnutrition. *Gastroenterology* **2013**, *144*, 369–380. [CrossRef] [PubMed]
115. Sukka-Ganesh, B.; Mohammed, K.A.; Kaye, F.; Goldberg, E.P.; Nasreen, N. Ephrin-A1 inhibits NSCLC tumor growth via induction of Cdx-2 a tumor suppressor gene. *BMC Cancer* **2012**, *12*, 309. [CrossRef] [PubMed]
116. Ahmad, R.; Chaturvedi, R.; Olivares-Villagomez, D.; Habib, T.; Asim, M.; Shivesh, P.; Polk, D.B.; Wilson, K.T.; Washington, M.K.; Van Kaer, L.; et al. Targeted colonic claudin-2 expression renders resistance to epithelial injury, induces immune suppression, and protects from colitis. *Mucosal Immunol.* **2014**, *7*, 1340–1353. [CrossRef] [PubMed]
117. Sonoki, H.; Tanimae, A.; Endo, S.; Matsunaga, T.; Furuta, T.; Ichihara, K.; Ikari, A. Kaempherol and Luteolin Decrease Claudin-2 Expression Mediated by Inhibition of STAT3 in Lung Adenocarcinoma A549 Cells. *Nutrients* **2017**, *9*, 597. [CrossRef] [PubMed]
118. Buchert, M.; Papin, M.; Bonnans, C.; Darido, C.; Raye, W.S.; Garambois, V.; Pelegrin, A.; Bourgaux, J.F.; Pannequin, J.; Joubert, D.; et al. Symplekin promotes tumorigenicity by up-regulating claudin-2 expression. *Proc. Natl. Acad. Sci. USA* **2010**, *107*, 2628–2633. [CrossRef] [PubMed]
119. Maruhashi, R.; Akizuki, R.; Sato, T.; Matsunaga, T.; Endo, S.; Yamaguchi, M.; Yamazaki, Y.; Sakai, H.; Ikari, A. Elevation of sensitivity to anticancer agents of human lung adenocarcinoma A549 cells by knockdown of claudin-2 expression in monolayer and spheroid culture models. *Biochim. Biophys. Acta Mol. Cell Res.* **2018**, *1865*, 470–479. [CrossRef] [PubMed]
120. Ikari, A.; Sato, T.; Takiguchi, A.; Atomi, K.; Yamazaki, Y.; Sugatani, J. Claudin-2 knockdown decreases matrix metalloproteinase-9 activity and cell migration via suppression of nuclear Sp1 in A549 cells. *Life Sci.* **2011**, *88*, 628–633. [CrossRef] [PubMed]

121. Behrens, A.; Sibilia, M.; Wagner, E.F. Amino-terminal phosphorylation of c-Jun regulates stress-induced apoptosis and cellular proliferation. *Nat. Genet.* **1999**, *21*, 326–329. [CrossRef] [PubMed]
122. Paquet-Fifield, S.; Koh, S.L.; Cheng, L.; Beyit, L.M.; Shembrey, C.; Molck, C.; Behrenbruch, C.; Papin, M.; Gironella, M.; Guelfi, S.; et al. Tight Junction Protein Claudin-2 Promotes Self-Renewal of Human Colorectal Cancer Stem-like Cells. *Cancer Res.* **2018**, *78*, 2925–2938. [CrossRef] [PubMed]
123. Larre, I.; Castillo, A.; Flores-Maldonado, C.; Contreras, R.G.; Galvan, I.; Munoz-Estrada, J.; Cereijido, M. Ouabain modulates ciliogenesis in epithelial cells. *Proc. Natl. Acad. Sci. USA* **2011**, *108*, 20591–20596. [CrossRef] [PubMed]
124. Sanchez, I.; Dynlacht, B.D. Cilium assembly and disassembly. *Nat. Cell Biol.* **2016**, *18*, 711–717. [CrossRef] [PubMed]
125. Mima, S.; Takehara, M.; Takada, H.; Nishimura, T.; Hoshino, T.; Mizushima, T. NSAIDs suppress the expression of claudin-2 to promote invasion activity of cancer cells. *Carcinogenesis* **2008**, *29*, 1994–2000. [CrossRef] [PubMed]
126. Dreymueller, D.; Theodorou, K.; Donners, M.; Ludwig, A. Fine Tuning Cell Migration by a Disintegrin and Metalloproteinases. *Mediators Inflamm.* **2017**, *2017*, 9621724. [CrossRef] [PubMed]
127. Ikari, A.; Takiguchi, A.; Atomi, K.; Sato, T.; Sugatani, J. Decrease in claudin-2 expression enhances cell migration in renal epithelial Madin-Darby canine kidney cells. *J. Cell Physiol.* **2011**, *226*, 1471–1478. [CrossRef] [PubMed]
128. Takehara, M.; Nishimura, T.; Mima, S.; Hoshino, T.; Mizushima, T. Effect of claudin expression on paracellular permeability, migration and invasion of colonic cancer cells. *Biol. Pharm Bull.* **2009**, *32*, 825–831. [CrossRef] [PubMed]
129. Du, W.; Xu, X.; Niu, Q.; Zhang, X.; Wei, Y.; Wang, Z.; Zhang, W.; Yan, J.; Ru, Y.; Fu, Z.; et al. Spi-B-Mediated Silencing of Claudin-2 Promotes Early Dissemination of Lung Cancer Cells from Primary Tumors. *Cancer Res.* **2017**, *77*, 4809–4822. [CrossRef] [PubMed]
130. Zhang, X.; Wang, H.; Li, Q.; Li, T. CLDN2 inhibits the metastasis of osteosarcoma cells via down-regulating the afadin/ERK signaling pathway. *Cancer Cell Int.* **2018**, *18*, 160. [CrossRef] [PubMed]
131. Campbell, K.; Casanova, J. A common framework for EMT and collective cell migration. *Development* **2016**, *143*, 4291–4300. [CrossRef] [PubMed]
132. Chaffer, C.L.; San Juan, B.P.; Lim, E.; Weinberg, R.A. EMT, cell plasticity and metastasis. *Cancer Metastasis Rev.* **2016**, *35*, 645–654. [CrossRef] [PubMed]
133. Che, J.; Yue, D.; Zhang, B.; Zhang, H.; Huo, Y.; Gao, L.; Zhen, H.; Yang, Y.; Cao, B. Claudin-3 Inhibits Lung Squamous Cell Carcinoma Cell Epithelial-mesenchymal Transition and Invasion via Suppression of the Wnt/beta-catenin Signaling Pathway. *Int. J. Med. Sci.* **2018**, *15*, 339–351. [CrossRef] [PubMed]
134. Suh, Y.; Yoon, C.H.; Kim, R.K.; Lim, E.J.; Oh, Y.S.; Hwang, S.G.; An, S.; Yoon, G.; Gye, M.C.; Yi, J.M.; et al. Claudin-1 induces epithelial-mesenchymal transition through activation of the c-Abl-ERK signaling pathway in human liver cells. *Oncogene* **2013**, *32*, 4873–4882. [CrossRef] [PubMed]
135. Nishida, M.; Yoshida, M.; Nishiumi, S.; Furuse, M.; Azuma, T. Claudin-2 Regulates Colorectal Inflammation via Myosin Light Chain Kinase-Dependent Signaling. *Dig. Dis. Sci.* **2013**, *58*, 1546–1559. [CrossRef] [PubMed]
136. Guillemot, L.; Paschoud, S.; Pulimeno, P.; Foglia, A.; Citi, S. The cytoplasmic plaque of tight junctions: A scaffolding and signalling center. *Biochim. Biophys. Acta* **2008**, *1778*, 601–613. [CrossRef] [PubMed]
137. Balda, M.S.; Matter, K. Tight junctions as regulators of tissue remodelling. *Curr. Opin. Cell Biol.* **2016**, *42*, 94–101. [CrossRef] [PubMed]
138. Hartsock, A.; Nelson, W.J. Adherens and tight junctions: Structure, function and connections to the actin cytoskeleton. *Biochim. Biophys. Acta* **2008**, *1778*, 660–669. [CrossRef] [PubMed]
139. Herve, J.C.; Derangeon, M.; Sarrouilhe, D.; Bourmeyster, N. Influence of the scaffolding protein Zonula Occludens (ZOs) on membrane channels. *Biochim. Biophys. Acta* **2014**, *1838*, 595–604. [CrossRef] [PubMed]
140. Yano, T.; Matsui, T.; Tamura, A.; Uji, M.; Tsukita, S. The association of microtubules with tight junctions is promoted by cingulin phosphorylation by AMPK. *J. Cell Biol.* **2013**, *203*, 605–614. [CrossRef] [PubMed]
141. Yano, T.; Torisawa, T.; Oiwa, K.; Tsukita, S. AMPK-dependent phosphorylation of cingulin reversibly regulates its binding to actin filaments and microtubules. *Sci. Rep.* **2018**, *8*, 15550. [CrossRef] [PubMed]
142. Pollack, V.; Sarkozi, R.; Banki, Z.; Feifel, E.; Wehn, S.; Gstraunthaler, G.; Stoiber, H.; Mayer, G.; Montesano, R.; Strutz, F.; et al. Oncostatin M-induced effects on EMT in human proximal tubular cells: Differential role of ERK signaling. *Am. J. Physiol. Renal Physiol.* **2007**, *293*, F1714–F1726. [CrossRef] [PubMed]

143. Kinugasa, T.; Sakaguchi, T.; Gu, X.; Reinecker, H.C. Claudins regulate the intestinal barrier in response to immune mediators. *Gastroenterology* **2000**, *118*, 1001–1011. [CrossRef]
144. Kinugasa, T.; Huo, Q.; Higashi, D.; Shibaguchi, H.; Kuroki, M.; Tanaka, T.; Futami, K.; Yamashita, Y.; Hachimine, K.; Maekawa, S.; et al. Selective up-regulation of claudin-1 and claudin-2 in colorectal cancer. *Anticancer Res.* **2007**, *27*, 3729–3734. [CrossRef]
145. Moldvay, J.; Jackel, M.; Paska, C.; Soltesz, I.; Schaff, Z.; Kiss, A. Distinct claudin expression profile in histologic subtypes of lung cancer. *Lung Cancer* **2007**, *57*, 159–167. [CrossRef] [PubMed]
146. Xin, S.; Huixin, C.; Benchang, S.; Aiping, B.; Jinhui, W.; Xiaoyan, L.; Yu, W.B.; Minhu, C. Expression of Cdx2 and claudin-2 in the multistage tissue of gastric carcinogenesis. *Oncology* **2007**, *73*, 357–365. [CrossRef] [PubMed]
147. Lin, Z.; Zhang, X.; Liu, Z.; Liu, Q.; Wang, L.; Lu, Y.; Liu, Y.; Wang, M.; Yang, M.; Jin, X.; et al. The distinct expression patterns of claudin-2, -6, and -11 between human gastric neoplasms and adjacent non-neoplastic tissues. *Diagn. Pathol.* **2013**, *8*, 133. [CrossRef] [PubMed]
148. Virman, J.; Soini, Y.; Kujala, P.; Luukkaala, T.; Salminen, T.; Sunela, K.; Kellokumpu-Lehtinen, P.L. Claudins as prognostic factors for renal cell cancer. *Anticancer Res.* **2014**, *34*, 4181–4187. [PubMed]
149. Song, X.; Li, X.; Tang, Y.; Chen, H.; Wong, B.; Wang, J.; Chen, M. Expression of claudin-2 in the multistage process of gastric carcinogenesis. *Histol. Histopathol.* **2008**, *23*, 673–682. [PubMed]
150. Jung, H.; Jun, K.H.; Jung, J.H.; Chin, H.M.; Park, W.B. The expression of claudin-1, claudin-2, claudin-3, and claudin-4 in gastric cancer tissue. *J. Surg. Res.* **2011**, *167*, e185–e191. [CrossRef] [PubMed]
151. Chen, S.Y.; Zhang, R.G.; Duan, G.C. Pathogenic mechanisms of the oncoprotein CagA in H. pylori-induced gastric cancer (Review). *Oncol. Rep.* **2016**, *36*, 3087–3094. [CrossRef] [PubMed]
152. Basu, S.; Haase, G.; Ben-Ze'ev, A. Wnt signaling in cancer stem cells and colon cancer metastasis. *F1000Research* **2016**, *5*. [CrossRef] [PubMed]
153. Weber, C.R.; Nalle, S.C.; Tretiakova, M.; Rubin, D.T.; Turner, J.R. Claudin-1 and claudin-2 expression is elevated in inflammatory bowel disease and may contribute to early neoplastic transformation. *Lab. Investig.* **2008**, *88*, 1110–1120. [CrossRef] [PubMed]
154. Mezheyeuski, A.; Strell, C.; Hrynchyk, I.; Guren, T.K.; Dragomir, A.; Doroshenko, T.; Pashkova, O.; Gorgun, J.; Ruksha, K.; Pfeiffer, P.; et al. Treatment-related survival associations of claudin-2 expression in fibroblasts of colorectal cancer. *Virchows Arch.* **2018**, *472*, 395–405. [CrossRef] [PubMed]
155. Soini, Y. Claudins 2, 3, 4, and 5 in Paget's disease and breast carcinoma. *Hum. Pathol.* **2004**, *35*, 1531–1536. [CrossRef] [PubMed]
156. Kim, T.H.; Huh, J.H.; Lee, S.; Kang, H.; Kim, G.I.; An, H.J. Down-regulation of claudin-2 in breast carcinomas is associated with advanced disease. *Histopathology* **2008**, *53*, 48–55. [CrossRef] [PubMed]
157. Tabaries, S.; Dong, Z.; Annis, M.G.; Omeroglu, A.; Pepin, F.; Ouellet, V.; Russo, C.; Hassanain, M.; Metrakos, P.; Diaz, Z.; et al. Claudin-2 is selectively enriched in and promotes the formation of breast cancer liver metastases through engagement of integrin complexes. *Oncogene* **2011**, *30*, 1318–1328. [CrossRef] [PubMed]
158. Kimbung, S.; Kovacs, A.; Bendahl, P.O.; Malmstrom, P.; Ferno, M.; Hatschek, T.; Hedenfalk, I. Claudin-2 is an independent negative prognostic factor in breast cancer and specifically predicts early liver recurrences. *Mol. Oncol.* **2014**, *8*, 119–128. [CrossRef] [PubMed]
159. Barmeyer, C.; Fromm, M.; Schulzke, J.D. Active and passive involvement of claudins in the pathophysiology of intestinal inflammatory diseases. *Pflugers Arch.* **2017**, *469*, 15–26. [CrossRef] [PubMed]
160. Landy, J.; Ronde, E.; English, N.; Clark, S.K.; Hart, A.L.; Knight, S.C.; Ciclitira, P.J.; Al-Hassi, H.O. Tight junctions in inflammatory bowel diseases and inflammatory bowel disease associated colorectal cancer. *World J. Gastroenterol.* **2016** *22*, 3117–3126. [CrossRef] [PubMed]
161. Amasheh, M.; Fromm, A.; Krug, S.M.; Amasheh, S.; Andres, S.; Zeitz, M.; Fromm, M.; Schulzke, J.D. TNFalpha-induced and berberine-antagonized tight junction barrier impairment via tyrosine kinase, Akt and NFkappaB signaling. *J. Cell Sci.* **2010**, *123*, 4145–4155. [CrossRef] [PubMed]
162. Zeissig, S.; Burgel, N.; Gunzel, D.; Richter, J.; Mankertz, J.; Wahnschaffe, U.; Kroesen, A.J.; Zeitz, M.; Fromm, M.; Schulzke, J.D. Changes in expression and distribution of claudin 2, 5 and 8 lead to discontinuous tight junctions and barrier dysfunction in active Crohn's disease. *Gut* **2007**, *56*, 61–72. [CrossRef] [PubMed]
163. Luettig, J.; Rosenthal, R.; Barmeyer, C.; Schulzke, J.D. Claudin-2 as a mediator of leaky gut barrier during intestinal inflammation. *Tissue Barriers* **2015**, *3*, e977176. [CrossRef] [PubMed]
164. Aghdassi, A.A.; Weiss, F.U.; Mayerle, J.; Lerch, M.M.; Simon, P. Genetic susceptibility factors for alcohol-induced chronic pancreatitis. *Pancreatology* **2015**, *15*, S23–S31. [CrossRef] [PubMed]

165. Gong, Y.; Hou, J. Claudins in barrier and transport function-the kidney. *Pflugers Arch.* **2017**, *469*, 105–113. [CrossRef] [PubMed]
166. Fromm, M.; Piontek, J.; Rosenthal, R.; Gunzel, D.; Krug, S.M. Tight junctions of the proximal tubule and their channel proteins. *Pflugers Arch.* **2017**, *469*, 877–887. [CrossRef] [PubMed]
167. Szaszi, K.; Amoozadeh, Y. New insights into functions, regulation, and pathological roles of tight junctions in kidney tubular epithelium. *Int. Rev. Cell Mol. Biol.* **2014**, *308*, 205–271. [PubMed]
168. Balkovetz, D.F.; Chumley, P.; Amlal, H. Downregulation of claudin-2 expression in renal epithelial cells by metabolic acidosis. *Am. J. Physiol. Renal Physiol.* **2009**, *297*, F604–F611. [CrossRef] [PubMed]
169. Gonzalez, J.E.; DiGeronimo, R.J.; Arthur, D.E.; King, J.M. Remodeling of the tight junction during recovery from exposure to hydrogen peroxide in kidney epithelial cells. *Free Radic Biol. Med.* **2009**, *47*, 1561–1569. [CrossRef] [PubMed]
170. Singh, A.B.; Sugimoto, K.; Dhawan, P.; Harris, R.C. Juxtacrine activation of EGFR regulates claudin expression and increases transepithelial resistance. *Am. J. Physiol. Cell Physiol.* **2007**, *293*, C1660–C1668. [CrossRef] [PubMed]
171. Maggiorani, D.; Dissard, R.; Belloy, M.; Saulnier-Blache, J.S.; Casemayou, A.; Ducasse, L.; Gres, S.; Belliere, J.; Caubet, C.; Bascands, J.L.; et al. Shear Stress-Induced Alteration of Epithelial Organization in Human Renal Tubular Cells. *PLoS ONE* **2015**, *10*, e0131416. [CrossRef] [PubMed]
172. Gamero-Estevez, E.; Andonian, S.; Jean-Claude, B.; Gupta, I.; Ryan, A.K. Temporal Effects of Quercetin on Tight Junction Barrier Properties and Claudin Expression and Localization in MDCK II Cells. *Int. J. Mol. Sci.* **2019**, *20*, 4889. [CrossRef] [PubMed]
173. Martin-Martin, N.; Dan, Q.; Amoozadeh, Y.; Waheed, F.; McMorrow, T.; Ryan, M.P.; Szaszi, K. RhoA and Rho kinase mediate cyclosporine A and sirolimus-induced barrier tightening in renal proximal tubular cells. *Int. J. Biochem. Cell Biol.* **2012**, *44*, 178–188. [CrossRef] [PubMed]
174. Trujillo, J.; Molina-Jijon, E.; Medina-Campos, O.N.; Rodriguez-Munoz, R.; Reyes, J.L.; Loredo, M.L.; Tapia, E.; Sanchez-Lozada, L.G.; Barrera-Oviedo, D.; Pedraza-Chaverri, J. Renal tight junction proteins are decreased in cisplatin-induced nephrotoxicity in rats. *Toxicol. Mech. Methods* **2014**, *24*, 520–528. [CrossRef] [PubMed]
175. Bialik, J.F.; Ding, M.; Speight, P.; Dan, Q.; Miranda, M.Z.; Di Ciano-Oliveira, C.; Kofler, M.M.; Rotstein, O.D.; Pedersen, S.F.; Szaszi, K.; et al. Profibrotic epithelial phenotype: A central role for MRTF and TAZ. *Sci. Rep.* **2019**, *9*, 4323. [CrossRef] [PubMed]
176. Takigawa, M.; Iida, M.; Nagase, S.; Suzuki, H.; Watari, A.; Tada, M.; Okada, Y.; Doi, T.; Fukasawa, M.; Yagi, K.; et al. Creation of a Claudin-2 Binder and Its Tight Junction-Modulating Activity in a Human Intestinal Model. *J. Pharmacol. Exp. Ther.* **2017**, *363*, 444–451. [CrossRef] [PubMed]
177. Hashimoto, Y.; Hata, T.; Tada, M.; Iida, M.; Watari, A.; Okada, Y.; Doi, T.; Kuniyasu, H.; Yagi, K.; Kondoh, M. Safety evaluation of a human chimeric monoclonal antibody that recognizes the extracellular loop domain of claudin-2. *Eur. J. Pharm. Sci.* **2018**, *117*, 161–167. [CrossRef] [PubMed]

2

IL-13 Impairs Tight Junctions in Airway Epithelia

Hanna Schmidt [1,†], Peter Braubach [2,3,†], Carolin Schilpp [1], Robin Lochbaum [1], Kathrin Neuland [1], Kristin Thompson [1], Danny Jonigk [2,3], Manfred Frick [1], Paul Dietl [1] and Oliver H. Wittekindt [1,*]

1 Institute of General Physiology, Ulm University, Albert-Einstein-Allee 11, 89081 Ulm, Germany
2 Institute of Pathology, Hannover Medical School, Carl-Neuberg-Str. 130625 Hannover, Germany
3 German Center of Lung Research (DZL), Partnersite BREATH, 306245 Hannover, Germany
* Correspondence: oliver.wittekindt@uni-ulm.de
† These authors contributed equally to this work.

Abstract: Interleukin-13 (IL-13) drives symptoms in asthma with high levels of T-helper type 2 cells (T_h2-cells). Since tight junctions (TJ) constitute the epithelial diffusion barrier, we investigated the effect of IL-13 on TJ in human tracheal epithelial cells. We observed that IL-13 increases paracellular permeability, changes claudin expression pattern and induces intracellular aggregation of the TJ proteins zonlua occludens protein 1, as well as claudins. Furthermore, IL-13 treatment increases expression of ubiquitin conjugating E2 enzyme UBE2Z. Co-localization and proximity ligation assays further showed that ubiquitin and the proteasomal marker PSMA5 co-localize with TJ proteins in IL-13 treated cells, showing that TJ proteins are ubiquitinated following IL-13 exposure. UBE2Z upregulation occurs within the first day after IL-13 exposure. Proteasomal aggregation of ubiquitinated TJ proteins starts three days after IL-13 exposure and transepithelial electrical resistance (TEER) decrease follows the time course of TJ-protein aggregation. Inhibition of JAK/STAT signaling abolishes IL-13 induced effects. Our data suggest that that IL-13 induces ubiquitination and proteasomal aggregation of TJ proteins via JAK/STAT dependent expression of UBE2Z, resulting in opening of TJs. This may contribute to barrier disturbances in pulmonary epithelia and lung damage of patients with inflammatory lung diseases.

Keywords: lung; epithelia; interleukin 13; tight junction; UBE2Z; ubiquitin

---

1. Introduction

Asthma is a complex disease that involves environmental interactions, genetic risk factors and chronic airway inflammation. According to the underlying pathophysiological mechanisms, asthma syndrome was proposed to be subdivided into several endotypes [1]. Depending on the dominating inflammatory response, the endotypes can be grouped in those with low levels of T-helper type 2 cells (T_h2-cells), the T_h2 low endotype, and those with high T_h2-cell levels, described as T_h2 high endotype [2]. The latter subsumes allergic asthma phenotypes with mainly T_h2 cell driven symptoms. T_h2 cells were referred to as T-helper cells that produce and release high amounts of interleukins (IL) -3, -4, -5, -6, -10 and -13 [3]. In particular, IL-13 plays a pivotal role in allergic asthma [4,5]. Novel treatment strategies follow approaches, where IL-13 signaling is inhibited either by immuno-inhibition of IL-13 dependent receptor signaling [6] or by immuno-neutralization of IL-13 itself [6–10]. Both strategies can improve asthma symptoms and reduce risk of exacerbations.

Cigarette smoke is another trigger of pulmonary inflammation, which causes damage and barrier dysfunction [11]. Moreover, cigarette smoke induced inflammatory and metabolic conditions are associated with lung cancer development [12,13]. In this process, IL-13 signaling is considered to be involved via the IL-13 receptor α2 that promotes malignant transformation [14].

IL-13 has multiple effects on pulmonary epithelia. It induces airway hyperresponsiveness [15,16], mucus hypersecretion and airway remodeling [4,17–20], all of which are the hallmarks of allergic asthma. Additionally, lateral intercellular contacts and tight junctions (TJ) of the pulmonary epithelium are disturbed in patients with allergic asthma [21]. T_h2 cytokines IL-4 and IL-13 have been found to compromise barrier function of the pulmonary epithelium [22–26].

The barrier function of an epithelium can be described as its ability to limit the diffusion-dependent exchange of solvent and solute between apical and basolateral compartments. Airway epithelial cells are linked by several intercellular protein complexes that connect lateral cell membranes of neighboring cells. TJ form the most apically localized junction complex. They function as the major limiting diffusion barrier as well as a fence that separates the apical and lateral membranes of epithelial cells [27]. TJs are comprised of different claudin (cldn) isoforms and the isoform composition determines the specific functional properties of TJs, specifically their permeability, reflection coefficient and permselectivity [27]. The cldn composition of TJs in the pulmonary epithelium changes along the respiratory tree and adjusts epithelial function to the segment specific requirements [28].

T_h2 cytokines, particularly IL-4 and IL-13, perturb the protein composition of TJ in pulmonary epithelia via different mechanisms. First, they alter expression levels of several TJ proteins. This has been reported for patients with allergic rhinosinusitis where expression levels of TJ proteins occludin (ocldn) and junctional adhesion molecule A (JAM-A) were reduced and expression of cldn2, a cldn that increases TJ's permeability in sinonasal epithelial cells, was increased [29]. Similarly, Cldn18 expression levels were significantly reduced in brush samples of human epithelial cells from patients with T_h2-high asthma endotype that was linked to IL-13 signaling [24]. Second, T_h2 cytokines decrease protein density at the TJs without affecting expression levels of TJ proteins [24,25].

Many studies revealed that IL-13 is the major effector for perturbation of TJ permeability in T_h2-high asthma patients [21–26]. In addition, breakdown of the pulmonary barrier function is considered to be a hall mark in developing respiratory distress [30]. However, the IL-13 mediated perturbations of TJ are poorly understood. Our study aims at elucidating the effect of prolonged IL-13 exposure on primary human bronchial epithelial cells cultivated at air-liquid interface (ALI) conditions. We demonstrate that IL-13 alters expression levels of cldn8, cldn9 and cldn16, upregulates the E2 ubiquitin conjugating enzyme UBE2Z and induces proteasomal aggregation of TJ proteins together with ubiquitin (UBQ) and UBE2Z. This reduces protein density at the TJ and increases TJ permeability for ions and small molecules and might be an important contributor to patients' predisposition to asthma exacerbations and subsequent respiratory distress.

2. Results

2.1. *Effect of IL-13 on hTEpC Epithelia*

Human primary tracheal epithelial cells (hTEpC) were cultivated at air-liquid interface (ALI) for 21 days. Exposure of hTEpC epithelia to 10 ng/mL IL-13 or IL-4 for the entire ALI cultivation time resulted in a transepithelial electrical resistance (TEER) decrease (Figure 1A). Both cytokines are characteristic for T_H2 high asthma endotype [31]. IL-13 reduced TEER by approximately 85%. The TEER decrease caused by IL-4 was lower and only the effect of IL-13 was statistically significant. Hence, we focused on the epithelial response to IL-13 in all subsequent experiments. The reduction of TEER by IL-13 was in line with an increased apparent permeability coefficient (P_{app}) for sodium fluorescein (Figure 1B). This result indicated that IL-13 affects the paracellular diffusion barrier. Increased mucin 5ac (Muc5ac) expression levels (Figure 1C) confirm the biological activity and efficiency of IL-13. Inhibition of JAK/STAT signaling by Tofacitinib [32,33] or S-Ruxolitinib [34,35] (both added at concentration of 2.5 μM) attenuated IL-13 effects on TEER and paracellular permeability (Figure 1D,E). This indicates that IL-13 acts via JAK/STAT signaling on paracellular permeability.

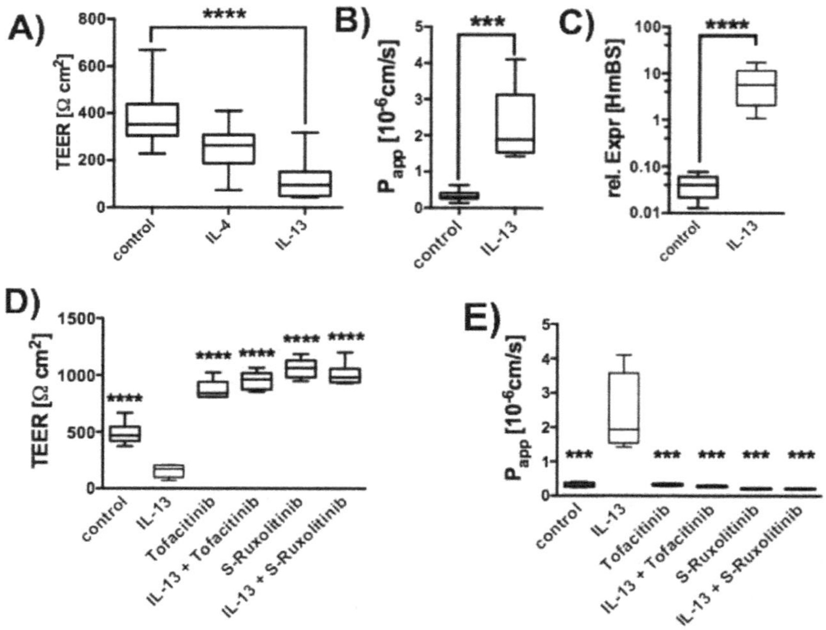

Figure 1. IL-13 increased paracellular permeability. Human tracheal epithelial cells were cultivated at air-liquid interface conditions in the absence (control) and presence of 10 ng/mL IL-13 (IL-13). Box plots summarize data as median values, the boxes represent percentiles, the whiskers indicate the minimum/maximum. (**A**) Response of human tracheal epithelial cells to T_H2 cytokines IL-4 and IL-13 (10 ng/mL each). Both cytokines reduced TEER. However, TEER reduction by IL-4 did not reach significance (Kruska-Wallis test with Dunn's correction for multiple testing, $p = 0.086$, IL-4 ($N = 8$) versus control cells (control), $N = 20$). IL-13 reduced TEER significantly (Kruska-Wallis test with Dunn's correction for multiple testing, $p < 0.0001$, IL-13, N = 17 versus control, $N = 20$). (**B**) IL-13 increases apparent permeability coefficient (P_{app}) for the small molecular marker sodium fluorescein (Mann-Whitney test, $p = 0.0002$, $N = 7$). Inhibition of JAK/STAT pathway by Tofacitinib and S-Ruxolitinib abolishes IL-13 induced effects on (**C**) Relative Expression of Muc5ac in human tracheal epithelial cells. Upregulation of mucus production was confirmed by detecting Muc5ac upregulation in IL-13 treated epithelia. (**D**) TEER of control epithelia (control), epithelia cultivated in the presence of IL-13 (IL-13), 2.5 µM Tofacitinib (Tofacitinib) or S- Ruxolitinib (S-Ruxolitinib) and in the presence of IL-13 + Tofactinib or IL-13 + S- Ruxolitinib. (all versus IL-13, ANOVA with Holm-Sidak correction for multiple testing, $p < 0.0001$, $N = 6$) and on (**E**) P_{app} (all versus IL-13, ANOVA with Holm-Sidak correction for multiple testing, $N = 4$, control $p = 0.0003$, Tofacitinib $p = 0.0003$, Tofacitinib + IL-13 $p = 0.0002$, S-Ruxolitinib $p = 0.0002$ and S-Ruxolitinib + IL-13 $p = 0.0002$). *** and **** indicate $p < 0.001$ and $p < 0.0001$, respectively.

Paracellular permeability depends on cldn composition of the tight junctions (TJ) [28]. Hence, we investigated the effect of IL-13 on expression levels of claudins and also on occludin (ocldn) as a TJ protein that seals the paracellular pathway. Semi quantitative reverse transcriptase PCR (RT-PCR) experiments revealed a significant reduction of expression levels in IL-13 treated versus control epithelia for cldn8, cldn9 and cldn16 by a factor of 7.0, 3.5 and 6.3, respectively (Figure 2A). Hence, we performed additional experiments to further investigate the effect of IL-13 on cldn8, cldn9, cldn16 expression. Cldn4, although not affected by IL-13, was included in subsequent experiments, since it interacts with cldn8 to form paracellular Cl^- pores in TJs [36] (Figure 2B–E). In line with the initial screening, IL-13 treatment reduces cldn8, cldn9 and cldn16 expression levels. Inhibition of JAK/STAT signaling by Tofacitinib [32,33] and S-Ruxolitinib [34,35] abolished the down-regulation of these cldns. Cldn4 expression was not affected neither by IL-13 nor by JAK/STAT inhibition. These experiments confirm that IL-13 specifically changes expression of cldn8, 9 and 16 via JAK/STAT signaling.

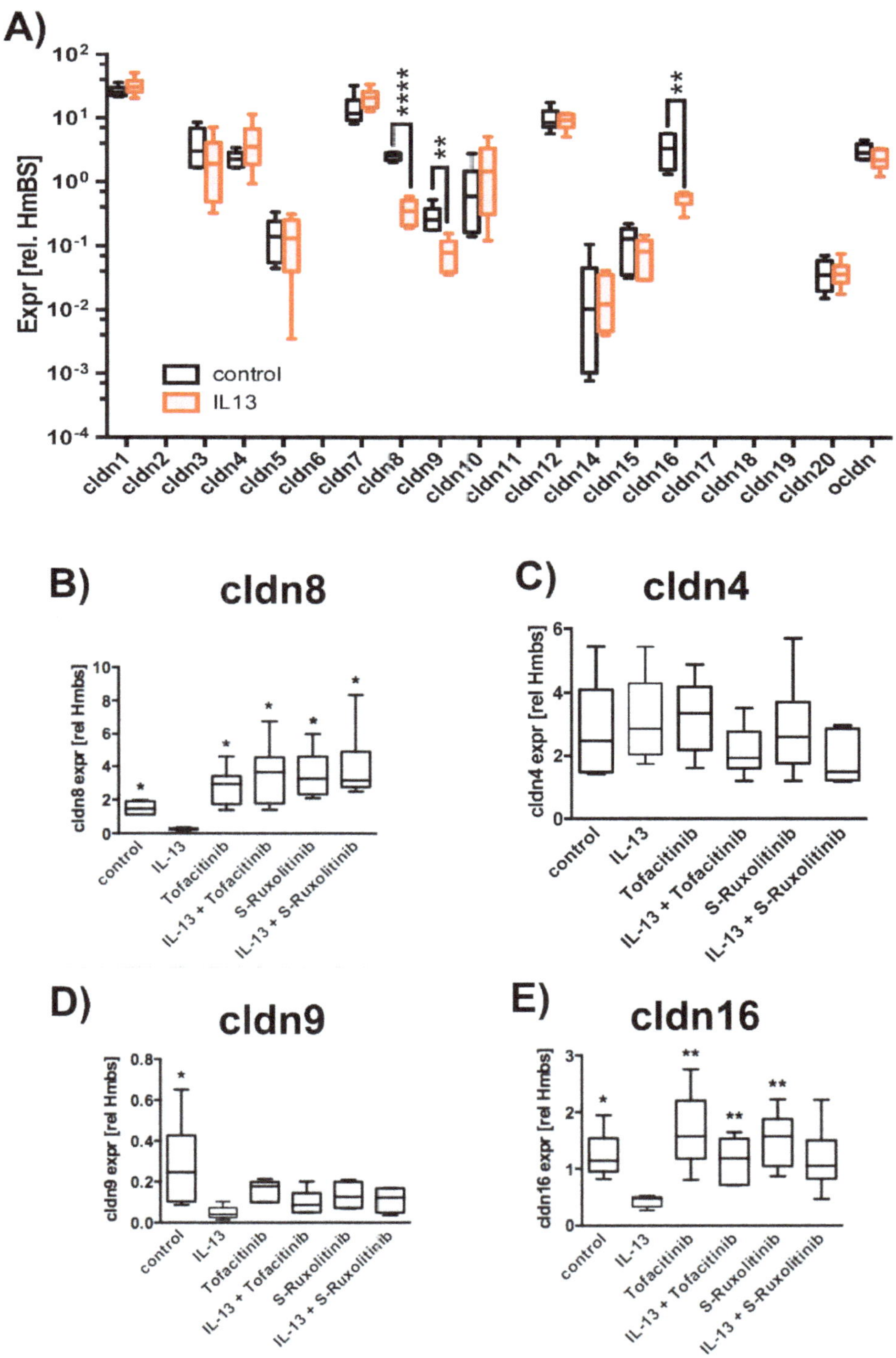

Figure 2. Effect of IL-13 on claudin and occludin expression levels. Human tracheal epithelial cells were cultivated at air-liquid interface conditions. Expression levels were normalized to HmBS. The box plot summarizes data as median values, the boxes represent percentiles, the whiskers indicate minimum/maximum. (**A**) Claudin expression levels in epithelia cultivated in the absence (control) or

presence of 10 ng/mL IL-13 (IL-13). Multiple T-test, all IL-13 versus control, cldn8 $p < 0.0001$, cldn9 $p = 0.005$ and cldn16 $p = 0.005$, $N = 6$). Transcripts encoding claudins 2, 6, 11, 17, 18 and 19 were not detected. Changes in cldn expression levels were confirmed and the effect of inhibition of JAK/STAT pathway was tested in subsequent independent experiments (control epithelia = control, epithelia cultivated in the presence of IL-13 = IL-13, 2.5 μM Tofacitinib = Tofacitinib or S-Ruxolitinib = S-Ruxolitinib and in the presence of IL-13 = IL-13 + Tofactinib or IL-13 + S- Ruxolitinib, respectively) (**B**) cldn8 expression (Mann Whitney test with Bonferroni correction all versus IL-13 and $N = 6$, control $p = 0.02$, Tofacitinib $p = 0.02$, IL-13 + Tofacitinib $p = 0.02$, Ruxolitinib $p = 0.02$ and IL-13 + Ruxolitinib $p = 0.02$) (**C**) cldn4 expression was not affected neither by IL-13 nor by Tofacitinib or Ruxolitinib. (**D**) cldn9 expression (Mann-Whitney test with Bonferroni correction, $N = 6$, control $p = 0.04$, Tofacitinib $p = 0.08$, IL-13 + Tofacitinib $p = 0.13$, Ruxolitinib $p = 0.07$ and IL-13 + Ruxolitinib $p = 0.2$) and (**E**) cldn16 expression (Mann-Whitney test with Bonferroni correction, $N = 6$, control $p = 0.02$, Tofacitinib $p = 0.01$, IL-13 + Tofacitinib $p = 0.01$, Ruxolitinib $p = 0.01$ and IL-13 + Ruxolitinib $p = 0.07$). *, ** and **** indicate $p < 0.05$, $p < 0.01$ and $p < 0.0001$, respectively.

To further elucidate the mechanisms underlying the IL-13-induced modulation of paracellular permeability, we investigated the cellular localization of TJ proteins (Figure 3A–C). We focused on cldn8, since we recently identified cldn8 as a claudin that regulates protein assembly at the TJs [37]. Cldn4 was chosen as the control, as its expression level is not modulated by neither IL-13 treatment nor inhibitors of JAK/STAT signaling (see above). Finally we investigated ocldn localization, since there is evidence that its localization at TJ in pulmonary epithelia is modulated by cldn8 [37]. In control epithelia, all the investigated TJ proteins colocalized with tight junction marker zonula occludens 1 (ZO1) (Figure 3A–C). Staining for TJ proteins revealed an almost continuous contouring of the cells, indicating their lateral assembly at the TJs. IL-13 treatment resulted in a highly diminished lateral localization of all TJ proteins, including ZO1. In these epithelia, the staining of TJ proteins resulted in a patchy cell contouring that indicates massively impaired protein assembly at the TJ. Moreover, TJ proteins were observed to form intracellular aggregates. This effect was not restricted to IL-13 regulated cldn8 but to all investigated TJ proteins, including ZO1. Inhibition of JAK/STAT signaling using 2.5 μM Tofacitinib [32,33] or 2.5 μM S-Ruxolitinib [34,35] abolished IL-13 induced intracellular localization of TJ proteins. Whereas TJ protein localization was affected by IL-13 treatment, the lateral localization of the adherens junction protein E-cadherin (E-cad) [38] remained almost unaffected by IL-13 (Figure 3D,E). This indicates that the IL-13 effect on cell–cell contacts is restricted to the TJ. Furthermore, IL-13 affects all TJ proteins but not exclusively those proteins, which are regulated at transcriptional levels by IL-13 treatment.

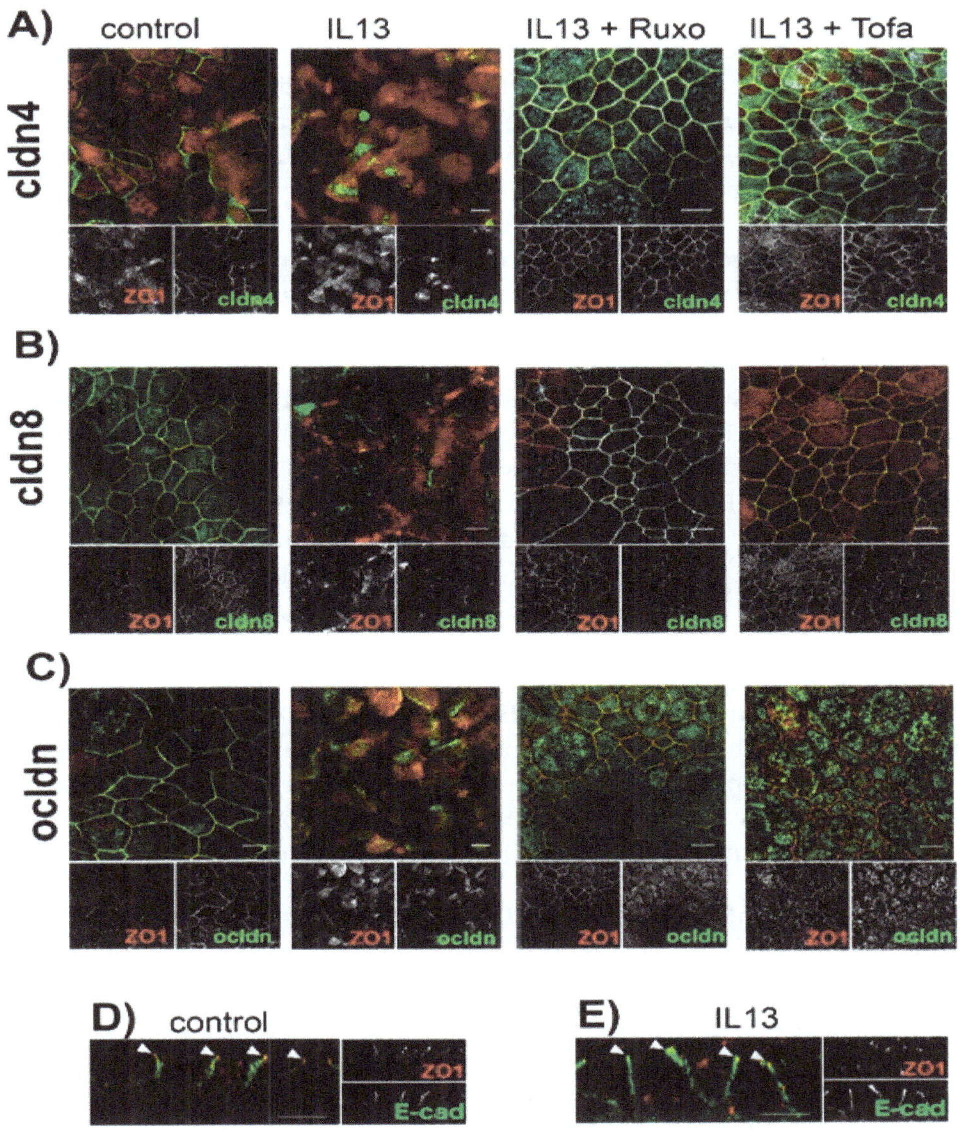

Figure 3. Effect of IL-13 on protein localization at TJ. Human tracheal epithelial cells were cultivated at air-liquid interface conditions. Epithelia were cultivated in the absence of compounds (control), in the presence of IL-13 (IL-13), in the presence of IL-13 + 2.5 μM S-Ruxolitinib (IL-13 + Ruxo) or in the presence of 2.5 μM Tofacitinib (IL-13 + Tofa). Scale bars represent 10 μm. (**A**) Intracellular localization of claudin 4 (cldn4, green channel). Zonula occludens 1 (ZO1, red channel) was used for counterstaining for tight junctions. (**B**) Intracellular localization of claudin 8 (cldn8, green channel). Counterstaining for tight junctions by ZO1 (red channel). (**C**) Intracellular localization of occludin (ocldn, green channel). Counterstaining for tight junctions by ZO1 (red channel). (**D**,**E**) Intracellular localization of E-cadherin (E-cad, green channel) as a marker for lateral adherens junction. ZO1 (red channel) was used for counterstaining of tight junctions.

2.2. IL-13 Activates the Ubiquitin-Proteasome Pathway

There is emerging evidence that the ubiquitin-proteasome pathway [39] controls TJ protein turnover and TJ permeability [40–45]. The observed localization of TJ proteins at intracellular granules upon IL-13 treatment may hint at the possibility that internalized proteins accumulate at proteasomal compartments. To address the question of whether IL-13 activates the ubiquitin-proteasome pathway, we initially screened for modulation of ubiquitin E1 activating and ubiquitin E2 conjugating enzymes. Semi-quantitative RT-PCR experiments revealed that IL-13 treatment specifically up-regulates expression of the ubiquitin E2 conjugating enzyme, UBE2Z (Figure 4A) 1.6 ± 0.16 fold (mean ± SD, $N = 3$; T-test $p = 0.01$). The detected upregulation of Muc5AC expression by IL-13 confirms its efficiency on the tested epithelia (Figure 4B). Subsequent experiments confirmed the upregulation of UBE2Z by IL-13, which was attenuated by inhibition of the JAK/STAT pathway (Figure 4C). The number of cells was limited due to the ALI cultivation on filters and hence limits the applicability of methods like pull down assays. Therefore, to further evaluate ubiquitination of TJ proteins, we performed proximity ligation assays (PLA, Figure 4D–G). In these experiments we investigated if cldn4 (Figure 4C), cldn8 (Figure 4D) and ZO1 (Figure 4E) form protein complexes together with ubiquitin (UBQ) in IL-13 treated cells. Indeed, compared to control cells, segmental analysis (Figure 4F) of cells revealed that IL-13 treatment results in increased number of fluorescent granules that reflect positive PLA signals. This indicates that ubiquitinated proteins and TJ-proteins localize within the same protein complex and suggests that IL-13 induces the ubiquitination of TJ-proteins. Staining for protein complexes that were positive for E-cadherin and UBQ revealed positive PLA signals in control as well as in IL-13 treated cells (Figure 4H). Also, abundance of PLA positive structures did not differ in control versus IL-13 treated cells (Figure 4I). Evidently, E-cad ubiquitination occurs even at basal conditions, presumably as part of protein turnover to maintain proteostasis. This basal E-cad ubiquitination is not affected by IL-13 and agrees with the observation that subcellular E-cad localization remained unaffected by IL-13 treatment shown above, given the results of the immunostaining experiments.

We investigated the intracellular localization of TJ proteins and markers of the ubiquitin-proteasome pathway in subsequent immunocytochemical experiments (Figure 5) and studied their colocalization using van Steensel's cross-correlation analysis [46]. In control cells, the investigated TJ proteins cldn8 and ZO1 colocalized (Figure 5A,C), resulting in a sharp peak when cross-correlation coefficients were plotted against lateral translocation. Hardly any colocalization could be observed in control cells for the TJ proteins with UBE2Z (Figure 5A) or UBQ (Figure 5C). The exposure of epithelia to IL-13 resulted in an overlapping localization of TJ proteins with UBE2Z (Figure 5B) as well as with UBQ (Figure 5D). In these cases, the van Steensel's plots revealed increases in the cross-correlation coefficients for all invested proteins. However, the broadening of peak when cross-correlation coefficients are plotted against the lateral translocation reflects the increased and inhomogeneous size of structures where TJ proteins, UBE2Z and UBQ colocalize. No colocalization of ZO1 with the proteasomal marker PSMA5 nor with UBQ or PSMA5 with UBQ could be observed in control epithelia (Figure 5E). In IL-13 treated epithelia (Figure 5F) ZO1 colocalized with UBQ and PSMA5. In these epithelia UBQ colocalizes with PSMA5 as well. Again van Steensel's plot revealed enlarged non-homogenous size of structures at which colocalization is detected.

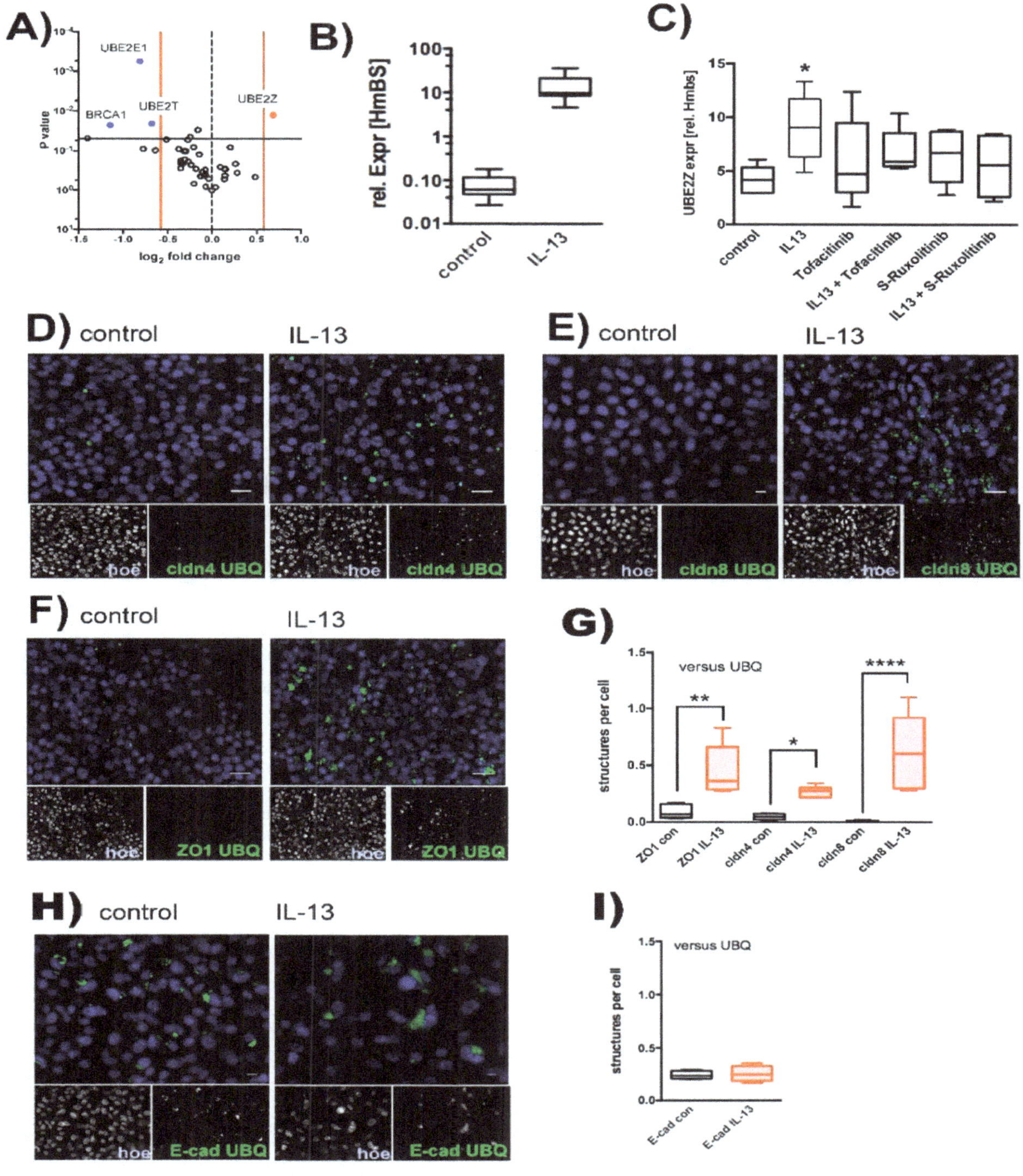

Figure 4. Ubiquitination of TJ proteins. Human tracheal epithelial cells were cultivated at air-liquid interface conditions in the absence (control) and presence of IL-13 (IL-13). (**A**) Screening for ubiquitin activating E1 and ubiquitin conjugating E2 enzyme by semi-quantitative RT-PCR. Volcano plot summarizes changes in expression levels (control vs. IL-13, each $N = 3$). Log2 fold changes are given as mean values. Vertical red lines give threshold of ±1.5 fold change. Horizontal black line gives threshold of $p = 0.05$. Genes that were considered to be up regulated are depicted in red (UBE2Z) and those to be down regulated are given in blue (BRCA1, UBE2T, UBE2E1). (**B**) Upregulation of UBE2Z by IL-13 was confirmed and effect of JAK/STAT pathway inhibition was tested in subsequent independent experiments. Relative expression of UBE2Z to HmBS of: control epithelia (control, $p = 0.021$), epithelia cultivated in the presence of IL-13 (IL-13), 2.5 μM Tofacitinib (Tofacitinib, $p = 0.035$) or S- Ruxolitinib (S-Ruxolitinib, $p = 0.16$) and in the presence of IL-13 + Tofactinib ($p = 0.16$) or IL-13 + S- Ruxolitinib ($p = 0.09$). All versus IL-13 and $N = 6$, ANOVA with Holm-Sidak correction for multiple testing. (**C**) IL-13

stimulation of epithelia used in "B" was confirmed by detecting Muc5ac upregulation (Mann-Whitney test $N = 6$, $p = 0.004$). Ubiquitination of TJ proteins were investigated in proximity ligation assays for (**D**) ubiquitination of claudin 4 (cldn4 UBQ, green channel, (**E**) ubiquitination of claudin 8 (cldn8 UBQ, green channel) and (**F**) ubiquitination of ZO1 (ZO1 UBQ, green channel). Nuclei were counterstained by Hoechst 33342 (blue channels). (**G**) PLA positive structures were identified as green fluorescent particles with an area of 75 μm^2 and above. Box plot summarizes quantification of PLA positive structures per cell as median, boxes represent percentile and whiskers give minimum/maximum (IL-13 versus control, Anova test with Holm-Sidaks correction for multiple testing, all $N = 4$, ZO1 $p = 0.0014$, cldn4 $p = 0.03$ and cldn8 $p < 0.0001$). (**H**) Ubiquitination of E-cadherin (E-cad UBQ, green channel). (**I**) PLA positive structures per cell summarized as median values, boxes represent percentiles and whiskers provide minimum/maximum values (E-cadherin = E-cad, ubiquitin = UBQ). No difference in PLA signal abundance could be detected in control versus IL-13 treated cells ($N = 4$, Mann-Whitney test $p > 0.9999$). *, ** and **** indicate $p < 0.05$, $p < 0.01$ and $p < 0.0001$, respectively. Scale bars in (**D**–**F**) and (**H**) represent 10 μM.

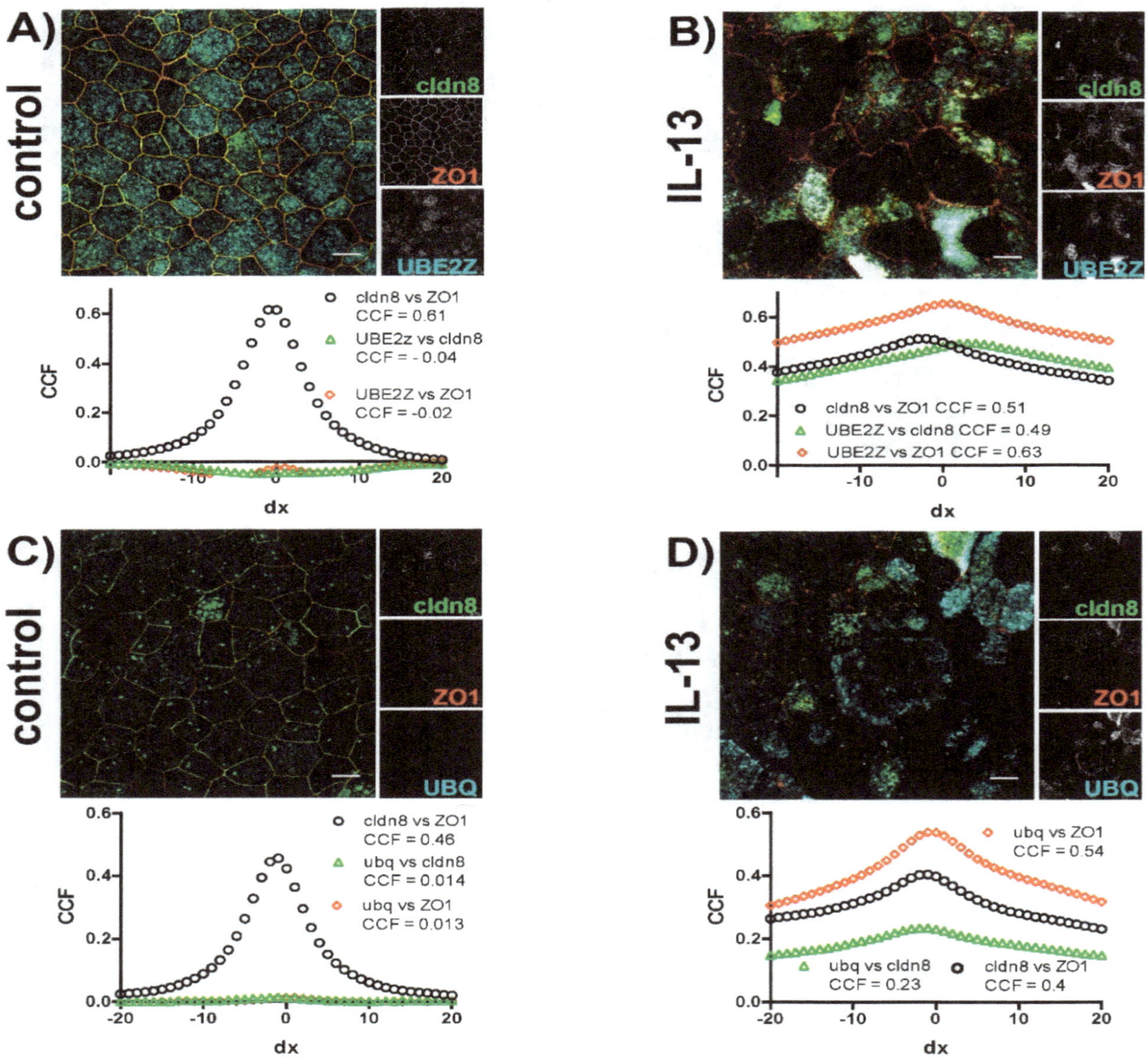

Figure 5. *Cont.*

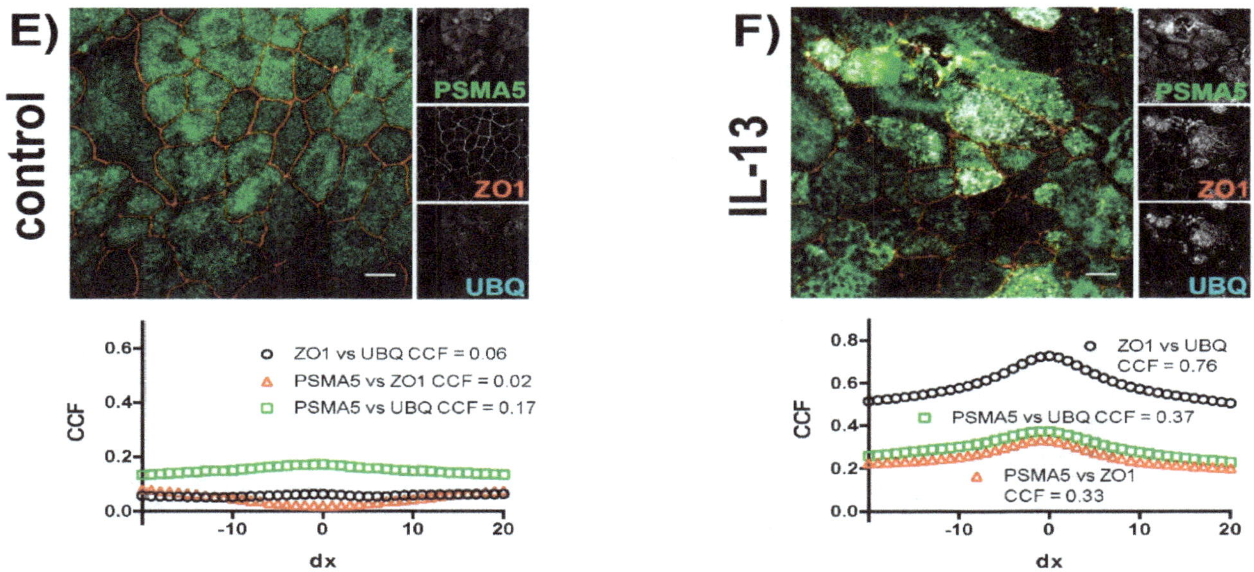

Figure 5. Colocalization of TJ proteins and proteins of the ubiquitin-proteasome machinery. Human tracheal epithelial cells were cultivated at air-liquid interface in the absence (control) and in the presence of IL-13 (IL-13). Images show merged fluorescence channels. Insets at the right give gray scale image for each channel. Van Steensel's plot below each image gives a cross-correlation coefficient (CCF) plotted against lateral delocalization (dx = 0 to ±20 pixel). Maxima of cross-correlation coefficient plot are given in the legend for each plot. (**A**) Control epithelia and (**B**) IL-13 treated epithelia immune-stained for Claudin 8 (cldn8, green channel), zonula occludens 1 (ZO1, red channel) and UBE2Z (cyan channel). (**C**) Control epithelia and (**D**) IL-13 treated epithelia immune-stained for Claudin 8 (cldn8, green channel), zonula occludens 1 (ZO1, red channel) and Ubiquitin (UBQ, cyan channel). (**E**) Control epithelia and (**F**) IL-13 treated epithelia immune-stained for proteasome subunit α Type 5 (PSMA5, green channel), zonula occludens 1 (ZO1, red channel) and ubiquitin (UBQ, cyan channel). Scale bars in (**A**–**F**) represent 10 μM.

2.3. *Effect of IL-13 on Mature hTEpC Epithelia*

In order to test if IL-13 also acts on mature epithelia, we added IL-13 at a final concentration of 10 ng/mL to hTEpC epithelia on day 18 after ALI establishment. Measurements were performed on the following days as indicated (Figure 6). In mature hTEpC epithelia, IL-13 reduced TEER from day 21 onwards (Figure 6A). The drop in TEER reached statistical significance on day 28 (control vs. IL-13 treated cells, $N = 12$, Mann-Whitney test $p = 0.0011$). The time course of TEER decrease correlated with time course of cldn8 expression levels, which decreased from day 21 onwards, three days after IL-13 exposure (Figure 6B). The decrease in cldn8 expression levels reached statistical significance from day 24 onwards. The increase in UBE2Z expression levels preceded TEER decrease and occurred already within the first day after IL-13 exposure (Figure 6C). It reached significant levels from day 19 onwards.

Time course of ubiquitination of TJ proteins was investigated in immunocytochemical experiments (Figure 6D–I). In these experiments, we used van Steensel's cross-correlation analysis [46] to quantify colocalization of ZO1 with cldn8 as well as of UBQ with ZO1 and cldn8. Time course of colocalization was visualized by plotting the maximum Pearson cross correlation coefficient detected from the van Steensel's plot against cultivation time.

In control cells cross-correlation coefficient for ZO1 and cldn8 colocalization increased from day 19 onwards (Figure 6H). This may reflect gradual tightening of TJ in mature epithelia during ALI cultivation, as it correlates with the slightly increasing TEER (compare Figure 6A). Furthermore, in these control cells no change in cross-correlation over time was observed for ZO1 and cldn8 with UBQ (Figure 6H).

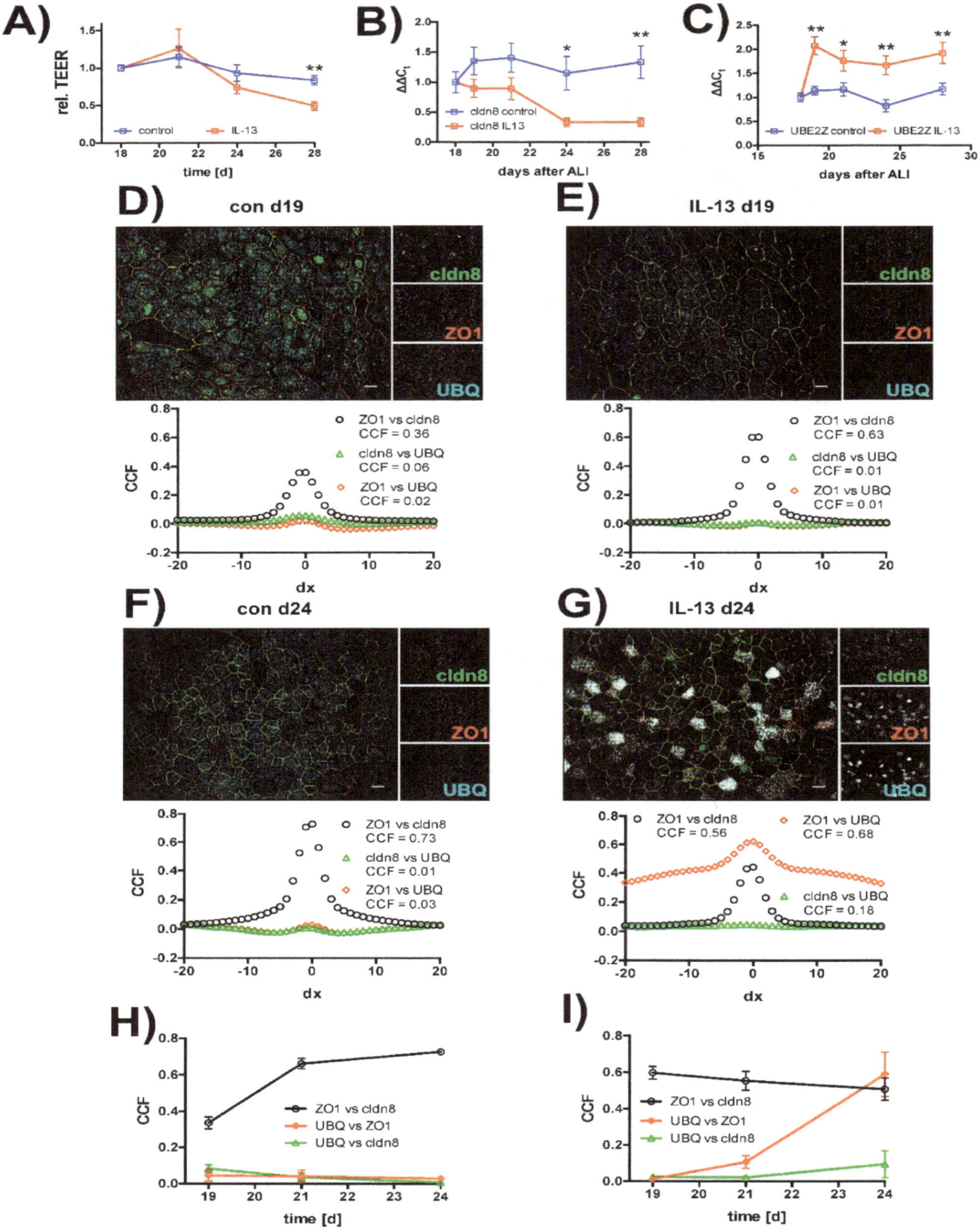

Figure 6. Time course of IL-13 induced effects on human tracheal epithelia. Human tracheal epithelial cells were seeded on Transwell filters and proliferated for 4 days. Afterwards, apical volume was removed and cells were cultivated for additional 18 days at air-liquid interface conditions. On day 18, IL-13 was added to the medium (IL-13). Control cells were not exposed to IL-13 (control). (**A**) Changes

of transepithelial electrical resistance (TEER) over time. TEER was normalized to mean TEER measured on day 18 directly before IL-13 exposure and relative TEER was plotted against cultivation time. Rel. TEER decreases from day 21 onwards. Decrease reached statistical significance on day 28. Data points summarize measurements as mean ±SEM (all $N = 10$ to 12, Mann-Whitney test $p = 0.001$). (**B**) Changes of claudin 8 (cldn8) expression levels over time. Changes in cldn8 expression are given as ΔΔCt values relative to the mean expression level on day 18. Data points give mean ΔΔCt ± SEM. Cldn8 expression drops from day 21 onwards. It reached significance on days 24 and 28 (all $N = 9$, Mann-Whitney test $p = 0.008$ and $p = 0.0002$, day 24 and 28, respectively). (**C**) Time course of UBE2Z expression. Changes in UBE2Z expression are given as ΔΔCt values relative to mean expression level on day 18 as mean ΔΔCt ± SEM. UBE2Z expression levels rose directly upon IL-13 exposure an reached significance from day 19 onwards (all $N = 9$, Mann-Whitney test $p = 0.0012$ day 19, $p = 0.039$ day 21, $p = 0.0027$ day 24 and $p = 0.0019$ day 28). Colocalization of tight junction proteins claudin 8 (cldn8, green channel) and zonula occludens 1 (ZO1, red channel) with ubiquitin (UBQ, cyan channel) was analyzed in van Steensel's plot (below each image). Examples are given for (**D**) control cells on day 19 (con d19) and (**E**) for cells exposed to IL-13 on day 19 (IL-13 d19) as well as (**F**) for control cells on day 24 (con d24) and (**G**) for cells exposed to IL-13 on day 24 (IL-13 d24). Changes of colocalization over time are analyzed by plotting the maximum of cross-correlation coefficients (CCF) from van Steensel's plot against the cultivation time. Changes of maximum CCF over time (**H**) in control cells and (**I**) in IL-13 treated cells. Black curves give CCF for zonula occludens 1 and claudin 8 (ZO1 vs. cldn8), red curves show CCF for ubiquitin and ZO1 (UBQ vs. ZO1) and green curves show CCF for UBQ and cldn8 (UBQ vs. CLDN8). Data points summarize CCF as mean ± SD all $N = 3$). * and ** indicate $p < 0.05$ and $p < 0.01$, respectively. Scale bars in (**D–G**) represent 10 μM.

In case of IL-13 treated epithelia, the cross-correlation coefficient for the colocalization of cldn8 with ZO1 did not change over time. In contrast to that and to control cells, cross-correlation coefficients for UBQ colocalization with ZO1 and cldn8 both increase, starting three days after IL-13 exposure, from day 21 onwards. However, the change in colocalization for cldn8 with UBQ was smaller compared to ZO1 with UBQ (Figure 6I) and most likely reflects a weaker staining efficiency and/or protein abundance of cldn8.

The onset of UBQ colocalization with TJ proteins correlates with onset of TEER decrease. The formation of protein aggregates does not start in all cells simultaneously. Instead, we observed an increasing amount of cells over time that were positive for granules containing tight junction proteins colocalizing with ubiquitin. On day 19 almost no cells were positive for protein aggregates. On day 21, when TEER starts to drop and the cross correlation coefficient starts to rise, 9.3 ± 5.1% of IL-13 treated cells formed aggregates (mean ± SD, $N = 3$). The amount of aggregate containing cells further increased to 23.6 ± 4.6% (mean ± SD, $N = 3$) on day 24. Thus, the time course of protein aggregate formation correlated with time course of epithelial permeability increase. This gives evidence that protein aggregate formation is the TJ damaging step in IL-13 exposed epithelia. These experiments indicate that IL-13 activates the ubiquitin-proteasome pathway that also involves UBE2Z in mature hTEpC epithelia.

2.4. *Ubiquitination of TJ Proteins in Sinunasal Epithelia*

To test whether the detected IL-13 induced ubiquitination and proteasomal aggregation also occurs in human airway epithelia, we investigated sinunasal biopsies from patients suffering from chronic inflammation with and without eosinophilia using PLA (Figure 7). These experiments revealed the formation of protein complexes that contain ZO1, cldn4 and cldn8 together with UBQ and UBE2Z. Especially in the absence of bacterial infection, eosinophilic inflammation is a fairly good and specific surrogate for high IL-13 levels [47]. Since measuring IL-13 levels in paraffin embedded tissue biopsies is erroneous, we used eosinophilia as a marker for elevated IL-13 tissue concentrations instead. There was no difference in number of PLA-positive particles detectable in eosinophilic inflammation in comparison to non-eosinophilic inflammation (all as mean ±SD in counts/μm^2 and $N = 3$ patients.

Eosinophilic rhinosinositis: UBQ/cldn4: 0.64 ± 0.18, UBQ/cldn8: 0.78 ± 0.41, UBQ/ZO1: 0.29 ± 0.14 and UBE2Z/cldn4: 0.38 ± 0.02, UBE2Z/cldn8: 0.4 ± 0.19, UBE2Z/ZO1: 0.51 ± 0.26 and non-eosinophilic rhinosinostis: UBQ/cldn4: 0.83 ± 0.28, UBQ/cldn8: 0.56 ± 0.14, UBQ/ZO1: 0.48 ± 0.33 and UBE2Z/cldn4: 0.41 ± 0.06, UBE2Z/cldn8: 1.15 ± 0.89, UBE2Z/ZO1: 0.3 ± 0.05). However, these experiments confirm that UBE2Z dependent ubiquitination is not restricted to in vitro models of airway epithelia but also occurs in vivo.

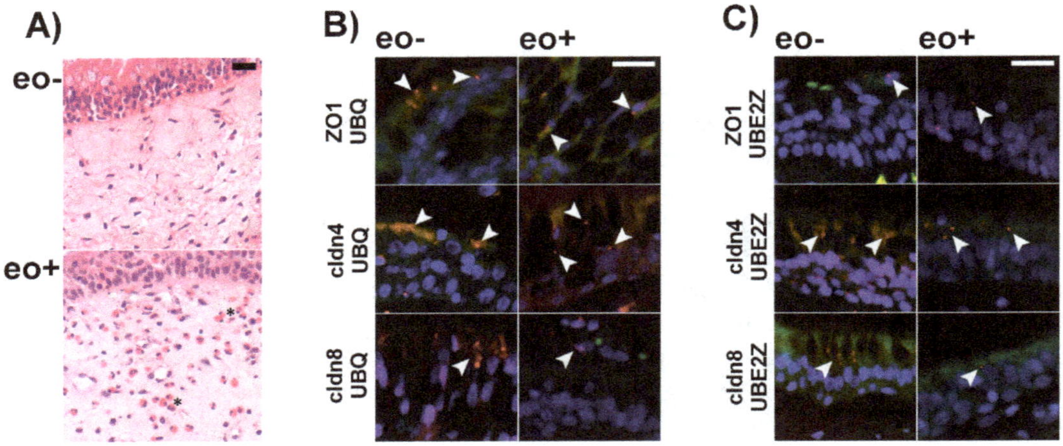

Figure 7. Proximity ligation assays (PLA) were performed on formalin fixed and paraffin embedded tissue from the sinunasal tract containing respiratory type epithelium of patients with and without eosinophilic inflammation. (**A**) shows representative Hematoxilin and Eosin stained images from nasal mucosa with (eo+) and without (eo−) eosinophilic inflammation. Eosinophilic granulocytes can be detected as reddish stained cells in eo+ mucosa (examples are highlighted by asterix) but were not detected in eo- mucosa. For the PLA merge images of nuclear counterstain (blue), autofluorescence (green) and PLA signals (red) are shown. Antibodies specific for ZO1, Claudin 4 and Claudin 8 were combined with antibodies detecting Ubiquitin (**A**) and Ube2Z (**B**). Specific PLA reaction signals (depicted by arrow heads) can be detected in the epithelium under all conditions, suggesting association between ZO1, Claudin 4 and Claudin 8 with Ubiquitin and UBE2Z respectively. Scale bars are 20 μm in panels (**A**–**C**).

3. Discussion

Early studies on transgenic mice revealed the dominating role of IL-13 in allergic asthma [4]. A more recent study identified IL-13 to modulate TJ permeability via altering claudin expression levels [48]. In accordance with these studies, we demonstrated that long-term exposure of epithelial cells to IL-13 results in perturbed claudin expression pattern and reduced protein densities at the TJs itself. A reduced protein density at the apico-lateral junctions of airway epithelia as a result of T_h2 cytokine exposure was already described [25,48] and was recently demonstrated to be linked to IL-13/IL-4 induced activation of the JAK/STAT signaling pathway [25], as we also demonstrated herein. Inhibition of either cytokines or their receptors have beneficial effects in patients with allergic asthma (summarized in reference [49]). However, most studies measured the reduction of inflammatory markers or parameters of lung function to quantify the therapeutic effect of IL-4/IL-13 inhibition but not changes in tightness of TJ. Hence these studies cannot be directly compared with the herein reported observations. In inflammatory bowel diseases, IL-13 acts predominantly on transcriptional levels. It seems to increase intestinal epithelial permeability by increasing cldn2 expression levels via activating STAT6, PI3K and MAPK pathways and in addition via reduction of tricellulin expression as a result of JNK, AP1 and ERK1/2 signaling [50].

Herein we demonstrate that IL-13 activates the ubiquitin-proteasome system, which involves UBE2Z and finally results in proteasomal aggregation of TJ proteins.

The ubiquitin-proteasome system was formerly described as a system that regulates non-lysosomal protein degradation [39]. It involves the marking of proteins that are targeted for degradation by E1-activating enzymes, E2 conjugases and finally E3 ubiquitin ligases. This sequential process results in a highly regulated and substrate specific degradation of proteins [39]. Later on, investigations revealed that UBQ dependent mechanisms also modulate the abundance of membrane spanning proteins at the plasma membrane and controls their lysosomal as well as non-lysosomal degradation [51–53]. Nowadays, it is commonly accepted that the ubiquitin-proteasome system is pivotal in maintaining cellular protein homeostasis via controlling protein turnover, in quality control of protein folding and in guiding of proteins to different subcellular destinations [54]. The ubiquitin-proteasome system has been shown to regulate protein abundance at TJ of cldn1, cldn2 and cldn4 in Madin-Darby Kidney cells (MDCK) cells [55], of cldn5 in endothelial cells [43], of cldn8 in kidney epithelia [40], of cldn4 in rat salivary gland epithelial cells [56], of ocldn in porcine epithelial cells [45], in vascular endothelial cells [57] and in Sertoli cells [42] and of zonula occludens protein 2 in MDCK cells [58]. These studies unequivocally demonstrate that ubiquitin dependent mechanisms are involved in regulating protein composition of TJ and thereby adjust their functional properties. However, the function of ubiquitination of TJ proteins is complex. On one hand, ubiquitination has been shown to label cldn1, cldn2 and ocldn for removal from the TJ via endocytosis and their transport along the endosomal and lysosomal pathway [55,57]. On the other hand, ubiquitination targets cldn5, cldn4 and also ocldn for proteasomal degradation [42,43,45,56]. Thereby, ubiquitination does not only act as a control mechanism to ensure appropriate protein folding or to adjust the protein half-life. It also becomes activated in response to environmental factors to adjust functional properties of TJs, as has been demonstrated for the cAMP dependent cldn4 regulation in Sertoli cells [42] and for ocldn modulation upon muscarinic acetylcholine receptor stimulation [56]. We herein identified IL-13 as another factor that activates the ubiquitin-proteasome system to increase the permeability of TJs. PLA revealed that upon IL-13 exposure, ubiquitinated TJ proteins localized intracellularly rather than at the TJ or anywhere else at the plasma membrane. Sorting and translocation of internalized TJ proteins from endosomal compartments towards the proteasome upon ubiquitination was recently reported for cldn5 [43]. In this study, the authors proposed a derlin1 dependent pathway [43] that presumably involves retrograde transport along endosomal compartments to the endoplasmic reticulum and employs endoplasmic reticulum quality control machinery to translocate the proteins to the proteasome [59,60]. Similar mechanisms could be involved in IL-13 induced proteasomal accumulation of TJ proteins in lung epithelia.

We observed an instantaneous upregulation of the E2 conjugase UBE2Z. The cDNA of UBE2Z was recently cloned and identified to encode an ubiquitin conjugating enzyme [61]. Later reports suggested using UBE2Z to regulate ubiquitin charging of ubiquitin E3 ligases and provide substrate specificity of protein ubiquitination [62,63]. Our experiments strongly support the hypothesis that IL-13 induces ubiquitination and proteasomal accumulation of TJ proteins that are involved in modulation of the paracellular diffusion barrier. Subcellular localization of E-cadherin that forms adherens junctions via homotypic trans-interaction and mediates the dynamic connection of adjacent lateral membranes [64] remained almost unaffected by IL-13. This indicates that lateral inter-cellular adhesion and integrity of epithelial cell layer is not preliminarily disturbed by IL-13. This specificity to TJ is possibly mediated by UBE2Z.

Time course of ubiquitin-mediated regulation of protein abundance at the TJs is reported as a rather fast process, operating within a few hours after stimulation [43,45,56–58]. In line with these studies, we observed an instantaneous upregulation of UBE2Z expression, indicating that lung epithelial cells rapidly respond to IL-13 exposure. However, measurable effects on TJ permeability were delayed and correlated with the reduction of protein abundance at the TJ as well as their accumulation at the proteasome. As recently reviewed, aggregation of poly-ubiquitinated proteins at the proteasome occurs in post-ischemic neurons and results from down-regulation of proteasome activity [65]. Indeed, proteasomal hyperactivity reduces aggregation of poly-ubiqutinated proteins [66]. Proteasomal activity depends on intra-cellular ATP [67]. A drop in intra-cellular ATP levels as a result of inflammatory stress conditions (as was introduced herein by IL-13 exposure) in combination with elevated protein transport to the proteasome and proteasomal degradation could exhaust proteasome activity to such an extent that proteins accumulate and aggregate. Impaired proteasomal activity is discussed to affect cellular proteostasis and therefore cause cellular stress and/or damage and also interfere with the location of proteins and the composition of polymeric protein complexes [68]. Such an assumption is in line with a recent study providing evidence that proteasome inhibition in mouse models of inflammatory bowel disease interrupts epithelial barrier function in colonic mucosa [69] and would also explain our observation of the delayed TJ damage as a consequence of impaired intracellular proteostasis.

Because of its distinct function in controlling intra-cellular proteostasis, the ubiquitin proteasome system is believed to play a pivotal role in inflammatory lung diseases and lung injury [70]. Genes involved in protein ubiquitination are up-regulated in the lung tissue of patients with chronic obstructive pulmonary disease (COPD) [71]. This goes in line with the observations that protein ubiquitination is increased in the lung tissue of COPD patients with emphysema [72]. Furthermore, cigarette smoke extract also induces aggregation of poly-ubiquitinated proteins in bronchial epithelial cells [73] and increases protein ubiquitination in lung endothelial cells [74]. These findings underscore the involvement of dysregulation of the ubiquitin proteasome system in the pathogenesis of chronic inflammatory lung diseases like COPD. Our findings are in line with these former observations. They demonstrate that IL-13 results in an incomplete proteasomal degradation of ubiquitinated proteins that coincide with TJ disturbance. They further point towards a yet unknown function of UBE2Z as an effector in IL-13 driven lung inflammation acting on TJs and thereby on paracellular barrier function in pulmonary epithelia. Opening of TJs of the pulmonary epithelium is a hallmark in the pathogenesis of alveolar edema and exudate formation and subsequent respiratory distress [30]. Disturbances of TJs are a predisposing condition for the development of acute respiratory distress syndrome [75] and lung injury for instance following mechanical ventilation [76,77].

Our results gives evidence that UBE2Z dependent ubiquitination of TJ-proteins also occurs in vivo. We used chronic inflammation with eosinophilic granulocytes as surrogate marker for an IL-13 high phenotype [47]. However, there was no significant difference in levels TJ-protein ubiquitination between samples with and without eosinophilic inflammation. Detection of eosinophilia does not measure cytokine concentrations. IL-13 concentrations for eosinophilic non-bacterial lung inflammation was recently measured to reach concentrations of 10.4 pg/mL in the sputum of humans [47] and around 100 ng/mL in bronchoalveolar lavage of IL-13 overexpressing transgenic mice [78]. These measurements may give an estimation of the range of IL-13 tissue levels during eosinophilic lung inflammation. However, it must be considered that eosinophilia reflects chronic inflammation, while a commonly used nasal steroid therapy in patients with chronic rhinosinusitis potentially reduces inflammatory response and cytokine levels. Under these conditions, IL-13 concentrations are probably below the levels that are needed to initiate the observed TJ protein ubiqutination in the herein investigated in vitro model. However, IL-13 tissue levels might increase and become effective to induce the observed TJ protein ubiquitination during exacerbation episodes. Nonetheless, our results demonstrate that TJ protein ubiquitination that involves UBE2Z also occurs in native tissue.

4. Materials and Methods

4.1. Compounds

IL-4 and IL-13 were obtained from Merck Millipore, Darmstadt, Germany. S-.Ruxolitinib and Tofacitinib, both were obtained from Selleckchem, distributed by Absource Diagnostics, Munich, Germany. If not mentioned otherwise, IL-13 was added at a concentration of 10 ng/mL. S-Ruxolitinib and Tofacitinib was used at concentrations of 2.5 μM each.

4.2. Cell Culture

Transwell permeable supports (Polyester membrane diameter 6.5 mm, 0.4 μm pore size, Costar distributed by Omnilab, Bremen, Germany) were overlaid with 100 μL Collagen I and II solution (StemCell Technologies, Cologne, Germany) diluted 1:100 in phosphate buffered Saline (PBS, Gibco distributed by Thermo Fisher, Darmstadt, Germany) and dried over night at room temperature. Dried filters were UV irradiated and stored until use at 4 °C.

Cryo-preserved passage 1 primary human tracheal epithelial cells (Promocell, Heidelberg Germany and Epithelix, Genève Switzerland) were proliferated in T75 flasks (Sarstedt, Nümbrecht, Germany) using Airway Epithelial Growth Medium (Promocell, Heidelberg, Germany) at 37 °C, 5% CO_2 and 95% humidity. Upon reaching 80% confluence, cells were detached using DetachKit (Promocell, Heidelberg, Germany) and suspended in Airway Epithelial Growth medium. 3.5×10^4 cells in 200 μL Airway Epithelial Growth Medium were seeded onto collagen-coated filters. Filters were placed in 24 well plates filled with 500 μL Airway Epithelial Growth Medium and incubated at 37 °C, 5% CO_2, 95% humidity for two days. Afterwards, air-liquid interface (ALI) conditions were established by removing the medium from the apical surface and replacing Airway Epithelial Growth Medium by 500 μL of fresh ALI-Medium. ALI-medium contained DMEM-H and LHC basal medium (both from Gibco via Thermo Fisher, Darmstadt, Germany) at a 1:1 ratio and was supplemented with: insulin 0.87 μM, hydrocortisone 0.21 μM, epidermal growth factor 0.5 ng/mL, triiodothyronine 0.01 μM, transferrin 0.125 μM, epinephrine 2.5 μM, bovine pituitary extract 10 μg/mL, bovine serum albumin 0.5 mg/mL (all from Promocell, Heidelberg, Germany), phosphorylethanolamine 0.5 μM, ethanolamine 0.5 μM, zinc sulfate 3 μM, retinoic acid 0.05 μM, ferrous sulfate 1.5 nM, calcium chloride 0.6 μM, magnesium chloride 0.11 μM, sodium selenite 30 μM, manganese chloride 1 μM, sodium silicate 0.5 nM, ammonium molybdate tetrahydrate 1 μM, ammonium metavanadate 5 μM, nickel sulfate 1 μM, tin chloride 0,5 μM (all from Sigma-Aldrich GmbH, Steinheim, Germany), 100 U/mL Penicillin and 100 μg/mL Streptomycin (both from Gibco via Thermo Fisher, Darmstadt, Germany). Medium was replaced every second day. Cells were incubated at 37 °C, 5% CO_2 and 95% humidity for the time given in the text.

4.3. TEER Measurement

Transepithelial electrical resistance (TEER) was measured by impedance spectroscopy using the cellZscope (NanoAnalytics, Münster, Germany). For measurements, the basal electrode was overlaid by 500 μL equilibrated ALI medium. After inserting filters, 100 μL of ALI medium were added to the apical surface and the apical electrode was placed into the apical liquid. The measurements were performed immediately after positioning of apical electrodes., The software package provided with the instrument (NanoAnalytics, Münster, Germany) was used for data acquisition and analysis.

4.4. Paracellular Permeability

The paracellular permeability was measured as the apparent permeability coefficient (P_{app}) of Na^+-fluorescein (Sigma-Aldrich GmbH, Steinheim, Germany). Na^+-fluorescein was added to the basolateral medium at a concentration of 150 μM. The Apical surface was overlaid with 100 μL of isotonic NaCl-solution. After incubation for 30 min at 37 °C, 95% humidity and 5% CO_2, the Na^+-fluorescein concentration was measured in the apical solution using an Infinite 200 plate reader with iControl software package (Tecan, Männedorf, Switzerland). P_{app} was calculated according to the equation: $P_{app} = (c_{ap} \times V_{ap})/(c_0 \times t \times A)$ with c_{ap} = Na^+-fluorescein concentration measured in the apical compartment after incubation, V_{ap} = Volume of isotonic NaCl-solution added to the apical side, C_0 = Na^+-fluorescein concentration in the basolateral compartment at time point t = 0, t = incubation time and A = epithelial surface area.

4.5. *qRT-PCR*

Total RNA was isolated from cells grown at ALI conditions in the absence or presence of IL-13, as indicated. Cell lysis was performed using my-Budget RNA Mini Kit (Biobudget, Krefeld, Germany) without prior cell suspension using the lysis buffer provided with the kit. RNA was isolated according to manufacturers' protocol. First strand synthesis was carried out on 0.6 to 0.8 µg total RNA using the VILO SuperScript cDNA Synthesis Kit and qRT-PCR was performed using SYBR GreenER qPCR Supermix (all from Thermo Fisher, Darmstadt, Germany) according to manufactures' protocols with Quantitect RT-PCR primer assays (Qiagen, Hilden, Germany). Quantitect primer assays: Cldn1 QT00225764, Cldn2 QT00089481, Cldn3 QT00201376, Cldn4 QT00241073, Cldn5 QT01681232, Cldn6 QT00235193, Cldn7 QT00236061, Cldn8 QT00212268, Cldn9 QT00209482, Cldn10 QT00031101, Cldn11 QT00008085, Cldn12 QT01012186, Cldn14 QT00234731, Cldn15 QT01017723, Cldn16 QT00039655, Cldn17 QT01018507, Cldn18 QT00039550, Cldn19 QT00083475, Cldn20 QT00218057, Occludin QT00081844, Hmbs QT00014462, UBE2Z QT0082516 (all from Qiagen, Hilden, Germany). To screen for transcripts encoding ubiquitin activating enzymes E1 and ubiquitin conjugating enzymes E2 ubiquitination RT^2 Profiler PCR array and RT^2 SYBR Green qPCR Mastermix were used (all from Qiagen, Hilden, Germany). RealPlex2 thermocycler with latest version of Realplex data acquisition and analysis software were used (Eppendorf, Hamburg, Germany). PCR cycling protocol: Initial denaturation and Taq activation 95 °C for 2 min, 40 cycles with denaturing 95 °C for 15 s, primer annealing and amplification 60 °C 75 s. PCR products were verified by determining melting curves, final denaturation 95 °C for 15 s, annealing at 60 °C for 15 sec and heat ramping from 60 to 95 °C with a ramp duration of 20 min. Crossing points were determined using the CalPlex method provided with the data acquisition and analysis software. Each target gene expression was determined as a triplet. PCR efficiency, correction for efficiency and relative expression were calculated as previously described [79].

4.6. *Immunocytochemistry*

Epithelial cells cultivated as ALI with and without exposure to IL-13 were washed with ice cold PBS and immediately fixed with ice-cold methanol for 1 min. After methanol removal and washing with PBS, filters and cells were incubated in blocking buffer (PBS with 1% Triton X-100 and 5% BSA) for 5 min at room temperature (RT). After removal of blocking buffer, primary antibody incubation was performed (1 h, RT). Antibodies were diluted in Diluent buffer (PBS with 0.2% Triton X-100 and 0.5% BSA). The following antibodies were used: anti-claudin 4 (ab53156), anti-ZO1 (ab99462), anti-ubiquitin (ab7254), anti-proteasome α-subunit type 5 (PSMA5, ab189855), anti-E-cadherin (ab1416) (all from abcam, Cambridge, UK), anti-claudin 8 (LS-B5505) (Life Span Bioscience, Seattle, WA, USA), anti-UBE2Z (H00065264-M01) (Abnova, Taipei, Taiwan) and anti-occludin (MAB7074) (R&D Systems, Minneapolis, MN, USA). Antibodies were removed by three-times washing with PBS buffer and filters were incubated for 30 min at RT with secondary antibodies diluted in Diluent buffer. Secondary antibodies were: Alexa Fluor 488 donkey anti-rabbit IgG, Alexa Fluor 568 donkey anti-goat IgG and Alexa Fluor 647 donkey anti-mouse IgG (all from Invitrogen, Karlsruhe, Germany). Hoechst 33342 (Invitrogen, Karlsruhe, Germany) was used for staining of the nuclei. Afterwards, filters were washed with PBS and mounted on coverslips using ibidi mounting medium (ibidi, Martinsried, Germany). Images of individually cultivated epithelia were taken from a randomly selected area close to the filter center on an inverted confocal microscope (Leica TCS SP5, Leica, Germany) using a 40× lens (Leica HCX PL APO CS 40× 1.25 oil). Images for the blue (DAPI), green (AlexaFluor 488), red (AlexaFluor 568) and far-red (AlexaFluor 647) channels were taken in sequential mode using appropriate excitation and emission settings. Images represent sections with 90 to 150 cells each. Image analysis was performed using latest version of ImageJ software (NIH, Bethesda, MD, USA) [80]. Colocalization analysis was performed according to the method established by van Steensel [46]. In brief: Cross-correlation of a dual labeling image was calculated by shifting one channel image pixelwise over a distance of ±20 pixel with respect to the other channel. Pearson's correlation coefficient was calculated and the CCF-value was obtained by plotting Pearson's correlation coefficient against the translocation of channels, Δx. A maximum of this function at $\Delta x = 0$ indicates overlapping fluorescence signals, a minimum of this function at $\Delta x = 0$ excludes overlapping fluorescence signals. A randomly distribution of fluorescence signals was indicated by a function without any minima/maxima.

4.7. *Proximity Ligation Assay*

Adjacent proteins were identified by proximity ligation assays (PLA) using the Duolink kit (Sigma Aldrich, Steinheim, Germany) on methanol fixed and permeabilized epithelia (primary antibodies see above). Secondary antibodies: Anti-Mouse MINUS DUO92004, Anti-Rabbit PLUS DUO92002, Goat Plus 82023 (supplied with the Duolink kit). PLA was performed according to the manufacturer's protocol. IMIC digital microscope (Till Photonics, Muenchen, Germany) equipped with UApo/340 40× /1.35 oil objective (Olympus, Hamburg, Germany) and latest version of Live Acquisition were used for microscopy, (Till Photonics, Muenchen, Germany). Images were taken for each individually cultivated filter from a single randomly selected region close to the filter center. PLA signals were quantified using latest version of ImageJ software [80] after binarization of particles with an area of 75 μm^2 and above.

4.8. *Immunohistochemistry*

Formalin-fixed paraffin embedded tissue of sinunasal biopsies with chronic inflammation with and without eosinophilia were selected randomly from the archives of the Institute for Pathology (Hannover Medical School). For evaluation 4 μm thick sections were taken and stained with the proximity ligation assay Duolink (Sigma-Aldrich GmbH, Steinheim, Germany). Primary and secondary antibodies are given above. PLA was performed according to manufacturer's protocol. Afterwards, the reaction to specimens was embedded with a medium containing DAPI. Representative microscopy images of respiratory epithelia were taken on a Keyence fluorescence microscope (BZ9000) with a 20× magnification lens using the appropriate filter sets to detect DAPI (Excitation 377/50 nm, Beamsplitter 409 nm, Emission 447/60 nm), green (autofluorescence) (Excitation 472/50 nm, Beamsplitter 495 nm, Emission 520/35 nm) and far red (Excitation 628/40 nm, Beamsplitter 660 nm, Emission 692/40 nm) signals. Microscopy images were fused using ImageJ. After linear intensity adjustment, a region of interest containing in-focus and intact respiratory epithelium was marked. PLA signals within this area were manually counted. Recorded values are given as counts per area. All experiments were performed with approval of the ethics committee of the Hannover Medical School (Project no. 2699-2015, approved April 2015).

4.9. *Statistical Analysis*

GraphPad Prims6 software (GraphPad, La Jolla, CA, USA) was used for statistical analysis (tests are given with the text). For each experiment cells were obtained from 2 to 4 individual donors. Number of experiments (N) gives number of individual cultivated filters. Differences were considered to be significant if calculated p-values were $p \leq 0.05$. Significance levels are given within the diagrams as: $0.05 \geq p > 0.01$ = *, $0.01 \geq p > 0.001$ = **, $0.001 \geq .001$ = *, p-levels are given as 0.00001 = ****. Used test methods and exact p-values are given within the figure legends.

Author Contributions: Conceptualization, H.S. and O.H.W.; Data curation, H.S., P.B. and O.H.W.; Formal analysis, P.B., C.S., R.L. and O.H.W.; Funding acquisition, M.F., P.D. and O.H.W.; Investigation, H.S., P.B., C.S., R.L., K.N., K.T. and O.H.W.; Methodology, H.S.; Resources, D.J., M.F., P.D. and O.H.W.; Supervision, P.D.; Writing—original draft, H.S. and P.B.; Writing—review & editing, K.T., D.J., M.F., P.D. and O.H.W.

Abbreviations

ALI	air-liquid interface
CCF	cross-correlation coefficient
cldn	claudin
E-cad	E-cadherin
hTEpC	human tracheal epithelial cells
IL	interleukin
JAK	janus kinase
JAM-A	junctional adhesion molecule A
MDCK	Madin-Darby kidney cells
muc5ac	mucin 5AC
ocldn	ocludin
P_{app}	apparent permeability coefficient
PLA	proximity ligation assay
PSMA5	proteasomal subunit alpha type 5
RT-PCR	reverse transcriptase PCR
STAT	signal transducer and activator of transcription
TEER	transepithelial electrical resistance
T_h2	T-helper cell type 2
TJ	tight junction
UBE2Z	ubiquitin conjugating enzyme 2Z
UBQ	ubiquitin
ZO1	zonula occludens 1 protein

References

1. Lötvall, J.; Akdis, C.A.; Bacharier, L.B.; Bjermer, L.; Casale, T.B.; Custovic, A.; Lemanske, R.F., Jr.; Wardlaw, A.J.; Wenzel, S.E.; Greenberger, P.A. Asthma endotypes: A new approach to classification of disease entities within the asthma syndrome. *J. Allergy Clin. Immunol.* **2011**, *127*, 355–360. [CrossRef] [PubMed]
2. Robinson, D.; Humbert, M.; Buhl, R.; Cruz, A.A.; Inoue, H.; Korom, S.; Hanania, N.A.; Nair, P. Revisiting Type 2-high and Type 2-low airway inflammation in asthma: Current knowledge and therapeutic implications. *Clin. Exp. Allergy* **2017**, *47*, 161–175. [CrossRef] [PubMed]
3. Del Prete, G. The concept of type-1 and type-2 helper T cells and their cytokines in humans. *Int. Rev. Immunol.* **1998**, *16*, 427–455. [CrossRef] [PubMed]
4. Zhu, Z.; Homer, R.J.; Wang, Z.; Chen, Q.; Geba, G.P.; Wang, J.; Zhang, Y.; Elias, J.A. Pulmonary expression of interleukin-13 causes inflammation, mucus hypersecretion, subepithelial fibrosis, physiologic abnormalities, and eotaxin production. *J. Clin. Investig.* **1999**, *103*, 779–788. [CrossRef] [PubMed]
5. Wills-Karp, M.; Luyimbazi, J.; Xu, X.; Schofield, B.; Neben, T.Y.; Karp, C.L.; Donaldson, D.D. Interleukin-13: Central mediator of allergic asthma. *Science* **1998**, *282*, 2258–2261. [CrossRef] [PubMed]
6. Hanania, N.A.; Noonan, M.; Corren, J.; Korenblat, P.; Zheng, Y.; Fischer, S.K.; Cheu, M.; Putnam, W.S.; Murray, E.; Scheerens, H.; et al. Lebrikizumab in moderate-to-severe asthma: Pooled data from two randomised placebo-controlled studies. *Thorax* **2015**, *70*, 748–756. [CrossRef] [PubMed]
7. Brightling, C.E.; Chanez, P.; Leigh, R.; O'Byrne, P.M.; Korn, S.; She, D.; May, R.D.; Streicher, K.; Ranade, K.; Piper, E. Efficacy and safety of tralokinumab in patients with severe uncontrolled asthma: A randomised, double-blind, placebo-controlled, phase 2b trial. *Lancet Respir. Med.* **2015**, *3*, 692–701. [CrossRef]
8. Baverel, P.G.; White, N.; Vicini, P.; Karlsson, M.O.; Agoram, B. Dose-exposure-response relationship of the investigational anti-interleukin-13 monoclonal antibody tralokinumab in patients with severe, uncontrolled asthma. *Clin. Pharmacol. Ther.* **2018**, *103*, 826–835. [CrossRef]
9. Blease, K.; Jakubzick, C.; Westwick, J.; Lukacs, N.; Kunkel, S.L.; Hogaboam, C.M. Therapeutic effect of IL-13 immunoneutralization during chronic experimental fungal asthma. *J. Immunol.* **2001**, *166*, 5219–5224. [CrossRef]

10. Tripp, C.S.; Cuff, C.; Campbell, A.L.; Hendrickson, B.A.; Voss, J.; Melim, T.; Wu, C.; Cherniack, A.D.; Kim, K. RPC4046, a novel anti-interleukin-13 antibody, blocks IL-13 binding to IL-13 α1 and α2 receptors: A randomized, double-blind, placebo-controlled, dose-escalation first-in-human study. *Adv. Ther.* **2017**, *34*, 1364–1381. [CrossRef]
11. Aghapour, M.; Raee, P.; Moghaddam, S.J.; Hiemstra, P.S.; Heijink, I.H. Airway epithelial barrier dysfunction in chronic obstructive pulmonary disease: Role of cigarette smoke exposure. *Am. J. Respir. Cell Mol. Biol.* **2018**, *58*, 157–169. [CrossRef] [PubMed]
12. Pezzuto, A.; Carico, E. Role of HIF-1 in cancer progression: Novel insights. A review. *Curr. Mol. Med.* **2018**, *18*, 343–351. [CrossRef] [PubMed]
13. Tonini, G.; D'Onofrio, L.; Dell'Aquila, E.; Pezzuto, A. New molecular insights in tobacco-induced lung cancer. *Future Oncol.* **2013**, *9*, 649–655. [CrossRef] [PubMed]
14. Meng, M.; Liao, H.; Zhang, B.; Pan, Y.; Kong, Y.; Liu, W.; Yang, P.; Huo, Z.; Cao, Z.; Zhou, Q. Cigarette smoke extracts induce overexpression of the proto-oncogenic gene interleukin-13 receptor α2 through activation of the PKA-CREB signaling pathway to trigger malignant transformation of lung vascular endothelial cells and angiogenesis. *Cell. Signal.* **2017**, *31*, 15–25. [CrossRef] [PubMed]
15. Venkayya, R.; Lam, M.; Willkom, M.; Grünig, G.; Corry, D.B.; Erle, D.J. The Th2 lymphocyte products IL-4 and IL-13 rapidly induce airway hyperresponsiveness through direct effects on resident airway cells. *Am. J. Respir. Cell Mol. Biol.* **2002**, *26*, 202–208. [CrossRef] [PubMed]
16. Morse, B.; Sypek, J.P.; Donaldson, D.D.; Haley, K.J.; Lilly, C.M. Effects of IL-13 on airway responses in the guinea pig. *Am. J. Physiol. Lung Cell. Mol. Physiol.* **2002**, *282*, L44–L49. [CrossRef] [PubMed]
17. Winkelmann, V.E.; Thompson, K.E.; Neuland, K.; Jaramillo, A.-M.; Fois, G.; Schmidt, H.; Wittekindt, O.H.; Han, W.; Tuvim, M.J.; Dickey, B.F.; et al. Inflammation-induced upregulation of P2X4 expression augments mucin secretion in airway epithelia. *Am. J. Physiol. Lung Cell. Mol. Physiol.* **2019**, *316*, L58–L70. [CrossRef]
18. Kanoh, S.; Tanabe, T.; Rubin, B.K. IL-13-induced MUC5AC production and goblet cell differentiation is steroid resistant in human airway cells. *Clin. Exp. Allergy* **2011**, *41*, 1747–1756. [CrossRef]
19. Kistemaker, L.E.M.; Hiemstra, P.S.; Bos, I.S.T.; Bouwman, S.; van den Berge, M.; Hylkema, M.N.; Meurs, H.; Kerstjens, H.A.M.; Gosens, R. Tiotropium attenuates IL-13-induced goblet cell metaplasia of human airway epithelial cells. *Thorax* **2015**, *70*, 668–676. [CrossRef]
20. Zhang, F.-Q.; Han, X.-P.; Zhang, F.; Ma, X.; Xiang, D.; Yang, X.-M.; Ou-Yang, H.-F.; Li, Z. Therapeutic efficacy of a co-blockade of IL-13 and IL-25 on airway inflammation and remodeling in a mouse model of asthma. *Int. Immunopharmacol.* **2017**, *46*, 133–140. [CrossRef]
21. Ohashi, Y.; Motojima, S.; Fukuda, T.; Makino, S. Airway hyperresponsiveness, increased intracellular spaces of bronchial epithelium, and increased infiltration of eosinophils and lymphocytes in bronchial mucosa in asthma. *Am. Rev. Respir. Dis.* **1992**, *145*, 1469–1476. [CrossRef] [PubMed]
22. Wise, S.K.; Laury, A.M.; Katz, E.H.; Den Beste, K.A.; Parkos, C.A.; Nusrat, A. Interleukin-4 and interleukin-13 compromise the sinonasal epithelial barrier and perturb intercellular junction protein expression. *Int. Forum Allergy Rhinol.* **2014**, *4*, 361–370. [CrossRef] [PubMed]
23. Wawrzyniak, P.; Wawrzyniak, M.; Wanke, K.; Sokolowska, M.; Bendelja, K.; Rückert, B.; Globinska, A.; Jakiela, B.; Kast, J.I.; Idzko, M.; et al. Regulation of bronchial epithelial barrier integrity by type 2 cytokines and histone deacetylases in asthmatic patients. *J. Allergy Clin. Immunol.* **2017**, *139*, 93–103. [CrossRef] [PubMed]
24. Sweerus, K.; Lachowicz-Scroggins, M.; Gordon, E.; LaFemina, M.; Huang, X.; Parikh, M.; Kanegai, C.; Fahy, J.V.; Frank, J.A. Claudin-18 deficiency is associated with airway epithelial barrier dysfunction and asthma. *J. Allergy Clin. Immunol.* **2016**, *139*, 1–10. [CrossRef] [PubMed]
25. Saatian, B.; Rezaee, F.; Desando, S.; Emo, J.; Chapman, T.; Knowlden, S.; Georas, S.N. Interleukin-4 and interleukin-13 cause barrier dysfunction in human airway epithelial cells. *Tissue Barriers* **2013**, *1*, e24333. [CrossRef] [PubMed]
26. Ahdieh, M.; Vandenbos, T.; Youakim, A. Lung epithelial barrier function and wound healing are decreased by IL-4 and IL-13 and enhanced by IFN-gamma. *Am. J. Physiol. Cell Physiol.* **2001**, *281*, C2029–C2038. [CrossRef] [PubMed]
27. Günzel, D. Claudins: Vital partners in transcellular and paracellular transport coupling. *Pflug. Arch. Eur. J. Physiol.* **2017**, *469*, 35–44. [CrossRef]

28. Schlingmann, B.; Molina, S.A.; Koval, M. Claudins: Gatekeepers of lung epithelial function. *Semin. Cell Dev. Biol.* **2015**, *42*, 47–57. [CrossRef]
29. Den Beste, K.A.; Hoddeson, E.K.; Parkos, C.A.; Nusrat, A.; Wise, S.K. Epithelial permeability alterations in an in vitro air-liquid interface model of allergic fungal rhinosinusitis. *Int. Forum Allergy Rhinol.* **2013**, *3*, 19–25. [CrossRef]
30. Holter, J.F.; Weiland, J.E.; Pacht, E.R.; Gadek, J.E.; Davis, W.B. Protein permeability in the adult respiratory distress syndrome. Loss of size selectivity of the alveolar epithelium. *J. Clin. Invest.* **1986**, *78*, 1513–1522. [CrossRef]
31. Ritchie, A.I.; Jackson, D.J.; Edwards, M.R.; Johnston, S.L. Airway epithelial orchestration of innate immune function in response to virus infection. A focus on asthma. *Ann. Am. Thorac. Soc.* **2016**, *13* (Suppl. 1), S55–S63.
32. Meyer, D.M.; Jesson, M.I.; Li, X.; Elrick, M.M.; Funckes-Shippy, C.L.; Warner, J.D.; Gross, C.J.; Dowty, M.E.; Ramaiah, S.K.; Hirsch, J.L.; et al. Anti-inflammatory activity and neutrophil reductions mediated by the JAK1/JAK3 inhibitor, CP-690,550, in rat adjuvant-induced arthritis. *J. Inflamm.* **2010**, *7*, 41. [CrossRef] [PubMed]
33. Gehringer, M.; Pfaffenrot, E.; Bauer, S.; Laufer, S.A. Design and synthesis of tricyclic jak3 inhibitors with picomolar affinities as novel molecular probes. *ChemMedChem* **2014**, *9*, 277–281. [CrossRef] [PubMed]
34. Fridman, J.S.; Scherle, P.A.; Collins, R.; Burn, T.C.; Li, Y.; Li, J.; Covington, M.B.; Thomas, B.; Collier, P.; Favata, M.F.; et al. Selective inhibition of JAK1 and JAK2 is efficacious in rodent models of arthritis: Preclinical characterization of INCB028050. *J. Immunol.* **2010**, *184*, 5298–5307. [CrossRef] [PubMed]
35. Quintás-Cardama, A.; Vaddi, K.; Liu, P.; Manshouri, T.; Li, J.; Scherle, P.A.; Caulder, E.; Wen, X.; Li, Y.; Waeltz, P.; et al. Preclinical characterization of the selective JAK1/2 inhibitor INCB018424: Therapeutic implications for the treatment of myeloproliferative neoplasms. *Blood* **2010**, *115*, 3109–3117. [CrossRef] [PubMed]
36. Hou, J.; Renigunta, A.; Yang, J.; Waldegger, S. Claudin-4 forms paracellular chloride channel in the kidney and requires claudin-8 for tight junction localization. *Proc. Natl. Acad. Sci. USA* **2010**, *107*, 18010–18015. [CrossRef]
37. Kielgast, F.; Schmidt, H.; Braubach, P.; Winkelmann, V.E.; Thompson, K.E.; Frick, M.; Dietl, P.; Wittekindt, O.H. Glucocorticoids regulate tight junction permeability of lung epithelia by modulating claudin 8. *Am. J. Respir. Cell Mol. Biol.* **2016**, *54*, 707–717. [CrossRef]
38. Nelson, W.J.; Shore, E.M.; Wang, A.Z.; Hammerton, R.W. Identification of a membrane-cytoskeleton complex containing the cell adhesion molecule uvomorulin (E-Cadherin), ankyrin, and fodrin in Madin-Darby canine kidney epithelial cells. *J. Cell Biol.* **1990**, *110*, 349–357. [CrossRef]
39. Ciechanover, A. The ubiquitin-proteasome proteolytic pathway. *Cell* **1994**, *79*, 13–21. [CrossRef]
40. Gong, Y.; Wang, J.; Yang, J.; Gonzales, E.; Perez, R.; Hou, J. KLHL3 regulates paracellular chloride transport in the kidney by ubiquitination of claudin-8. *Proc. Natl. Acad. Sci. USA* **2015**, *112*, 4340–4345. [CrossRef]
41. Lui, W.Y.; Lee, W.M. Regulation of junction dynamics in the testis—Transcriptional and post-translational regulations of cell junction proteins. *Mol. Cell. Endocrinol.* **2006**, *250*, 25–35. [CrossRef] [PubMed]
42. Wing, Y.L.; Lee, W.M. cAMP perturbs inter-Sertoli tight junction permeability barrier in vitro via its effect on proteasome-sensitive ubiquitination of occludin. *J. Cell. Physiol.* **2005**, *203*, 564–572.
43. Mandel, I.; Paperna, T.; Volkowich, A.; Merhav, M.; Glass-Marmor, L.; Miller, A. The ubiquitin-proteasome pathway regulates claudin 5 degradation. *J. Cell. Biochem.* **2012**, *113*, 2415–2423. [CrossRef]
44. Van Campenhout, C.A.; Eitelhuber, A.; Gloeckner, C.J.; Giallonardo, P.; Gegg, M.; Oller, H.; Grant, S.G.N.; Krappmann, D.; Ueffing, M.; Lickert, H. Dlg3 Trafficking and apical tight junction formation is regulated by Nedd4 and Nedd4-2 E3 Ubiquitin ligases. *Dev. Cell* **2011**, *21*, 479–491. [CrossRef] [PubMed]
45. Traweger, A.; Fang, D.; Liu, Y.C.; Stelzhammer, W.; Krizbai, I.A.; Fresser, F.; Bauer, H.C.; Bauer, H. The tight junction-specific protein occludin is a functional target of the E3 ubiquitin-protein ligase itch. *J. Biol. Chem.* **2002**, *277*, 10201–10208. [CrossRef] [PubMed]
46. Van Steensel, B.; van Binnendijk, E.P.; Hornsby, C.D.; van der Voort, H.T.; Krozowski, Z.S.; de Kloet, E.R.; van Driel, R. Partial colocalization of glucocorticoid and mineralocorticoid receptors in discrete compartments in nuclei of rat hippocampus neurons. *J. Cell Sci.* **1996**, *109*, 787–792. [PubMed]
47. Ghebre, M.A.; Bafadhel, M.; Desai, D.; Cohen, S.E.; Newbold, P.; Rapley, L.; Woods, J.; Rugman, P.; Pavord, I.D.; Newby, C.; et al. Biological clustering supports both "dutch" and "british" hypotheses of asthma and chronic obstructive pulmonary disease. *J. Allergy Clin. Immunol.* **2015**, *135*, 63–72. [CrossRef] [PubMed]

48. Sugita, K.; Steer, C.A.; Martinez-Gonzalez, I.; Altunbulakli, C.; Morita, H.; Castro-Giner, F.; Kubo, T.; Wawrzyniak, P.; Rückert, B.; Sudo, K.; et al. Type 2 innate lymphoid cells disrupt bronchial epithelial barrier integrity by targeting tight junctions through IL-13 in asthmatic patients. *J. Allergy Clin. Immunol.* **2018**, *141*, 300–310. [CrossRef] [PubMed]
49. May, R.D.; Fung, M. Strategies targeting the IL-4/IL-13 axes in disease. *Cytokine* **2015**, *75*, 89–116. [CrossRef]
50. Krug, S.M.; Bojarski, C.; Fromm, A.; Lee, I.M.; Dames, P.; Richter, J.F.; Turner, J.R.; Fromm, M.; Schulzke, J.-D. Tricellulin is regulated via interleukin-13-receptor α2, affects macromolecule uptake, and is decreased in ulcerative colitis. *Mucosal Immunol.* **2018**, *11*, 345–356. [CrossRef]
51. Skieterska, K.; Rondou, P.; Van Craenenbroeck, K. Regulation of G protein-coupled receptors by ubiquitination. *Int. J. Mol. Sci.* **2017**, *18*, 923. [CrossRef] [PubMed]
52. Hicke, L. Ubiquitin-dependent internalization regulation of plasma membrane proteins. *FASEB J.* **1997**, *11*, 1215–1226. [CrossRef] [PubMed]
53. Widagdo, J.; Guntupalli, S.; Jang, S.E.; Anggono, V. Regulation of AMPA receptor trafficking by protein ubiquitination. *Front. Mol. Neurosci.* **2017**, *10*, 1–10. [CrossRef] [PubMed]
54. Ciechanover, A. Intracellular protein degradation: From a vague idea thru the lysosome and the ubiquitin-proteasome system and onto human diseases and drug targeting. *Best Pract. Res. Clin. Haematol.* **2017**, *30*, 341–355. [CrossRef] [PubMed]
55. Takahashi, S.; Iwamoto, N.; Sasaki, H.; Ohashi, M.; Oda, Y.; Tsukita, S.; Furuse, M. The E3 ubiquitin ligase LNX1p80 promotes the removal of claudins from tight junctions in MDCK cells. *J. Cell Sci.* **2009**, *122*, 985–994. [CrossRef]
56. Cong, X.; Zhang, Y.; Li, J.; Mei, M.; Ding, C.; Xiang, R.-L.; Zhang, L.-W.; Wang, Y.; Wu, L.-L.; Yu, G.-Y. Claudin-4 is required for modulation of paracellular permeability by muscarinic acetylcholine receptor in epithelial cells. *J. Cell Sci.* **2015**, *128*, 2271–2286. [CrossRef]
57. Murakami, T.; Felinski, E.A.; Antonetti, D.A. Occludin phosphorylation and ubiquitination regulate tight junction trafficking and vascular endothelial growth factor-induced permeability. *J. Biol. Chem.* **2009**, *284*, 21036–21046. [CrossRef]
58. Wetzel, F.; Mittag, S.; Cano-Cortina, M.; Wagner, T.; Krämer, O.H.; Niedenthal, R.; Gonzalez-Mariscal, L.; Huber, O. SUMOylation regulates the intracellular fate of ZO-2. *Cell. Mol. Life Sci.* **2016**, *74*, 373–392. [CrossRef]
59. Dang, H.; Klokk, T.I.; Schaheen, B.; McLaughlin, B.M.; Thomas, A.J.; Durns, T.A.; Bitler, B.G.; Sandvig, K.; Fares, H. Derlin-dependent retrograde transport from endosomes to the golgi apparatus. *Traffic* **2011**, *12*, 1417–1431. [CrossRef]
60. Schaheen, B.; Dang, H.; Fares, H. Derlin-dependent accumulation of integral membrane proteins at cell surfaces. *J. Cell Sci.* **2009**, *122*, 2228–2239. [CrossRef]
61. Gu, X.; Zhao, F.; Zheng, M.; Fei, X.; Chen, X.; Huang, S.; Xie, Y.; Mao, Y. Cloning and characterization of a gene encoding the human putative ubiquitin conjugating enzyme E2Z (UBE2Z). *Mol. Biol. Rep.* **2007**, *34*, 183–188. [CrossRef] [PubMed]
62. Liu, X.; Zhao, B.; Sun, L.; Bhuripanyo, K.; Wang, Y.; Bi, Y.; Davuluri, R.V.; Duong, D.M.; Nanavati, D.; Yin, J.; et al. Orthogonal ubiquitin transfer identifies ubiquitination substrates under differential control by the two ubiquitin activating enzymes. *Nat. Commun.* **2017**, *8*, 1–12. [CrossRef] [PubMed]
63. Jin, J.; Li, X.; Gygi, S.P.; Harper, J.W. Dual E1 activation systems for ubiquitin differentially regulate E2 enzyme charging. *Nature* **2007**, *447*, 1135–1138. [CrossRef] [PubMed]
64. Mehta, S.; Nijhuis, A.; Kumagai, T.; Lindsay, J.; Silver, A. Defects in the adherens junction complex (E-cadherin/ β-catenin) in inflammatory bowel disease. *Cell Tissue Res.* **2015**, *360*, 749–760. [CrossRef] [PubMed]
65. Caldeira, M.V.; Salazar, I.L.; Curcio, M.; Canzoniero, L.M.T.; Duarte, C.B. Role of the ubiquitin-proteasome system in brain ischemia: Friend or foe? *Prog. Neurobiol.* **2014**, *112*, 50–69. [CrossRef] [PubMed]
66. Choi, W.H.; De Poot, S.A.H.; Lee, J.H.; Kim, J.H.; Han, D.H.; Kim, Y.K.; Finley, D.; Lee, M.J. Open-gate mutants of the mammalian proteasome show enhanced ubiquitin-conjugate degradation. *Nat. Commun.* **2016**, *7*, 1–12. [CrossRef] [PubMed]
67. Kanayama, H.O.; Tamura, T.; Ugai, S.; Kagawa, S.; Tanahashi, N.; Yoshimura, T.; Tanaka, K.; Ichihara, A. Demonstration that a human 26S proteolytic complex consists of a proteasome and multiple associated protein components and hydrolyzes ATP and ubiquitin-ligated proteins by closely linked mechanisms. *Eur. J. Biochem.* **1992**, *206*, 567–578. [CrossRef]

68. Balch, W.E.; Morimoto, R.I. Adapting proteostatis for disease intervention. *Science* **2008**, *319*, 916–919. [CrossRef]
69. Inoue, S.; Nakase, H.; Matsuura, M.; Mikami, S.; Ueno, S.; Uza, N.; Chiba, T. The effect of proteasome inhibitor MG132 on experimental inflammatory bowel disease. *Clin. Exp. Immunol.* **2009**, *156*, 172–182. [CrossRef]
70. Vadász, I.; Weiss, C.H.; Sznajder, J.I. Ubiquitination and proteolysis in acute lung injury. *Chest* **2012**, *141*, 763–771. [CrossRef]
71. Stepaniants, S.; Wang, I.M.; Boie, Y.; Mortimer, J.; Kennedy, B.; Elliott, M.; Hayashi, S.; Luo, H.; Wong, J.; Loy, L.; et al. Genes related to emphysema are enriched for ubiquitination pathways. *BMC Pulm. Med.* **2014**, *14*, 1–11. [CrossRef] [PubMed]
72. Min, T.; Bodas, M.; Mazur, S.; Vij, N. Critical role of proteostasis-imbalance in pathogenesis of COPD and severe emphysema. *J. Mol. Med.* **2011**, *89*, 577–593. [CrossRef] [PubMed]
73. Tran, I.; Ji, C.; Ni, I.; Min, T.; Tang, D.; Vij, N. Role of cigarette smoke-induced aggresome formation in chronic obstructive pulmonary disease-emphysema pathogenesis. *Am. J. Respir. Cell Mol. Biol.* **2015**, *53*, 159–173. [CrossRef] [PubMed]
74. Kim, S.Y.; Kim, H.J.; Park, M.K.; Huh, J.W.; Park, H.Y.; Ha, S.Y.; Shin, J.H.; Lee, Y.S. Mitochondrial E3 ubiquitin protein ligase 1 mediates cigarette smoke-induced endothelial cell death and dysfunction. *Am. J. Respir. Cell Mol. Biol.* **2016**, *54*, 284–296. [CrossRef] [PubMed]
75. Ware, L.B.; Matthay, M.A. The acute respiratory distress syndrome. *N. Engl. J. Med.* **2000**, *342*, 1334–1349. [CrossRef]
76. Li, G.; Flodby, P.; Luo, J.; Kage, H.; Sipos, A.; Gao, D.; Ji, Y.; Beard, L.M.L.; Marconett, C.N.; DeMaio, L.; et al. Knockout mice reveal key roles for Claudin 18 in alveolar barrier properties and fluid homeostasis. *Am. J. Respir. Cell Mol. Biol.* **2014**, *51*, 210–222. [CrossRef] [PubMed]
77. Kage, H.; Flodby, P.; Gao, D.; Kim, Y.H.; Marconett, C.N.; DeMaio, L.; Kim, K.-J.; Crandall, E.D.; Borok, Z. Claudin 4 knockout mice: Normal physiological phenotype with increased susceptibility to lung injury. *Am. J. Physiol. Lung Cell. Mol. Physiol.* **2014**, *307*, L524–L536. [CrossRef]
78. Fulkerson, P.C.; Fischetti, C.A.; Hassman, L.M.; Nikolaidis, N.M.; Rothenberg, M.E. Persistent effects induced by IL-13 in the lung. *Am. J. Respir. Cell Mol. Biol.* **2006**, *35*, 337–346. [CrossRef]
79. Pfaffl, M.W. A new mathematical model for relative quantification in real-time RT-PCR. *Nucleic Acids Res.* **2001**, *29*, e45. [CrossRef]
80. Schindelin, J.; Rueden, C.T.; Hiner, M.C.; Eliceiri, K.W. The ImageJ ecosystem: An open platform for biomedical image analysis. *Mol. Reprod. Dev.* **2015**, *82*, 518–529. [CrossRef]

3

Detailed Clinical Features of Deafness Caused by a Claudin-14 Variant

Tomohiro Kitano [1], Shin-ichiro Kitajiri [1,*], Shin-ya Nishio [1,2] and Shin-ichi Usami [1,2]

1 Department of Otorhinolaryngology, Shinshu University School of Medicine, 3-1-1 Asahi, Matsumoto, Nagano 390-8621, Japan; tomokitano@shinshu-u.ac.jp (T.K.); nishio@shinshu-u.ac.jp (S.N.); usami@shinshu-u.ac.jp (S.U.)

2 Department of Hearing Implant Sciences, Shinshu University School of Medicine, 3-1-1 Asahi, Matsumoto, Nagano 390-8621, Japan

* Correspondence: kitajiri@shinshu-u.ac.jp.

Abstract: Tight junctions are cellular junctions that play a major role in the epithelial barrier function. In the inner ear, claudins, occludin, tricellulin, and angulins form the bicellular or tricellular binding of membrane proteins. In these, one type of claudin gene, *CLDN14*, was reported to be responsible for human hereditary hearing loss, DFNB29. Until now, nine pathogenic variants have been reported, and most phenotypic features remain unclear. In the present study, genetic screening for 68 previously reported deafness causative genes was carried out to identify *CLDN14* variants in a large series of Japanese hearing loss patients, and to clarify the prevalence and clinical characteristics of DFNB29 in the Japanese population. One patient had a homozygous novel variant (c.241C>T: p.Arg81Cys) (0.04%: 1/2549). The patient showed progressive bilateral hearing loss, with post-lingual onset. Pure-tone audiograms indicated a high-frequency hearing loss type, and the deterioration gradually spread to other frequencies. The patient showed normal vestibular function. Cochlear implantation improved the patient's sound field threshold levels, but not speech discrimination scores. This report indicated that claudin-14 is essential for maintaining the inner ear environment and suggested the possible phenotypic expansion of DFNB29. This is the first report of a patient with a tight junction variant receiving a cochlear implantation.

Keywords: tight junction; Claudin-14; *CLDN14*; hearing loss; vestibular function; cochlear implantation

---

1. Introduction

1.1. Hearing Loss

Hearing loss (HL) is the most common sensory impairment and is diagnosed in approximately two in every 1000 children [1]. At least 60% of all childhood nonsyndromic sensorineural hearing loss is caused by genetic factors [2]. The inheritance patterns of this form of HL include autosomal recessive, autosomal dominant, X-linked, and mitochondrial. Autosomal recessive nonsyndromic hereditary HL (ARNSHL) is typically prelingual, and accounts for approximately 70% of nonsyndromic hereditary HL patients [3]. Thus far, 75 causative genes for ARNSHL have been identified [4]. One form of ARNSHL is DFNB29 (OMIM #: 614035), which is caused by variants in the *CLDN14* gene.

1.2. Tight Junctions in the Inner Ear

For our sound receiving process, it is necessary to convert sound vibration to nerve action potential (mechano-transduction) [5]. The center for this mechano-transduction is the cochlea, located in the inner ear. The cochlea is filled with two types of lymph fluid, the endolymph and perilymph. These fluids are completely different in their chemical composition: the perilymph resembles extracellular fluids in general [6,7] but the endolymph has the characteristics of an intracellular fluid in that it has high K^+ and low Na^+ concentrations [8]. Furthermore, the electrical potential of the endolymph, i.e., endocochlear potential (EP) is positive by approximately 80 to 90 mV relative to the perilymph [9,10]. It is now widely accepted that these characteristics of the endolymph (high K^+ concentration and EP) are indispensable for cochlear hair cells to transduce acoustic stimuli into electrical signals [11]. The barrier function of the epithelial cell sheet prevents paracellular permeability, and the separation of these two fluids is essential for the maintenance of their differences in composition [12]. Tight junctions (TJs) are cellular junctions that play a major role in epithelial barrier function. In the inner ear, claudins, occludin, tricellulin, and the angulin family proteins (angulin-1/*LSR*, angulin-2/*ILDR1*, and angulin-3/*ILDR2*) form the bicellular or tricellular binding of membrane proteins. Among these, claudin-14, tricellulin, and angulin-2 are reported to be responsible for human deafness DFNB29, DFNB49, and DFNB42, respectively [13–15].

The claudins are a family of proteins that play a major role in epithelial barrier function, especially in bicellular junctions. There are 24 claudins that have been identified in humans thus far [13,16–18], with at least 10 of these reported to be expressed in the inner ear [19]. Their expression can be categorized into three types: (1) claudin-1, claudin-2, claudin-3, claudin-9, claudin-10, claudin-12, claudin-14, and claudin-18 that are expressed in multiple sites such as the organ of Corti, Reissner's membrane, the spiral limbus, and the marginal cells of the stria vascularis; (2) claudin-8 that is expressed in all the aforementioned sites except for the organ of Corti; and (3) claudin-11 that is expressed only in the basal cells of the stria vascularis [19]. The barrier functions are speculated to be dependent on the combinations of these claudins, and previous reports have shown claudin-9, claudin-11, and claudin-14 to be critical for hearing functions [20–25], with mutations in these proteins causing deafness in humans and mice. Murine studies have indicated that claudin-11 (cldn-11) knockout mice demonstrate hearing loss due to reduced EP [20,21]. Other studies have indicated that, as a result of the rapid degeneration of cochlear hair cells shortly after birth, *Cldn9* mutant mice and *Cldn14* mutant mice demonstrate hearing loss. However, these two animal models do not display reduced EP [22,23]. The variations within these phenotypes are thought to be correlated with the multiple functions of the TJs in the inner ear. Thus, the TJs in the epithelial barriers are significantly involved in inner ear function.

1.3. CLDN14 Variants in Deafness

To date, nine different variants in *CLDN14* [13,26–29] have been reported to cause HL in the Pakistani, Greek, and Canadian populations. Seven of the nine variants were reported from Pakistani consanguineous families. The *CLDN14* variants are a relatively common cause of recessive hearing loss, which is responsible for 2.25% of HL patients in a Pakistani study cohort [28], whereas no pathogenic variants were reported from east Asian populations [30,31]. Although previously reported papers have shown some clinical characteristics of patients with *CLDN14* variants, the audiovestibular findings, such as progression and details of vestibular function, remain unclear. In addition, there are no reports on patients with cochlear implantation (CI), thus the outcome of CI is also unknown. In the present study, we used massively parallel DNA sequencing (MPS) to detect pathogenic variants in *CLDN14* among a large series of Japanese HL patients.

Here, we present a patient with a novel variant in the *CLDN14* gene identified from a group of non-consanguineous HL patients and discuss additional clinical features including CI outcomes. This is the first report of a diagnosis of hearing caused by *CLDN14* in an east Asian population and in such a patient receiving a cochlear implant as intervention.

2. Results

2.1. Detected Variant

One possible disease-causing variant, *CLDN14*: NM_144492: c.241C>T, leading to p.Arg81Cys, was homozygously detected in one of 1577 probands with autosomal recessive HL (Table 1, Figure 1), whereas no pathogenic variants were found in the other 972 probands with autosomal dominant or inheritance pattern of unknown HL. At the same amino acid residue, which is within the first extracellular loop, a different missense change c.242G>A, leading to p.Arg81His, has been reported (Figure 2) [27]. No candidate variants in the other 67 deafness genes were identified in the proband. In detail, the variants in two other genes, *PTPRQ* and *EYA4*, were also detected from the proband but were not segregated. As shown in Figure 1, family segregation was confirmed by using Sanger sequencing. The variant was not identified in ExAC, gnomAD, 3.5KJPN or the 1208 Japanese exome variants, in addition to the 333 in-house Japanese normal hearing control databases.

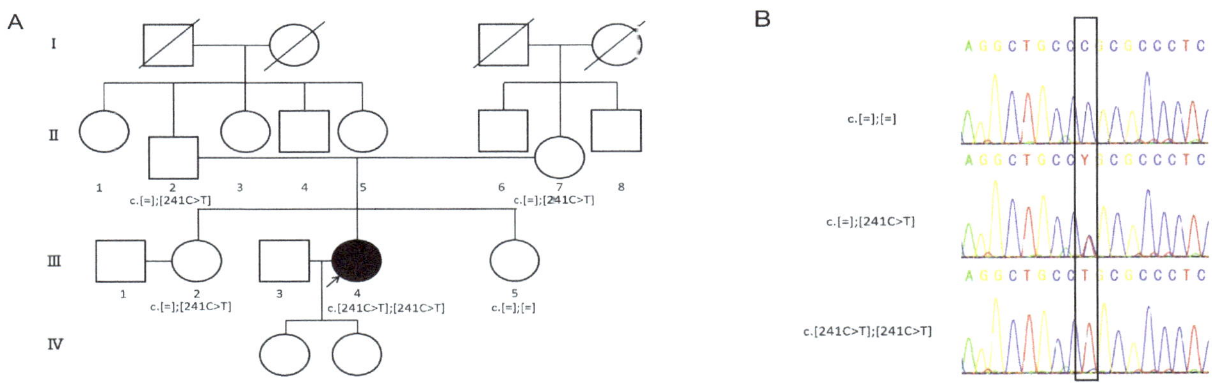

Figure 1. Pedigree and *CLDN14* variants of the family. (**A**) Pedigree shows autosomal recessive inherited hearing loss (HL), (**B**) the electropherograms of this family. Target genome enrichment for 68 previously reported deafness causative genes and massively parallel DNA sequencing are carried out for this proband (III-4). Sanger sequencing is used for family segregation analysis. Genetic analysis results are shown under the proband and family members.

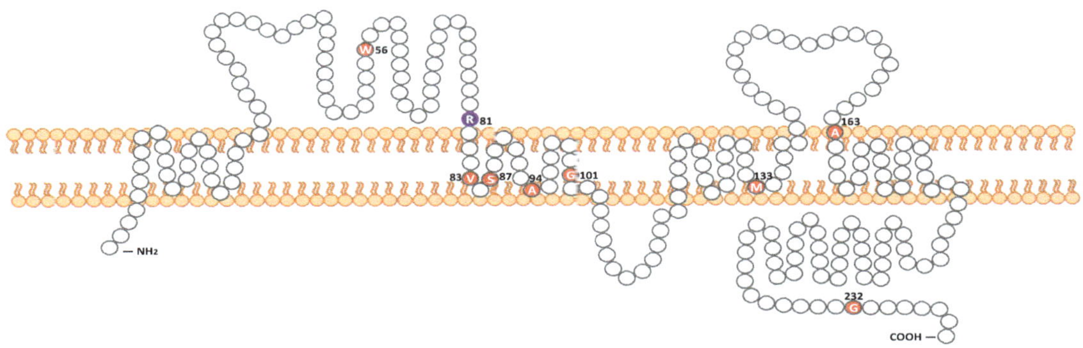

Figure 2. Location of pathogenic variants in Claudin-14. Red colored amino acid residues indicate previously reported claudin-14 variants. The blue colored residue indicates the positions of *CLDN14* p.Arg81His and p.Arg81Cys.

We employed in silico software (SIFT, polyphen-2, LRT, Mutation Taster, etc.), and almost all scores indicated "damaging" (Table 1). The corresponding amino acid was well conserved across species (170/170 (100%) in vertebrates). Taken together, according to the American College of Medical Genetics (ACMG) guideline, the variant was classified as "likely pathogenic" (PS4, PM2, and PM5).

2.2. Clinical Findings

The proband was a 37-year-old female. The newborn hearing screening program had not yet started at birth and there were no particular complications in the perinatal period. The female proband also passed an elementary school health checkup at six years of age, but was suspected of hearing loss at age nine and was referred to an otolaryngology clinic at a local general hospital. Pure-tone audiograms showed steep high-frequency sensorineural HL (SNHL) with an average of 77 dBHL in both ears (Figure 3A). It appeared to progress slowly, and the proband began to wear hearing aids bilaterally at 11 years of age. Over a period of 20 years, their residual hearing in the lower frequencies gradually deteriorated and hearing aids became ineffective. At the age of 34, the proband consulted our hospital for further examination. Otoscopic examination revealed a normal tympanic membrane. Computed tomography (CT) and magnetic resonance imaging (MRI) of the temporal bones showed no malformations, hearing level was approximately 100 dBHL in both ears, but the left ear showed moderate residual hearing in the lower frequencies. We performed vestibular assessment (caloric test and cervical-ocular vestibular evoked myogenic potentials, i.e., c/oVEMPs) as preoperative examinations prior to CI. The caloric test, cVEMPs, and oVEMPs represent the function of the semicircular canal, the saccule and inferior vestibular nerve, and the utricle and superior vestibular nerve, respectively. As shown in Figure 3B, all vestibular testing showed normal vestibular functions.

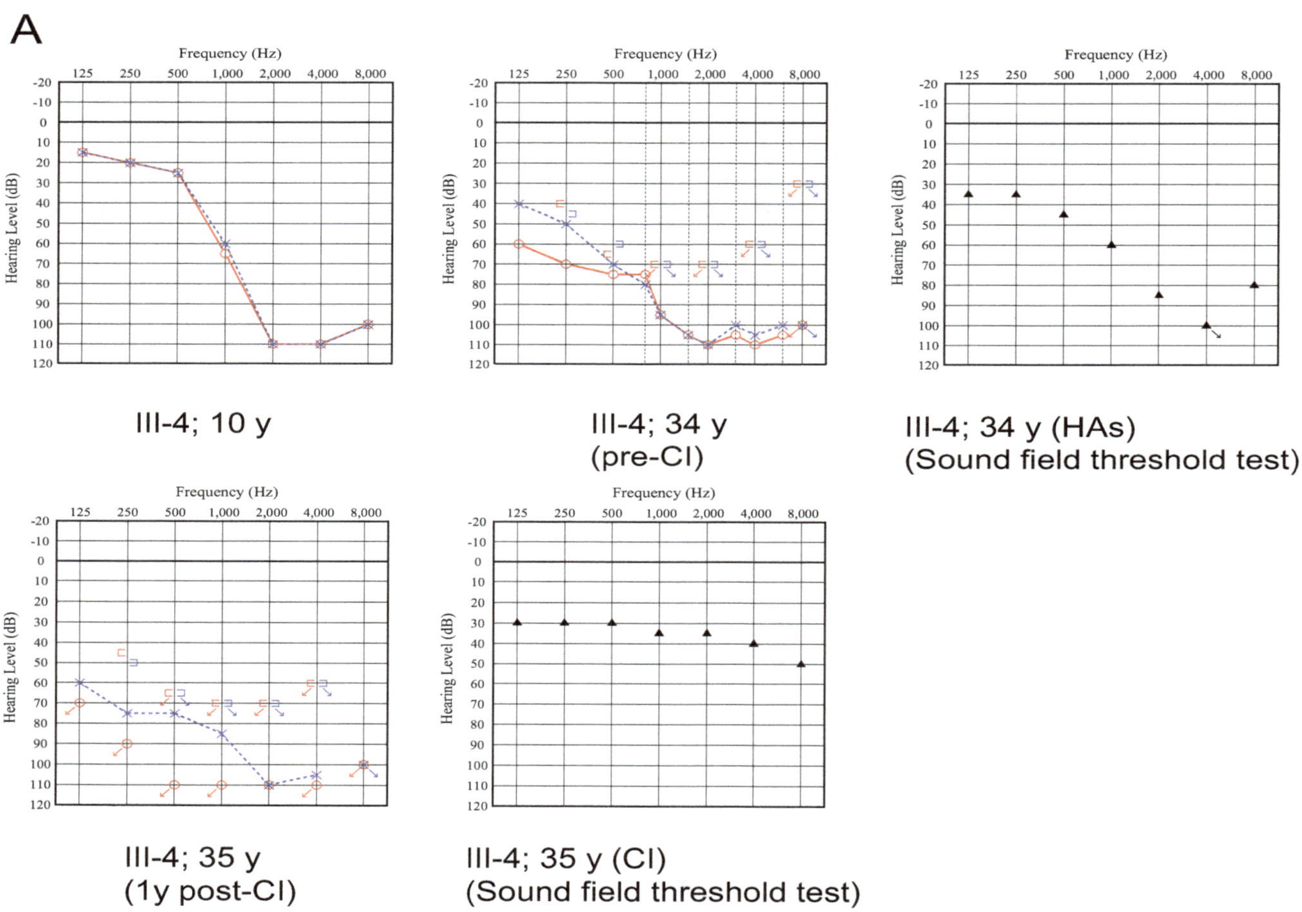

Figure 3. *Cont.*

B

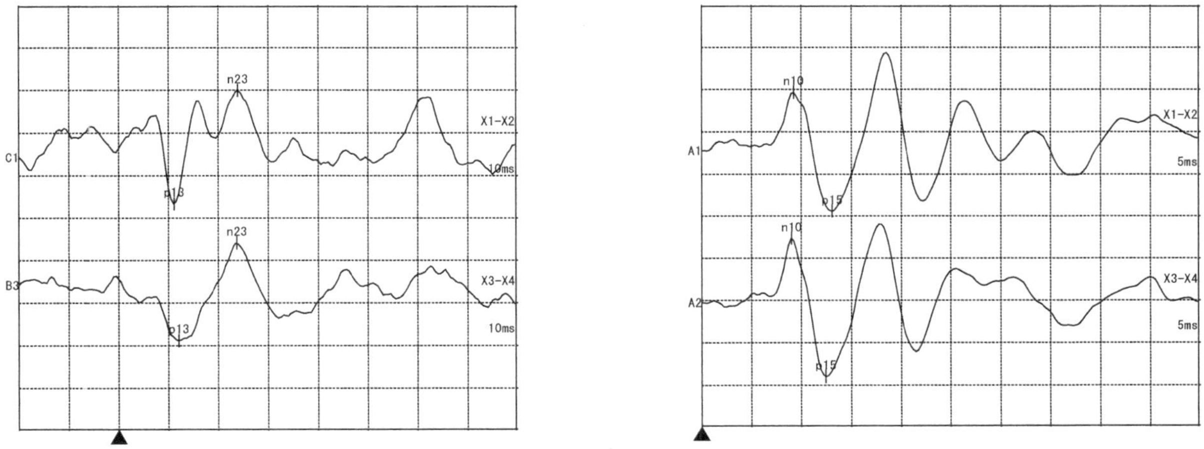

Figure 3. (**A**) Pure-tone audiometry (PTA) shows bilateral progressive sensorineural hearing loss in the proband. After a cochlear implantation (CI) in the right ear, sound field hearing threshold tests with a CI show 40 dBSPL. Red: right ear, Blue: left ear, Circle and Cross: Right and left ear air conduction hearing level respectively, Square bracket: Bone conduction hearing level, Triangle: both ears with intervention, Arrow: over measurement limit (**B**) There are no obvious differences between each ear in the cervical vestibular evoked myogenic potentials (right), and ocular vestibular evoked myogenic potentials (left).

The proband underwent a CI (MED-EL FLEX28) in the right ear at the age 35. Sound field threshold levels were improved from 77.5 dBSPL (with hearing aids) to 37.5 dBSPL (with the cochlear implant) (Figure 3A). To evaluate speech perception outcomes, the proband underwent speech discrimination testing (using the 67S Japanese monosyllable test, presenting in 65 dBSPL). Contrary to the good improvement in sound field threshold levels, the proband's speech perception outcomes improved from 12% to only 32% two years after receiving a CI.

3. Discussion

3.1. Frequency of CLDN14-Associated HL in the Japanese Population

We discovered a novel causative variant in the *CLDN14* gene as a cause of progressive SNHL in a Japanese patient. To date, a total of nine pathogenic variants in *CLDN14* have been reported (Table 2, Figure 2). Among them, seven were reported from Pakistan, and one each from Greece and Canada. Otherwise, according to the previous reports, no pathogenic variants of *CLDN14* have been detected in the Korean and Chinese populations [30,31]. Hence, this is the first report of a causative *CLDN14* variant in an east Asian population.

The incidence of *CLDN14* variants was 0.04% (1/2549) among the Japanese HL patients, and 0.06% (1/1577) among the families with autosomal recessive HL in the Japanese population. In clear contrast to this, variants of *CLDN14* account for 2.25% of autosomal recessive HL in the Pakistani population. In this study, the patient was identified from a non-consanguineous family and the parents of the proband were from a geographically remote area, whereas almost all of the previous reports were from consanguineous families [13,27–29]. As no pathogenic *CLDN14* variants were found in our cohort, but one HL patient with a homozygous *CLDN14* variant was identified, it is necessary to consider such rare HL cases.

Table 1. Possible causative variant identified in this study.

		Prediction Score								Allele Frequency in Controls		
Nucleotide Change	Amino Acid Change	SIFT *	PolyPhen2_HVAR *	LRT *	Mut_Taster *	Mut_Assessor *	REVEL *	CADD	Evolutional Conservation **	ExAC	gnomAD	3.5kJPN
c.241C > T	p.R81C	0.78	0.88	0.84	0.81	0.98	0.89	27.6	Yes (100%)	0.000042	0.000037	0

* The prediction scores of each algorithm included on the ANNOVAR software were converted from the original scoring system. Scores closer to 1.0 indicated the mutation was more damaging, and those closer to 0 indicated they were more tolerant. ** Conserved in 170 vertebrates from the HGMD professional.

Table 2. Summary of clinical features associated with *CLDN14* variants.

			HL		Pure-Tone Audiometry						
Nucleotide Change	Genotype	Amino Acid Change	Onset	Progression	Severity of HL	Audiometric Configuration	Consanguineous Marriage	Vestibuler Function	Hearing Intervention	Family Origin	Reference
c.167G > A	homozygote	p.Trp56 *	NA	No	severe to profound	gently sloping	Yes	NA	NA	Pakistan	Lee et al. 2011 [27]
c.241C > T	homozygote	p.Arg81Cys	9 y.o.	Yes	Severe	steeply sloping	No	Normal	CI	Japan	This study
c.242G > A	homozygote	p.Arg81His	NA	No	severe to profound	gently sloping	Yes	NA	NA	Pakistan	Lee et al. 2011 [27]
c.254T > A	homozygote	p.Val85Asp	prelingual	No	moderate to profound	HF	Yes	NA	NA	Pakistan	Wilcox et al. 2001 [13], Bashir et al. 2013 [28]
c.259_260TC > AT	homozygote	p.Ser87Ile	prelingual	No	severe to profound	flat	Yes	NA	NA	Pakistan	Bashir et al. 2013 [28]
c.281C > T	homozygote	p.Ala94Val	NA	No	moderate to profound	flat, DE	Yes	NA	NA	Pakistan	Bashir et al. 2013 [28]
c.301G > A	homozygote	p.Gly101Arg	NA	NA	NA	NA	NA	NA	NA	Greece	Wattenhofer et al. 2005 [26]
c.398delT	homozygote	p.Met133Argfs *24	congenital	NA	moderate to profound	gently sloping	Yes	NA	NA	Pakistan	Wilcox et al. 2001 [13], Bashir et al. 2013 [28]
c.488C > T	homozygote	p.Ala163Val	5–7 y.o.	Yes	mild to severe	steeply sloping	Yes	Normal	HA	Canada	Pater et al. 2016 [29]
c.694G > A	homozygote	p.Gly232Arg	NA	No	severe to profound	gently sloping	Yes	NA	NA	Pakistan	Lee et al. 2011 [27]

Abbreviations: HF, high-frequency hearing loss; DE, deaf; CI, cochlear implant; HA, hearing aid; N/A, not available.

3.2. Clinical Characteristics

3.2.1. Progression and Onset of DFNB29

Until Pater et al. reported the progression of HL [29], the phenotype of DFNB29 was associated with non-progressive, congenital or prelingual HL with a variable degree of severity [13,26–28]. In this report, serial audiograms showed hearing deterioration especially at low frequencies (125 to 500 Hz). The average threshold progression rate of HL in the low frequencies was about 1.6 dB/year. Moreover, in this case, HL in the high frequencies was already impaired from the first decade, but there may have been no problem at birth as there was no problem in articulation. According to Pater's report, the phenotype of DFNB29 showed no hearing impairment until three years of age, and its onset occurred after age four especially in the higher frequencies. After that it progressed to almost deaf at above 2 kHz by nine years of age, therefore, it might be that our patient had a similar course. Indeed, in our case, no hearing impairment was detected at the elementary school health checkup at six years of age and it was only suspected at nine years of age. Lee et al. reported HL patients with another amino acid change, p.Arg81His [27], at the same amino acid residue. In that report, the phenotype was described as non-progressive prelingual HL. The audiogram indicated bilateral severe to profound HL with a gentle slope at 20 years of age. From our study and the previous reports, the hearing deterioration in patients with *CLDN14* remains unclear, and further study including serial audiograms is required to clarify this.

3.2.2. Vestibular Examination

We could not identify any dysfunction on the vestibular examinations. According to past reports, claudin-14 is also expressed in the vestibule of mice (such as in the sensory epithelia) [19]. Vestibular function in *Cldn14* knockout model mice has not been reported and the details remain unknown. This is the first report to mention in detail the vestibular function, including caloric testing, cVEMPs, and oVEMPs. Endolymph in cochlear is characterized by extremely high resting potential called EP, in addition to high K^+ concentration. In contrast, in the vestibular endolymph, the K^+ concentration is high, but no EP can be detected. Claudin-14 may be necessary for maintaining a barrier to resist EP, but not K^+.

3.2.3. Outcome of Cochlear Implantation

This is the first report showing the outcome for a CI for a *CLDN14*-associated HL patient. A CI represents the most successful neural prosthesis in clinical cases [32]. For patients with severe to profound SNHL, CI has been established as the standard therapy [33]. The implant is surgically implanted and works by transducing acoustic energy into an electrical signal, with an electrode array in the cochlea used to stimulate the surviving spiral ganglion cells of the auditory nerve [34].

As mentioned above, the patient showed relatively poor improvement in speech discrimination scores, whereas good improvement in sound field threshold levels was observed. Since the cochlear implant directly stimulates the cochlear nerve, the influence of the inner ear should be excluded. Thus, it is likely that the unsatisfactory speech discrimination was due to deficiencies in the spiral ganglion or central auditory pathway. In our previous study, *CLDN14* was expressed at the same level in the organ of Corti and spiral ganglions [35]. Including this report, there have been no cases with intellectual disability indicating central involvement; therefore, it is possible that the spiral ganglion may be involved in claudin-14 associated HL.

4. Materials and Methods

4.1. Subjects

All procedures were approved by the Shinshu University Ethical Committee (No. 387—4 September 2012 and No. 576—2 May 2017) as well as the respective Ethical Committees of the other participating institutions described elsewhere [36] and were carried out after obtaining written informed consent from all subjects (or from their next of kin, caretaker, or guardian in the case of minors or children). A total of 2549 probands from unrelated Japanese HL families were enrolled from the 67 otolaryngology departments across Japan participating in the present study from May 2012 to September 2016. The age of the probands ranged from 0 to 79 years (mean ± SD: 22.1 ± 19.7). The hereditary patterns of the HL in the families of the probands were autosomal dominant in 602, autosomal recessive in 1577, and unknown inheritance mode in 370.

4.2. Variant Analysis

For the genetic analysis for this proband, we performed target genome enrichment for 68 previously reported genetic causes for deafness and MPS analysis described elsewhere [37]. In brief, amplicon libraries were prepared using an Ion AmpliSeq™ Custom Panel (Applied Biosystems, Life Technologies, Carlsbad, CA, USA), in accordance with the manufacturer's instructions, for 68 genes reported to cause nonsyndromic hereditary HL [37]. Emulsion PCR and sequencing was performed according to the manufacturer's instructions. The detailed protocol has been described elsewhere [38,39]. MPS was performed with an Ion Proton™ system using the Ion PI™ Hi-Q™ Sequencing 200 Kit and Ion PI™ Chip (ThermoFisher Scientific, Waltham, MA, USA) according to the manufacturers' instructions. The sequence data were mapped against the human genome sequence (build GRCh37/hg19) with a Torrent Mapping Alignment Program. The mean depth of coverage of 68 target genes was 363.0. The percentage of each region with more than 20 times coverage (indicating the percentage of each region sequenced 20 times or more by MPS) was 95.31%. After sequence mapping, the DNA variant regions were piled up with Torrent Variant Caller plug-in software. After variant detection, their effects were analyzed using ANNOVAR software [40,41].

The missense, nonsense, insertion/deletion, and splicing variants were selected from among the identified variants. Variants were further selected as less than 1% of (1) the ExAC [42,43], (2) gnomAD [44], (3) 3.5KJPN [45], (4) the Human Genetic Variation Database (dataset for 1208 Japanese exome variants) [46], and (5) the 333 in-house Japanese normal hearing controls. Direct sequencing was utilized to confirm the selected variants. The pathogenicity of a variant was evaluated by ACMG (American College of Medical Genetics, Bethesda, MD, USA) standards and guidelines [47]. For missense variants, in particular, functional prediction software, including Sorting Intolerant from Tolerant (SIFT) [48], Polymorphism Phenotyping (PolyPhen2) [49], Likelihood Ratio Test (LRT) [50], Mutation Taster [51], Mutation Assessor [52], Rare Exome Variant Ensemble Learner (REVEL) [53], and Combined Annotation Dependent Depletion (CADD) [54] were used on the ANNOVAR software. Conservation of the variant site was also evaluated in 170 vertebrates from the HGMD professional [55]. Segregation analysis was performed for the proband and family members.

4.3. Clinical Evaluations

The age of onset of HL, the incidence of progressive HL, and episodes of vertigo and dizziness were analyzed based on the medical charts of the probands with the *CLDN14* variant.

Pure-tone audiometry was performed to evaluate HL. The pure-tone average (PTA) was calculated from the audiometric thresholds at four frequencies (0.5, 1, 2, and 4 kHz). If an individual did not respond to the maximum hearing level at a frequency, 5 dB was added to the maximum hearing level. The severity of HL was classified into mild (PTA: 20–40 dBHL), moderate (41–70 dBHL), severe (71–95 dBHL), and profound (>95 dBHL). The audiometric configurations were categorized as low-frequency, mid-frequency, high-frequency (gently or steeply sloping), flat, and deaf [56].

The vestibular examination findings, including caloric testing and the measurement of cervical and ocular vestibular evoked myogenic potentials (c/oVEMPs), were analyzed. Caloric testing involved the measurement of the maximum slow phase velocity (SPV) by cold water irrigation (20 °C, 5 mL, 20 s). We defined a maximum SPV value below 10 deg/s as representing areflexia and a value between 10 and 20 deg/s as representing hyporeflexia. For cVEMPs testing, electromyography (EMG) was performed using a pair of surface electrodes mounted on the upper half and sternal head of the sternocleidomastoid muscle, respectively. The electrographic signal was recorded using a Neuropack evoked potential recorder (Nihon Kohden Co Ltd., Tokyo, Japan). Clicks lasting for 0.1 ms at 105 dBnHL were presented through a headphone. The stimulation rate was 5 Hz, the bandpass filter intensity was 20 to 2000 Hz, and the analysis time was 50 ms. The responses to 100 stimuli were averaged twice. The oVEMPs testing was measured by bone-conductive vibration (BCV). The BCV was delivered in 4 ms tone bursts of 500 Hz vibration (rise and fall time = 1 ms and plateau time = 2 ms) by using a hand-held 4810 mini-shaker (Bruel and Kjaer, Naerum, Denmark), which was placed on the midline (Fz) of the forehead. The active electrode was located over the inferior orbital margin and a reference electrode was placed 2 cm below the active electrode. The ground electrode was placed on the chin. The patients laid in a supine position on a bed and looked up at an angle of approximately 30 degrees above straight ahead during the recording. The signals were amplified and bandpass filtered between 20 and 2000 Hz. The stimulus intensity was 115 dB force level, 500 Hz with an analysis time of 40 ms, and 50 responses were averaged for each run. The VEMPs asymmetry was calculated as follows: asymmetry ratio (AR) = (larger amplitude − smaller amplitude) × 100/(larger amplitude + smaller amplitude). In this study, an asymmetry ratio of >30% was defined as a decreased reaction and no reaction in amplitude VEMPs as absent.

Intervention for HL, including the use of hearing aids or cochlear implants, was investigated. To evaluate speech perception outcomes, speech discrimination scores (using the 67S Japanese monosyllable test) were used. The subjects sat 1 m away from the sound source facing zero-degree azimuth, and recorded monosyllable words in quiet were presented in the sound field at 65 dBSPL.

5. Conclusions

We present a patient with a novel variant in the *CLDN14* gene identified from a non-consanguineous family. This is the first report of *CLDN14*-associated HL in an east Asian population. Serial audiograms indicated high-frequency hearing loss type, and the deterioration gradually spread to other frequencies, finally resulting in deafness. The patient showed normal vestibular function for caloric testing, cVEMPs, and oVEMPs. This is also the first report of an HL patient with a tight junction variant receiving a CI. The CI improved the proband's sound field threshold levels, but not their speech discrimination scores. This information contributes to our understanding, diagnosis, and treatment of HL caused by TJ disorders.

Author Contributions: Conceptualization, S.U.; methodology, S.N.; software, S.N.; validation, T.K., S.K., S.N. and S.U.; formal analysis, S.N.; investigation, T.K.; resources, T.K. and S.N.; data curation, S.N.; writing—original draft preparation, T.K.; writing—review and editing, S.K. and S.U.; visualization, T.K.; supervision, S.U.; project administration, S.U.; funding acquisition, S.U.

Acknowledgments: The authors thank the proband and their family members who participated in this study. We would also like to thank Sachiko Matsuda and Fumiko Tomioka for their technical assistance, and Ayano Koide for her assistance with visualization.

Abbreviations

HL	Hearing loss
ARNSHL	Autosomal recessive nonsyndromic hereditary hearing loss
EP	Endocochlear potential
TJ	Tight junction
CI	Cochlear implantation
MPS	Massively parallel DNA sequencing
PTA	Pure-tone average

cVEMPs	Cervical vestibular evoked myogenic potentials
oVEMPs	Ocular vestibular evoked myogenic potentials
SNHL	Sensorineural hearing loss
dBSPL	Decibel sound pressure level
dBHL	Decibel hearing level

References

1. Morton, C.C.; Nance, W.E. Newborn Hearing Screening—A Silent Revolution. *N. Engl. J. Med.* **2006**, *354*, 2151–2164. [CrossRef] [PubMed]
2. Smith, R.J.H.; Bale, J.F.; White, K.R. Sensorineural hearing loss in children. *Lancet* **2005**, *365*, 879–890. [CrossRef]
3. Hilgert, N.; Smith, R.J.H.; Van Camp, G. Forty-six genes causing nonsyndromic hearing impairment: Which ones should be analyzed in DNA diagnostics? *Mutat. Res.* **2009**, *681*, 189–196. [CrossRef] [PubMed]
4. Hereditary Hearing Loss Homepage. Available online: https://hereditaryhearingloss.org/ (accessed on 13 May 2019).
5. Richardson, G.P.; de Monvel, J.B.; Petit, C. How the Genetics of Deafness Illuminates Auditory Physiology. *Annu. Rev. Physiol.* **2011**, *73*, 311–334. [CrossRef] [PubMed]
6. Wangemann, P.; Schacht, J. Homeostatic Mechanisms in the Cochlea. In *The Cochlea*; Dallos, P., Popper, A.N., Fay, R.R., Eds.; Springer: New York, NY, USA, 1996; Volume 8, pp. 130–185.
7. Ferrary, E.; Sterkers, O. Mechanisms of endolymph secretion. *Kidney Int.* **1998**, *65*, S98–S103.
8. Sterkers, O.; Ferrary, E.; Amiel, C. Production of inner ear fluids. *Phys. Rev.* **1988**, *68*, 1083–1128. [CrossRef] [PubMed]
9. Tasaki, I.; Spyroupoulos, C.S. Stria vascularis as source of endocochlear potential. *J. Neurophysiol.* **1959**, *22*, 149–155. [CrossRef]
10. Konishi, T.; Hamrick, P.E.; Walsh, P.J. Ion transport in guinea pig cochlea. I. Potassium and sodium transport. *Acta Otolaryngol.* **1978**, *86*, 22–34. [CrossRef]
11. Hudspeth, A.J. How the ear's works work. *Nature* **1989**, *341*, 397–404. [CrossRef]
12. Nunes, F.D.; Lopez, L.N.; Lin, H.W.; Davies, C.; Azevedo, R.B.; Gow, A.; Kachar, B. Distinctsubdomain organization and molecular composition of a tight junction with adherens junction features. *J. Cell Sci.* **2006**, *119*, 4819–4827. [CrossRef]
13. Wilcox, E.R.; Burton, Q.L.; Naz, S.; Riazuddin, S.; Smith, T.N.; Ploplis, B.; Belyantseva, I.; Ben-Yosef, T.; Liburd, N.A.; Morell, R.J.; et al. Mutations in the gene encoding tight junction claudin-14 cause autosomal recessive deafness DFNB29. *Cell* **2001**, *104*, 165–172. [CrossRef]
14. Riazuddin, S.; Ahmed, Z.M.; Fanning, A.S.; Lagziel, A.; Kitajiri, S.; Ramzan, K.; Khan, S.N.; Chattaraj, P.; Friedman, P.L.; Anderson, J.M.; et al. Tricellulin Is a Tight-Junction Protein Necessary for Hearing. *Am. J. Hum. Genet.* **2006**, *79*, 1040–1051. [CrossRef]
15. Borck, G.; Rehman, A.U.; Lee, K.; Pogoda, H.; Kakar, N.; von Ameln, S.; Grillet, N.; Hildebrand, M.S.; Ansar, M.; Basit, S.; et al. Loss-of-Function Mutations of ILDR1 Cause Autosomal-Recessive Hearing Impairment DFNB42. *Am. J. Hum. Genet.* **2011**, *88*, 127–137. [CrossRef]
16. Furuse, M.; Fujita, K.; Hiiragi, T.; Fujimoto, K.; Tsukita, S. Claudin-1 and -2: Novel integral membrane proteins localizing at tight junctions with no sequence similarity to occludin. *J. Cell Biol.* **1998**, *141*, 1539–1550. [CrossRef] [PubMed]
17. Elkouby-Naor, L.; Ben-Yosef, T. Functions of claudin tight junction proteins and their complex interactions in various physiologicalsystems. *Int. Rev. Cell Mol. Biol.* **2010**, *279*, 1–32. [PubMed]
18. Gunzel, D.; Fromm, M. Claudins and other tight junction proteins. *Compr. Physiol.* **2012**, *2*, 1819–1852. [PubMed]
19. Kitajiri, S.-I.; Furuse, M.; Morita, K. Expression patterns of claudins, tight junction adhesion molecules, in the inner ear. *Hear. Res.* **2004**, *187*, 25–34. [CrossRef]
20. Gow, A.; Davies, C.; Southwood, C.M.; Frolenkov, G.; Chrustowski, M.; Ng, L.; Yamauchi, D.; Marcus, D.C.; Kachar, B. Deafness in Claudin 11-null mice reveals the critical contribution of basal cell tight junctions to

stria vascularis function. *J. Neurosci.* **2004**, *24*, 7051–7062. [CrossRef] [PubMed]

21. Kitajiri, S.; Miyamoto, T.; Mineharuetal, A. Compartmentalization established by claudin-11-based tight junctions in stria vascularis is required for hearing through generation of endocochlear potential. *J. Cell Sci.* **2004**, *117*, 5087–5096. [CrossRef]
22. Ben-Yosef, T.; Belyantseva, I.A.; Saunders, T.L.; Hughes, E.D.; Kawamoto, K.; Van Itallie, C.M.; Beyer, L.A.; Halsey, K.; Gardner, D.J.; Wilcox, E.R.; et al. Claudin 14 knockout mice, a model for autosomal recessive deafness DFNB29, are deaf due to cochlear hair cell degeneration. *Hum. Mol. Genet.* **2003**, *12*, 2049–2061. [CrossRef] [PubMed]
23. Nakano, Y.; Kim, S.H.; Kim, H.M.; Sanneman, J.D.; Zhang, Y.; Smith, R.J.; Marcus, D.C.; Wangemann, P.; Nessler, R.A.; Banfi, B. A claudin-9—Based ion permeability barrier is essential for hearing. *PLoS Genet.* **2009**, *5*, e1000610. [CrossRef] [PubMed]
24. Matter, K.; Balda, M.S. Signalling to and from tight junctions. *Nat. Rev. Mol. Cell Biol.* **2003**, *4*, 225–237. [CrossRef]
25. Schneeberger, E.E.; Lynch, R.D. The tight junction: A multifunctional complex. *Am. J. Physiol. Cell Physiol.* **2004**, *286*, C1213–C1228. [CrossRef] [PubMed]
26. Wattenhofer, M.; Reymond, A.; Falciola, V.; Charollais, A.; Caille, D.; Borel, C.; Lyle, R.; Estivill, X.; Petersen, M.B.; Meda, P.; et al. Different mechanisms preclude mutant CLDN14 proteins from forming tight junctions in vitro. *Hum. Mutat.* **2005**, *25*, 543–549. [CrossRef] [PubMed]
27. Lee, K.; Ansar, M.; Andrade, P.B.; Khan, B.; Santos-Cortez, R.L.P.; Ahmad, W.; Leal, S.M. Novel CLDN14 mutations in Pakistani families with autosomal recessive non-syndromic hearing loss. *Am. J. Med. Genet. A* **2012**, *158*, 315–321. [CrossRef] [PubMed]
28. Bashir, Z.-E.-H.; Latief, N.; Belyantseva, I.A.; Iqbal, F.; Amer Riazuddin, S.; Khan, S.N.; Friedman, T.B.; Riazuddin, S.; Riazuddin, S. Phenotypic variability of CLDN14 mutations causing DFNB29 hearing loss in the Pakistani population HHS Public Access. *J. Hum. Genet.* **2013**, *58143*, 102–108. [CrossRef] [PubMed]
29. Pater, J.A.; Benteau, T.; Griffin, A.; Penney, C.; Stanton, S.G.; Predham, S.; Kielley, B.; Squires, J.; Zhou, J.; Li, Q.; et al. A common variant in CLDN14 causes precipitous, prelingual sensorineural hearing loss in multiple families due to founder effect. *Hum. Genet.* **2017**, *136*, 107–118. [CrossRef] [PubMed]
30. Kim, M.-A.; Kim, Y.-R.; Sagong, B.; Cho, H.-J.; Bae, J.W.; Kim, J.; Lee, J.; Park, H.-J.; Choi, J.Y.; Lee, K.-Y.; et al. Genetic Analysis of Genes Related to Tight Junction Function in the Korean Population with Non-Syndromic Hearing Loss. *PLoS ONE* **2014**, *9*, e95646. [CrossRef] [PubMed]
31. Lu, Y.; Yao, J.; Wei, Q.; Xu, J.; Xing, G.; Cao, X. Genetic analysis of CLDN14 in the Chinese population affected with non-syndromic hearing loss. *Int. J. Pediatr. Otorhinolaryngol.* **2018**, *105*, 6–11. [CrossRef] [PubMed]
32. Roche, J.P.; City, I.; Hansen, M.R. On the Horizon: Cochlear Implant Technology. *Otolaryngol. Clin. North Am.* **2016**, *48*, 1097–1116. [CrossRef] [PubMed]
33. Wilson, B.S. Getting a decent (but sparse) signal to the brain for users of cochlear implants. *Hear. Res.* **2014**, *322*, 24–38. [CrossRef] [PubMed]
34. Deep, N.L.; Dowling, E.M.; Jethanamest, D.; Carlson, M.L. Cochlear Implantation: An Overview. *J. Neurol. Surg. Part B* **2019**, *80*, 169–177. [CrossRef]
35. Nishio, S.; Takumi, Y.; Usami, S. Laser-capture micro dissection combined with next-generation sequencing analysis of cell type-specific deafness gene expression in the mouse cochlea. *Hear. Res.* **2017**, *348*, 87–97. [CrossRef]
36. Nishio, S.Y.; Usami, S.I. Deafness Gene Variations in a 1120 Nonsyndromic Hearing Loss Cohort: Molecular Epidemiology and Deafness Mutation Spectrum of Patients in Japan. *Ann. Otol. Rhinol. Laryngol.* **2015**, *124*, 49S–60S. [CrossRef] [PubMed]
37. Kitano, T.; Miyagawa, M.; Nishio, S.-Y.; Moteki, H.; Oda, K.; Ohyama, K.; Miyazaki, H.; Hidaka, H.; Nakamura, K.-I.; Murata, T.; et al. POU4F3 mutation screening in Japanese hearing loss patients: Massively parallel DNA sequencing-based analysis identified novel variants associated with autosomal dominant hearing loss. *PLoS ONE* **2017**, *12*, 1–15. [CrossRef] [PubMed]
38. Miyagawa, M.; Nishio, S.Y.; Ikeda, T.; Fukushima, K.; Usami, S.I. Massively Parallel DNA Sequencing Successfully Identifies New Causative Mutations in Deafness Genes in Patients with Cochlear Implantation and EAS. *PLoS ONE* **2013**, *8*, e75793. [CrossRef] [PubMed]
39. Nishio, S.Y.; Moteki, H.; Usami, S.I. Simple and efficient germline copy number variant visualization method for the Ion AmpliSeq^{TM} custom panel. *Mol. Genet. Genomic Med.* **2018**, *6*, 678–686. [CrossRef] [PubMed]
40. Chang, X.; Wang, K. wANNOVAR: Annotating genetic variants for personal genomes via the web. *J. Med. Genet.* **2012**, *49*, 433–436. [CrossRef]

41. Wang, K.; Li, M.; Hakonarson, H. ANNOVAR: Functional annotation of genetic variants from high-throughput sequencing data. *Nucleic Acids Res.* **2010**, *38*, e164. [CrossRef]
42. Abboud, H.E.; Abecasis, G.; Aguilar-Salinas, C.A.; Arellano-Campos, O.; Atzmon, G.; Aukrust, I.; Barr, C.L.; Bell, G.I.; Bergen, S.; Bjørkhaug, L.; et al. Analysis of protein-coding genetic variation in 60,706 humans. *Nature* **2016**, *536*, 285–291.
43. The Exome Aggregation Consortium Database (ExAC). Available online: http://exac.broadinstitute.org/ (accessed on 13 May 2019).
44. The Genome Aggregation Database (gnomAD). Available online: https://gnomad.broadinstitute.org/ (accessed on 13 May 2019).
45. Integrative Japanese Genome Variation Database (3.5KJPN). Available online: https://ijgvd.megabank.tohoku.ac.jp/statistics/statistics-3.5kjpn-all (accessed on 13 May 2019).
46. Narahara, M.; Higasa, K.; Nakamura, S.; Tabara, Y.; Kawaguchi, T.; Ishii, M.; Matsubara, K.; Matsuda, F.; Yamada, R. Large-scale East-Asian eQTL mapping reveals novel candidate genes for LD mapping and the genomic landscape of transcriptional effects of sequence variants. *PLoS ONE* **2014**, *9*, e100924. [CrossRef] [PubMed]
47. Richards, S.; Aziz, N.; Bale, S.; Bick, D.; Das, S.; Gastier-Foster, J.; Grody, W.W.; Hegde, M.; Lyon, E.; Spector, E.; et al. Standards and guidelines for the interpretation of sequence variants: A joint consensus recommendation of the American College of Medical Genetics and Genomics and the Association for Molecular Pathology. *Genet. Med.* **2015**, *17*, 405–424. [CrossRef] [PubMed]
48. Kumar, P.; Henikoff, S.; Ng, P.C. Predicting the effects of coding non-synonymous variants on protein function using the SIFT algorithm. *Nat. Protoc.* **2009**, *4*, 1073–1081. [CrossRef] [PubMed]
49. Adzhubei, I.A.; Schmidt, S.; Peshkin, L.; Ramensky, V.E.; Gerasimova, A.; Bork, P.; Kondrashov, A.S.; Sunyaev, S.R. A method and server for predicting damaging missense mutations. *Nat. Methods* **2010**, *7*, 248–249. [CrossRef] [PubMed]
50. Chun, S.; Fay, J.C. Identification of deleterious mutations within three human genomes. *Genome Res.* **2009**, *19*, 1553–1561. [CrossRef]
51. Schwarz, J.M.; Rödelsperger, C.; Schuelke, M.; Seelow, D. MutationTaster evaluates disease-causing potential of sequence alterations. *Nat. Methods* **2010**, *7*, 575–576. [CrossRef]
52. Reva, B.; Antipin, Y.; Sander, C. Predicting the functional impact of protein mutations: Application to cancer genomics. *Nucleic Acids Res.* **2011**, *39*, 37–43. [CrossRef]
53. Ioannidis, N.M.; Rothstein, J.H.; Pejaver, V.; Middha, S.; McDonnell, S.K.; Baheti, S.; Musolf, A.; Li, Q.; Holzinger, E.; Karyadi, D.; et al. REVEL: An Ensemble Method for Predicting the Pathogenicity of Rare Missense Variants. *Am. J. Hum. Genet.* **2016**, *99*, 877–885. [CrossRef]
54. Kircher, M.; Witten, D.M.; Jain, P.; O'Roak, B.J.; Cooper, G.M.; Shendure, J. A general framework for estimating the relative pathogenicity of human genetic variants. *Nat. Genet.* **2014**, *46*, 310–315. [CrossRef]
55. The Human Gene Mutation Database Professional (HGMD). Available online: http://www.hgmd.cf.ac.uk/ (accessed on 13 May 2019).
56. Mazzoli, M.; Van Camp, G.; Newton, V.; Giarbini, N.; Declau, F.; Parving, A. Recommendations for the Description of Genetic and Audiological Data for Families with Nonsyndromic Hereditary Hearing Impairment. *Audiol. Med.* **2003**, *1*, 148–150.

4

Reciprocal Association between the Apical Junctional Complex and AMPK: A Promising Therapeutic Target for Epithelial/Endothelial Barrier Function?

Kazuto Tsukita [1,2], Tomoki Yano [3], Atsushi Tamura [1,4] and Sachiko Tsukita [1,4,*]

1 Laboratory of Biological Science, Graduate School of Frontier Biosciences, Osaka University, Suita, Osaka 565-0811, Japan; kazusan@kuhp.kyoto-u.ac.jp (K.T.); atamura@biosci.med.osaka-u.ac.jp (A.T.)
2 Department of Neurology, Graduate School of Medicine, Kyoto University, Kyoto 606-8507, Japan
3 Laboratory of Biological Science, Graduate School of Medicine, Osaka University, Suita, Osaka 565-0811, Japan; t-yano@biosci.med.osaka-u.ac.jp
4 Strategic Innovation and Research Center, Teikyo University, Tokyo 173-8605, Japan
* Correspondence: atsukita@biosci.med.osaka-u.ac.jp

Abstract: Epithelial/endothelial cells adhere to each other via cell–cell junctions including tight junctions (TJs) and adherens junctions (AJs). TJs and AJs are spatiotemporally and functionally integrated, and are thus often collectively defined as apical junctional complexes (AJCs), regulating a number of spatiotemporal events including paracellular barrier, selective permeability, apicobasal cell polarity, mechano-sensing, intracellular signaling cascades, and epithelial morphogenesis. Over the past 15 years, it has been acknowledged that adenosine monophosphate (AMP)-activated protein kinase (AMPK), a well-known central regulator of energy metabolism, has a reciprocal association with AJCs. Here, we review the current knowledge of this association and show the following evidences: (1) as an upstream regulator, AJs activate the liver kinase B1 (LKB1)–AMPK axis particularly in response to applied junctional tension, and (2) TJ function and apicobasal cell polarization are downstream targets of AMPK and are promoted by AMPK activation. Although molecular mechanisms underlying these phenomena have not yet been completely elucidated, identifications of novel AMPK effectors in AJCs and AMPK-driven epithelial transcription factors have enhanced our knowledge. More intensive studies along this line would eventually lead to the development of AMPK-based therapies, enabling us to manipulate epithelial/endothelial barrier function.

Keywords: tight junction; adherens junction; apical junctional complex; AMP-activated protein kinase (AMPK); paracellular barrier

1. Introduction

Epithelial/endothelial cells adhere to each other via cell–cell junctions, thereby collectively functioning as a permselective barrier between the internal and external environments of the body to define the framework of every biological compartment of our body [1–6]. Because the size and function of each compartment and their inter-relationships are critical for biological systems, the specific and general principles regulating the permselective barrier function of epithelial/endothelial cells are of considerable interest. Additionally, it is attractive from a therapeutic point of view to investigate methods to manipulate barrier function, given that its disruption can lead to various diseases [7–17]. For example, changes in paracellular barrier function can initiate and accelerate inflammation (both locally and systemically) [10–12,15–17], contribute to tumor formation [14,18], initiate infection [19], and disrupt ion homeostasis [9,13,20], and have even been suggested as being potentially involved in the initiation and progression of psychiatric and neurodegenerative diseases [21–26].

Tight junctions (TJs) are vertebrate-specific junctions that are located most apically in epithelial cell–cell junctions. It is generally accepted that adherens junctions (AJs), which are located immediately at the basal side of TJs, are essential for epithelial cell sheet formation, on the basis of which TJs create a paracellular permselective barrier [6,27,28]. Therefore, TJs and AJs are considered to be spatiotemporally and functionally integrated, and are thus often collectively known as apical junctional complexes (AJCs). In addition to creating a paracellular barrier, AJCs play essential roles in apicobasal cell polarity, mechano-sensing, intracellular signaling cascades, and epithelial morphogenesis [1,6,28–32]. A number of signaling networks are involved in AJC-related events and, over the past 15 years, adenosine monophosphate (AMP)-activated protein kinase (AMPK), a well-known central regulator of energy metabolism signaling pathway, has been shown to have a reciprocal association with AJCs [33–40]. More specifically, AJs activate AMPK particularly in response to applied junctional tension, whereas TJ barrier function and apicobasal cell polarization are promoted by AMPK activation. Herein, we review the critical studies regarding this reciprocal association and discuss the therapeutic potential of AMPK to manipulate epithelial/endothelial barrier function.

2. Overview of Function and Regulation of AMPK

In the 1970s and 1980s, a series of biochemical experiments identified the presence of a 3-hydroxy-3-methylglutaryl-coenzyme A (HMG-CoA) reductase kinase, which was found to be activated by AMP [41–43]. Meanwhile, an acetyl-coenzyme A (acetyl-CoA) carboxylase kinase was purified from rat liver in the mid-1980s and was also found to be activated by AMP [44]. Later, the independently identified HMG-CoA reductase kinase and acetyl-CoA carboxylase kinase were shown to be the same protein, and this protein was subsequently given the name "AMPK" after its characteristic feature of activation by AMP [45,46].

AMPK is a heterotrimeric serine/threonine kinase composed of one catalytic α-subunit and two regulatory β- and γ-subunits, and is highly conserved in all eukaryotes from protozoa to vertebrates (Figure 1) [47–56].

AMPK functions as a central regulator of energy metabolism and, once activated, it turns on catabolic processes as well as turns off anabolic processes, leading to upregulation of adenosine triphosphate (ATP) concentrations (Figure 2). Activation of AMPK is determined by the balance of the following three factors (Figure 2) [48,49,57]: (1) the AMP/ATP ratio [48,57], (2) the activity of upstream kinases to phosphorylate α-subunit Thr-172 including liver kinase B1 (LKB1) and Ca^{2+}/calmodulin-dependent protein kinase kinase β (CaMKKβ) [58,59], and (3) the activity of upstream protein phosphatases that dephosphorylate α-subunit Thr-172 such as protein phosphatase 2A and 2C [49,60].

Notably, binding of AMP to the γ-subunit can activate AMPK by three mechanisms that can be antagonized by ATP (Figure 3): (1) AMP drives direct allosteric activation of AMPK [61], (2) AMP renders AMPK susceptible to α-subunit Thr-172 phosphorylation by upstream kinases [62,63], and (3) AMP protects phosphorylated α-subunit Thr-172 from dephosphorylation by upstream phosphatases [64]. It should also be noted that the effect of direct allosteric activation by AMP on AMPK activity is modest (approximately a fivefold increase in activity) [61] compared with that of α-subunit Thr-172 phosphorylation (more than 100-fold increase in activity) [64]; therefore, the phosphorylation status of α-subunit Thr-172 is essentially a molecular on–off switch for AMPK activity [57] (Figure 3). Because AMPK activation has potential to mediate a variety of different cellular consequences, AMPK activation is controlled, not only temporally by the phosphorylation status of α-subunit Thr-172, but also spatially [47,57]. Spatially, AMPK is mobilized to various subcellular organelles only in response to appropriate signals. As examples, glucose starvation leads to AMPK targeting to lysosomes, increased autophagic influx results in AMPK targeting to autophagosomes, and mitochondrial damage causes AMPK targeting to mitochondria [65].

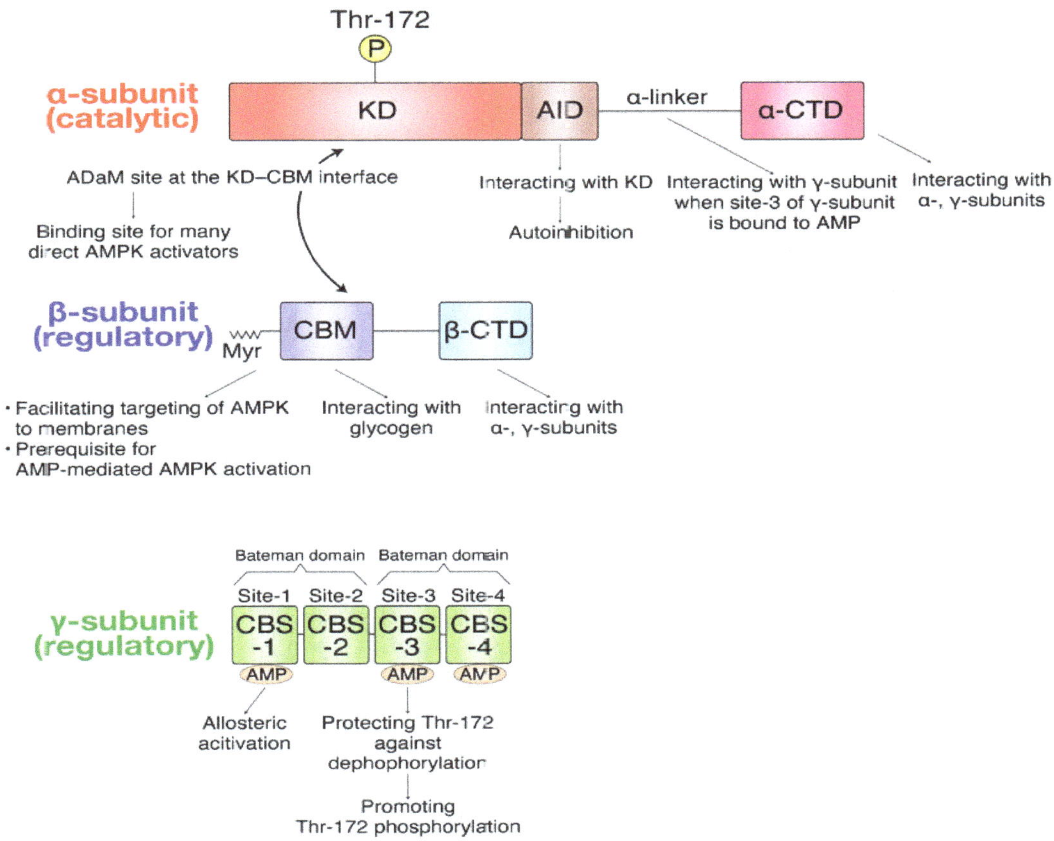

Figure 1. Domains of adenosine monophosphate (AMP)-activated protein kinase (AMPK) subunits. The kinase domain (KD) in the catalytic α-subunit exerts AMPK kinase activity, and Thr-172 phosphorylation of the KD is critical for its activity. The autoinhibitory domain (AID) binds to the backside of KD, constraining its mobility, thereby inhibiting kinase activity [52]. When AMP is bound at site-3 in the γ-subunit, the α-linker binds to the γ-subunit, thereby pulling AID away from the KD [53]. The C-terminal domain of the α-subunit (α-CTD) is responsible for interacting with the β- and γ-subunits. The allosteric drug and metabolite (ADaM) site at the KD–carbohydrate-binding module (CBM) interface is a binding site for many direct AMPK activators that induces allosteric activation, including A-769662 and compound 991 (see Table 1) [56]. The N-terminal of the regulatory β-subunit has been shown to be myristoylated, which facilitates targeting of AMPK to membranes and is prerequisite for AMP-mediated α-subunit Thr-172 phosphorylation [54]. The CBM in the β-subunit interacts with glycogen, which facilitates targeting of AMPK to glycogen for glycogen turnover regulation [55]. The C-terminal domain of the β-subunit (β-CTD) interacts with both α- and γ- subunits and serves as a connection between the α- and γ- subunits [51]. The regulatory γ-subunits contain four cystathionine beta synthase (CBS) domains, packing together to generate two Bateman domains, thereby providing four potential AMP-binding sites (site-1 to site-4) [48]. However, in mammalian AMPK, only site-1, site-3, and site-4 bind to AMP because aspartic acid residues in a conserved potential binding site in site-2 are changed to arginine [48]. Site-4 contains a permanently-bound AMP molecule, whereas AMP transiently binds to site-1 and site-3 (site-1 has a much stronger binding affinity for AMP than site-3). AMP binding to site-1 causes allosteric activation of AMPK, and AMP binding to site-3 promotes α-subunit Thr-172 phosphorylation by protecting AMPK from upstream phosphatases [49]. Myr, myristoylation.

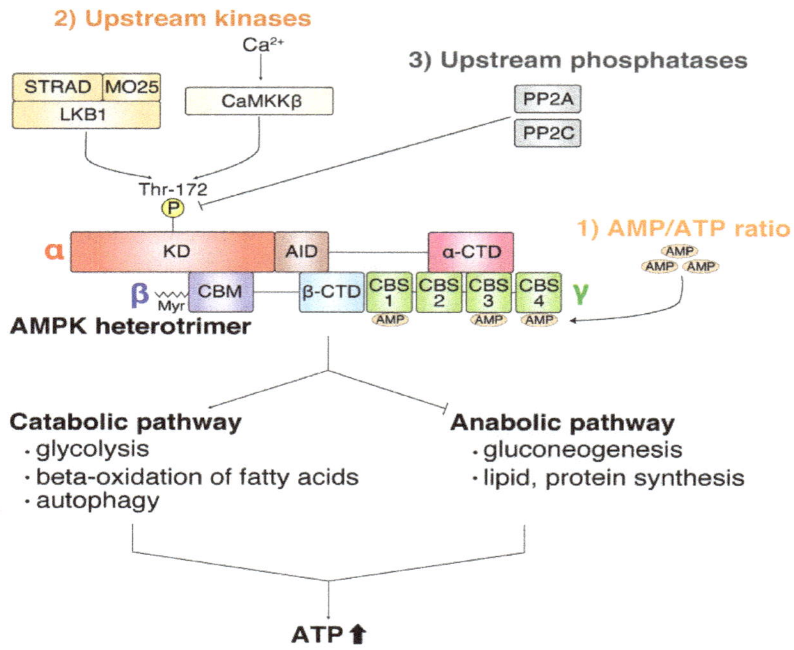

Figure 2. Overview of adenosine monophosphate (AMP)-activated protein kinase (AMPK) activation and the biological consequences. Activation of AMPK is determined by the balance of (1) AMP/adenosine triphosphate (ATP) ratio, (2) upstream kinase activity, and (3) upstream phosphatase activity. Once activated, AMPK activates catabolic pathways such as glycolysis, beta-oxidation of fatty acids, and autophagy, as well as suppresses anabolic pathway such as gluconeogenesis and lipid/protein synthesis to increase ATP levels. LKB1, liver kinase B1; STRAD, ste-20 related adaptor; MO25, mouse protein 25; CaMKKβ, CaMKK Ca^{2+}/calmodulin-dependent protein kinase kinase β; PP2A, protein phosphatase 2A; PP2C, protein phosphatase 2C; KD, kinase domain; AID, auto-inhibitory domain; CTD, C-terminal domain; CBM, carbohydrate-binding module; Myr, myristoylation; CBS, cystathionine beta synthase.

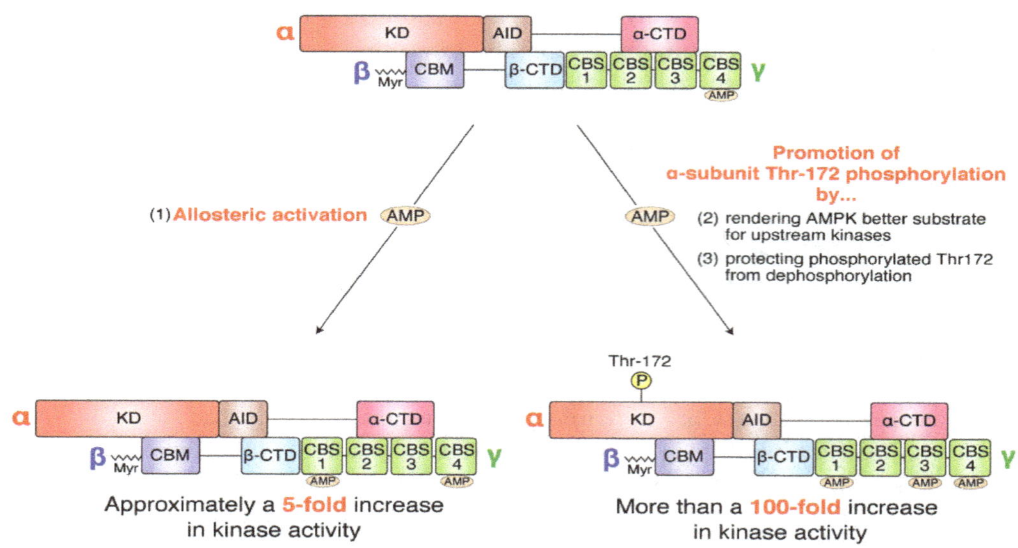

Figure 3. Adenosine monophosphate (AMP) activates AMP-activated protein kinase (AMPK) in several different ways. Allosteric activation results in approximately a fivefold increase in kinase activity, whereas α-subunit Thr-172 phosphorylation causes more than a 100-fold increase in kinase activity; therefore, α-subunit Thr-172 phosphorylation is essentially considered to be a molecular on–off switch for AMPK. KD, kinase domain; AID, auto-inhibitory domain; CTD, C-terminal domain; CBM, carbohydrate-binding module; Myr, myristoylation; CBS, cystathionine beta synthase.

3. Notable Findings in the Mid-2000s Established the Important Role of the LKB1–AMPK Axis in Apicobasal Epithelial Cell Polarity

In 2004, Clevers and colleagues shocked the epithelial cell research community with their seminal paper [33]. Until then, it was believed that apicobasal epithelial cell polarity was established after cells formed cell–cell junctions [66]. However, using intestinal epithelial cell lines with inducible LKB1 activity, they showed that, even in the absence of junctional cell–cell contacts, activation of LKB1 alone can fully establish apicobasal cell polarity in a cell-autonomous manner with the three major aspects of cell polarity all fulfilled, these being the appropriate sorting of apical and basolateral plasma membrane markers, formation of a full brush border, and correct junctional localization of junctional proteins. LKB1 forms a complex with ste-20 related adaptor (STRAD) and mouse protein 25 (MO25) that exerts many functions through the phosphorylation of many substrates [67]. In 2005, the downstream LKB1 effector responsible for establishing apicobasal cell polarity was suggested to be AMPK by investigating the functional consequences of a missense mutation identified in Peutz–Jeghers syndrome (PJS) that is caused by germline mutations of the LKB1 gene [34]. It was found that there is a certain LKB1-C-terminal missense mutation found in PJS patients that abolish the kinase activity responsible for almost all effects of the LKB1/STRAD/MO25 complex but still impair AMPK activity and apicobasal polarity establishment [34]. Subsequently, in 2007, it was found that not only *Drosophila* LKB1-null mutants but also *Drosophila* AMPK-null mutants display severe abnormalities in apicobasal cell polarity [36]. These findings collectively showed the crucial role of the LKB1–AMPK axis in apicobasal epithelial cell polarity establishment and have certainly paved the way for subsequent elucidation of the intimate association between AMPK and AJCs.

4. AJs and Junctional Tension Promote the LKB1–AMPK Axis to Provide Energy to Resist Applied Force

Interestingly, in 2009, it was found that AJs can activate LKB1 complexes [38]. In this study, it was shown that active LKB1 complexes co-localize with E-cadherin in polarized Caco-2 and Madin–Darby canine kidney (MDCK) II cells, that E-cadherin-dependent junctional maturation gradually mobilizes LKB1 complexes to AJs, and that AJ localization of LKB1 complexes is critical for the activation of AMPK. However, the physiological relevance of the LKB1–AMPK axis regulation at AJs remained unknown until 2017 when this regulation was shown to be critical for epithelial cells to resist against applied force. In epithelial cells, mechano-sensing, especially at AJs, triggers robust actin cytoskeletal rearrangements to resist externally applied forces [68–70]. From an energetic point of view, these changes are very costly, consuming about half of the total ATP in a cell. In 2017, it was found that junctional tension directly applied on E-cadherin, as well as shear stress on MDCK II cells, facilitates AJ-mediated LKB1–AMPK activation. Activated AMPK subsequently increases ATP levels and activates RhoA to support peri-junctional actomyosin filament rearrangement, thereby contributing to the cell stiffness critical for resisting against applied force [39] (Figure 4). In vivo importance of this AJ-mediated LKB1–AMPK regulation is suggested by the study of the tumor suppressor gene folliculin (FLCN), the causative gene for a rare autosomal-dominant disorder named Birt–Hogg–Dubé (BHD) syndrome. It was shown that FLCN deletion in lung epithelium results in downregulation of E-cadherin and LKB1 expression, which leads to the impairment of AMPK activation and alveolar collapse [71]. Collectively, these findings established the critical role of AJs and junctional tension in promoting the LKB1–AMPK axis to resist applied force.

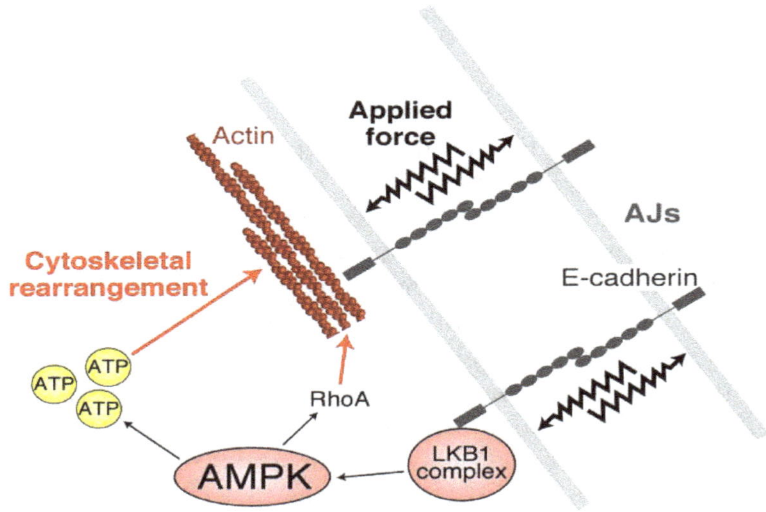

Figure 4. Adherens junctions (AJs) activate the liver kinase B1 (LKB1)–AMP-activated protein kinase (AMPK) axis. AJs can promote LKB1 complex activation [38], which is facilitated by applied force (zigzag arrows) [39]. Activated LKB1 complexes subsequently activate AMPK, which increases ATP levels and promotes RhoA activation to rearrange cytoskeletons, providing stiffness to resist against applied force.

5. TJ Function Is Reinforced by AMPK

The first report that showed the association of AMPK with TJs was published in 2006 [35]. At that time, AMPK was shown to facilitate apicobasal polarity establishment [34,36] and TJs have long been believed to function as a "fence" to prevent the diffusion of proteins and lipids, thereby having a role in apicobasal cell polarity [72] (later, "fence" function of TJs has been basically denied [73,74], although some controversial discussions appeared again recently [75]). Therefore, the role of AMPK in TJ assembly was investigated using a calcium switch experiment, a prevalent experiment to study TJ assembly. In a calcium switch experiment, calcium depletion from the medium of confluent epithelial cells destabilizes AJs, which triggers TJ disassembly with zonula occludens-1 (ZO-1) translocation from cell–cell junctions to the cytoplasm. Calcium restoration causes TJ assembly with ZO-1 redistribution from the cytoplasm to cell–cell junctions to form nascent ZO-1 puncta, which later forms continuous ZO-1 structures. Using this experimental procedure, it was shown that AMPK is activated during TJ assembly in MDCK II cells and that 5-aminoimidazole-4-carboxamide ribonucleoside (AICAR), a potent AMPK activating reagent, facilitates ZO-1 redistribution from the cytoplasm to cell–cell junctions, whereas expression of a dominant-negative AMPK (a kinase-dead form of AMPK) significantly delays ZO-1 redistribution with impaired establishment of the paracellular barrier. It was also noted that AMPK activation alone does not lead to the formation of continuous ZO-1 structures from puncta, suggesting that AMPK is required in the initiation of TJ assembly but other factors are needed to support further TJ maturation [35]. Shortly after this, it was also shown that AMPK activation by AICAR partially protects TJs from disassembly induced by calcium depletion [37]. Altogether, these studies established that TJs are a downstream target of AMPK and are reinforced by its activation.

6. Molecular Basis of the Effects of AMPK on AJCs: Implication of Novel AMPK Effectors in AJCs and AMPK-Driven Epithelial Transcription Factors

Since these early studies, there has been a general consensus that AMPK has a role in reinforcing TJ function and/or establishing apicobasal polarity in a variety of epithelial cells both in vitro and in vivo [76–85]. Although the molecular mechanisms underlying these phenomena are not yet completely elucidated, recent identification of novel AMPK effectors in AJCs and AMPK-driven epithelial transcription factors have improved our knowledge.

To date, myosin regulatory light chain (MRLC), afadin, cingulin, claudins, and G-alpha interacting vesicle associated protein (GIV)/Girdin were identified as AMPK effectors in AJCs. The first one identified was MRLC, which was originally shown in 2007 to be directly phosphorylated at Thr-21 by AMPK using in vitro phosphorylation assays [36]. In that report, it was shown that this phosphorylation of MRLC is critical in the phenotype of AMPK-null mutants in *Drosophila* which displays severe abnormalities in apicobasal cell polarity [36]. However, it should be noted that this finding has since been questioned because this phosphorylation site does not match the optimal AMPK substrate motif found in all other established AMPK substrates [86]. In 2011, afadin was identified as the second AMPK substrate in AJCs by the study investigating adhesion-related molecules, which are important in AMPK-dependent TJ assembly [87]. Prior to that study, afadin, together with another cell adhesion molecule nectin, was known to constitute an important adhesion systems in epithelial cells [88], and nectin–afadin-based adhesion was also known to appear during the earliest stages in the process of epithelial morphogenesis prior to the expression of TJ proteins [89]. In that study, it was found that afadin is a direct substrate for AMPK using in vitro phosphorylation assays. It was also found that AMPK activation by AICAR treatment increases the interaction of afadin with ZO-1 and that knocking down afadin expression prevented AMPK-mediated facilitation of TJ assembly in MDCK II cells, suggesting the possibility that AMPK reinforces nectin–afadin-based adhesion at an early stage of cell–cell junction establishment to facilitate TJ assembly [87]. The third AMPK effector in AJCs was identified in 2013, during the search for microtubule-binding proteins in the AJC fraction, when we found that cingulin is a direct AMPK substrate and AMPK-mediated phosphorylation of cingulin facilitates cingulin-mediated linkages of apical microtubule networks to TJs in epithelial cells [80,90]. Cingulin is a specific TJ protein [91] and has two potential evolutionally conserved AMPK target motifs at residues Ser-132 and Ser-150, and we showed that wild-type cingulin can be directly phosphorylated by AMPK at both sites using in vitro phosphorylation assays. Furthermore, we showed that AMPK-mediated cingulin phosphorylation strengthens the interaction of cingulin with microtubules in mouse mammary gland epithelial Eph4 cells and is important in epithelial three-dimensional morphogenesis [80,92]. The fourth group of AMPK effectors in AJCs is the claudins. In 2014, it was shown that AMPK activation by AICAR increases phosphorylation of claudin-4 at Ser-199 in an extracellular signal-regulated kinase 1/2-dependent manner in SV40 immortalized rat submandibular acinar cell lines [93]. Shortly after this, it was shown that claudin-1 can also be directly phosphorylated at Thr-191 by AMPK using in vitro phosphorylation assays and that this phosphorylation may potentially be important in TJ reinforcement in epithelial Eph4 cells [94]. The last AMPK effector in AJCs discovered so far was GIV/Girdin, which was identified as a direct AMPK substrate in 2016 [95]. Prior to that work, GIV/Girdin was shown to be a guanine-nucleotide-exchange factor (GEF) for $G\alpha i3$ in trimeric G proteins and to play an important role in TJ formation and establishment of apicobasal epithelial cell polarity [96]. In that work, it was found that GIV/Girdin has an evolutionally conserved, potential AMPK phosphorylation motif at Ser-245 in the N-terminus of GIV/Girdin and that AMPK directly phosphorylates GIV/Girdin at Ser-245 using in vitro phosphorylation assays. It was also shown that, during apicobasal cell polarity establishment in MDCK II cells, GIV/Girdin becomes phosphorylated by AMPK and subsequently localizes at TJs through its binding affinity for the TJ-linked apical microtubule network. Further, it was shown that this phosphorylation of GIV/Girdin is essential in AMPK-dependent TJ formation and apicobasal epithelial cell polarity establishment, and that an oncogenic mutant of GIV which cannot be phosphorylated by AMPK increases tumor cell growth through the defects in establishing apicobasal cell polarity [95]. These results collectively show that there are several AMPK substrates in AJCs (MRLC, cingulin, claudins, and GIV/Girdin) and that the phosphorylation of these substrates impacts TJ function and apicobasal cell polarity, although how AMPK coordinates these different pathways merits future investigation (Figure 5).

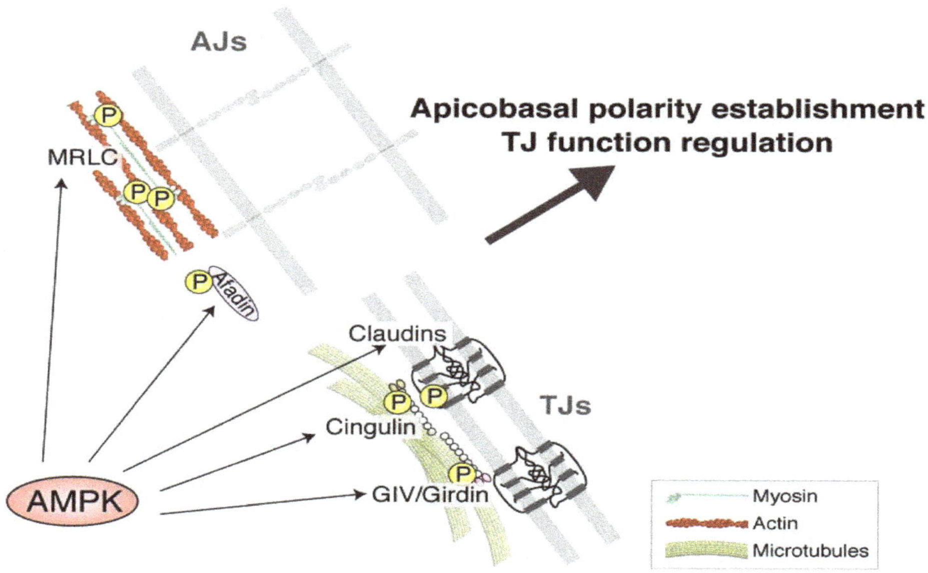

Figure 5. AMP-activated protein kinase (AMPK)-mediated regulation of apicobasal polarity establishment and tight junction (TJ) function are at least partially dependent on its phosphorylation of its substrates in AJCs. AMPK has been shown to phosphorylate myosin regulatory light chain (MRLC), cingulin, G-alpha interacting vesicle associated protein (GIV)/Girdin, claudins, and afadin, which can explain the role of AMPK in apicobasal polarity establishment and tight junction function regulation, at least partially. P, phosphate; AJs, adherens junctions.

Another important mechanism explaining the association between AMPK and TJs was described in 2017 [97]. Prior studies suggested the possibility that AMPK upregulates the expression of TJ proteins, thereby exerting a positive effect on TJ function; however, the underlying mechanism remained unknown [81,84]. In that study, it was found that, in intestinal epithelial Caco-2 cells, AMPK activation by AICAR upregulates expression of caudal type homeobox 2 (CDX2), a key transcription factor governing the differentiation of intestinal epithelial cells. It was also shown that this upregulation is caused by AMPK-dependent recruitment of methylase PRC2 and demethylase LSD1 to induce histone modifications of the CDX2 promoter. Furthermore, it was shown that intestine-conditional AMPK-knockout mice displayed impairment of intestinal epithelial differentiation and barrier function with a significant reduction in CDX2 expression [97]. Subsequent research found that AMPK activation by purple potato extract (PPE) in Caco-2 cells upregulates TJ protein expression in correlation with CDX2 expression and that AMPK-knockout abolishes the positive effects of PPE on CDX2 expression and TJ protein expression [98]. These findings suggest that, in addition to the AMPK substrates in AJCs, an AMPK-driven epithelial transcription factor plays an important role in AMPK-dependent TJ reinforcement via upregulation of TJ protein content (Figure 6).

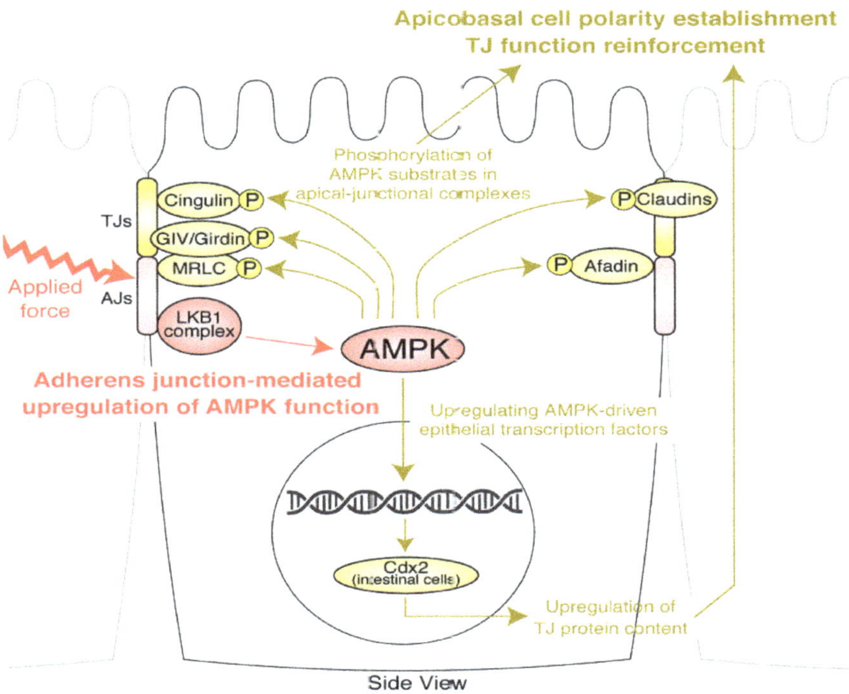

Figure 6. Overview of the reciprocal association between apical junctional complexes and AMP-activated kinase (AMPK). Adherens junctions (AJs) activate the liver kinase B1 (LKB1)–AMPK axis particularly in response to applied junctional tension. AMPK reinforces tight junction (TJ) function and promotes apicobasal cell polarization establishment through AMPK effectors in AJCs and AMPK-driven epithelial transcription factors.

7. Role of AMPK in Endothelial Barrier Function

Thus far, we have mainly focused on the role of AMPK in epithelial AJCs. Another intriguing aspect of AMPK is that it also has been shown to enhance endothelial TJ function [99–102]. An endothelial cell monolayer lines the inside of the intima of all blood and lymphatic vessels, forming a semipermeable barrier between the interstitial fluids within tissues and circulating fluids [103]. In general, TJs together with AJs form pericellular zipper-like structures between endothelial cells to restrict free diffusion of molecules, but the specific architecture and composition varies in different vascular beds. For example, TJs are more sophisticated in small arterioles, especially in highly specialized vascular beds like the blood–brain barrier (BBB) and inner blood–retinal barrier, whereas AJs are more prominent in postcapillary venules [103–105].

The overall molecular composition of endothelial TJs and AJs are similar to those of epithelial cells, with some small differences [103]. For example, the cadherin-family protein in epithelial AJs is E-cadherin, whereas it is vascular endothelial-cadherin (VE-cadherin) in endothelial AJs. It is noteworthy that N-cadherin is another major cadherin-family protein expressed in endothelial cells, but is excluded from endothelial AJs for an unknown reason [106,107]. N-cadherin has been shown to be important in the interactions of endothelial cells with pericytes and smooth muscle cells, which are thought to play an essential role in the maturation and stabilization of endothelial cells [108,109]. Interestingly, in 2011, using a pulmonary microvascular endothelial cell (PMVEC) wound healing model, it was found that AMPK-mediated TJ reinforcement is critical for PMVEC repair and that AMPK colocalizes with N-cadherin in PMVEC [99]. Following this finding, the biological importance of the AMPK–N-cadherin interaction was investigated using the same PMVEC wound healing model, and it was found that a short-hairpin RNA which decreases the expression of N-cadherin causes translocation of AMPK away from the membrane and inhibits AMPK-mediated barrier repair [110].

These findings collectively suggest that the N-cadherin–AMPK axis may be important for endothelial barrier assembly, but further studies are warranted for validation of these intriguing findings.

Table 1. List of commonly-used AMPK activators and inhibitors.

	Compound Name	Mechanism	Isoform Selectivity	Reference
Indirect activators	Metformin (biguanides)	Inhibiting the mitochondrial respiratory chain Complex I → Increase in AMP/ATP ratio	No	[111,112]
	Troglitazone (thiazolidinediones)	Inhibiting the mitochondrial respiratory chain Complex I → Increase in AMP/ATP ratio	No	[112,113]
	Resveratrol (polyphenols)	Inhibiting F1F0–ATPase/ATP synthase → Increase in AMP/ATP ratio	No	[112,114]
	Quercetin (polyphenols)	Inhibiting F1F0–ATPase/ATP synthase → Increase in AMP/ATP ratio	No	[112,115]
	α-Lipoic acid	Increasing calcium level?	No	[116]
Direct activators	5-Aminoimidazole-4-carboxamide ribonucleoside (AICAR)	Prodrug, converted to 5-amino-1-β-d-ribofuranosylimidazole-4-carboxamide-5′-monophosphate (ZMP) → ZMP binds to site-3 in the γ-subunit	No	[112,117]
	Compound-13 (C-13)	Prodrug, converted to Compound-2 (C-2) → C-2 binds to the γ-subunit	α1 > α2	[118]
	A-769662	Binding to the allosteric drug and metabolite (ADaM) site → Allosteric activation	β1 > β2	[56,119]
	Compound 991	Binding to the ADaM site → Allosteric activation	β1 > β2	[56]
	Salicylate	Binding to the ADaM site → Allosteric activation	β1	[120]
	MT63-78	Binding to the ADaM site → Allosteric activation	β1	[121]
Inhibitors	Compound C (dorsomorphin)	Binding to the α-subunit → Competitive inhibition of ATP	No	[111,122]
	MT47-100	Binding to the β2-subunit → Allosteric inhibition	β2 (β1 activation)	[123]
	SBI-0206965	Binding to the α-subunit → Mixed-type inhibition of ATP	No	[65]

8. Therapeutic Potential of Targeting AMPK for Manipulation of Epithelial Barrier Function

One of the fascinating features of AMPK is that it is "druggable". Indeed, we can artificially activate AMPK either directly (direct activator) or indirectly through AMP or calcium accumulation (indirect activator) using preexisting food compounds such as polyphenols and α-lipoic acid, as well as drugs such as metformin, thiazolidinediones, AICAR, and A79662 (summarized in Table 1). This leads to the question of whether these agents have any potential as therapeutics for diseases caused by barrier dysfunction. The answer may be "yes".

By improving intestinal epithelial barrier dysfunction, metformin-induced activation of AMPK can alleviate inflammatory bowel disease in mouse models induced by interleukin-10 deficiency, dextran sulfate sodium, and heat stress [84,85,124]. Metformin-induced intestinal barrier reinforcement was also shown to alleviate fructose-induced non-alcoholic fatty liver disease by blocking translocation of bacterial endotoxin [78]. Moreover, it was shown that metformin-induced AMPK activation may inhibit respiratory infection via reinforcement of airway epithelial barriers [79], attenuate the severity of ischemia-induced renal tubular injury by facilitating TJ assembly in renal epithelial cells [77], and facilitate recovery from acute ischemic stroke by attenuating BBB disruption [102,125] (metformin is an indirect AMPK activator that has other AMPK-independent effects; however, these protective effects of metformin on epithelial barrier function was shown to be dependent on AMPK [77–79,84,85,102,124,125]). Furthermore, AICAR-induced AMPK activation was also shown to alleviate lipopolysaccharide (LPS)-induced lung injury via restoring pulmonary endothelial barrier integrity [110], and LPS-induced cardiac dysfunction by attenuating left ventricular wall edema [100]. Therefore, these results collectively show that AMPK is a promising therapeutic target for epithelial/endothelial barrier manipulation. However, it should be noted that AMPK has wide variety of function in wide variety of organs with its activity being spatiotemporally regulated. Therefore, systemic and uniform AMPK activation would cause too many off-target effects. AMPK subunits (α, β, or γ) occur as multiple isoforms encoded by distinct genes (α1 or α2; β1 or β2; γ1, γ2, or γ3), giving rise to 12 possible AMPK α/β/γ heterotrimers that are differently expressed in different tissues and exert different effects [47]. For example, the α1/β2/γ1 heterotrimer is predominant in intestinal epithelial cells [126] and, while α1/β2/γ1, α2/β2/γ1, and α1/β2/γ3 heterotrimers are present in skeletal muscles [127,128], only the α2/β2/γ3 heterotrimer is activated during high intensity exercise [128,129]. Most AMPK activators/inhibitors that are commonly used today are not isoform-specific; therefore, further elucidation of isoform specific localization and/or function, as well as further development of isoform-specific activators and inhibitors, is important for AMPK-based organ- and/or barrier-specific therapy.

9. Conclusions

AMPK is a well-known central regulator of energy metabolism. As we have reviewed here, it has also been shown over 15 years that AMPK has a reciprocal association with AJCs, and fine-tunes TJ function and apicobasal cell polarity establishment in epithelial/endothelial cells. The complete molecular mechanisms underlying these phenomena have not yet been elucidated; however, identifications of novel AMPK effectors in AJCs and AMPK-driven epithelial transcription factors have enhanced our knowledge, and recent technological advances will contribute to better understanding. For example, recent development of transgenic mice expressing a highly sensitive fluorescence resonance energy transfer-based biosensor would visualize dynamic AMPK responses in vivo [130,131]. The formation and regulation of AJC-mediated epithelial cell–cell adhesion and paracellular barrier are critical for the biological systems with its disruption being implicated in various diseases; therefore, further elucidation of molecular details underlying an association between AMPK and AJCs will contribute to the better understanding of epithelial cell physiology as well as the development of therapies, enabling us to manipulate epithelial/endothelial barrier function.

Author Contributions: K.T., writing—original draft preparation; T.Y., A.T. and S.T., writing—review and editing.

Acknowledgments: We thank all members of Tsukita Lab for their useful comments and discussions.

Abbreviations

TJs	Tight junctions
AJs	Adherens junctions
AMP	Adenosine monophosphate
AMPK	AMP-activated protein kinase
HMG-CoA	3-hydroxy-3-methylglutaryl-coenzyme A
acetyl-CoA	Acetyl-coenzyme A
ATP	Adenosine triphosphate
LKB1	Liver kinase B1
CaMKKβ	Ca^{2+}/calmodulin-dependent protein kinase kinase β
PJS	Peutz–Jeghers syndrome
MDCK	Madin–Darby canine kidney
FLCN	Folliculin
BHD	Birt–Hogg–Dubé
AICAR	5-Aminoimidazole-4-carboxamide ribonucleoside
MRLC	Myosin regulatory light chain
GIV	G-alpha interacting vesicle associated protein
GEF	Guanine-nucleotide-exchange factor
CDX2	Caudal type homeobox 2

PPE	Purple potato extract
BBB	Blood–brain barrier
VE-cadherin	vascular endothelial-cadherin
PMVEC	Pulmonary microvascular endothelial cell
ZMP	5-amino-1-β-d-ribofuranosylimidazole- 4-carboxamide-5′-monophosphate
LPS	Lipopolysaccharide

References

1. Tsukita, S.; Furuse, M.; Itoh, M. Multifunctional strands in tight junctions. *Nat. Rev. Mol. Cell Biol.* **2001**, *2*, 285–293. [CrossRef] [PubMed]
2. Tamura, A.; Yamazaki, Y.; Hayashi, D.; Suzuki, K.; Sentani, K.; Yasui, W.; Tsukita, S. Claudin-based paracellular proton barrier in the stomach. *Ann. N. Y. Acad. Sci.* **2012**, *1258*, 108–114. [CrossRef] [PubMed]
3. Tamura, A.; Tsukita, S. Paracellular barrier and channel functions of TJ claudins in organizing biological systems: Advances in the field of barriology revealed in knockout mice. *Semin. Cell Dev. Biol.* **2014**, *36*, 177–185. [CrossRef] [PubMed]
4. Suzuki, H.; Tani, K.; Tamura, A.; Tsukita, S.; Fujiyoshi, Y. Model for the architecture of claudin-based paracellular ion channels through tight junctions. *J. Mol. Biol.* **2015**, *427*, 291–297. [CrossRef]
5. Nakamura, S.; Irie, K.; Tanaka, H.; Nishikawa, K.; Suzuki, H.; Saitoh, Y.; Tamura, A.; Tsukita, S.; Fujiyoshi, Y. Morphologic determinant of tight junctions revealed by claudin-3 structures. *Nat. Commun.* **2019**, *10*, 816. [CrossRef]
6. Tsukita, S.; Tanaka, H.; Tamura, A. The Claudins: From Tight Junctions to Biological Systems. *Trends Biochem. Sci.* **2019**, *44*, 141–152. [CrossRef]
7. Tanaka, H.; Tamura, A.; Suzuki, K.; Tsukita, S. Site-specific distribution of claudin-based paracellular channels with roles in biological fluid flow and metabolism. *Ann. N. Y. Acad. Sci.* **2017**, *1405*, 44–52. [CrossRef]
8. Wada, M.; Tamura, A.; Takahashi, N.; Tsukita, S. Loss of claudins 2 and 15 from mice causes defects in paracellular Na+ flow and nutrient transport in gut and leads to death from malnutrition. *Gastroenterology* **2013**, *144*, 369–380. [CrossRef]
9. Matsumoto, K.; Imasato, M.; Yamazaki, Y.; Tanaka, H.; Watanabe, M.; Eguchi, H.; Nagano, H.; Hikita, H.; Tatsumi, T.; Takehara, T.; et al. Claudin 2 deficiency reduces bile flow and increases susceptibility to cholesterol gallstone disease in mice. *Gastroenterology* **2014**, *147*, 1134–1145.e10. [CrossRef]
10. Tanaka, H.; Takechi, M.; Kiyonari, H.; Shioi, G.; Tamura, A.; Tsukita, S. Intestinal deletion of Claudin-7 enhances paracellular organic solute flux and initiates colonic inflammation in mice. *Gut* **2015**, *64*, 1529–1538. [CrossRef]
11. Tokumasu, R.; Yamaga, K.; Yamazaki, Y.; Murota, H.; Suzuki, K.; Tamura, A.; Bando, K.; Furuta, Y.; Katayama, I.; Tsukita, S. Dose-dependent role of claudin-1 in vivo in orchestrating features of atopic dermatitis. *Proc. Natl. Acad. Sci. USA* **2016**, *113*, E4061–E4068. [CrossRef] [PubMed]
12. Citi, S. Intestinal barriers protect against disease. *Science* **2018**, *359*, 1097–1098. [CrossRef] [PubMed]
13. Tanaka, H.; Imasato, M.; Yamazaki, Y.; Matsumoto, K.; Kunimoto, K.; Delpierre, J.; Meyer, K.; Zerial, M.; Kitamura, N.; Watanabe, M.; et al. Claudin-3 regulates bile canalicular paracellular barrier and cholesterol gallstone core formation in mice. *J. Hepatol.* **2018**, *69*, 1308–1316. [CrossRef] [PubMed]
14. Suzuki, K.; Sentani, K.; Tanaka, H.; Yano, T.; Suzuki, K.; Oshima, M.; Yasui, W.; Tamura, A.; Tsukita, S. Deficiency of Stomach-Type Claudin-18 in Mice Induces Gastric Tumor Formation Independent of H pylori Infection. *Cell. Mol. Gastroenterol. Hepatol.* **2019**, *8*, 119–142. [CrossRef]
15. Nalle, S.C.; Zuo, L.; Ong, M.L.D.M.; Singh, G.; Worthylake, A.M.; Choi, W.; Manresa, M.C.; Southworth, A.P.; Edelblum, K.L.; Baker, G.J.; et al. Graft-versus-host disease propagation depends on increased intestinal epithelial tight junction permeability. *J. Clin. Investig.* **2019**, *129*, 902–914. [CrossRef]
16. Graham, W.V.; He, W.; Marchiando, A.M.; Zha, J.; Singh, G.; Li, H.-S.; Biswas, A.; Ong, M.L.D.M.; Jiang, Z.-H.; Choi, W.; et al. Intracellular MLCK1 diversion reverses barrier loss to restore mucosal homeostasis. *Nat. Med.* **2019**, *25*, 690–700. [CrossRef]
17. Kuo, W.-T.; Shen, L.; Zuo, L.; Shashikanth, N.; Ong, M.L.D.M.; Wu, L.; Zha, J.; Edelblum, K.L.; Wang, Y.; Wang, Y.; et al. Inflammation-induced Occludin Downregulation Limits Epithelial Apoptosis by Suppressing Caspase-3 Expression. *Gastroenterology* **2019**, *157*, 1323–1337. [CrossRef]

18. Hagen, S.J.; Ang, L.-H.; Zheng, Y.; Karahan, S.N.; Wu, J.; Wang, Y.E.; Caron, T.J.; Gad, A.P.; Muthupalani, S.; Fox, J.G. Loss of Tight Junction Protein Claudin 18 Promotes Progressive Neoplasia Development in Mouse Stomach. *Gastroenterology* **2018**, *155*, 1852–1867. [CrossRef]
19. Shah, J.; Rouaud, F.; Guerrera, D.; Vasileva, E.; Popov, L.M.; Kelley, W.L.; Rubinstein, E.; Carette, J.E.; Amieva, M.R.; Citi, S. A Dock-and-Lock Mechanism Clusters ADAM10 at Cell-Cell Junctions to Promote α-Toxin Cytotoxicity. *Cell Rep.* **2018**, *25*, 2132–2147.e7. [CrossRef]
20. King, A.J.; Siegel, M.; He, Y.; Nie, B.; Wang, J.; Koo-McCoy, S.; Minassian, N.A.; Jafri, Q.; Pan, D.; Kohler, J.; et al. Inhibition of sodium/hydrogen exchanger 3 in the gastrointestinal tract by tenapanor reduces paracellular phosphate permeability. *Sci. Transl. Med.* **2018**, *10*, eaam6474. [CrossRef]
21. Clairembault, T.; Leclair-Visonneau, L.; Coron, E.; Bourreille, A.; Le Dily, S.; Vavasseur, F.; Heymann, M.-F.; Neunlist, M.; Derkinderen, P. Structural alterations of the intestinal epithelial barrier in Parkinson's disease. *Acta Neuropathol. Commun.* **2015**, *3*, 12. [CrossRef]
22. Montagne, A.; Zhao, Z.; Zlokovic, B.V. Alzheimer's disease: A matter of blood-brain barrier dysfunction? *J. Exp. Med.* **2017**, *214*, 3151–3169. [CrossRef] [PubMed]
23. Greene, C.; Kealy, J.; Humphries, M.M.; Gong, Y.; Hou, J.; Hudson, N.; Cassidy, L.M.; Martiniano, R.; Shashi, V.; Hooper, S.R.; et al. Dose-dependent expression of claudin-5 is a modifying factor in schizophrenia. *Mol. Psychiatry* **2018**, *23*, 2156–2166. [CrossRef] [PubMed]
24. Tsukita, K.; Taguchi, T.; Sakamaki-Tsukita, H.; Tanaka, K.; Suenaga, T. The vagus nerve becomes smaller in patients with Parkinson's disease: A preliminary cross-sectional study using ultrasonography. *Parkinsonism Relat. Disord.* **2018**, *55*, 148–149. [CrossRef] [PubMed]
25. Tsukita, K.; Sakamaki-Tsukita, H.; Tanaka, K.; Suenaga, T.; Takahashi, R. Value of in vivo α-synuclein deposits in Parkinson's disease: A systematic review and meta-analysis. *Mov. Disord.* **2019**, *34*, 1452–1463. [CrossRef] [PubMed]
26. Van IJzendoorn, S.C.D.; Derkinderen, P. The Intestinal Barrier in Parkinson's Disease: Current State of Knowledge. *J. Parkinsons Dis.* **2019**, *9*, S323–S329. [CrossRef] [PubMed]
27. Rosenthal, R.; Günzel, D.; Theune, D.; Czichos, C.; Schulzke, J.-D.; Fromm, M. Water channels and barriers formed by claudins. *Ann. N. Y. Acad. Sci.* **2017**, *1397*, 100–109. [CrossRef] [PubMed]
28. Konishi, S.; Yano, T.; Tanaka, H.; Mizuno, T.; Kanoh, H.; Tsukita, K.; Namba, T.; Tamura, A.; Yonemura, S.; Gotoh, S.; et al. Vinculin is critical for the robustness of the epithelial cell sheet paracellular barrier for ions. *Life Sci. Alliance* **2019**, *2*, e201900414. [CrossRef]
29. Yonemura, S.; Wada, Y.; Watanabe, T.; Nagafuchi, A.; Shibata, M. alpha-Catenin as a tension transducer that induces adherens junction development. *Nat. Cell Biol.* **2010**, *12*, 533–542. [CrossRef]
30. Zihni, C.; Mills, C.; Matter, K.; Balda, M.S. Tight junctions: From simple barriers to multifunctional molecular gates. *Nat. Rev. Mol. Cell Biol.* **2016**, *17*, 564–580. [CrossRef]
31. Yano, T.; Kanoh, H.; Tamura, A.; Tsukita, S. Apical cytoskeletons and junctional complexes as a combined system in epithelial cell sheets. *Ann. N. Y. Acad. Sci.* **2017**, *1405*, 32–43. [CrossRef] [PubMed]
32. Citi, S. The mechanobiology of tight junctions. *Biophys. Rev.* In press. [CrossRef] [PubMed]
33. Baas, A.F.; Kuipers, J.; van der Wel, N.N.; Batlle, E.; Koerten, H.K.; Peters, P.J.; Clevers, H.C. Complete polarization of single intestinal epithelial cells upon activation of LKB1 by STRAD. *Cell* **2004**, *116*, 457–466. [CrossRef]
34. Forcet, C.; Etienne-Manneville, S.; Gaude, H.; Fournier, L.; Debilly, S.; Salmi, M.; Baas, A.; Olschwang, S.; Clevers, H.; Billaud, M. Functional analysis of Peutz-Jeghers mutations reveals that the LKB1 C-terminal region exerts a crucial role in regulating both the AMPK pathway and the cell polarity. *Hum. Mol. Genet.* **2005**, *14*, 1283–1292. [CrossRef] [PubMed]
35. Zhang, L.; Li, J.; Young, L.H.; Caplan, M.J. AMP-activated protein kinase regulates the assembly of epithelial tight junctions. *Proc. Natl. Acad. Sci. USA* **2006**, *103*, 17272–17277. [CrossRef] [PubMed]
36. Lee, J.H.; Koh, H.; Kim, M.; Kim, Y.; Lee, S.Y.; Karess, R.E.; Lee, S.-H.; Shong, M.; Kim, J.-M.; Kim, J.; et al. Energy-dependent regulation of cell structure by AMP-activated protein kinase. *Nature* **2007**, *447*, 1017–1020. [CrossRef]
37. Zheng, B.; Cantley, L.C. Regulation of epithelial tight junction assembly and disassembly by AMP-activated protein kinase. *Proc. Natl. Acad. Sci. USA* **2007**, *104*, 819–822. [CrossRef]
38. Sebbagh, M.; Santoni, M.-J.; Hall, B.; Borg, J.-P.; Schwartz, M.A. Regulation of LKB1/STRAD localization and function by E-cadherin. *Curr. Biol.* **2009**, *19*, 37–42. [CrossRef]

39. Bays, J.L.; Campbell, H.K.; Heidema, C.; Sebbagh, M.; DeMali, K.A. Linking E-cadherin mechanotransduction to cell metabolism through force-mediated activation of AMPK. *Nat. Cell Biol.* **2017**, *19*, 724–731. [CrossRef]
40. Rowart, P.; Wu, J.; Caplan, M.J.; Jouret, F. Implications of AMPK in the Formation of Epithelial Tight Junctions. *Int. J. Mol. Sci.* **2018**, *19*, 2040. [CrossRef]
41. Beg, Z.H.; Allmann, D.W.; Gibson, D.M. Modulation of 3-hydroxy-3-methylglutaryl coenzyme A reductase activity with cAMP and wth protein fractions of rat liver cytosol. *Biochem. Biophys. Res. Commun.* **1973**, *54*, 1362–1369. [CrossRef]
42. Carlson, C.A.; Kim, K.H. Regulation of hepatic acetyl coenzyme A carboxylase by phosphorylation and dephosphorylation. *J. Biol. Chem.* **1973**, *248*, 378–380. [CrossRef]
43. Ferrer, A.; Caelles, C.; Massot, N.; Hegardt, F.G. Activation of rat liver cytosolic 3-hydroxy-3-methylglutaryl coenzyme A reductase kinase by adenosine 5′-monophosphate. *Biochem. Biophys. Res. Commun.* **1985**, *132*, 497–504. [CrossRef]
44. Carling, D.; Zammit, V.A.; Hardie, D.G. A common bicyclic protein kinase cascade inactivates the regulatory enzymes of fatty acid and cholesterol biosynthesis. *FEBS Lett.* **1987**, *223*, 217–222. [CrossRef]
45. Sim, A.T.; Hardie, D.G. The low activity of acetyl-CoA carboxylase in basal and glucagon-stimulated hepatocytes is due to phosphorylation by the AMP-activated protein kinase and not cyclic AMP-dependent protein kinase. *FEBS Lett.* **1988**, *233*, 294–298. [CrossRef]
46. Grahame Hardie, D.; Carling, D.; Sim, A.T.R. The AMP-activated protein kinase: A multisubstrate regulator of lipid metabolism. *Trends Biochem. Sci.* **1989**, *14*, 20–23. [CrossRef]
47. Carling, D. The AMP-activated protein kinase cascade—A unifying system for energy control. *Trends Biochem. Sci.* **2004**, *29*, 18–24. [CrossRef]
48. Xiao, B.; Heath, R.; Saiu, P.; Leiper, F.C.; Leone, P.; Jing, C.; Walker, P.A.; Haire, L.; Eccleston, J.F.; Davis, C.T.; et al. Structural basis for AMP binding to mammalian AMP-activated protein kinase. *Nature* **2007**, *449*, 496–500. [CrossRef]
49. Xiao, B.; Sanders, M.J.; Underwood, E.; Heath, R.; Mayer, F.V.; Carmena, D.; Jing, C.; Walker, P.A.; Eccleston, J.F.; Haire, L.F.; et al. Structure of mammalian AMPK and its regulation by ADP. *Nature* **2011**, *472*, 230–233. [CrossRef]
50. Langendorf, C.G.; Ngoei, K.R.W.; Scott, J.W.; Ling, N.X.Y.; Issa, S.M.A.; Gorman, M.A.; Parker, M.W.; Sakamoto, K.; Oakhill, J.S.; Kemp, B.E. Structural basis of allosteric and synergistic activation of AMPK by furan-2-phosphonic derivative C2 binding. *Nat. Commun.* **2016**, *7*, 10912. [CrossRef]
51. Salt, I.P.; Hardie, D.G. AMP-Activated Protein Kinase: An Ubiquitous Signaling Pathway With Key Roles in the Cardiovascular System. *Circ. Res.* **2017**, *120*, 1825–1841. [CrossRef] [PubMed]
52. Chen, L.; Jiao, Z.-H.; Zheng, L.-S.; Zhang, Y.-Y.; Xie, S.-T.; Wang, Z.-X.; Wu, J.-W. Structural insight into the autoinhibition mechanism of AMP-activated protein kinase. *Nature* **2009**, *459*, 1146–1149. [CrossRef] [PubMed]
53. Li, X.; Wang, L.; Zhou, X.E.; Ke, J.; de Waal, P.W.; Gu, X.; Tan, M.H.E.; Wang, D.; Wu, D.; Xu, H.E.; et al. Structural basis of AMPK regulation by adenine nucleotides and glycogen. *Cell Res.* **2015**, *25*, 50–66. [CrossRef] [PubMed]
54. Oakhill, J.S.; Chen, Z.-P.; Scott, J.W.; Steel, R.; Castelli, L.A.; Ling, N.; Macaulay, S.L.; Kemp, B.E. β-Subunit myristoylation is the gatekeeper for initiating metabolic stress sensing by AMP-activated protein kinase (AMPK). *Proc. Natl. Acad. Sci. USA* **2010**, *107*, 19237–19241. [CrossRef]
55. Oligschlaeger, Y.; Miglianico, M.; Chanda, D.; Scholz, R.; Thali, R.F.; Tuerk, R.; Stapleton, D.I.; Gooley, P.R.; Neumann, D. The recruitment of AMP-activated protein kinase to glycogen is regulated by autophosphorylation. *J. Biol. Chem.* **2015**, *290*, 11715–11728. [CrossRef]
56. Xiao, B.; Sanders, M.J.; Carmena, D.; Bright, N.J.; Haire, L.F.; Underwood, E.; Patel, B.R.; Heath, R.B.; Walker, P.A.; Hallen, S.; et al. Structural basis of AMPK regulation by small molecule activators. *Nat. Commun.* **2013**, *4*, 3017. [CrossRef]
57. Carling, D.; Thornton, C.; Woods, A.; Sanders, M.J. AMP-activated protein kinase: New regulation, new roles? *Biochem. J.* **2012**, *445*, 11–27. [CrossRef]
58. Hawley, S.A.; Boudeau, J.; Reid, J.L.; Mustard, K.J.; Udd, L.; Mäkelä, T.P.; Alessi, D.R.; Hardie, D.G. Complexes between the LKB1 tumor suppressor, STRAD alpha/beta and MO25 alpha/beta are upstream kinases in the AMP-activated protein kinase cascade. *J. Biol.* **2003**, *2*, 28. [CrossRef]
59. Hawley, S.A.; Pan, D.A.; Mustard, K.J.; Ross, L.; Bain, J.; Edelman, A.M.; Frenguelli, B.G.; Hardie, D.G.

Calmodulin-dependent protein kinase kinase-beta is an alternative upstream kinase for AMP-activated protein kinase. *Cell Metab.* **2005**, *2*, 9–19. [CrossRef]

60. Davies, S.P.; Helps, N.R.; Cohen, P.T.; Hardie, D.G. 5′-AMP inhibits dephosphorylation, as well as promoting phosphorylation, of the AMP-activated protein kinase. Studies using bacterially expressed human protein phosphatase-2C alpha and native bovine protein phosphatase-2AC. *FEBS Lett.* **1995**, *377*, 421–425.
61. Cheung, P.C.; Salt, I.P.; Davies, S.P.; Hardie, D.G.; Carling, D. Characterization of AMP-activated protein kinase gamma-subunit isoforms and their role in AMP binding. *Biochem. J.* **2000**, *346*, 659–669. [CrossRef] [PubMed]
62. Hong, S.-P.; Leiper, F.C.; Woods, A.; Carling, D.; Carlson, M. Activation of yeast Snf1 and mammalian AMP-activated protein kinase by upstream kinases. *Proc. Natl. Acad. Sci. USA* **2003**, *100*, 8839–8843. [CrossRef] [PubMed]
63. Oakhill, J.S.; Steel, R.; Chen, Z.-P.; Scott, J.W.; Ling, N.; Tam, S.; Kemp, B.E. AMPK is a direct adenylate charge-regulated protein kinase. *Science* **2011**, *332*, 1433–1435. [CrossRef] [PubMed]
64. Sanders, M.J.; Grondin, P.O.; Hegarty, B.D.; Snowden, M.A.; Carling, D. Investigating the mechanism for AMP activation of the AMP-activated protein kinase cascade. *Biochem. J.* **2007**, *403*, 139–148. [CrossRef] [PubMed]
65. Dite, T.A.; Langendorf, C.G.; Hoque, A.; Galic, S.; Rebello, R.J.; Ovens, A.J.; Lindqvist, L.M.; Ngoei, K.R.W.; Ling, N.X.Y.; Furic, L.; et al. AMP-activated protein kinase selectively inhibited by the type II inhibitor SBI-0206965. *J. Biol. Chem.* **2018**, *293*, 8874–8885. [CrossRef] [PubMed]
66. Wodarz, A. Establishing cell polarity in development. *Nat. Cell Biol.* **2002**, *4*, E39–E44. [CrossRef] [PubMed]
67. Lizcano, J.M.; Göransson, O.; Toth, R.; Deak, M.; Morrice, N.A.; Boudeau, J.; Hawley, S.A.; Udd, L.; Mäkelä, T.P.; Hardie, D.G.; et al. LKB1 is a master kinase that activates 13 kinases of the AMPK subfamily, including MARK/PAR-1. *EMBO J.* **2004**, *23*, 833–843. [CrossRef]
68. Daniel, J.L.; Molish, I.R.; Robkin, L.; Holmsen, H. Nucleotide exchange between cytosolic ATP and F-actin-bound ADP may be a major energy-utilizing process in unstimulated platelets. *Eur. J. Biochem.* **1986**, *156*, 677–684. [CrossRef]
69. Bernstein, B.W.; Bamburg, J.R. Actin-ATP hydrolysis is a major energy drain for neurons. *J. Neurosci.* **2003**, *23*, 1–6. [CrossRef]
70. Borghi, N.; Sorokina, M.; Shcherbakova, O.G.; Weis, W.I.; Pruitt, B.L.; Nelson, W.J.; Dunn, A.R. E-cadherin is under constitutive actomyosin-generated tension that is increased at cell-cell contacts upon externally applied stretch. *Proc. Natl. Acad. Sci. USA* **2012**, *109*, 12568–12573. [CrossRef]
71. Goncharova, E.A.; Goncharov, D.A.; James, M.L.; Atochina-Vasserman, E.N.; Stepanova, V.; Hong, S.-B.; Li, H.; Gonzales, L.; Baba, M.; Linehan, W.M.; et al. Folliculin controls lung alveolar enlargement and epithelial cell survival through E-cadherin, LKB1, and AMPK. *Cell Rep.* **2014**, *7*, 412–423. [CrossRef] [PubMed]
72. Diamond, J.M. Twenty-first Bowditch lecture. The epithelial junction: Bridge, gate, and fence. *Physiologist* **1977**, *20*, 10–18. [PubMed]
73. Umeda, K.; Ikenouchi, J.; Katahira-Tayama, S.; Furuse, K.; Sasaki, H.; Nakayama, M.; Matsui, T.; Tsukita, S.; Furuse, M.; Tsukita, S. ZO-1 and ZO-2 independently determine where claudins are polymerized in tight-junction strand formation. *Cell* **2006**, *126*, 741–754. [CrossRef] [PubMed]
74. Ikenouchi, J.; Umeda, K.; Tsukita, S.; Furuse, M.; Tsukita, S. Requirement of ZO-1 for the formation of belt-like adherens junctions during epithelial cell polarization. *J. Cell Biol.* **2007**, *176*, 779–786. [CrossRef] [PubMed]
75. Otani, T.; Nguyen, T.P.; Tokuda, S.; Sugihara, K.; Sugawara, T.; Furuse, K.; Miura, T.; Ebnet, K.; Furuse, M. Claudins and JAM-A coordinately regulate tight junction formation and epithelial polarity. *J. Cell Biol.* **2019**, *218*, 3372–3396. [CrossRef]
76. Peng, L.; Li, Z.-R.; Green, R.S.; Holzman, I.R.; Lin, J. Butyrate enhances the intestinal barrier by facilitating tight junction assembly via activation of AMP-activated protein kinase in Caco-2 cell monolayers. *J. Nutr.* **2009**, *139*, 1619–1625. [CrossRef]
77. Seo-Mayer, P.W.; Thulin, G.; Zhang, L.; Alves, D.S.; Ardito, T.; Kashgarian, M.; Caplan, M.J. Preactivation of AMPK by metformin may ameliorate the epithelial cell damage caused by renal ischemia. *Am. J. Physiol. Renal Physiol.* **2011**, *301*, F1346–F1357. [CrossRef]
78. Spruss, A.; Kanuri, G.; Stahl, C.; Bischoff, S.C.; Bergheim, I. Metformin protects against the development of fructose-induced steatosis in mice: Role of the intestinal barrier function. *Lab. Investig.* **2012**, *92*, 1020–1032. [CrossRef]
79. Garnett, J.P.; Baker, E.H.; Naik, S.; Lindsay, J.A.; Knight, G.M.; Gill, S.; Tregoning, J.S.; Baines, D.L. Metformin

reduces airway glucose permeability and hyperglycaemia-induced Staphylococcus aureus load independently of effects on blood glucose. *Thorax* **2013**, *68*, 835–845. [CrossRef]
80. Yano, T.; Matsui, T.; Tamura, A.; Uji, M.; Tsukita, S. The association of microtubules with tight junctions is promoted by cingulin phosphorylation by AMPK. *J. Cell Biol.* **2013**, *203*, 605–614. [CrossRef]
81. Park, H.-Y.; Kunitake, Y.; Hirasaki, N.; Tanaka, M.; Matsui, T. Theaflavins enhance intestinal barrier of Caco-2 Cell monolayers through the expression of AMP-activated protein kinase-mediated Occludin, Claudin-1, and ZO-1. *Biosci. Biotechnol. Biochem.* **2015**, *79*, 130–137. [CrossRef] [PubMed]
82. Wang, B.; Wu, Z.; Ji, Y.; Sun, K.; Dai, Z.; Wu, G. L-Glutamine Enhances Tight Junction Integrity by Activating CaMK Kinase 2-AMP-Activated Protein Kinase Signaling in Intestinal Porcine Epithelial Cells. *J. Nutr.* **2016**, *146*, 501–508. [CrossRef] [PubMed]
83. Patkee, W.R.A.; Carr, G.; Baker, E.H.; Baines, D.L.; Garnett, J.P. Metformin prevents the effects of Pseudomonas aeruginosa on airway epithelial tight junctions and restricts hyperglycaemia-induced bacterial growth. *J. Cell. Mol. Med.* **2016**, *20*, 758–764. [CrossRef] [PubMed]
84. Xue, Y.; Zhang, H.; Sun, X.; Zhu, M.-J. Metformin Improves Ileal Epithelial Barrier Function in Interleukin-10 Deficient Mice. *PLoS ONE* **2016**, *11*, e0168670. [CrossRef] [PubMed]
85. Chen, L.; Wang, J.; You, Q.; He, S.; Meng, Q.; Gao, J.; Wu, X.; Shen, Y.; Sun, Y.; Wu, X.; et al. Activating AMPK to Restore Tight Junction Assembly in Intestinal Epithelium and to Attenuate Experimental Colitis by Metformin. *Front. Pharmacol.* **2018**, *9*, 761. [CrossRef] [PubMed]
86. Shackelford, D.B.; Shaw, R.J. The LKB1-AMPK pathway: Metabolism and growth control in tumour suppression. *Nat. Rev. Cancer* **2009**, *9*, 563–575. [CrossRef] [PubMed]
87. Zhang, L.; Jouret, F.; Rinehart, J.; Sfakianos, J.; Mellman, I.; Lifton, R.P.; Young, L.H.; Caplan, M.J. AMP-activated protein kinase (AMPK) activation and glycogen synthase kinase-3β (GSK-3β) inhibition induce Ca2+-independent deposition of tight junction components at the plasma membrane. *J. Biol. Chem.* **2011**, *286*, 16879–16890. [CrossRef]
88. Yamada, A.; Fujita, N.; Sato, T.; Okamoto, R.; Ooshio, T.; Hirota, T.; Morimoto, K.; Irie, K.; Takai, Y. Requirement of nectin, but not cadherin, for formation of claudin-based tight junctions in annexin II-knockdown MDCK cells. *Oncogene* **2006**, *25*, 5085–5102. [CrossRef]
89. Brakeman, P.R.; Liu, K.D.; Shimizu, K.; Takai, Y.; Mostov, K.E. Nectin proteins are expressed at early stages of nephrogenesis and play a role in renal epithelial cell morphogenesis. *Am. J. Physiol. Renal Physiol.* **2009**, *296*, F564–F574. [CrossRef]
90. Yano, T.; Torisawa, T.; Oiwa, K.; Tsukita, S. AMPK-dependent phosphorylation of cingulin reversibly regulates its binding to actin filaments and microtubules. *Sci. Rep.* **2018**, *8*, 15550. [CrossRef]
91. Citi, S.; Sabanay, H.; Jakes, R.; Geiger, B.; Kendrick-Jones, J. Cingulin, a new peripheral component of tight junctions. *Nature* **1988**, *333*, 272–276. [CrossRef] [PubMed]
92. Mangan, A.J.; Sietsema, D.V.; Li, D.; Moore, J.K.; Citi, S.; Prekeris, R. Cingulin and actin mediate midbody-dependent apical lumen formation during polarization of epithelial cells. *Nat. Commun.* **2016**, *7*, 12426. [CrossRef] [PubMed]
93. Xiang, R.-L.; Mei, M.; Cong, X.; Li, J.; Zhang, Y.; Ding, C.; Wu, L.-L.; Yu, G.-Y. Claudin-4 is required for AMPK-modulated paracellular permeability in submandibular gland cells. *J. Mol. Cell Biol.* **2014**, *6*, 486–497. [CrossRef] [PubMed]
94. Shiomi, R.; Shigetomi, K.; Inai, T.; Sakai, M.; Ikenouchi, J. CaMKII regulates the strength of the epithelial barrier. *Sci. Rep.* **2015**, *5*, 13262. [CrossRef] [PubMed]
95. Aznar, N.; Patel, A.; Rohena, C.C.; Dunkel, Y.; Joosen, L.P.; Taupin, V.; Kufareva, I.; Farquhar, M.G.; Ghosh, P. AMP-activated protein kinase fortifies epithelial tight junctions during energetic stress via its effector GIV/Girdin. *Elife* **2016**, *5*, e20795. [CrossRef]
96. Sasaki, K.; Kakuwa, T.; Akimoto, K.; Koga, H.; Ohno, S. Regulation of epithelial cell polarity by PAR-3 depends on Girdin transcription and Girdin-Gαi3 signaling. *J. Cell Sci.* **2015**, *128*, 2244–2258. [CrossRef]
97. Sun, X.; Yang, Q.; Rogers, C.J.; Du, M.; Zhu, M.-J. AMPK improves gut epithelial differentiation and barrier function via regulating Cdx2 expression. *Cell Death Differ.* **2017**, *24*, 819–831. [CrossRef]
98. Sun, X.; Du, M.; Navarre, D.A.; Zhu, M.-J. Purple Potato Extract Promotes Intestinal Epithelial Differentiation and Barrier Function by Activating AMP-Activated Protein Kinase. *Mol. Nutr. Food Res.* **2018**, *62*. [CrossRef]
99. Creighton, J.; Jian, M.; Sayner, S.; Alexeyev, M.; Insel, P.A. Adenosine monophosphate-activated kinase alpha1 promotes endothelial barrier repair. *FASEB J.* **2011**, *25*, 3356–3365. [CrossRef]
100. Castanares-Zapatero, D.; Bouleti, C.; Sommereyns, C.; Gerber, B.; Lecut, C.; Mathivet, T.; Horckmans, M.;

Communi, D.; Foretz, M.; Vanoverschelde, J.-L.; et al. Connection between cardiac vascular permeability, myocardial edema, and inflammation during sepsis: Role of the α1AMP-activated protein kinase isoform. *Crit. Care Med.* **2013**, *41*, e411–e422. [CrossRef]

101. Takata, F.; Dohgu, S.; Matsumoto, J.; Machida, T.; Kaneshima, S.; Matsuo, M.; Sakaguchi, S.; Takeshige, Y.; Yamauchi, A.; Kataoka, Y. Metformin induces up-regulation of blood-brain barrier functions by activating AMP-activated protein kinase in rat brain microvascular endothelial cells. *Biochem. Biophys. Res. Commun.* **2013**, *433*, 586–590. [CrossRef] [PubMed]
102. Liu, Y.; Tang, G.; Li, Y.; Wang, Y.; Chen, X.; Gu, X.; Zhang, Z.; Wang, Y.; Yang, G.-Y. Metformin attenuates blood-brain barrier disruption in mice following middle cerebral artery occlusion. *J. Neuroinflammation* **2014**, *11*, 177. [CrossRef] [PubMed]
103. Komarova, Y.A.; Kruse, K.; Mehta, D.; Malik, A.B. Protein Interactions at Endothelial Junctions and Signaling Mechanisms Regulating Endothelial Permeability. *Circ. Res.* **2017**, *120*, 179–206. [CrossRef] [PubMed]
104. Simionescu, M.; Simionescu, N.; Palade, G.E. Segmental differentiations of cell junctions in the vascular endothelium. The microvasculature. *J. Cell Biol.* **1975**, *67*, 863–885. [CrossRef] [PubMed]
105. Schneeberger, E.E. Structure of intercellular junctions in different segments of the intrapulmonary vasculature. *Ann. N. Y. Acad. Sci.* **1982**, *384*, 54–63. [CrossRef] [PubMed]
106. Navarro, P.; Ruco, L.; Dejana, E. Differential localization of VE- and N-cadherins in human endothelial cells: VE-cadherin competes with N-cadherin for junctional localization. *J. Cell Biol.* **1998**, *140*, 1475–1484. [CrossRef] [PubMed]
107. Gentil-dit-Maurin, A.; Oun, S.; Almagro, S.; Bouillot, S.; Courçon, M.; Linnepe, R.; Vestweber, D.; Huber, P.; Tillet, E. Unraveling the distinct distributions of VE- and N-cadherins in endothelial cells: A key role for p120-catenin. *Exp. Cell Res.* **2010**, *316*, 2587–2599. [CrossRef]
108. Kim, J.H.; Kim, J.H.; Yu, Y.S.; Kim, D.H.; Kim, K.-W. Recruitment of pericytes and astrocytes is closely related to the formation of tight junction in developing retinal vessels. *J. Neurosci. Res.* **2009**, *87*, 653–659. [CrossRef]
109. Hellström, M.; Gerhardt, H.; Kalén, M.; Li, X.; Eriksson, U.; Wolburg, H.; Betsholtz, C. Lack of pericytes leads to endothelial hyperplasia and abnormal vascular morphogenesis. *J. Cell Biol.* **2001**, *153*, 543–553. [CrossRef]
110. Jian, M.-Y.; Liu, Y.; Li, Q.; Wolkowicz, P.; Alexeyev, M.; Zmijewski, J.; Creighton, J. N-cadherin coordinates AMP kinase-mediated lung vascular repair. *Am. J. Physiol. Lung Cell Mol. Physiol.* **2016**, *310*, L71–L85. [CrossRef]
111. Zhou, G.; Myers, R.; Li, Y.; Chen, Y.; Shen, X.; Fenyk-Melody, J.; Wu, M.; Ventre, J.; Doebber, T.; Fujii, N.; et al. Role of AMP-activated protein kinase in mechanism of metformin action. *J. Clin. Investig.* **2001**, *108*, 1167–1174. [CrossRef] [PubMed]
112. Hawley, S.A.; Ross, F.A.; Chevtzoff, C.; Green, K.A.; Evans, A.; Fogarty, S.; Towler, M.C.; Brown, L.J.; Ogunbayo, O.A.; Evans, A.M.; et al. Use of cells expressing gamma subunit variants to identify diverse mechanisms of AMPK activation. *Cell Metab.* **2010**, *11*, 554–565. [CrossRef] [PubMed]
113. Fryer, L.G.D.; Parbu-Patel, A.; Carling, D. The Anti-diabetic drugs rosiglitazone and metformin stimulate AMP-activated protein kinase through distinct signaling pathways. *J. Biol. Chem.* **2002**, *277*, 25226–25232. [CrossRef] [PubMed]
114. Ahn, J.; Lee, H.; Kim, S.; Park, J.; Ha, T. The anti-obesity effect of quercetin is mediated by the AMPK and MAPK signaling pathways. *Biochem. Biophys. Res. Commun.* **2008**, *373*, 545–549. [CrossRef] [PubMed]
115. Baur, J.A.; Pearson, K.J.; Price, N.L.; Jamieson, H.A.; Lerin, C.; Kalra, A.; Prabhu, V.V.; Allard, J.S.; Lopez-Lluch, G.; Lewis, K.; et al. Resveratrol improves health and survival of mice on a high-calorie diet. *Nature* **2006**, *444*, 337–342. [CrossRef]
116. Shen, Q.W.; Zhu, M.J.; Tong, J.; Ren, J.; Du, M. Ca^{2+}/calmodulin-dependent protein kinase kinase is involved in AMP-activated protein kinase activation by α-lipoic acid in C2C12 myotubes. *Am. J. Physiol. Cell Physiol.* **2007**, *293*, C1395–C1403. [CrossRef]
117. Corton, J.M.; Gillespie, J.G.; Hawley, S.A.; Hardie, D.G. 5-aminoimidazole-4-carboxamide ribonucleoside. A specific method for activating AMP-activated protein kinase in intact cells? *Eur. J. Biochem.* **1995**, *229*, 558–565. [CrossRef]
118. Gómez-Galeno, J.E.; Dang, Q.; Nguyen, T.H.; Boyer, S.H.; Grote, M.P.; Sun, Z.; Chen, M.; Craigo, W.A.; van Poelje, P.D.; MacKenna, D.A.; et al. A Potent and Selective AMPK Activator That Inhibits de Novo Lipogenesis. *ACS Med. Chem. Lett.* **2010**, *1*, 478–482. [CrossRef]
119. Cool, B.; Zinker, B.; Chiou, W.; Kifle, L.; Cao, N.; Perham, M.; Dickinson, R.; Adler, A.; Gagne, G.; Iyengar, R.; et al. Identification and characterization of a small molecule AMPK activator that treats key components of

type 2 diabetes and the metabolic syndrome. *Cell Metab.* **2006** *3*, 403–416. [CrossRef]

120. Calabrese, M.F.; Rajamohan, F.; Harris, M.S.; Caspers, N.L.; Magyar, R.; Withka, J.M.; Wang, H.; Borzilleri, K.A.; Sahasrabudhe, P.V.; Hoth, L.R.; et al. Structural Basis for AMPK Activation: Natural and Synthetic Ligands Regulate Kinase Activity from Opposite Poles by Different Molecular Mechanisms. *Structure* **2014**, *22*, 1161–1172. [CrossRef]
121. Zadra, G.; Photopoulos, C.; Tyekucheva, S.; Heidari, P.; Weng, Q.P.; Fedele, G.; Liu, H.; Scaglia, N.; Priolo, C.; Sicinska, E.; et al. A novel direct activator of AMPK inhibits prostate cancer growth by blocking lipogenesis. *EMBO Mol. Med.* **2014**, *6*, 519–538. [CrossRef] [PubMed]
122. Handa, N.; Takagi, T.; Saijo, S.; Kishishita, S.; Takaya, D.; Toyama, M.; Terada, T.; Shirouzu, M.; Suzuki, A.; Lee, S.; et al. Structural basis for compound C inhibition of the human AMP-activated protein kinase α2 subunit kinase domain. *Acta Crystallogr. D Biol. Crystallogr.* **2011**, *67*, 480–487. [CrossRef] [PubMed]
123. Scott, J.W.; Galic, S.; Graham, K.L.; Foitzik, R.; Ling, N.X.Y.; Dite, T.A.; Issa, S.M.A.; Langendorf, C.G.; Weng, Q.P.; Thomas, H.E.; et al. Inhibition of AMP-Activated Protein Kinase at the Allosteric Drug-Binding Site Promotes Islet Insulin Release. *Chem. Biol.* **2015**, *22*, 705–711. [CrossRef] [PubMed]
124. Xia, Z.; Huang, L.; Yin, P.; Liu, F.; Liu, Y.; Zhang, Z.; Lin, J.; Zou, W.; Li, C. L-Arginine alleviates heat stress-induced intestinal epithelial barrier damage by promoting expression of tight junction proteins via the AMPK pathway. *Mol. Biol. Rep.*. in press. [CrossRef] [PubMed]
125. Farbood, Y.; Sarkaki, A.; Khalaj, L.; Khodagholi, F.; Badavi, M.; Ashabi, G. Targeting Adenosine Monophosphate-Activated Protein Kinase by Metformin Adjusts Post-Ischemic Hyperemia and Extracellular Neuronal Discharge in Transient Global Cerebral Ischemia. *Microcirculation* **2015**, *22*, 534–541. [CrossRef]
126. Harmel, E.; Grenier, E.; Bendjoudi Ouadda, A.; El Chebly, M.; Ziv, E.; Beaulieu, J.F.; Sané, A.; Spahis, S.; Laville, M.; Levy, E. AMPK in the small intestine in normal and pathophysiological conditions. *Endocrinology* **2014**, *155*, 873–888. [CrossRef]
127. Wojtaszewski, J.F.P.; Birk, J.B.; Frøsig, C.; Holten, M.; Pilegaard, H.; Dela, F. 5′AMP activated protein kinase expression in human skeletal muscle: Effects of strength training and type 2 diabetes. *J. Physiol.* **2005**, *564*, 563–573. [CrossRef]
128. Birk, J.B.; Wojtaszewski, J.F.P. Predominant alpha2/beta2/gamma3 AMPK activation during exercise in human skeletal muscle. *J. Physiol.* **2006**, *577*, 1021–1032. [CrossRef]
129. Thomson, D.M. The Role of AMPK in the Regulation of Skeletal Muscle Size, Hypertrophy, and Regeneration. *Int. J. Mol. Sci.* **2018**, *19*, 3125. [CrossRef]
130. Tsou, P.; Zheng, B.; Hsu, C.-H.; Sasaki, A.T.; Cantley, L.C. A Fluorescent Reporter of AMPK Activity and Cellular Energy Stress. *Cell Metab.* **2011**, *13*, 476–486. [CrossRef]
131. Konagaya, Y.; Terai, K.; Hirao, Y.; Takakura, K.; Imajo, M.; Kamioka, Y.; Sasaoka, N.; Kakizuka, A.; Sumiyama, K.; Asano, T.; et al. A Highly Sensitive FRET Biosensor for AMPK Exhibits Heterogeneous AMPK Responses among Cells and Organs. *Cell Rep.* **2017**, *21*, 2628–2638. [CrossRef] [PubMed]

5

Regulation of Epithelial Cell Functions by the Osmolality and Hydrostatic Pressure Gradients: A Possible Role of the Tight Junction as a Sensor

Shinsaku Tokuda [1,2,*] and Alan S. L. Yu [2]

1 Department of Respiratory Medicine, Graduate School of Medicine, Kyoto University, Kyoto 606-8507, Japan
2 Division of Nephrology and Hypertension, Department of Internal Medicine, University of Kansas Medical Center, Kansas City, KS 66160, USA
* Correspondence: tokku@kuhp.kyoto-u.ac.jp

Abstract: Epithelia act as a barrier to the external environment. The extracellular environment constantly changes, and the epithelia are required to regulate their function in accordance with the changes in the environment. It has been reported that a difference of the environment between the apical and basal sides of epithelia such as osmolality and hydrostatic pressure affects various epithelial functions including transepithelial transport, cytoskeleton, and cell proliferation. In this paper, we review the regulation of epithelial functions by the gradients of osmolality and hydrostatic pressure. We also examine the significance of this regulation in pathological conditions especially focusing on the role of the hydrostatic pressure gradient in the pathogenesis of carcinomas. Furthermore, we discuss the mechanism by which epithelia sense the osmotic and hydrostatic pressure gradients and the possible role of the tight junction as a sensor of the extracellular environment to regulate epithelial functions.

Keywords: tight junction; osmolality; hydrostatic pressure; cancer; sensor

---

1. Introduction

In multicellular organisms, epithelia act as a barrier to the external environment and contribute to maintain the homeostasis in the internal environment. The environmental conditions including osmolality and hydrostatic pressure (HP) constantly changes with the biological activity [1–3]. The epithelia are required to regulate their functions including transepithelial transport in accordance with the changes in the environmental condition to maintain the homeostasis in the internal environment. It has been reported that differences of the osmolality and HP between the apical and basal sides of the epithelia affects various epithelial functions as reviewed in this paper.

In this paper, we review these reports and illustrate the effects of osmotic and HP gradients in the regulation of epithelial functions. Furthermore, we examine the significance of the regulation in pathological conditions and discuss the mechanism by which epithelia sense the gradients of osmolality and HP.

2. Effects of Osmolality on the Epithelia and Endothelia

2.1. Regulation of Osmolality in the Body

When there is a concentration difference of a certain substance between a semipermeable membrane which is permeable to water but not to the substance, the substance exerts a driving force for the water movement and the driving force is called osmotic pressure or osmolality. Epithelia act as a barrier to the external environment and osmolality in the apical side of the epithelia constantly changes with the biological activity. For example, the osmolality in the apical side of an intestine dramatically and dynamically changes with diet and its digestion. The osmolality in the surface of an airway epithelium changes with secretion and moisture, and urine osmolality fluctuates in accordance with the regulation of osmolality in a body (see below).

In contrast, the osmolality inside the body is strictly regulated within a narrow range (275–295 mOsm/kg) [4,5]. The regulation of osmolality is mainly performed in the kidney except in aquatic animals. The excretion of free water in urine is regulated in the kidney, which is further regulated by hormones such as antidiuretic hormone (ADH) [6]. In the interstitium of the kidney, the osmotic gradient is formed from medulla toward cortex (high osmolality in medulla) [5,7]. The glomerular filtrate flows through renal tubules, and renal tubules go down to medulla and then make a sharp loop (Henle's loop) and return to cortex. The permeability of water in renal tubules after the Henle's loop is very low and the filtrate grows hypotonic as sodium chloride is reabsorbed. Then the renal tubules gather into collecting ducts and again go down to medulla [5,7]. In collecting ducts, water is absorbed in accordance with the amount of aquaporin (AQP) in the cell membranes and ADH regulates the amount of AQP in the apical cell membrane [6]. As such, the amount of free water excretion in urine is regulated and osmolality in the body is kept within a narrow range. The osmolality in urine ranges from 30 to 1200 mOsm/kg in humans [5,7].

Therefore, osmolality in the apical side constantly changes whereas osmolality in the basal side is kept in a narrow range in most epithelia. Then, what kind of effects does the osmotic change have on the epithelia?

2.2. *Effect of Osmolality on Cell Volume and Other Cell Functions*

The effects of osmolality on cells received attention as early as 1930s. In the 1960s, the regulation of cell volume after the osmotic changes in the extracellular environment was actively investigated by using erythrocytes. When the extracellular osmolality is lowered, water flows into the cells as most cell membranes of animal cells have high water permeability and results in cell swelling. Then, cells start to restore their volume to their original size by excreting potassium chloride (regulatory volume decrease: RVD). In contrast, when the extracellular osmolality is raised, water flows out from the cells and results in cell shrinkage. Then cells start to restore their volume by the uptake of sodium chloride (regulatory volume increase: RVI). The mechanism of osmotic cell volume regulation has been intensively investigated, and transporters and channels involved in RVD and RVI have been identified [8,9]. Furthermore, cell volume alteration by the osmotic changes have been known to have an impact on various cell functions including cell proliferation, cell apoptosis, metabolism, epithelial transport and migration [1,10].

2.3. *Effects of the Osmotic Gradient Between Apical and Basal Sides on the Epithelia*

As noted above, osmolality in the apical side changes constantly whereas osmolality in the basal side is kept in a narrow range. This explains why there are often differences of the osmolality between apical and basal sides in most epithelia. Interestingly, the difference of the osmolality (the osmotic gradient across the epithelia) is known to affect various epithelial functions in a different manner from osmotic volume changes. In this paper, we focus on the studies which clearly show that the osmotic gradient between apical and basal sides, but not increase or decrease of osmolality in both sides, affects epithelial functions.

2.3.1. Jejunum

The osmolality in the apical side of the intestinal epithelium drastically changes with the diet, and the osmotic gradient affects transepithelial transport of a jejunal epithelium in a guinea pig [11]. Madara has reported that increase of the osmolality in the apical side with mannitol (up to 600 mOsm) elevates transepithelial electrical resistance (TER) whereas increase of the osmolality in the basal side has no effect on TER and increase of the osmolality in both sides slightly reduces TER [11].

Thus, the osmotic gradient from the apical to basal side is thought to elevate TER. TER reflects ion permeability across the epithelia. There are two pathways for ions across the epithelia: a transcellular pathway via apical and basal cell membranes and a paracellular pathway across intercellular space. The permeability of the paracellular pathway is regulated by tight junctions (TJs) [12–14]. To study the effect of the osmotic gradient on TJs, Madara investigated morphological changes in TJs by freeze-fracture electron microscopy. Apical hyperosmolality induces increase in number and depth of TJ strands (Figure 1A, [11]). Thus, the osmotic gradient from apical to basal side increases transepithelial ion permeability with the morphological changes in TJs in the jejunum. These changes may contribute the regulation of ion absorption in the jejunum.

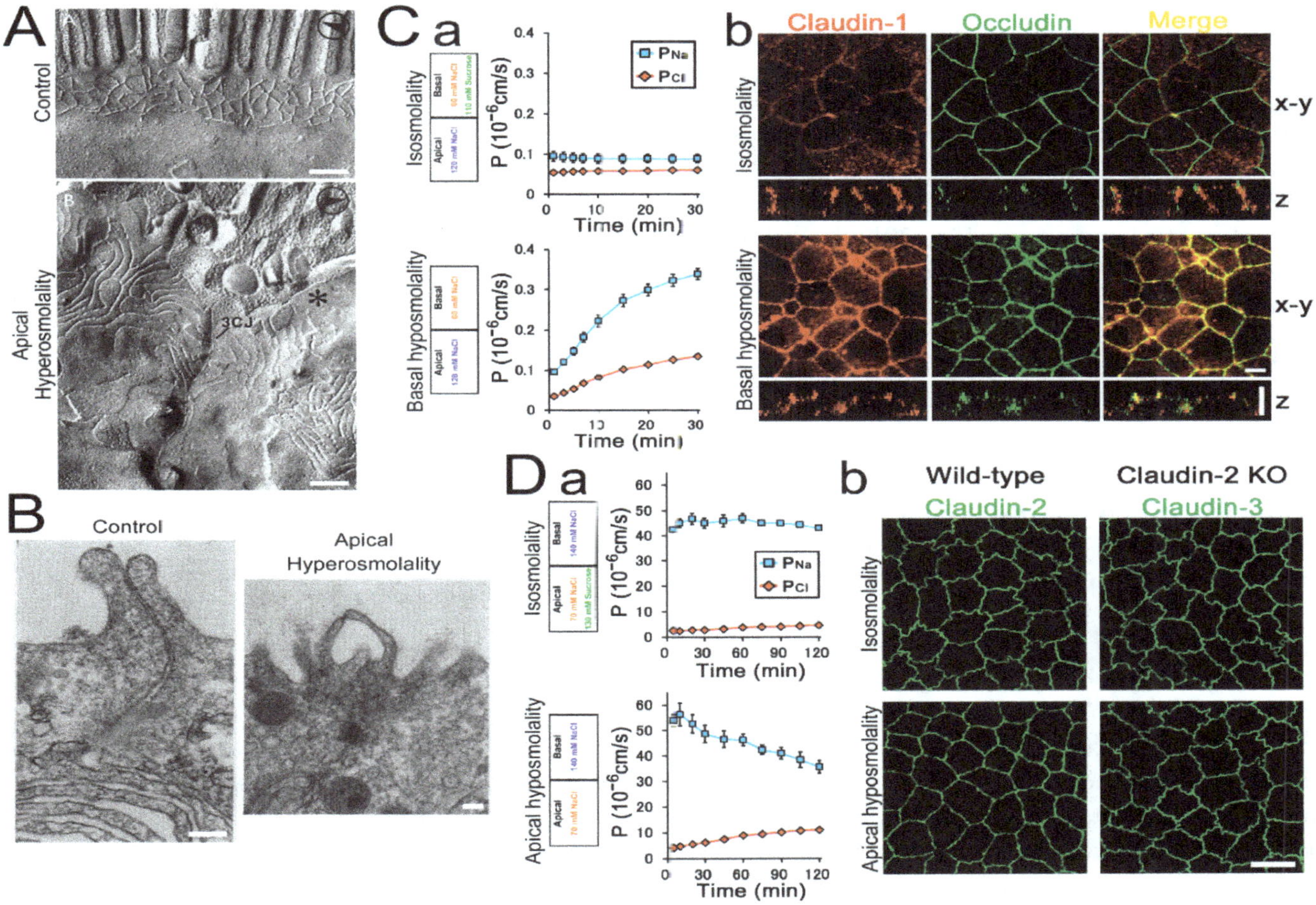

Figure 1. Effects of the osmotic gradient in various epithelia. (**A**) Freeze-fracture electron micrographs in the jejunal epithelium. Apical osmolality was increased to 600 mOsm with mannitol and cells were fixed 20 min after the osmotic changes. Apical hyperosmolality increased tight junction (TJ) strand number and depth. Scale bar = 200 nm. From Madara. *J. Cell Biol.* 1983 [11] with permission. (**B**) Transmission electron micrographs in the bladder epithelium. Apical osmolality was increased with 240 mM urea and cells were fixed 10 min after the osmotic changes. Apical hyperosmolality induced bleb formation between TJ strands. Scale bar = 200 nm. From Wade et al., *Am. J. Physiol.* 1973 [15] with permission. (**C**) Effects of basal hypoosmolality on *Xenopus* A6 cells. (**a**) Basal osmolality was decreased by the reduction of NaCl concentration or counterbalanced by the addition of sucrose, and permeability of sodium and chloride (P_{Na} and P_{Cl}) were calculated from transepithelial resistance and dilution potentials in the presence of Na^+, K^+ and Cl^- channel blockers. Basal hypoosmolality increased P_{Na} and P_{Cl} with the selective increase of P_{Na}. (**b**) Immunofluorescence of claudin-1 and occludin.

Cells were fixed 30 min after the osmotic changes. Basal hypoosmolality altered claudin-1 localization to the apical end and claudin-1 showed colocalization with occludin. Scale bar = 5 μm. Modified from Tokuda et al., *Biochem. Biophys. Res. Commun.* 2008 [16] and Tokuda et al., *Biochem. Biophys. Res. Commun.* 2010 [17] with permission. (**D**) Effects of apical hypoosmolality on Madin–Darby canine kidney (MDCK) II cells. (**a**) Apical osmolality was decreased by the reduction of NaCl concentration or counterbalanced by the addition of sucrose. Apical hypoosmolality induced the reduction of cation selectivity. (**b**) Immunofluorescence of claudin-2 or claudin-3 in wild-type and claudin-2 knockout cells. Cells were fixed 30 min after the osmotic changes. Apical hypoosmolality altered the shape of cell–cell contact from zigzag to more straight shape in wild-type cells but not in claudin-2 knockout cells. Scale bar = 10 μm. Modified from Tokuda et al., *PLoS ONE.* 2016 [18] with permission.

2.3.2. Skin

In the skin, osmolality in the apical side changes with the condition of sweat and drying, and the osmotic gradient affects TER in a frog skin [19,20]. The increase of osmolality in the apical side with mannitol, acetamide, or thiourea reduces TER. In contrast, increase of osmolality in the basal side elevates TER. Increase of osmolality in both sides has almost no effect on TER [19]. The decrease of TER with apical hyperosmolality and the increase of TER with basal hyperosmolality is also reported in another study in a frog skin [20]. Thus, the osmotic gradient from apical to basal side reduces TER whereas that from basal to apical side elevates TER in the skin.

2.3.3. Retina

In the retina, a retinal pigment epithelium (RPE) separates retinal (apical) and choroidal (basal) environment and contributes to a blood–retinal barrier (BRB), which provides proper environment for photoreceptor cells. The osmolality in the choroidal side is higher than retinal side in the physiological condition. When the BRB is disrupted in the pathological conditions such as diabetic retinopathy, the osmolality in the retinal side increases and results in the accumulation of water and macular edema [21].

The osmotic gradient is reported to affect the electrophysiological property of the RPE [22]. To measure the electrophysiological property, a microelectrode was placed inside of and across the RPE and electrophysiological measurement including electroretinogram (ERG) was performed in chick retinas. Apical hyperosmolality with 25 mM mannitol induces depolarization of the basal cell membrane with decrease of membrane resistance and amplifies light-evoked c-wave in ERG. Basal hyperosmolality has opposite effects on these measurements and hyperosmolality in both sides has no effect on c-wave in ERG. Thus, the osmotic gradient affects electrophysiological property of the RPE, which may have a role in the regulation of BRB in the physiological and pathological conditions.

2.3.4. Vascular Endothelium

Endothelia in the brain form the blood–brain barrier (BBB) and restrict the permeation of substances into the brain. Infusion of osmotic agents such as mannitol into the carotid artery is known to induce transient increase of TJ permeability in the endothelia in the BBB [23,24], which is considered as the method to improve the drug delivery into the brain for the treatment of brain tumors and other brain diseases [25–27]. Interestingly, the osmotic gradient causes increase of permeability in the endothelia in bovine major cerebral artery [28]. Apical hyperosmolality with 20% mannitol increases albumin permeability in the endothelia. Basal hyperosmolality also increases albumin permeability. In contrast, hyperosmolality in both sides has no effect on albumin permeability. Thus, the osmotic gradient in both directions increases the albumin permeability in the brain endothelia, and the same mechanism may be involved in the increase of BBB permeability by the infusion of mannitol into the carotid artery.

2.3.5. Bladder

As noted above, the urine osmolality fluctuates in accordance with the amount of free water excretion in the urine, thus the osmolality in the apical side of bladder changes with this fluctuation. The osmotic gradient affects the transepithelial transport of bladder epithelium. In toad bladder, apical hyperosmolality with urea increases the permeability of sucrose and water. Basal hyperosmolality or hyperosmolality in both sides has almost no effect on the permeability of sucrose and reduces water permeability [29]. Interestingly, apical hyperosmolality induces bleb formation between TJ strands (Figure 1B, [15]). The blebs are also formed in basal hypoosmolality but not in basal hyperosmolality, hyperosmolality in both sides, apical hypoosmolality or hypoosmolality in both sides. Thus, the osmotic gradient from apical to basal side increases permeability of sucrose and water with structural changes in TJs, which is thought to contribute the regulation of transepithelial transport in the bladder.

2.3.6. Kidney, Distal Tubule

As noted above, the osmolality of apical and basal sides in renal tubules varies in each segment of nephron, and the osmotic gradient is known to affect transepithelial transport in *Xenupus* A6 cells, a model of distal tubule cells [16,17]. In A6 cells, basal hypoosmolality reduces TER. The permeability of sodium (P_{Na}) is selectively increased than that of chloride (P_{Cl}) (Figure 1C). A counterbalance of the osmotic gradient by adding sucrose eliminates the reduction of TER. Apical hyperosmolality also reduces TER with the selective increase of P_{Na} [16]. The osmotic gradient from apical to basal side also affects the localization of the claudin, a family protein of an integral membrane protein in TJs. Claudins are major constituent of tight junction strands and thought to be a major determinant of TJ permeability [30–32]. In A6 cells, claudin-1 is mainly localized in the entire lateral cell membrane and shows no colocalization with occludin (other TJ protein mainly localized at TJs). The osmotic gradient from apical to basal side alters the localization of claudin-1 to the apical end of the lateral membrane and claudin-1 shows colocalization with occludin (Figure 1C) [17]. In contrast, osmotic gradient from basal to apical side reduces TER with an equal increase in P_{Na} and P_{Cl} [16,17]. The selective increase of P_{Na} by the osmotic gradient from apical to basal side is also observed in Madin–Darby canine kidney (MDCK) I cells, a model of distal tubule cells [18]. Thus, the osmotic gradient from apical to basal side reduce TER with selective increase of P_{Na} in distal tubule cell models, and the regulation of transepithelial transport by the osmotic gradient may have a role in the regulation of ion reabsorption in the distal nephron.

2.3.7. Kidney, Proximal Tubule

Sodium chloride is filtered in the glomerulus and approximately 70% of sodium chloride in the glomerular filtrate is absorbed in proximal tubules. The proximal tubules have high water permeability and the osmotic gradient created by the reabsorption of sodium generates the driving force for water absorption. However, the filtrate in the apical side of the proximal tubules flows before the osmolality in the apical and basal sides become equal and thus the osmolality in the basal side is slightly higher than the apical side in proximal tubules [33–35]. In addition, fluctuation of plasma osmolality in the physiological conditions such as the changes in sodium chloride or blood glucose concentration in the plasma affects the osmolality of the filtrate in the proximal tubules. The osmotic gradient is known to affect the permeability of TJs in MDCK II cells, a model of proximal tubule cells [18]. TJs of MDCK II cells have high ion permeability with cation selectivity. Claudin-2 is expressed in MDCK II cells as well as proximal tubule cells in vivo, and claudin-2 is known to form highly conductive channels with cation selectivity in TJ strands and to be a major determinant of the permeability property of TJs in MDCK II cells [36–38]. Interestingly, apical hypoosmolality induces a reduction of cation selectivity in MDCK II cells (Figure 1D). A counterbalance of the osmotic gradient by adding sucrose or mannitol eliminates the reduction of cation selectivity. Basal hyperosmolality also reduces cation selectivity and basal hypoosmolality, apical hyperosmolality, or hyperosmolality in both sides does not affect cation selectivity. The expression level of claudin-2 is not changed two hours after the apical hypoosmolality. Thus, the osmotic gradient from basal to apical side is thought to cause the reduction of cation selectivity in MDCK II cells. In addition, the osmotic gradient from basal to apical side induces bleb formation between TJ strands with the changes in actin filaments. Furthermore, the osmotic gradient from basal to apical side alters the shape of cell–cell contact in MDCK II cells from zigzag to more straight shape (Figure 1D). Interestingly, the reduction of cation selectivity, the bleb formation and the alteration in the shape of cell–cell contact are not observed in claudin-2 knockout MDCK II cells, suggesting that claudin-2 mediates these changes. Since claudin-2 is expressed in proximal tubules in vivo, the regulation of charge selectivity in TJs by the osmotic gradient may have a role in the regulation of ion reabsorption in the proximal tubules in the physiological conditions.

2.3.8. Summary

Effects of the osmotic gradient on epithelia and endothelia are summarized in Table 1. The osmotic gradient is involved in the regulation of various cell functions in various epithelia and endothelia. The cell functions regulated by the osmotic gradient include transepithelial transport (paracellula transport in many cases), polarization and resistance of cell membrane, actin filaments and shape of cell–cell contact. Interestingly, the direction of the osmotic gradient which induces cell responses and the regulated cell functions are different among the types of epithelia. Therefore, it is speculated that epithelia distinguish the direction of the osmotic gradient and each epithelium has a different mechanism to regulate cell functions, which is likely to reflect a difference of physiological roles in each epithelium.

Table 1. Effects of the osmotic gradient on epithelia and endothelia.

Organ	Type of the Osmotic Gradient	Cell Response	References
Jejunum	Apical hyper (600 mOsm with mannitol)	TER, TJ strand number↑	[11]
Skin	Apical hyper (210 mM mannitol, acetamide, thiourea)	TER↓	[19,20]
	Basal hyper (210 mM mannitol, acetamide, thiourea)	TER↑	
Retina (RPE)	Apical hyper (25 mM mannitol)	depolarization, Rt↓ in basal cell membrane	[22]
	Basal hyper (25 mM mannitol)	polarization, Rt↑ in basal cell membrane	
Brain (Endothelia)	Apical hyper (1100 mM mannitol)	$P_{albumin}$↑	[28]
	Basal hyper (1100 mM mannitol)	$P_{albumin}$↑	
Bladder	Apical hyper (240 mM Urea)	$P_{sucrose}$↑, P_{H2O}↑	[15,29]
	Apical hyper (240 mM Urea) or Basal hypo	Bleb formation between TJ strands	
Kidney (Distal tubule)	Basal hypo or Apical hyper (120 mM NaCl)	TER↓, Claudin-1 localization to TJs	[16–18]
	Apical hypo or Basal hyper (120 mM NaCl)	TER↓	
Kidney (Proximal tubule)	Apical hypo or Basal hyper (70 mM NaCl)	Cation selectivity↓, Bleb formation between TJ strands, Changes in cell–cell contact shape	[18]
Gallbladder	Apical hyper (100 mM sucrose)	TER↑, $P_{sucrose}$↓, $P_{1,4\text{-butanediol}}$↓	[39]

RPE = retinal pigment epithelium; Hyper = hyperosmolality; Hypo = hypoosmolality; TER = transepithelial electrical resistance; ↑ = increase; ↓ = decrease; TJ = tight junction; P_X = permeability of X; Rt = resistance.

3. Effects of HP on the Epithelia and Endothelia

As shown in Table 1, the osmotic gradient is involved in the regulation of various cell functions in various epithelia and endothelia. If the permeability of the substances used in the studies in Table 1 across the epithelial sheets is lower than water permeability, the concentration difference of the substances is assumed to act as a driving force for the water movement across the epithelia. To put in perspective the magnitude of this, the concentration difference of just 1 mM osmotically active solute (fasting blood glucose level fluctuates from 3.9 to 6.1 mM in the physiological condition) is estimated to be 26 cmH_2O. Surprisingly, HP less than 26 cmH_2O has been reported to have great impact on epithelial functions as we will review here.

3.1. HP in the Body

As in the case of osmolality, HP in the apical side constantly changes in accordance with biological activity. For example, the HP in the apical side of intestine drastically changes with diet. HP in a gallbladder and a bladder changes with the amount of storage (bile and urine, respectively).

In the lung, respiration affects the pressure in the airway and the pressure in the apical side of alveoli periodically becomes negative in inspiration and positive in expiration. In contrast, HP in the basal side (interstitial fluid pressure: IFP) is regulated in a certain range. Most loose tissues have slightly negative IFP (−1 to −3 cmH_2O) and the negative value is relatively high in the lung IFP (−10 cmH_2O). In contrast, IFP in the encased tissues such as kidney shows slightly positive values (0 to 2 cmH_2O) [7,40,41]. Therefore, it is thought that HP is higher in the apical side in most epithelia in the physiological condition and the HP gradient between apical and basal sides fluctuates with the biological activity.

3.2. Effects of HP on Epithelia

3.2.1. Intestine

In the intestine, HP in the apical side dynamically and drastically changes with diet, and water transport in the intestine is affected by the HP gradient. In the isolated dog jejunum and ileum, water is absorbed from the apical to the basal side in normal conditions. The HP from the apical to the basal side up to 20 cmH_2O has no effect on water transport. In contrast, HP from basal to apical side reduces the water absorption dependent on the pressure and the direction of water movement is reversed at 4 cmH_2O [42]. Similar effects of the HP gradient on water transport is observed in a small intestine in a hamster and a colon in a rat [43,44]. Thus, the HP gradient from basal side suppresses water absorption in the intestinal epithelia.

3.2.2. Trachea

Chronic inflammation in the trachea such as asthma is known to increase permeability of airway epithelia. On the other hand, chronic inflammation is also known to increase IFP (see Section 3.4), and the HP gradient from the basal side affects the transepithelial transport in the tracheal epithelia. HP from basal side increases transepithelial conductance (reciprocal of electrical resistance) and permeability of mannitol and albumin in dog tracheal epithelium, transepithelial conductance and permeability of 70 kDa and 2000 kDa dextran in bovine tracheal epithelium, and permeability of 70 kDa dextran in guinea pig tracheal epithelium [45–47]. HP from apical from basal side has no effect on transepithelial transport in dog and bovine tracheal epithelia whereas it induces small increase in the permeability of 70 kDa dextran in guinea pig tracheal epithelium. Thus, the HP gradient from basal side increases the permeability of tracheal epithelium, which may be involved in the pathogenesis of asthma and other airway diseases.

3.2.3. Alveolus

The HP in alveoli periodically alters with respiration. In addition, the pressure is thought to be affected in various clinical conditions such as acute respiratory distress syndrome (ARDS), interstitial pneumonia, and mechanical ventilation. The HP gradient affects transcellular transport in alveolar epithelia. In *Xenopus* alveoli, sodium is absorbed and chloride and potassium are secreted via the epithelial sodium channel (ENaC), chloride, and potassium channels in the apical cell membrane, respectively. The HP from the apical side reduces short-circuit current (Isc) from apical to basal side, the reflection of net ion transport across the epithelia. The HP from the basal side also reduces Isc whereas the increase of the HP in both sides has no effect on Isc, suggesting both directions of the HP gradients reduce Isc. The blocker of ENaC or a chloride channel enhances the reduction of Isc induced by the HP from apical side whereas the potassium channel blocker decreases the reduction of Isc [48,49]. Thus, the HP gradient from apical side increases sodium and chloride secretion and reduces potassium secretion via the transcellular pathway in the alveolar epithelia.

3.2.4. Kidney, Distal Tubule

In the renal tubules, HP is applied from apical side by the flow of filtrate, and the HP gradient is known to affect various cell functions of *Xenopus* A6 cells [50,51]. The HP from basal side increases transepithelial conductance dependent on the degree of pressure. The HP from basal side also alters the structure of actin filaments and claudin-1 localization at the lateral side, increases cell height and stimulates transcellular chloride secretion (Figure 2A). The HP from apical side or the increase of the HP in both sides does not induce these changes. Thus, the HP gradient from the basal side affects various cell functions including transepithelial conductance, actin structure, claudin-1 localization, cell height, and transcellular transport in A6 cells.

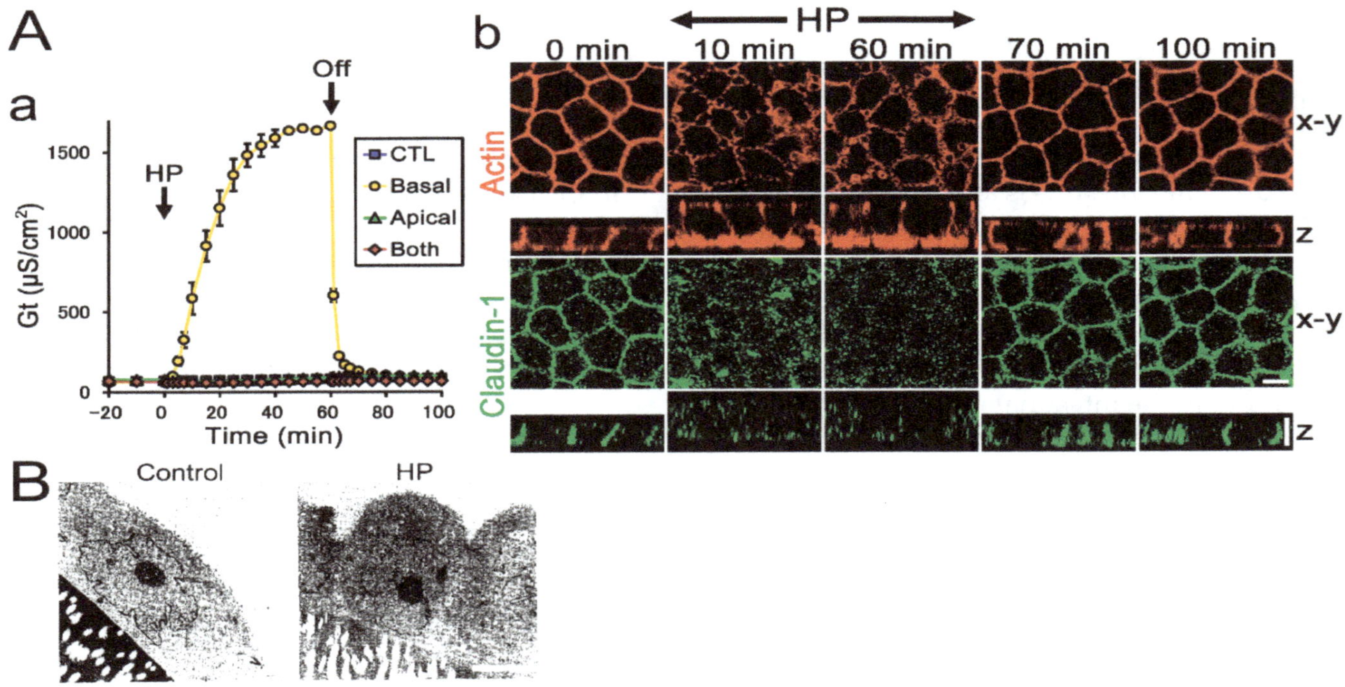

Figure 2. Effects of the hydrostatic pressure (HP) gradient in renal distal tubule cells and podocytes. (**A**) Effects of HP on *Xenopus* A6 cells. (**a**) Time course of transepithelial conductance. 8 cmH_2O HP was applied from apical, basal, or both sides from time 0 to 60 min. The HP from basal side increased transepithelial conductance with reversibility. (**b**) Immunofluorescence of F-actin and claudin-1. 8 cmH_2O HP from basal side was applied from time 0 to 60 min. HP from basal side increased cell height and altered actin structure and claudin-1 localization with reversibility. Modified from Tokuda et al., *Biochem. Biophys. Res. Commun.* 2009 [51] with permission. (**B**) Transmission electron micrographs of podocytes. 1 cmH_2O HP was applied from basal side for three days. Podocyte cells showed more round shape and had wide intercellular space when the HP was applied from basal side. Scale bar = 5 μm. From Coers et al., *Pathobiology.* 1996 [52] with permission.

3.2.5. Glomerulus

In the glomerulus, glomerular epithelial cells (podocytes) play an important role in glomerular filtration. The podocytes have TJs in fetal period, but they develop slit diaphragms, special structure for the glomerular filtration, during the development, and lose TJs [53]. When podocytes are isolated and cultured in vitro, they dedifferentiate and lose various characteristics within 24 h, and they form TJs again. The glomerular filtrate flows from basal to apical side in a podocyte cell sheet in vivo, and the HP gradient from basal side has been reported to affect various characteristics of the podocytes [52]. When podocyte cells from a rat cell line are cultured on a membrane filter and 1 cmH_2O HP is applied from basal side by changing the height of culture medium, the podocyte cell sheet forms a large whirl-like configuration compared with the control (no HP gradient) condition. Furthermore, podocyte cells show more round shape and have wide intercellular space and deep intercellular indentations (Figure 2B). The HP from basal side also induces the reduction of TER and the loss of keratine-18 expression, one of the dedifferentiation markers in the isolated podocytes, in some cells. Thus, the HP gradient from a the basal side affects various characteristics in podocytes, and it is speculated that these changes may be related to the redifferentiation of podocytes.

3.2.6. Bladder

In the bladder, the HP from an apical side fluctuates with the amount of urine storage. The HP gradient is known to affect transepithelial transport of the bladder epithelium [54–56]. The HP from an apical side increases Isc from apical to basal side whereas HP from a basal side decreases Isc in the bullfrog bladder [54]. The increase of Isc by the HP from apical side is also observed in the rabbit bladder, which is inhibited by ENaC, chloride and potassium channel blockers [55,56]. Thus, the HP gradient from an apical side affects sodium absorption and chloride and potassium secretion in the bladder epithelium, which is thought to contribute to the regulation of ion absorption and excretion in the bladder.

3.2.7. Mammary Gland

In mammary glands, HP in the apical side fluctuates with milk production and breastfeeding and the permeability of TJs in mammary epithelia is known to be affected by the status of milk storage in vivo. The HP gradient has been reported to affect transepithelial transport in cultured mammary epithelial HC11 cells [57,58]. The HP from basal side decreases TER and changes the direction of Isc. The increase of HP in both sides has no effect on TER. Thus, the HP gradient from a basal side affects TER and transcellular transport, which may be involved in the regulation of transepithelial transport in milk production and breastfeeding.

3.2.8. Summary

Effects of the HP gradient on epithelia are summarized in Table 2. There are several reports which show that HP from the apical or basal side affects transepithelial transport in gallbladder, hepatocyte, nasal epithelia, and proximal tubule [59–62]. However, these reports do not have clear data that shows that the gradient is a definitive cause of the changes due to the purpose of the studies and/or the difficulty in an experimental system, and we do not include these studies in the Table 2. Nevertheless, there is a possibility that the HP gradient is also involved in the regulation of cell functions in these epithelia.

As shown in Table 2, the HP gradient is involved in the regulation of various cell functions in various epithelia. The cell functions regulated by the HP gradient include transepithelial transport (transcellular and paracellular transport), cytoskeleton and cell shape. Surprisingly, only several to several tens cmH_2O HP gradients induce various cell responses, which are less than the osmotic pressure induced by the 1 mM concentration difference of the nonionized osmotic substance as a driving force for the water movement (26 cmH_2O). In addition, the HP gradient and the osmotic gradient which are assumed to generate the water movement of a same direction induce different cell responses in some cases (see the cases in the jejunum and the distal tubule in Tables 1 and 2). Therefore, the osmotic gradient and the HP gradient are thought to affect various cell functions through different mechanisms (see Section 4).

Furthermore, the direction of the HP gradient which induces cell responses and the regulated cell functions are different among the types of epithelia. Thus, it is speculated that epithelia distinguish the direction of the HP gradient and each epithelium has a different mechanism to regulate cell functions, which is likely to reflect the difference of physiological roles in each epithelium.

Table 2. Effects of the HP gradient on epithelia.

Organ	Type of HP Gradient	Cell Response	References
Jejunum, ileum	Basal (20 cmH_2O)	Water absorption↓	[42–44]
Trachea	Basal (5–20 cmH_2O)	TER↓, $P_{mannitol}$↑, P_{water}↑, $P_{70kDa\ or\ 2000kDa\ dextran}$↑, $P_{albumin}$↑	[45–47]
	Apical (20 cmH_2O)	$P_{70kDa\ dextran}$↑	
Alveolus	Apical (5 cmH_2O)	Isc↓, K^+ secretion↓, Na^+ absorption↑, Cl^- secretion↑	[48,49]
	Basal (5 cmH_2O)	Isc↓	
Kidney (Distal tubule)	Basal (8 cmH_2O)	TER↓, Cl^- secretion↑, Claudin-1 localization, Actin structure, Cell height	[50,51]
Kidney (Podocyte)	Basal (1 cmH_2O)	Cell shape, TER↓, Expression of keratin-18↓	[52]
Bladder	Apical (1–8 cmH_2O)	Na^+ absorption↑, Cl^- secretion↑, K^+ secretion↑	[54–56]
Mammary gland	Basal (10.2 cmH_2O)	TER↓, Isc↓	[57,58]
Cervical epithelium	Basal (2.1 cmH_2O)	TER↓, $P_{pyranine}$↑	[63]

HP = hydrostatic pressure; TER = transepithelial electrical resistance; ↑ = increase; ↓ = decrease; P_X = permeability of X; Isc = short-circuit current.

3.3. *Effects of HP on Endothelia*

Relatively high HP is applied to the endothelia from the apical side by the blood flow. The pressure is especially high in the artery (approximately 100 cmH_2O), which fluctuates with heartbeats. There are many reports which show that the HP from apical or the basal side affects various endothelial functions such as transendothelial transport, cell adhesion, cell shape, intracellular Ca^{2+} concentration, expression of cytokines and so on [64–72]. On the other hand, the increase of 50–270 cmH_2O HP to the whole cell by using specialized pressure chamber also affects various endothelial functions including transendothelial electrical resistance, cytoskeletons, cell proliferation, and expression of cytokines and cell adhesion proteins [73–77]. Furthermore, smaller HP (2 to 27 cmH_2O) is also shown to have effects on F-actin and cell proliferation [78–81], although there is a report which shows that 136 cmH_2O HP has no effect on F-actin and cell proliferation in endothelia [82]. Thus, it is important to distinguish the effect of the increase of the HP on whole cell and the effect of the HP gradient on endothelia. However, there is no study which clearly shows that the HP gradient has an effect on endothelial functions. Thus, there is a possibility that HP gradient affects endothelial cell functions, but currently we do not have enough evidence.

3.4. *HP in the Pathological Conditions*

The HP in the basal side of epithelia (IFP) is known to change in pathological conditions. For example, IFP is markedly decreased in burn to −27 to −42 cmH_2O [83]. Edema induced by raising vascular pressure increases IFP in a dog lung [84]. On the other hand, acute inflammation induces interstitial edema but the IFP is rather decreased transiently by the reduction of tension in the interstitium due to physical denaturation of collagen [85]. In contrast, in chronic inflammation the IFP is known to be increased; experimental inflammation by the ligation of first maxillary molar in rat gingiva and experimental pulpitis in cat dental pulp induce the increase of IFP [86,87]. Meanwhile, the IFP is also increased in cancer tissues. The increase of IFP was reported in xenograft tumor in animals as early as 1950 [88–90]. The IFP is also increased in most human tumors to 14–54.4 cmH_2O [91–96]. The precise mechanism of the increase of IFP in tumors is still incompletely understood, but fibrosis, fluid accumulation and increase of cell density is thought to be involved in the mechanism [97,98]. In addition, the increase of IFP is known to be the factor associated with poor prognosis in uterine cervical cancer [99,100].

It is worth noting that IFP is increased in both chronic inflammation and carcinomas. The causal association between chronic inflammation and carcinomas is reported in almost all organs [101]. The mechanism of the association is generally thought to be due to immune responses induced by the chronic inflammation; reactive oxygen species from immune cells contribute to the occurrence of gene mutation and cytokines secreted by the immune cells promote cell proliferation, inhibition of apoptosis, angiogenesis, metastasis and epithelial mesenchymal transition (EMT) of the carcinomas [101,102].

On the other hand, since IFP is increased in both chronic inflammation and carcinomas, it is thought that the increased IFP may also contribute the promotion of carcinomas in chronic inflammation.

3.4.1. Effects of HP on Tumor Cells

The increase of HP in the extracellular environment by using a specialized pressure chamber is known to affect the characteristics of cancer cells. In osteosarcoma, the increase of 27 to 68 cmH_2O HP induces promotion or suppression of cell proliferation dependent on cell lines. In the cell lines which show the promotion of cell proliferation by the HP, the HP increases sensitivity to cisplatin, elevates the expression of tissue plasminogen activator (TPA) and vascular endothelial growth factor C (VEGF-C), and suppresses the expression of VEGF-A [103–106]. In lung cancer cells (CL1-5 cells and A549 cells), 27 cmH_2O HP increases migration speed and cell volume with the elevation of the expression of various proteins including AQP1, Snail, vinculin and caveolin-1 [107,108]. In urothelial carcinoma cells, 102–1020 cmH_2O HP promotes apoptosis with the elevation of the expression of Fas ligand, toll-like receptor 6 and connective tissue growth factor in the presence of mitomycin C [109,110]. In addition, extremely high HP more than 1 million cmH_2O induces immunogenic cell death in tumor cells and is used for the method in cancer immunotherapy [111,112]. In contrast, small degree of the HP gradient from a basal side has been known to have carcinogenic effects on the epithelia.

3.4.2. Effects of the HP Gradient on Carcinogenic Properties of Epithelia

When epithelia are cultured on permeable filters and the HP gradient is applied from a basal side by changing the height of culture medium, the HP induces epithelial stratification in some epithelial cell lines including MDCK I cells, Caco-2 cells, a model of colon cancer cells, and EpH4 cells, a model of mammary epithelial cells (Figure 3, [113]). The stratification is not observed when the HP is applied from apical side or the HP in both sides are decreased or increased, suggesting the stratification is caused by the HP gradient from basal side. In MDCK II cells, the HP gradient from a basal side does not induce the stratification, indicating the responsiveness of epithelial stratification to the HP gradient is different among the cell types. The stratification continues to develop in the presence of the HP gradient from a basal side, and the elimination of the HP gradient restores the epithelia to an almost single layer in 2 days after the elimination.

Interestingly, cavities with the characteristics of apical cell polarity are formed within the stratified epithelia. Microvilli are observed at the surface of the cavities, and TJs with the functional barrier assessed by the biotin tracer experiment are formed between the cells surrounding the cavities. It is worth noting that a similar abnormality in cell polarity is observed in the epithelia expressing oncogene and in vivo cancer tissue. When K-ras is expressed in MDCK cells and the cells are culture on permeable filters, epithelial stratification is induced and cavities with microvilli and TJs are observed within the stratification [114]. Similar stratification is reported in rat salivary gland cells (Pa-4 cells) when Raf-1 is expressed and cultured on permeable filter [115]. Furthermore, small cavities with microvilli and TJs are observed within pulmonary metastases of the mammary adenocarcinoma induced by the infection of mammary tumor virus in mice, and the permeation of lanthanum nitrate is blocked by these TJs [116]. Therefore, the epithelial stratification and abnormal cell polarity induced by the HP gradient from a basal side are thought to be common characteristics observed in oncogene expressing cells and in vivo carcinomas.

In addition, the HP gradient from a basal side affects cell proliferation and apoptosis. The proportion of S phase cells assessed by the BrdU assay is elevated by the HP gradient from a basal side, indicating the acceleration of a cell cycle. The HP gradient from a basal side also reduces the amount of dead cells in culture supernatant, and the elimination of the HP gradient induces the emergence of numerous TUNEL positive cells from 6 h after the elimination, suggesting the suppression of apoptosis by the HP gradient. In addition, the HP gradient from a basal side decreases TER and increases P_{Na} with the increase of claudin-2 expression level. Furthermore, the epithelial stratification is suppressed by the activation of protein kinase A (PKA) and promoted by the inhibition of PKA [113]. Thus, the HP gradient from basal side induces abnormal cell polarity, accelerates cell proliferation and suppresses cell apoptosis, resulting in the epithelial stratification via the PKA pathway.

In contrast to the carcinogenic effects of the HP gradient from a basal side, the outcome of the treatment of the bladder carcinoma by the HP suggests the possibility that the HP gradient from an apical side may have an inhibitory effect on carcinomas.

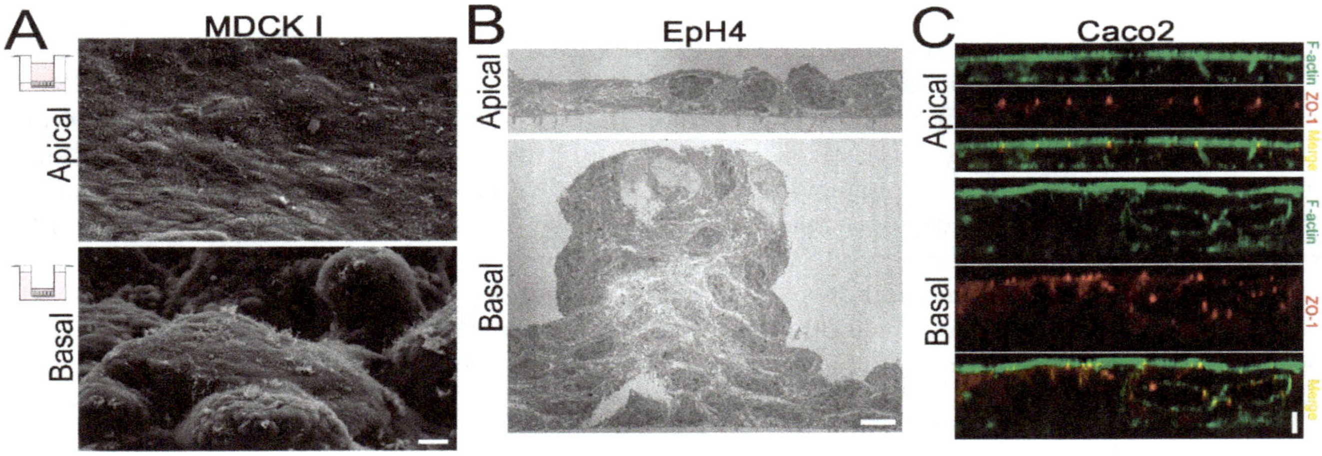

Figure 3. Effects of the HP gradient on carcinogenic properties of epithelia. (**A**) Scanning electron micrographs in MDCK I cells. 0.6 cmH_2O HP was applied from apical or basal side for four days. A bumpy surface with cell masses was observed when the HP was applied from a basal side. Scale bar = 10 µm. (**B**) Transmission electron micrographs in EpH4 cells. 0.6 cmH_2O HP was applied from apical or basal side for four days. HP from a basal side induced stratification. Cavities were observed within the stratification. Scale bar = 5 µm. (**C**) Immunofluorescence of F-actin and ZO-1 in Caco2 cells. 0.6 cmH_2O HP was applied from apical or basal side for eight days. HP from a basal side induced stratification. ZO-1 was localized at the cavities within the stratification. Scale bar = 5 µm. Modified from Tokuda et al., *PLoS ONE*. 2015 [113] with permission.

3.4.3. Treatment of the Bladder Carcinoma by HP from an Apical Side

In 1972, Helmstein developed and introduced a method for the treatment of the bladder carcinoma using HP. In this method, the pressure inside the bladder is kept at the level of diastolic blood pressure for 5–6 h under general anesthesia. Surprisingly, in the studies including the patients with pathological stage T4, the reduction of the tumor has been observed in 70%–100% and complete response has been achieved in 25%–45% [117–121]. The cause of antitumor effect in the HP treatment was speculated to be the ischemia induced by the HP; tumor cells were thought to be more sensitive to the ischemia by the HP than normal cells due to the vulnerability of tumor vessels and changes in metabolism in tumor cells, and the HP from apical side induced cell death only in tumor cells.

The current standard therapy of non-muscle invasive bladder carcinomas is transurethral resection of visible bladder tumor (TURBT) and intravesical instillation of chemotherapy or BCG dependent on the risk factors of recurrence, and the HP treatment for the bladder carcinoma is not performed in current practical clinic. The mechanism of the HP treatment of the bladder carcinoma is still poorly understood, but since the HP gradient from basal side has carcinogenic effects on epithelia and the elimination of the HP gradient induces cell apoptosis [113], it is likely that the HP gradient from apical side in the HP treatment itself may have suppressive effect on tumor cells and contribute to tumor reduction.

3.4.4. Intervention of IFP in Tumors

If the HP gradient from basal side promotes the growth of carcinomas and the elimination of the HP gradient contributes to the treatment of carcinomas, the intervention of the HP gradient in tumor tissues may be used as the treatment of carcinomas. In the HP treatment of bladder carcinomas, the HP from the apical side is applied under general anesthesia, but it is practically difficult to change the HP in the apical side in other organs. In contrast, various interventions are known to affect the IFP in tumor tissues. In xenograft tumors in mice, hamsters or rats, knockout of Neural/glial antigen 2 (NG2) proteoglycan, hyperthermia (43 °C), and the treatments of the antibody against VEGF receptor-2, ZD6126 (tubulin-binding agent) and inhibitor of platelet-derived growth factor (PDGF) receptor reduce IFP whereas the treatment of angiotensin II elevates IFP [122–127]. However, the mechanism of the increase of IFP in tumor tissues is still not fully understood, and it is difficult to constantly keep the IFP in normal level. Thus, it is currently not feasible to eliminate the HP gradient with the intervention except for the bladder carcinoma. On the other hand, the elucidation of the mechanism in the regulation of epithelial functions by the HP gradient is thought to provide a clue for the alternative method to develop the treatment of carcinomas. Then, how do the epithelia sense the HP gradient?

4. Theoretical Speculation of the Mechanism about How Epithelia Sense Osmotic and HP Gradients

In physiological conditions, there are various differences between the apical and basal environments of the epithelia including osmolality, HP, concentration of various substances, and electrical potential. Steep gradients of osmolality, HP, concentration of various substances, and electrical potential between apical and basal sides are formed at the sites that act as barriers to these differences, and the major barriers are the apical and basal cell membranes and the TJs. It is reasonable to speculate that the sensor(s) of these differences must be located at the site of these steep gradients (Figure 4A). Thus, next we discuss the possibility that either the cell membranes or the TJs act as the sensor of the gradients.

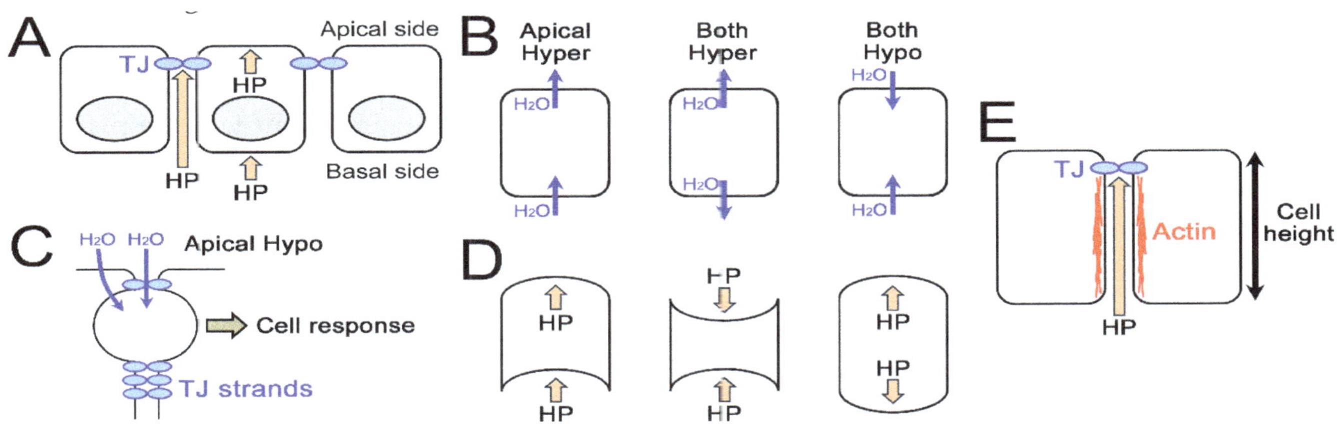

Figure 4. Theoretical speculation of the mechanism about how epithelia sense osmotic and HP gradients. (**A**) A model of epithelium in the HP from basal side. Steep HP gradients are formed at TJs and apical and basal cell membranes. (**B**) The water movement through apical and basal cell membranes in apical hyperosmolality (*left*), hyperosmolality in both sides (*middle*), and hypoosmolality in both sides (*right*). Hyper = hyperosmolality; Hypo = hypoosmolality. (**C**) Possible mechanism of bleb formation between TJ strands in apical hyperosmolality. (**D**) Effects of HP from basal side (*left*), increase of HP in both side (*middle*) and decrease of HP in both sides (*right*) on apical and basal cell membranes. (**E**) Possible effects of HP from basal side on F-actin in the lateral side and cell height.

4.1. Osmolality

The osmotic gradient induces various changes in epithelia (Table 1). It is shown in the studies in Table 1 that either apical or basal hyperosmolality induces cell responses but hyperosmolality in both sides does not induce these responses except for the study in gallbladder. Further, it is also shown that either apical or basal hypoosmolality induces cell response but hypoosmolality in both sides does not induce these responses in the bladder, proximal tubule and distal tubule cells [15–18]. In addition, the cell responses induced by the osmotic gradient are somewhat different from the responses induced by cell volume changes in hyperosmolality or hypoosmolality [1,10]. Therefore, epithelial cells are thought to sense and respond to the transepithelial osmotic gradient (i.e., the osmotic difference between the two sides) and not merely the overall extracellular osmolality.

First, we will consider the possibility that the apical cell membrane acts as a sensor of apical hyperosmolality. In the apical hyperosmotic condition, water flows out through the apical cell membrane (Figure 4B). In the condition of hyperosmolality in both sides, water also flows out through the apical cell membrane (Figure 4B). If the apical cell membrane is a sensor of the apical hyperosmolality and the water flow through the apical cell membrane induces cell responses, same cell responses should be induced by the hyperosmolality in both sides. However, the cell responses induced by the apical hyperosmolality are not induced by the hyperosmolality in both sides. Thus it is unlikely that the outflow of water through the apical cell membrane by itself is the trigger of cell responses induced by the apical hyperosmolality. Similarly, in the apical hyperosmotic condition, water flows in through the basal cell membrane. In the condition of hypoosmolality in both sides, water also flows in through the basal cell membrane. If the basal cell membrane is a sensor of the apical hyperosmolality and the water flow through the basal cell membrane induces cell responses, same cell responses should be induced by the hypoosmolality in both sides. However, the cell responses induced by the apical hyperosmolality in the bladder and proximal and distal tubule cells are not induced by the hypoosmolality in both sides. Thus it is unlikely that the inflow of water through the basal cell membrane by itself is the trigger of cell responses induced by the apical hyperosmolality (Figure 4B). There remains a possibility that apical and basal membranes act together as a sensor. If this is the case, the possible mechanism is that the water flow from both cell membranes induces the changes in cell membranes, cell shape and/or cytoskeleton and serves as a trigger. However, it may be difficult to distinguish between these changes in the osmotic gradient and those in cell swelling and shrinkage and provide enough sensitivity as a sensor. Therefore, it is not very likely that cell membranes are the sensor of the osmotic gradient.

In contrast, the possibility that TJs act as a sensor of the osmotic gradient is quite plausible. The movement of water and/or osmotic substances through the TJs is assumed to be sensed by TJs. One of the possible mechanisms in this case is that the movement of water and substances through the TJ strands act as shear stress and the shear stress is sensed by the TJ strands. Furthermore, the water movement may cause the accumulation of water between TJ strands because of differences in the water permeability between diffusion across TJ strands and diffusion through the paracellular space between strands, and this water accumulation may serve as a trigger of cell responses (Figure 4C). The bleb formation between TJ strands observed in the presence of the osmotic gradient is thought to support this possibility (Figure 1B, [15,18]). In addition, since TJ permeability is regulated by the osmotic gradient in many cases, TJs may constitute a feedback system in which TJs sense the extracellular environment and regulate the functions of themselves.

4.2. HP

The HP gradient also induces various changes in epithelia (Table 2 and [113]). Some studies in Table 1 show that these changes are not induced by the increase or decrease of HP in both sides [51,52,113]. In addition, the atmospheric pressure fluctuates more than 1 cmH_2O in a day and more than 5 cmH_2O from day to day, and it is unlikely that the fluctuation of atmospheric pressure induces the changes observed in Table 2. Thus, as in the case of osmolality, epithelial cells are thought to distinguish the HP gradient and the changes of HP in both sides.

In the condition of the HP gradient from basal side, it is thought that basal and apical cell membranes are pushed to apical side with the water movement (Figure 4D). In contrast, basal cell membrane is pushed to apical side and apical cell membrane is pushed to basal side with the water movement in the condition that the HP is increased in both sides, and the cell membranes are pushed to opposite direction with the water movement in the condition that HP is decreased in both sides.

Thus, if the cell membranes act as a sensor of the HP gradient, it is unlikely that the push of basal cell membrane to apical side with the water movement by itself or the push of apical cell membrane to apical side by with the water movement itself is the trigger of the responses induced by the HP gradient (Figure 4D). There remains a possibility that the push of apical cell membrane to apical side and basal cell membrane act together as a sensor. If this is the case, the possible mechanism is that as in the case of osmolality, the push of cell membranes induces the changes in cell membranes, cell shape and/or cytoskeleton and serves as a trigger. However, it may be difficult to distinguish between these changes in the HP gradient and those in the increase or decrease of HP in both sides and provide enough sensitivity as a sensor. Therefore, it is not very likely that cell membranes are the sensor of the HP gradient.

In contrast, it is simple to consider the possibility that TJs act as a sensor of the HP gradient. The push of TJ strands to apical or basal side and movement of water through the TJs is sensed by the TJs. One of the possible mechanisms in this case is that the pressure to the TJ strands changes the structure of the intracellular proteins in TJ complexes and the tension applied to the cytoskeleton (Figure 5). The structural changes of F-actin in the lateral side and the increase of cell height by the HP gradient from basal side in *Xenupus* A6 cells may reflect the effect of HP gradient on the tension in F-actin (Figures 2A and 4E, [51]).

In summary, it is likely that TJs act as a sensor of osmotic and HP gradients, although it is difficult to rule out the possibility that apical and basal cell membranes act together as a sensor. Lastly, we further discuss the possible role of TJs as a sensor

4.3. *Possible Role of TJs as a Sensor*

The TJs are one mode of the junctional complexes located in the most apical part of the complexes [12]. The TJs regulate the permeability of the paracellular pathway [13,14]. Claudins are a family protein of integral membrane proteins (26 members in human) in TJs and a major constituent of TJ strands [30,128,129]. Most epithelia express multiple claudins and expression patterns of claudins are thought to determine the permeability of TJs in each epithelium [31,32]. Most claudins have PDZ binding motif in carboxy-terminal tail and bind to scaffolding proteins in TJs which have PDZ motif such as ZO family proteins, multiple PDZ domain protein (MPDZ) and Pals1-associated tight junction protein (PATJ). These scaffolding proteins further bind to transmembrane proteins including claudins and other proteins, scaffolding proteins, F-actin and many other proteins involved in the regulation of various cell functions including cytoskeleton, transcription, signal transduction, and vesicular trafficking. Thus, TJs is a complex comprised of diverse proteins which are involved in the regulation of various cell functions including cell differentiation, proliferation and apoptosis [130–134].

As discussed above, the sensor of the environmental gradients between apical and basal sides of the epithelia is required to distinguish the direction of the gradient and the difference of the stimulation (such as osmolality and HP) to regulate various cell functions dependent on the physiological requirement in each epithelium. Since claudins are large family proteins and expression patterns of claudins in TJs are different among epithelial and endothelial cell types, TJ strands are thought to be quite diverse among cell types and likely to have the ability to distinguish the direction and the difference of the stimulation dependent on cell types. Furthermore, diverse proteins in TJ complex are thought to be a suitable system to transduce the stimulation sensed by the TJ strands to regulate various cell functions (Figure 5). So far, there is little evidence that shows that TJs act as a sensor of the environmental gradient between apical and basal sides. However, since the changes in the shape of cell–cell contact and F-actin induced by the osmotic gradient from apical to basal side in MDCK II cells are not observed in claudin-2 knockout cells (Figure 1D, [18]), claudin-2 is thought to mediate these changes by the osmotic gradient and likely supports the possibility that TJs act as a sensor. In addition, it is known that the expression pattern of claudins is changed in most carcinomas and changes in the expression level of some claudins affects cell functions such as cell proliferation [135,136]. Since the HP gradient from basal side has carcinogenic effects on epithelia, the altered expression pattern of claudins in carcinomas may contribute to transduce favorable signals induced by the HP gradient from basal side for cancer growth.

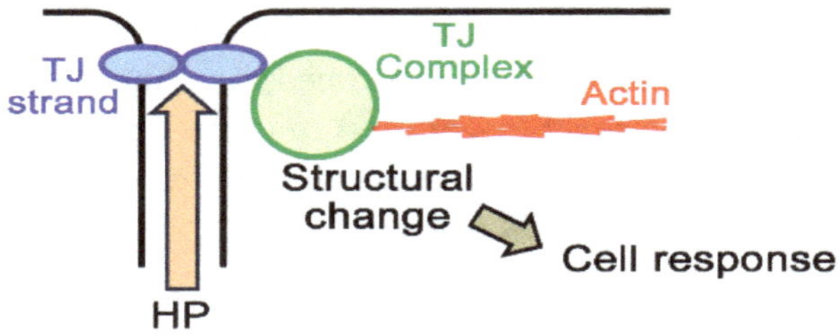

Figure 5. Possible mechanism of TJs as a sensor.

5. Conclusions

Osmotic and HP gradients have great impact on various cell functions in various epithelia. It is required to elucidate the mechanism about how epithelia sense the extracellular environment and regulate cell functions; the mechanism is thought to have physiological and cell biological significance and may contribute to the development of cancer therapy. Theoretical speculation suggests the possibility that TJs are involved in the sensing of osmotic and HP gradients, although obviously more evidence is required. Since TJs are intricate complexes composed of multiple proteins, it is speculated to be challenging to elucidate the mechanism due to the redundancy of many proteins and enormous protein interactions. We expect more examples that indicate the regulation of cell functions by the environmental gradient between apical and basal sides, which may provide a clue for the elucidation of the mechanism.

Abbreviations

HP	hydrostatic pressure
ADH	antidiuretic hormone
AQP	aquaporin
RVD	regulatory volume decrease
RVI	regulatory volume increase
TER	transepithelial electrical resistance
TJ	tight junction
P_{Na}	permeability of sodium
P_{Cl}	permeability of chloride
MDCK	Madin–Darby canine kidney
RPE	retinal pigment epithelium
BRB	blood–retinal barrier
ERG	electroretinogram
BBB	blood–brain barrier
IFP	interstitial fluid pressure
ARDS	acute respiratory distress syndrome
ENaC	epithelial sodium channel
Isc	short-circuit current
EMT	epithelial mesenchymal transition
TPA	tissue plasminogen activator
VEGF	vascular endothelial growth factor
TURBT	transurethral resection of visible bladder tumor

References

1. Hoffmann, E.K.; Lambert, I.H.; Pedersen, S.F. Physiology of cell volume regulation in vertebrates. *Physiol. Rev.* **2009**, *89*, 193–277. [CrossRef] [PubMed]
2. Swabb, E.A.; Hynes, R.A.; Donowitz, M. Elevated intraluminal pressure alters rabbit small intestinal transport in vivo. *Am. J. Physiol.* **1982**, *242*, 58–64. [CrossRef] [PubMed]
3. Loring, S.H.; Topulos, G.P.; Hubmayr, R.D. Transpulmonary Pressure: The Importance of Precise Definitions and Limiting Assumptions. *Am. J. Respir. Crit. Care Med.* **2016**, *194*, 1452–1457. [CrossRef] [PubMed]
4. Rasouli, M. Basic concepts and practical equations on osmolality: Biochemical approach. *Clin. Biochem.* **2016**, *49*, 936–941. [CrossRef] [PubMed]
5. Sands, J.M.; Layton, H.E. The physiology of urinary concentration: An update. *Semin. Nephrol.* **2009**, *29*, 178–195. [CrossRef] [PubMed]
6. Bankir, L.; Bichet, D.G.; Morgenthaler, N.G. Vasopressin: Physiology, assessment and osmosensation. *J. Intern. Med.* **2017**, *282*, 284–297. [CrossRef] [PubMed]
7. Guyton, A.C.; Hall, J.E. *Textbook of Medical Physiology*, 11th ed.; Elsevier Saunders: Philadelphia, PA, USA, 2006; pp. 187–188, 348–357.
8. Okada, Y.; Maeno, E.; Shimizu, T.; Dezaki, K.; Wang, J.; Morishima, S. Receptor-mediated control of regulatory volume decrease (RVD) and apoptotic volume decrease (AVD). *J. Physiol.* **2001**, *532*, 3–16. [CrossRef] [PubMed]
9. Hoffmann, E.K.; Pedersen, S.F. Cell volume homeostatic mechanisms: Effectors and signalling pathways. *Acta. Physiol.* **2011**, *202*, 465–485. [CrossRef] [PubMed]
10. Lang, F.; Busch, G.L.; Ritter, M.; Völkl, H.; Waldegger, S.; Gulbins, E.; Häussinger, D. Functional significance of cell volume regulatory mechanisms. *Physiol. Rev.* **1998**, *78*, 247–306. [CrossRef]
11. Madara, J.L. Increases in guinea pig small intestinal transepithelial resistance induced by osmotic loads are accompanied by rapid alterations in absorptive-cell tight-junction structure. *J. Cell Biol.* **1983**, *97*, 125–136. [CrossRef] [PubMed]
12. Farquhar, M.G.; Palade, G.E. Junctional complexes in various epithelia. *J. Cell Biol.* **1963**, *17*, 375–412. [CrossRef] [PubMed]
13. Claude, P.; Goodenough, D.A. Fracture faces of zonulae occludentes from "tight" and "leaky" epithelia. *J. Cell. Biol.* **1973**, *58*, 390–400. [CrossRef] [PubMed]
14. Powell, D.W. Barrier function of epithelia. *Am. J. Physiol.* **1981**, *241*, 275–288. [CrossRef] [PubMed]
15. Wade, J.B.; Revel, J.P.; DiScala, V.A. Effect of osmotic gradients on intercellular junctions of the toad bladder. *Am. J. Physiol.* **1973**, *224*, 407–415. [CrossRef] [PubMed]
16. Tokuda, S.; Niisato, N.; Nakajima, K.; Marunaka, Y. Regulation of the paracellular Na+ and Cl- conductances by the NaCl-generated osmotic gradient in a manner dependent on the direction of osmotic gradients. *Biochem. Biophys. Res. Commun.* **2008**, *366*, 464–470. [CrossRef] [PubMed]
17. Tokuda, S.; Miyazaki, H.; Nakajima, K.; Yamada, T.; Marunaka, Y. NaCl flux between apical and basolateral side recruits claudin-1 to tight junction strands and regulates paracellular transport. *Biochem. Biophys. Res. Commun.* **2010**, *393*, 390–396. [CrossRef] [PubMed]
18. Tokuda, S.; Hirai, T.; Furuse, M. Effects of Osmolality on Paracellular Transport in MDCK II Cells. *PLoS ONE* **2016**, *11*, e0166904. [CrossRef] [PubMed]
19. Lindley, B.D.; Hoshiko, T.; Leb, D.E. Effects of D2O and osmotic gradients on potential and resistance of the isolated frog skin. *J. Gen. Physiol.* **1964**, *47*, 773–793. [CrossRef] [PubMed]
20. Ussing, H.H. Relationship between osmotic reactions and active sodium transport in the frog skin epithelium. *Acta. Physiol. Scand.* **1965**, *63*, 141–155. [CrossRef]
21. Willermain, F.; Janssens, S.; Arsenijevic, T.; Piens, I.; Bolaky, N.; Caspers, L.; Perret, J.; Delporte, C. Osmotic stress decreases aquaporin-4 expression in the human retinal pigment epithelial cell line, ARPE-19. *Int. J. Mol. Med.* **2014**, *34*, 533–538. [CrossRef]
22. Shirao, Y.; Steinberg, R.H. Mechanisms of effects of small hyperosmotic gradients on the chick RPE. *Invest. Ophthalmol. Vis. Sci.* **1987**, *28*, 2015–2025. [PubMed]
23. Rapoport, S.I. Effect of concentrated solutions on blood–brain barrier. *Am. J. Physiol.* **1970**, *219*, 270–274. [CrossRef] [PubMed]
24. Nagy, Z.; Pappius, H.M.; Mathieson, G.; Hüttner, I. Opening of tight junctions in cerebral endothelium. I. Effect of hyperosmolar mannitol infused through the internal carotid artery. *J. Comp. Neurol.* **1979**

185, 569–578. [CrossRef] [PubMed]

25. Robinson, P.J. Facilitation of drug entry into brain by osmotic opening of the blood–brain barrier. *Clin. Exp. Pharmacol. Physiol.* **1987**, *14*, 887–901. [CrossRef] [PubMed]
26. Rapoport, S.I. Osmotic opening of the blood–brain barrier. *Ann. Neurol.* **1988**, *24*, 677–684. [CrossRef] [PubMed]
27. Kroll, R.A.; Neuwelt, E.A. Outwitting the blood–brain barrier for therapeutic purposes: Osmotic opening and other means. *Neurosurgery* **1998**, *42*, 1083–1100. [CrossRef] [PubMed]
28. Machi, T.; Kassell, N.F.; Scheld, M.W.; Lehmann, G.A. Effect of mannitol on the permeability of cultured endothelial cells. *Fukuoka Igaku Zasshi* **1996**, *87*, 178–183. [PubMed]
29. Urakabe, S.; Handler, J.S.; Orloff, J. Effect of hypertonicity on permeability properties of the toad bladder. *Am. J. Physiol.* **1970**, *218*, 1179–1187. [CrossRef]
30. Furuse, M.; Fujita, K.; Hiiragi, T.; Fujimoto, K.; Tsukita, S. Claudin-1 and -2: Novel integral membrane proteins localizing at tight junctions with no sequence similarity to occludin. *J. Cell Biol.* **1998**, *141*, 1539–1550. [CrossRef]
31. Van Itallie, C.M.; Anderson, J.M. Claudins and epithelial paracellular transport. *Annu. Rev. Physiol.* **2006**, *68*, 403–429. [CrossRef]
32. Günzel, D.; Yu, A.S. Claudins and the modulation of tight junction permeability. *Physiol. Rev.* **2013**, *93*, 525–569. [CrossRef]
33. Green, R.; Giebisch, G. Reflection coefficients and water permeability in rat proximal tubule. *Am. J. Physiol.* **1989**, *257*, 658–668. [CrossRef]
34. Williams, J.C., Jr.; Schafer, J.A. Cortical interstitium as a site for solute polarization during tubular absorption. *Am. J. Physiol.* **1988**, *254*, 813–823. [CrossRef] [PubMed]
35. Schafer, J.A. Transepithelial osmolality differences, hydraulic conductivities, and volume absorption in the proximal tubule. *Annu. Rev. Physiol.* **1990**, *52*, 709–726. [CrossRef] [PubMed]
36. Furuse, M.; Furuse, K.; Sasaki, H.; Tsukita, S. Conversion of zonulae occludentes from tight to leaky strand type by introducing claudin-2 into Madin–Darby canine kidney I cells. *J. Cell Biol.* **2001**, *153*, 263–272. [CrossRef] [PubMed]
37. Amasheh, S.; Meiri, N.; Gitter, A.H.; Schöneberg, T.; Mankertz, J.; Schulzke, J.D.; Fromm, M. Claudin-2 expression induces cation-selective channels in tight junctions of epithelial cells. *J. Cell Sci.* **2002**, *115*, 4969–4976. [CrossRef]
38. Tokuda, S.; Furuse, M. Claudin-2 knockout by TALEN-mediated gene targeting in MDCK cells: Claudin-2 independently determines the leaky property of tight junctions in MDCK cells. *PLoS ONE* **2015**, *10*, e0119869. [CrossRef]
39. Smulders, A.P.; Tormey, J.D.; Wright, E.M. The effect of osmotically induced water flows on the permeability and ultrastructure of the rabbit gallbladder. *J. Membr. Biol.* **1972**, *7*, 164–197. [CrossRef]
40. Scholander, P.F.; Hargens, A.R.; Miller, S.L. Negative pressure in the interstitial fluid of animals. Fluid tensions are spectacular in plants; in animals they are elusively small, but just as vital. *Science* **1968**, *161*, 321–328. [CrossRef]
41. Miserocchi, G.; Passi, A.; Negrini, D.; Del Fabbro, M.; De Luca, G. Pulmonary interstitial pressure and tissue matrix structure in acute hypoxia. *Am. J. Physiol. Lung Cell Mol. Physiol.* **2001**, *280*, 881–887. [CrossRef]
42. Hakim, A.A.; Lifson, N. Effects of pressure on water and solute transport by dog intestinal mucosa in vitro. *Am. J. Physiol.* **1969**, *216*, 276–284. [CrossRef] [PubMed]
43. Wanitschke, R.; Nell, G.; Rummel, W. Influence of hydrostatic pressure gradients on net transfer of sodium and water across isolated rat colonic mucosa. *Naunyn. Schmiedebergs Arch. Pharmacol.* **1977**, *297*, 191–194. [CrossRef] [PubMed]
44. Wilson, T.H. A modified method for study of intestinal absorption in vitro. *J. Appl. Physiol.* **1956**, *9*, 137–140. [CrossRef] [PubMed]
45. Kondo, M.; Finkbeiner, W.E.; Widdicombe, J.H. Changes in permeability of dog tracheal epithelium in response to hydrostatic pressure. *Am. J. Physiol.* **1992**, *262*, 176–182. [CrossRef] [PubMed]
46. Azizi, F.; Matsumoto, P.S.; Wu, D.X.; Widdicombe, J.H. Effects of hydrostatic pressure on permeability of airway epithelium. *Exp. Lung Res.* **1997**, *23*, 257–267. [CrossRef] [PubMed]
47. Gustafsson, B.G.; Persson, C.G. Asymmetrical effects of increases in hydrostatic pressure on macromolecular movement across the airway mucosa. A study in guinea-pig tracheal tube preparations. *Clin. Exp. Allergy* **1991**, *21*, 121–126. [CrossRef] [PubMed]
48. Bogdan, R.; Veith, C.; Clauss, W.; Fronius, M. Impact of mechanical stress on ion transport in native lung

epithelium (Xenopus laevis): Short-term activation of Na+, Cl (-) and K+ channels. *Pflugers Arch.* **2008**, *456*, 1109–1120. [CrossRef] [PubMed]

49. Richter, K.; Kiefer, K.P.; Grzesik, B.A.; Clauss, W.G.; Fronius, M. Hydrostatic pressure activates ATP-sensitive K+ channels in lung epithelium by ATP release through pannexin and connexin hemichannels. *FASEB J.* **2014**, *28*, 45–55. [CrossRef]
50. Tokuda, S.; Niisato, N.; Nagai, T.; Taruno, A.; Nakajima, K.; Miyazaki, H.; Yamada, T.; Hosogi, S.; Ohta, M.; Nishio, K.; et al. Regulation of paracellular Na+ and Cl(-) conductances by hydrostatic pressure. *Cell Biol. Int.* **2009**, *33*, 949–956. [CrossRef]
51. Tokuda, S.; Miyazaki, H.; Nakajima, K.; Yamada, T.; Marunaka, Y. Hydrostatic pressure regulates tight junctions, actin cytoskeleton and transcellular ion transport. *Biochem. Biophys. Res. Commun.* **2009**, *390*, 1315–1321. [CrossRef]
52. Coers, W.; Vos, J.T.; Huitema, S.; Dijk, F.; Weening, J.J. Biological alterations of rat podocytes cultured under basolateral hydrostatic pressure. *Pathobiology* **1996**, *64*, 222–232. [CrossRef] [PubMed]
53. Reeves, W.; Caulfield, J.P.; Farquhar, M.G. Differentiation of epithelial foot processes and filtration slits: Sequential appearance of occluding junctions, epithelial polyanion, and slit membranes in developing glomeruli. *Lab. Invest.* **1978**, *39*, 90–100. [PubMed]
54. Stiffler, D.F.; Thornburg, K.L.; Swanson, R.E. Structural and functional responses of the bullfrog urinary bladder to distension caused by hydrostatic pressure gradients. *Arch. Physiol. Biochem.* **2000**, *108*, 405–414. [CrossRef] [PubMed]
55. Ferguson, D.R.; Kennedy, I.; Burton, T.J. ATP is released from rabbit urinary bladder epithelial cells by hydrostatic pressure changes–a possible sensory mechanism? *J. Physiol.* **1997**, *505*, 503–511. [CrossRef] [PubMed]
56. Wang, E.C.; Lee, J.M.; Johnson, J.P.; Kleyman, T.R.; Bridges, R.; Apodaca, G. Hydrostatic pressure-regulated ion transport in bladder uroepithelium. *Am. J. Physiol. Renal Physiol.* **2003**, *285*, 651–663. [CrossRef] [PubMed]
57. Mießler, K.S.; Vitzthum, C.; Markov, A.G.; Amasheh, S. Basolateral pressure challenges mammary epithelial cell monolayer integrity, in vitro. *Cytotechnology* **2018**, *70*, 567–576. [CrossRef]
58. Mießler, K.S.; Markov, A.G.; Amasheh, S. Hydrostatic pressure incubation affects barrier properties of mammary epithelial cell monolayers, in vitro. *Biochem. Biophys. Res. Commun.* **2018**, *495*, 1089–1093. [CrossRef]
59. Eldrup, E.; Frederiksen, O.; Møllgård, K.; Rostgaard, J. Effects of a small serosal hydrostatic pressure on sodium and water transport and morphology in rabbit gall-bladder. *J. Physiol.* **1982**, *331*, 67–85. [CrossRef]
60. Toyota, N.; Miyai, K.; Hardison, W.G. Effect of biliary pressure versus high bile acid flux on the permeability of hepatocellular tight junction. *Lab. Invest.* **1984**, *50*, 536–542.
61. Berg, S.; Wollmer, P.; Andersson, M.; Persson, C.G.; Greiff, L. Effects of experimental changes in nasal airway pressure on mucosal output of plasma. *Clin. Physiol. Funct. Imaging.* **2003**, *23*, 155–158. [CrossRef]
62. Bank, N.; Koch, K.M.; Aynedjian, H.S.; Aras, M. Effect of changes in renal perfusion pressure on the suppression of proximal tubular sodium reabsorption due to saline loading. *J. Clin. Invest.* **1969**, *48*, 271–283. [CrossRef] [PubMed]
63. Gorodeski, G.I. The cultured human cervical epithelium: A new model for studying paracellular transport. *J. Soc. Gynecol. Investig.* **1996**, *3*, 267–280. [CrossRef] [PubMed]
64. DeMaio, L.; Tarbell, J.M.; Scaduto, R.C., Jr.; Gardner, T.W.; Antonetti, D.A. A transmural pressure gradient induces mechanical and biological adaptive responses in endothelial cells. *Am. J. Physiol. Heart Circ. Physiol.* **2004**, *286*, 731–741. [CrossRef] [PubMed]
65. Stamer, W.D.; Roberts, B.C.; Epstein, D.L. Hydraulic pressure stimulates adenosine 3′,5′-cyclic monophosphate accumulation in endothelialcells from Schlemm's canal. *Invest. Ophthalmol. Vis. Sci.* **1999**, *40*, 1983–1988. [PubMed]
66. Lei, Y.; Stamer, W.D.; Wu, J.; Sun, X. Cell senescence reduced the mechanotransduction sensitivity of porcine angular aqueous plexus cells to elevation of pressure. *Invest. Ophthalmol. Vis. Sci.* **2014**, *55*, 2324–2328. [CrossRef] [PubMed]
67. Larsson, B.; Skärby, T.; Edvinsson, L.; Hardebo, J.E.; Owman, C. Vincristine reduces damage of the blood–brain barrier induced by high intravascular pressure. *Neurosci. Lett.* **1980**, *17*, 155–159. [CrossRef]
68. Tsukimoto, K.; Mathieu-Costello, O.; Prediletto, R.; Elliott, A.R.; West, J.B. Ultrastructural appearances of pulmonary capillaries at high transmural pressures. *J. Appl. Physiol.* **1991**, *71*, 573–582. [CrossRef] [PubMed]

69. Patel, M.; Chignalia, A.Z.; Isbatan, A.; Bommakanti, N.; Dull, R.O. Ropivacaine inhibits pressure-induced lung endothelial hyperpermeability in models of acute hypertension. *Life Sci.* **2019**, *222*, 22–28. [CrossRef]
70. Kuebler, W.M.; Ying, X.; Bhattacharya, J. Pressure-induced endothelial Ca(2+) oscillations in lung capillaries. *Am. J. Physiol. Lung Cell Mol. Physiol.* **2002**, *282*, 917–923. [CrossRef]
71. Gan, L.; Doroudi, R.; Hägg, U.; Johansson, A.; Selin-Sjögren, L.; Jern, S. Differential immediate-early gene responses to shear stress and intraluminal pressure in intact human conduit vessels. *FEBS Lett.* **2000**, *477*, 89–94. [CrossRef]
72. Riou, S.; Mees, B.; Esposito, B.; Merval, R.; Vilar, J.; Stengel, D.; Ninio, E.; van Haperen, R.; de Crom, R.; Tedgui, A.; et al. High pressure promotes monocyte adhesion to the vascular wall. *Circ. Res.* **2007**, *100*, 1226–1233. [CrossRef]
73. Prystopiuk, V.; Fels, B.; Simon, C.S.; Liashkovich, I.; Pasrednik, D.; Kronlage, C.; Wedlich-Söldner, R.; Oberleithner, H.; Fels, J. A two-phase response of endothelial cells to hydrostatic pressure. *J. Cell Sci.* **2018**, *131*, jcs206920. [CrossRef]
74. Ohashi, T.; Sugaya, Y.; Sakamoto, N.; Sato, M. Hydrostatic pressure influences morphology and expression of VE-cadherin of vascular endothelial cells. *J. Biomech.* **2007**, *40*, 2399–2405. [CrossRef]
75. Thoumine, O.; Nerem, R.M.; Girard, P.R. Oscillatory shear stress and hydrostatic pressure modulate cell-matrix attachment proteins in cultured endothelial cells. *In Vitro Cell Dev. Biol. Anim.* **1995**, *31*, 45–54. [CrossRef]
76. Hasel, C.; Dürr, S.; Bauer, A.; Heydrich, R.; Brüderlein, S.; Tambi, T.; Bhanot, U.; Möller, P. Pathologically elevated cyclic hydrostatic pressure induces CD95-mediated apoptotic cell death in vascular endothelial cells. *Am. J. Physiol. Cell Physiol.* **2005**, *289*, 312–322. [CrossRef]
77. Shin, H.Y.; Smith, M.L.; Toy, K.J.; Williams, P.M.; Bizios, R.; Gerritsen, M.E. VEGF-C mediates cyclic pressure-induced endothelial cell proliferation. *Physiol. Genomics* **2002**, *11*, 245–251. [CrossRef]
78. Shin, H.Y.; Underwood, R.M.; Fannon, M.W. Fluid pressure is a magnitude-dependent modulator of early endothelial tubulogenic activity: Implications related to a potential tissue-engineering control parameter. *Tissue Eng. Part A* **2012**, *18*, 2590–2600. [CrossRef]
79. Salwen, S.A.; Szarowski, D.H.; Turner, J.N.; Bizios, R. Three-dimensional changes of the cytoskeleton of vascular endothelial cells exposed to sustained hydrostatic pressure. *Med. Biol. Eng. Comput.* **1998**, *36*, 520–527. [CrossRef]
80. Acevedo, A.D.; Bowser, S.S.; Gerritsen, M.E.; Bizios, R. Morphological and proliferative responses of endothelial cells to hydrostatic pressure: Role of fibroblast growth factor. *J. Cell Physiol.* **1993**, *157*, 603–614. [CrossRef]
81. Schwartz, E.A.; Bizios, R.; Medow, M.S.; Gerritsen, M.E. Exposure of human vascular endothelial cells to sustained hydrostatic pressure stimulates proliferation. Involvement of the alphaV integrins. *Circ. Res.* **1999**, *84*, 315–322. [CrossRef]
82. Tworkoski, E.; Glucksberg, M.R.; Johnson, M. The effect of the rate of hydrostatic pressure depressurization on cells in culture. *PLoS ONE* **2018**, *13*, e0189890. [CrossRef]
83. Lund, T.; Wiig, H.; Reed, R.K. Acute postburn edema: Role of strongly negative interstitial fluid pressure. *Am. J. Physiol.* **1988**, *255*, 1069–1074. [CrossRef]
84. Bhattacharya, J.; Gropper, M.A.; Shepard, J.M. Lung expansion and the perialveolar interstitial pressure gradient. *J. Appl. Physiol.* **1989**, *66*, 2600–2605. [CrossRef]
85. Reed, R.K.; Rubin, K. Transcapillary exchange: Role and importance of the interstitial fluid pressure and the extracellular matrix. *Cardiovasc. Res.* **2010**, *87*, 211–217. [CrossRef]
86. Fjaertoft, M.; Johannessen, A.C.; Heyeraas, K.J. Micropuncture measurements of interstitial fluid pressure in normal and inflamed gingiva in rats. *J. Periodontal Res.* **1992**, *27*, 534–538. [CrossRef]
87. Tønder, K.J.; Kvinnsland, I. Micropuncture measurements of interstitial fluid pressure in normal and inflamed dental pulp in cats. *J. Endod.* **1983**, *9*, 105–109. [CrossRef]
88. Young, J.S.; Lumsden, C.E.; Stalker, A.L. The significance of the tissue pressure of normal testicular and of neoplastic (Brown-Pearce carcinoma) tissue in the rabbit. *J. Pathol. Bacteriol.* **1950**, *62*, 313–333. [CrossRef]
89. Boucher, Y.; Baxter, L.T.; Jain, R.K. Interstitial pressure gradients in tissue-isolated and subcutaneous tumors: Implications for therapy. *Cancer Res.* **1990**, *50*, 4478–4484.
90. Boucher, Y.; Jain, R.K. Microvascular pressure is the principal driving force for interstitial hypertension in solid tumors: Implications for vascular collapse. *Cancer Res.* **1992**, *52*, 5110–5114.
91. Roh, H.D.; Boucher, Y.; Kalnicki, S.; Buchsbaum, R.; Bloomer, W.D.; Jain, R.K. Interstitial hypertension in carcinoma of uterine cervix in patients: Possible correlation with tumor oxygenation and radiation response.

Cancer Res. **1991**, *51*, 6695–6698.
92. Boucher, Y.; Kirkwood, J.M.; Opacic, D.; Desantis, M.; Jain, R.K. Interstitial hypertension in superficial metastatic melanomas in humans. *Cancer Res.* **1991**, *51*, 6691–6694.
93. Less, J.R.; Posner, M.C.; Boucher, Y.; Borochovitz, D.; Wolmark, N.; Jain, R.K. Interstitial hypertension in human breast and colorectal tumors. *Cancer Res.* **1992**, *52*, 6371–6374.
94. Gutmann, R.; Leunig, M.; Feyh, J.; Goetz, A.E.; Messmer, K.; Kastenbauer, E.; Jain, R.K. Interstitial hypertension in head and neck tumors in patients: Correlation with tumor size. *Cancer Res.* **1992**, *52*, 1993–1995.
95. Curti, B.D.; Urba, W.J.; Alvord, W.G.; Janik, J.E.; Smith, J.W., 2nd; Madara, K.; Longo, D.L. Interstitial pressure of subcutaneous nodules in melanoma and lymphoma patients: Changes during treatment. *Cancer Res.* **1993**, *53*, 2204–2207.
96. Nathanson, S.D.; Nelson, L. Interstitial fluid pressure in breast cancer, benign breast conditions, and breast parenchyma. *Ann. Surg. Oncol.* **1994**, *1*, 333–338. [CrossRef]
97. Heldin, C.H.; Rubin, K.; Pietras, K.; Ostman, A. High interstitial fluid pressure—An obstacle in cancer therapy. *Nat. Rev. Cancer* **2004**, *4*, 806–813. [CrossRef]
98. Lunt, S.J.; Fyles, A.; Hill, R.P.; Milosevic, M. Interstitial fluid pressure in tumors: Therapeutic barrier and biomarker of angiogenesis. *Future Oncol.* **2008**, *4*, 793–802. [CrossRef]
99. Milosevic, M.; Fyles, A.; Hedley, D.; Pintilie, M.; Levin, W.; Manchul, L.; Hill, R. Interstitial fluid pressure predicts survival in patients with cervix cancer independent of clinical prognostic factors and tumor oxygen measurements. *Cancer Res.* **2001**, *61*, 6400–6405.
100. Fyles, A.; Milosevic, M.; Pintilie, M.; Syed, A.; Levin, W.; Manchul, L.; Hill, R.P. Long-term performance of interstial fluid pressure and hypoxia as prognostic factors in cervix cancer. *Radiother. Oncol.* **2006**, *80*, 132–137. [CrossRef]
101. Coussens, L.M.; Werb, Z. Inflammation and cancer. *Nature* **2002**, *420*, 860–867. [CrossRef]
102. Hanahan, D.; Weinberg, R.A. Hallmarks of cancer: The next generation. *Cell* **2011**, *144*, 646–674. [CrossRef] [PubMed]
103. Diresta, G.R.; Nathan, S.S.; Manoso, M.W.; Casas-Ganem, J.; Wyatt, C.; Kubo, T.; Boland, P.J.; Athanasian, E.A.; Miodownik, J.; Gorlick, R.; et al. Cell proliferation of cultured human cancer cells are affected by the elevated tumor pressuresthat exist in vivo. *Ann. Biomed. Eng.* **2005**, *33*, 1270–1280. [CrossRef] [PubMed]
104. Nathan, S.S.; DiResta, G.R.; Casas-Ganem, J.E.; Hoang, B.H.; Sowers, R.; Yang, R.; Huvos, A.G.; Gorlick, R.; Healey, J.H. Elevated physiologic tumor pressure promotes proliferation and chemosensitivity in human osteosarcoma. *Clin. Cancer Res.* **2005**, *11*, 2389–2397. [CrossRef]
105. Nathan, S.S.; Huvos, A.G.; Casas-Ganem, J.E.; Yang, R.; Linkov, I.; Sowers, R.; DiResta, G.R.; Gorlick, R.; Healey, J.H. Tumor interstitial fluid pressure may regulate angiogenic factors in osteosarcoma. *J. Orthop. Res.* **2008**, *26*, 1520–1525. [CrossRef] [PubMed]
106. Aung, K.Z.; Pereira, B.P.; Tan, P.H.; Han, H.C.; Nathan, S.S. Interstitial fluid pressure as an alternate regulator of angiogenesis independent of hypoxia driven HIF-1α in solid tumors. *J. Orthop. Res.* **2012**, *30*, 2038–2045. [CrossRef]
107. Kao, Y.C.; Lee, C.H.; Kuo, P.L. Increased hydrostatic pressure enhances motility of lung cancer cells. *Conf. Proc. IEEE Eng. Med. Biol. Soc.* **2014**, *2014*, 2928–2931. [CrossRef]
108. Kao, Y.C.; Jheng, J.R.; Pan, H.J.; Liao, W.Y.; Lee, C.H.; Kuo, P.L. Elevated hydrostatic pressure enhances the motility and enlarges the size of the lung cancercells through aquaporin upregulation mediated by caveolin-1 and ERK1/2 signaling. *Oncogene* **2017**, *36*, 863–874. [CrossRef]
109. Chen, S.K.; Chung, C.A.; Cheng, Y.C.; Huang, C.J.; Ruaan, R.C.; Chen, W.Y.; Li, C.; Tsao, C.W.; Hu, W.W.; Chien, C.C. Hydrostatic pressure enhances mitomycin C induced apoptosis in urothelial carcinoma cells. *Urol. Oncol.* **2014**, *32*, 17–24. [CrossRef]
110. Chen, S.K.; Chung, C.A.; Cheng, Y.C.; Huang, C.J.; Chen, W.Y.; Ruaan, R.C.; Li, C.; Tsao, C.W.; Hu, W.W.; Chien, C.C. Toll-like receptor 6 and connective tissue growth factor are significantly upregulated in mitomycin-C-treated urothelial carcinoma cells under hydrostatic pressure stimulation. *Genet. Test Mol. Biomarkers* **2014**, *18*, 410–416. [CrossRef]
111. Adkins, I.; Hradilova, N.; Palata, O.; Sadilkova, L.; Palova-Jelinkova, L.; Spisek, R. High hydrostatic pressure in cancer immunotherapy and biomedicine. *Biotechnol. Adv.* **2018**, *36*, 577–582. [CrossRef]
112. Shinitzky, M.; Goldman, Y. Immunotherapy of cancer with pressure modified cells. *Isr. Med. Assoc. J.* **2000**, *2*, 615–620. [PubMed]
113. Tokuda, S.; Kim, Y.H.; Matsumoto, H.; Muro, S.; Hirai, T.; Mishima, M.; Furuse, M. Effects of Hydrostatic Pressure on Carcinogenic Properties of Epithelia. *PLoS ONE* **2015**, *10*, e0145522. [CrossRef] [PubMed]

114. Schoenenberger, C.A.; Zuk, A.; Kendall, D.; Matlin, K.S. Multilayering and loss of apical polarity in MDCK cells transformed with viral K-ras. *J. Cell Biol.* **1991**, *112*, 873–889. [CrossRef] [PubMed]
115. Li, D.; Mrsny, R.J. Oncogenic Raf-1 disrupts epithelial tight junctions via downregulation of occludin. *J. Cell Biol.* **2000**, *148*, 791–800. [CrossRef] [PubMed]
116. Pitelka, D.R.; Hamamoto, S.T.; Taggart, B.N. Epithelial cell junctions in primary and metastatic mammary tumors of mice. *Cancer Res.* **1980**, *40*, 1588–1599. [PubMed]
117. Helmstein, K. Treatment of bladder carcinoma by a hydrostatic pressure technique. Report on 43 cases. *Br. J. Urol.* **1972**, *44*, 434–450. [CrossRef] [PubMed]
118. Glashan, R.W. A critical review of the management of bladder neoplasia using a modified form of Helmstein's pressure therapy. *Br. J. Urol.* **1975**, *47*, 57–66. [CrossRef]
119. Hirose, K.; Seto, T.; Takayasu, H. Re-evaluation of hydrostatic pressure treatment for malignant bladder lesions. *J. Urol.* **1977**, *118*, 762–764. [CrossRef]
120. Debré, B.; Steg, A. Hydrostatic distension of the bladder: Technique and indications. 89 cases (author's transl). *Nouv. Presse. Med.* **1981**, *10*, 21–24.
121. Antonsen, H.K.; Lose, G.; Højensgård, J.C. The Helmstein bladder distension treatment for tumours and severe bleeding. *Int. Urol. Nephrol.* **1986**, *18*, 421–427. [CrossRef]
122. Tong, R.T.; Boucher, Y.; Kozin, S.V.; Winkler, F.; Hicklin, D.J.; Jain, R.K. Vascular normalization by vascular endothelial growth factor receptor 2 blockade induces a pressure gradient across the vasculature and improves drug penetration in tumors. *Cancer Res.* **2004**, *64*, 3731–3736. [CrossRef] [PubMed]
123. Skliarenko, J.V.; Lunt, S.J.; Gordon, M.L.; Vitkin, A.; Milosevic, M.; Hill, R.P. Effects of the vascular disrupting agent ZD6126 on interstitial fluid pressure and cell survival in tumors. *Cancer Res.* **2006**, *66*, 2074–2080. [CrossRef] [PubMed]
124. Pietras, K.; Stumm, M.; Hubert, M.; Buchdunger, E.; Rubin, K.; Heldin, C.H.; McSheehy, P.; Wartmann, M.; Ostman, A. STI571 enhances the therapeutic index of epothilone B by a tumor-selective increase of drug uptake. *Clin. Cancer Res.* **2003**, *9*, 3779–3787. [PubMed]
125. Leunig, M.; Goetz, A.E.; Dellian, M.; Zetterer, G.; Gamarra, F.; Jain, R.K.; Messmer, K. Interstitial fluid pressure in solid tumors following hyperthermia: Possible correlation with therapeutic response. *Cancer Res.* **1992**, *52*, 487–490. [PubMed]
126. Ozerdem, U.; Hargens, A.R. A simple method for measuring interstitial fluid pressure in cancer tissues. *Microvasc. Res.* **2005**, *70*, 116–120. [CrossRef] [PubMed]
127. Hori, K.; Suzuki, M.; Saito, S.; Tanda, S.; Zhang, Q.H.; Li, H.C. Changes in vessel pressure and interstitial fluid pressure of normal subcutis and subcutaneous tumor in rats due to angiotensin II. *Microvasc. Res.* **1994**, *48*, 246–256. [CrossRef] [PubMed]
128. Furuse, M.; Sasaki, H.; Fujimoto, K.; Tsukita, S. A single gene product, claudin-1 or -2, reconstitutes tight junction strands and recruits occludin in fibroblasts. *J. Cell Biol.* **1998**, *143*, 391–401. [CrossRef]
129. Mineta, K.; Yamamoto, Y.; Yamazaki, Y.; Tanaka, H.; Tada, Y.; Saito, K.; Tamura, A.; Igarashi, M.; Endo, T.; Takeuchi, K.; et al. Barrier function of epithelia. *FEBS Lett.* **2011**, *585*, 606–612. [CrossRef]
130. Tsukita, S.; Furuse, M.; Itoh, M. Multifunctional strands in tight junctions. *Nat. Rev. Mol. Cell Biol.* **2001**, *2*, 285–293. [CrossRef]
131. Schneeberger, E.E.; Lynch, R.D. The tight junction: A multifunctional complex. *Am. J. Physiol. Cell Physiol.* **2004**, *286*, 1213–1228. [CrossRef]
132. Van Itallie, C.M.; Anderson, J.M. Architecture of tight junctions and principles of molecular composition. *Semin. Cell Dev. Biol.* **2014**, *36*, 157–165. [CrossRef] [PubMed]
133. González-Mariscal, L.; Domínguez-Calderón, A.; Raya-Sandino, A.; Ortega-Olvera, J.M.; Vargas-Sierra, O.; Martínez-Revollar, G. Tight junctions and the regulation of gene expression. *Semin. Cell Dev. Biol.* **2014**, *36*, 213–223. [CrossRef] [PubMed]
134. Zihni, C.; Mills, C.; Matter, K.; Balda, M.S. Tight junctions: From simple barriers to multifunctional molecular gates. *Nat. Rev. Mol. Cell Biol.* **2016**, *17*, 564–580. [CrossRef] [PubMed]
135. Turksen, K.; Troy, T.C. Junctions gone bad: Claudins and loss of the barrier in cancer. *Biochim. Biophys. Acta* **2011**, *1816*, 73–79. [CrossRef] [PubMed]
136. Bhat, A.A.; Uppada, S.; Achkar, I.W.; Hashem, S.; Yadav, S.K.; Shanmugakonar, M.; Al-Naemi, H.A.; Haris, M.; Uddin, S. Tight Junction Proteins and Signaling Pathways in Cancer and Inflammation: A Functional Crosstalk. *Front. Physiol.* **2019**, *9*, 1942. [CrossRef] [PubMed]

6

ZO-2 is a Master Regulator of Gene Expression, Cell Proliferation, Cytoarchitecture and Cell Size

Lorenza González-Mariscal *, Helios Gallego-Gutiérrez, Laura González-González and Christian Hernández-Guzmán

Center for Research and Advanced Studies (Cinvestav), Department of Physiology, Biophysics and Neuroscience, Mexico City 07360, Mexico

* Correspondence: lorenza@fisio.cinvestav.mx

Abstract: ZO-2 is a cytoplasmic protein of tight junctions (TJs). Here, we describe ZO-2 involvement in the formation of the apical junctional complex during early development and in TJ biogenesis in epithelial cultured cells. ZO-2 acts as a scaffold for the polymerization of claudins at TJs and plays a unique role in the blood–testis barrier, as well as at TJs of the human liver and the inner ear. ZO-2 movement between the cytoplasm and nucleus is regulated by nuclear localization and exportation signals and post-translation modifications, while ZO-2 arrival at the cell border is triggered by activation of calcium sensing receptors and corresponding downstream signaling. Depending on its location, ZO-2 associates with junctional proteins and the actomyosin cytoskeleton or a variety of nuclear proteins, playing a role as a transcriptional repressor that leads to inhibition of cell proliferation and transformation. ZO-2 regulates cell architecture through modulation of Rho proteins and its absence induces hypertrophy due to inactivation of the Hippo pathway and activation of mTOR and S6K. The interaction of ZO-2 with viral oncoproteins and kinases and its silencing in diverse carcinomas reinforce the view of ZO-2 as a tumor regulator protein.

Keywords: tight junctions; ZO-2; cholestasis; gene transcription; hypertrophy; tumor suppressor; NLS; NES; CaSR; RhoA

---

1. Introduction

Zonula occludens proteins ZO-1, ZO-2, and ZO-3 belong to the membrane associated guanylate kinase homologue (MAGUK) protein family and are concentrated at the cytoplasmic face of tight junctions (TJs) in epithelial cells. Although these proteins share a core of common structural and functional domains and interacting partners, they also exhibit particular features that give rise to certain non-redundant functions among these proteins (for review see [[1,2]]). This review concentrates on ZO-2 and its non-canonical role in gene transcription, cell proliferation, and modulation of cell size, cytoarchitecture, and cancer.

ZO-2 is a 160 kDa protein, first identified by its co-immunoprecipitation with ZO-1 in epithelial cells [3]. ZO-2 is encoded by the *TJP2* gene located on human chromosome 9 q21.11 [4]. ZO-2 is present at TJs but in non-epithelial cells like fibroblasts that lack TJs, ZO-2 concentrates at adherens junctions (AJs) [5]. In cardiac muscle cells, the observations are contradictory. Some report the presence of ZO-2 in co-localization with ZO-1 at specialized intercellular junctions, known as fascia adherens or intercalated discs, which connect the opposing ends of cardiac muscle cells [5]. Others indicate that only ZO-1 is present at fascia adherens [6] and that ZO-2 has a diffuse cytoplasmic distribution in myocardium tissue [7].

ZO-2 is a scaffold protein, whose amino segment, containing PDZ1-3-SH3-GuK domains, binds to integral and peripheral proteins of the TJs, including occludin, claudins, JAM-A, cingulin and ZO-1, to proteins of the AJs, like α-catenin and β-catenin, and to gap junction connexins (for review see [8]).

Instead, the carboxyl segment of ZO-2, which exhibits the acidic and proline rich regions and ends with a motif that binds PDZ (PSD95, Dlg1 and ZO-1) domains, distributes when separately introduced into epithelial cells, along actin filaments [5] (Figure 1).

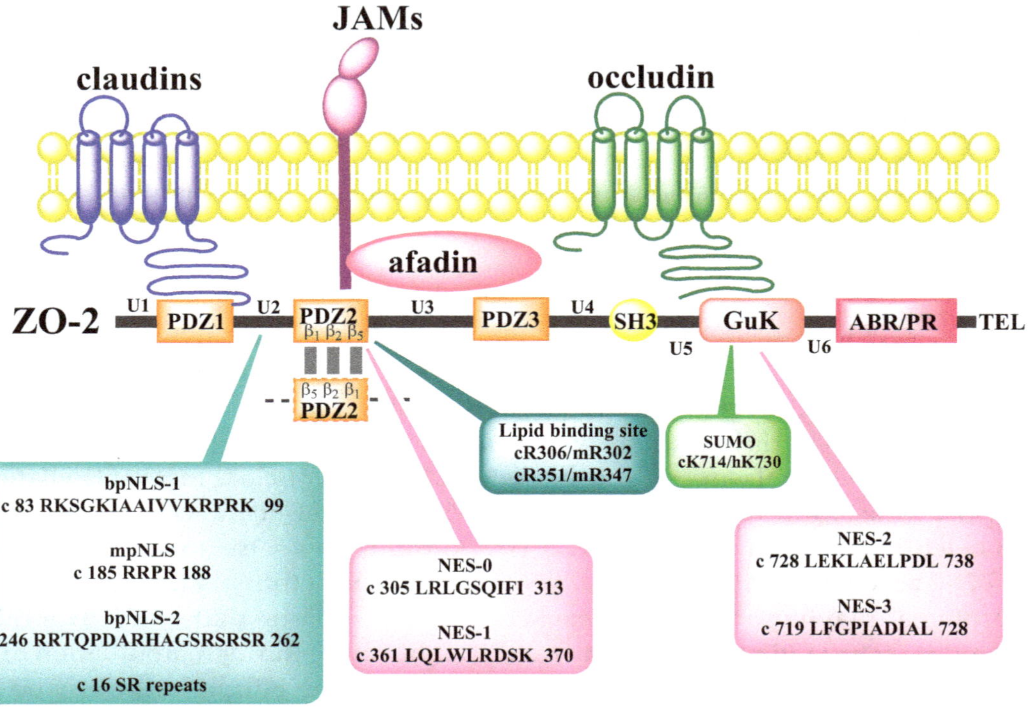

Figure 1. ZO-2 molecular organization and interactions with integral tight junction (TJ) proteins at the plasma membrane. ZO-2 domains (PDZ, SH3, and GuK), regions (U, unique; ABR, actin binding; PR, proline rich), and PDZ-binding motif (TEL) are indicated, as well as the nuclear localization signals (NLS) and exportation signals (NES), SUMOylation (SUMO) and lipid binding sites, and dimerization region. The ZO-2 sequence is identified by letters: c, canine; m, mouse; h, human. Numbers correspond to amino acids. Clusters of basic amino acids (K/R) in the bpNLS are shown in red. bp, bipartite; mp, monopartite.

PDZ1-3 modules, SH3 and GuK domains, and the acidic region of ZO-2 display a high percent of identity and similarity to those in other ZO proteins, with ZO-1 having a higher percent of both than with ZO-3 [9]. In situ ZO-2 is present as a ZO-1/ZO-2 complex, but not in ZO-2/ZO-3 or ZO-1/ZO-2/ZO-3 complexes [10]. The high conservation present between PDZ2 domains in ZO proteins allows these PDZ domains to dimerize via three-dimensional domain swapping, generating heterodimers of ZO-1-PDZ2/ZO-2-PDZ2 and ZO-1-PDZ2/ZO-3-PDZ2 domains only, as well as homodimers of ZO-1-PDZ2 and ZO-2-PDZ2 domains only [11]. Structural analysis of ZO-2 PDZ2 revealed that it has five β sheets and two α helices and that ZO-2-PDZ2 homodimers form by the interaction of three antiparallel β sheets, β1-β5′, β1′-β5, and β2-β2′, due to extensive inter-subunit hydrogen bonds and hydrophobic interactions. In addition, chemical crosslinking and dynamic laser light scatter experiments revealed that ZO-1-PDZ2 and ZO-2-PDZ2 form oligomers in solution. This oligomerization mediated by PDZ2 domains in ZO-1/ZO-2 proteins might provide a scaffold for the assembly of TJs.

Both ZO-1 and ZO-2 independently allow the polymerization of claudins and determine the site of TJ strand formation [12]. Thus, epithelial cells lack TJs when ZO-1 and ZO-2 expression is suppressed, and when either of these proteins is exogenously expressed, claudins polymerize and TJ filaments are again observed in freeze-fracture replicas. However, when a truncated segment of ZO-1 was introduced containing only the PDZ1-3 domains, it localized in the cytoplasm, not in the membrane, and the claudins did not polymerize. However, when a longer construct consisting of PDZ1-3 and SH3–GuK domains was introduced, TJ strands formed. The importance of the SH3–GuK segment is highlighted by the fact that it also plays a role in the dimerization of MAGUK proteins and binds to the AJ proteins afadin and α-catenin, allowing the recruitment of ZO-1 to the proximity of the plasma membrane. If in ZO-1 knock out (KO)/ZO-2 knock down (KD) cells the amino PDZ1-3 segment of ZO-1 was forcibly recruited to the membrane by the addition of a myristoylation signal, no claudin polymerization happened. Instead, if this construct was allowed to dimerize by the introduction of a FK506-binding protein (FKBPv) signal and a homodimerizer, claudin polymerization was abundant throughout the lateral membrane, highlighting the crucial importance of ZO-1 dimerization for claudin polymerization. Although the latter experiments were done with ZO-1, the same results were expected to occur with ZO-2, since their MAGUK domains are so similar. Taken together, these observations suggest that claudin binding through the PDZ1 domain of ZO-1 or ZO-2, as well as the SH3–GuK mediated recruitment of these proteins to the AJ and the dimerization of ZO-1 and ZO-2, allow the polymerization of claudins at the TJ region.

2. ZO-2 During Early Development

In mouse preimplantation embryos, ZO-2 assembles at blastomere junctional sites at the 16-cell stage, at the same time as cingulin but later than ZO-1α^-, rab13, and the Par-3/Par-6/aPKC complex, which reach the junctions after the compaction of 8-cell morulae. At the 16-cell stage, ZO-2 co-localizes with E-cadherin in a transient complex that contains proteins of both the TJ and the AJ. From the 32-cell stage onwards, ZO-2 and the rest of the TJ proteins, also including occludin and claudins, separate from those of the AJ and a blastocoel cavity is formed [13] (for review see [14]). Lack of ZO-2 in the trophectoderm of mouse preimplantation embryos delays the formation of the blastocoel cavity and induces an increased assembly of ZO-1, suggesting a compensatory mechanism of ZO proteins. In accordance, the double KD of ZO-1 and ZO-2 induces a more severe inhibition of blastocoel formation [13]. Early after implantation, KO of ZO-2 is lethal and mouse embryos die, showing decreased proliferation at embryonic day 6.5 (E6.5) and increased apoptosis at E7.5 [15]. However, embryonic lethality of ZO-2 KO mice was rescued by injecting ZO-2 KO embryonic stem cells into wild type blastocysts to generate viable ZO-2 chimeras, hence indicating that ZO-2 is crucial for extraembryonic tissue rather than for the development of the embryo *per se* [16].

3. ZO-2 Presence at TJs Is Crucial in the Mouse Testis and the Human Liver and Inner Ear

ZO-2 is present in the TJs of all epithelial cells and has a role on the barrier function of TJs, as ZO-2 KD cells achieve a much lower peak value of transepithelial electrical resistance (TER) than parental cells [17,18]. Therefore, the observation that the absence of ZO-2 causes severe barrier impairment in mouse testes and human liver and inner ear tissue suggests that compensatory mechanisms, like the over-expression of ZO-1, are not effective in these tissues, where ZO-2 plays a critical and non-redundant role (Table 1).

Table 1. ZO-2 mutations and consequences.

Species	Disease	Alteration	Location within ZO-2	Features of Disease	Reference
Human	ADNSHL	A112T	PDZ1		[19]
		T1188A	TEL		
		G694E	GuK		[20]
		Genomic duplication at chromosome 9p13.3-q21.13			[4]
	Familial hypercholanemia	V48A	PDZ1	Patients with itching, elevated bile acid concentration and fat malabsorption. TJP2 mutation combined with bile acid Coenzyme A (BAAT) mutation	[21]
	Intrahepatic cholestasis	I875T	GuK	71-year-old patient	[22]
		R322W	PDZ2	Disease since early infancy	
		A256T	ZO-2 deficiency		[23]
		S296A			
		Y261S			
		A454G			
		A664S			
		Q318G			
		A632fs			
		S1136A			
		T880S			[24]
		N814Q		Disease since early infancy that leads to hepatocellular carcinoma	[25]
		A273fs			[26]
		Exon 18 deletion			[27]
		T62M	PDZ1	Intrahepatic cholestasis of pregnancy (ICP)	[28]
		T626S	SH3		
Mouse		ZO-2 KO		Reduced male fertility	[16]

ADNSHL: Autosomal dominant non-syndromic hearing loss.

3.1. *ZO-2 and the Blood–Testis Barrier*

TJs form the blood–testis barrier (BTB) at the lowermost portion of the lateral membrane of Sertoli cells. This barrier divides the paracellular space in two compartments: The basal compartment, which contains diploid spermatogonia, and the adluminal compartment, which contains differentiating spermatocytes and haploid spermatids. The BTB, constituted by occludin, claudin-3 and -11, ZO-1, and ZO-2, is critical for the maintenance of a proper microenvironment for germinal cells and to prevent autoimmunity (for review see [29]). Male ZO-2 chimera mice generated from wild type blastocysts, in which ZO-2 KO embryonic stem cells substituted the normal stem cells, showed reduced fertility, smaller testes, degeneration of seminiferous tubules, and permeable BTB, even when the other TJ proteins were present at the BTB. This suggests that ZO-2 plays a unique and critical role in the BTB [16].

Moreover, the organophosphate pesticide methamidophos (MET), which is widely used in agriculture in developing countries and is known to cause adverse effects in human male reproductive function, opens the BTB of mice and perturbs spermatogenesis. These changes correlate with MET formation of covalent bonds with ZO-2 in serine, tyrosine, and lysine residues. MET bonds formed in ZO-2 are proposed to interfere with ZO-2 degradation and TJ sealing, since cultured Madin-Darby canine kidney (MDCK) cells transfected with ZO-2 mutated at MET target sites developed lower values of TER [30].

3.2. *ZO-2 in the Liver*

In the liver, TJs present in hepatocytes and cholangiocytes in intrahepatic canaliculi prevent leakage of biliary components through the paracellular space and into the liver parenchyma. Familial hypercholanemia is a rare disease characterized by elevated serum bile acid concentration, pruritus, and fat malabsorption. In Amish individuals with this condition, the missense mutation V48A in the first PDZ domain of TJP2 was identified. This mutation reduces PDZ1 domain stability and binding to the C-terminal domain peptides of claudin-1, -2, -3, -5, and -7. In some individuals, this mutation is associated with an additional missense mutation on the gene for bile acid Coenzyme A (BAAT), a bile-conjugating enzyme. In a patient homozygous for the *TJP2* mutation and heterozygous with respect to the *BAAT* mutation, TJ depth viewed by transmission electron microscopy was greater than in control individuals. Considering that greater TJ depth was observed in liver disease associated with greater TJ permeability, it was concluded that the missense mutation in *TJP2* generated leakier TJs in the liver, which affected bile acid localization [21].

In children with familial intrahepatic cholestasis, a condition where bile does not flow through the intrahepatic ducts of the liver, homozygous protein truncating mutations in *TJP2* leading to a complete absence of the protein were detected. In this case, the absence of ZO-2 is accompanied by a failure of localization of claudin-1, but not of claudin-2, to the canalicular membrane [31]. Another case of compound heterozygous mutations in *TJP2* predicted to abolish TJ protein translation was observed in a young child with familial intrahepatic cholestasis [24]. Additional cases of infants with progressive intrahepatic cholestasis that led to chronic cholestatic hepatitis with cirrhosis and hepatocellular carcinoma were also reported. In these cases, homozygous and heterozygous *TJP2* mutations were present, ZO-2 expression was absent, and claudin-1 expression was decreased [25–27].

Intrahepatic cholestasis of pregnancy, characterized by pruritus, hepatic impairment, and elevated serum bile acids, can lead to spontaneous preterm delivery and stillbirth. In women with this condition, heterozygous mutations in *TJP2* were discovered. These novel T62M and T626S mutations were respectively located in the PDZ1 and the SH3 domains of ZO-2; and no patient harbored more than one mutation [28]. Since only three patients were identified with these mutations, larger cohorts of pregnant patients should be studied to understand why these variants give rise to leaky TJs in the liver during pregnancy (Table 1).

3.3. *ZO-2 in the Inner Ear*

In the mammalian ear, the hair cells of the cochlea convert the vibrations induced by sound into electrical signals by K^+-mediated depolarization. A K^+-rich fluid, named endolymph and produced by the stria vascularis, bathes the apical surface of hair cells. The basolateral surface of a hair cell is surrounded by the perilymph, which has a low K^+ concentration. To prevent mixing endolymph and perilymph and to maintain a high resting potential in the endolymph ($\approx$90 mV), the cells surrounding the liquid are sealed with TJs, which are very tight in hair cells and supporting cells in the reticular lamina of the organ of Corti and in some cells of the stria vascularis. Accordingly, hearing loss associated with the degeneration of cochlear hair cells was observed in KO mice for claudin-14 [32], occludin [33], tricellulin [34], and ILDR-1/angulin-2 [35], and in a mice carrying a missense mutation of claudin-9 [36]. Claudin-11-null mice lacked TJs in the basal cells of the stria vascularis and their deafness was associated with a low endocochlear potential [37,38].

Autosomal dominant nonsyndromic hearing loss caused by mutations in the *TJP2* gene was observed in a Chinese family involving the missense mutation 2081G>A (G694E) [20], located within the GuK domain of ZO-2, and in the Korean population due to two pathogenic variations: 334G>A (A112T), which is present in the PDZ1 domain of ZO-2, and 3562A>G (T1188A), which is situated within the carboxyl terminal PDZ binding motif TEL of ZO-2 [19]. Since the latter, as well as the PDZ1 and GuK domains of ZO-2, are protein–protein binding regions, it is expected that mutations at these sites would compromise TJ stability and the interaction of ZO-2 with other TJ proteins and signaling molecules (Table 1).

4. Subcellular Localization and Traffic of ZO-2

ZO-2 can be found in the cytoplasm, nuclei, and TJs. In unfertilized mouse eggs, ZO-2 distributes diffusely in the cytoplasm. In the zygote, ZO-2 concentrates in specks around both pronuclei. After the first cell division these specks appear in the cytoplasm close to the plane of cleavage. From late 2-cell embryos until 8-cell compaction, ZO-2 concentrates at the nuclei of the blastomeres, but is excluded from the nucleoli. After 8-cell compaction, nuclear ZO-2 diminishes in intensity and instead concentrates at junctional sites from the 16-cell stage onward [13].

In confluent epithelial cells, ZO-2 concentrates at the TJs, whereas in sparse cultures, ZO-2 is present both at the nucleus and TJs [39], thus revealing that the localization of ZO-2 is regulated by the degree of cell–cell contact (Figure 2A). Nuclear recruitment of ZO-2 can also be induced at sites of monolayer wounding [39] or with environmental stresses, such as heat shock (42 °C) and chemical insult ($CdCl_2$) [40].

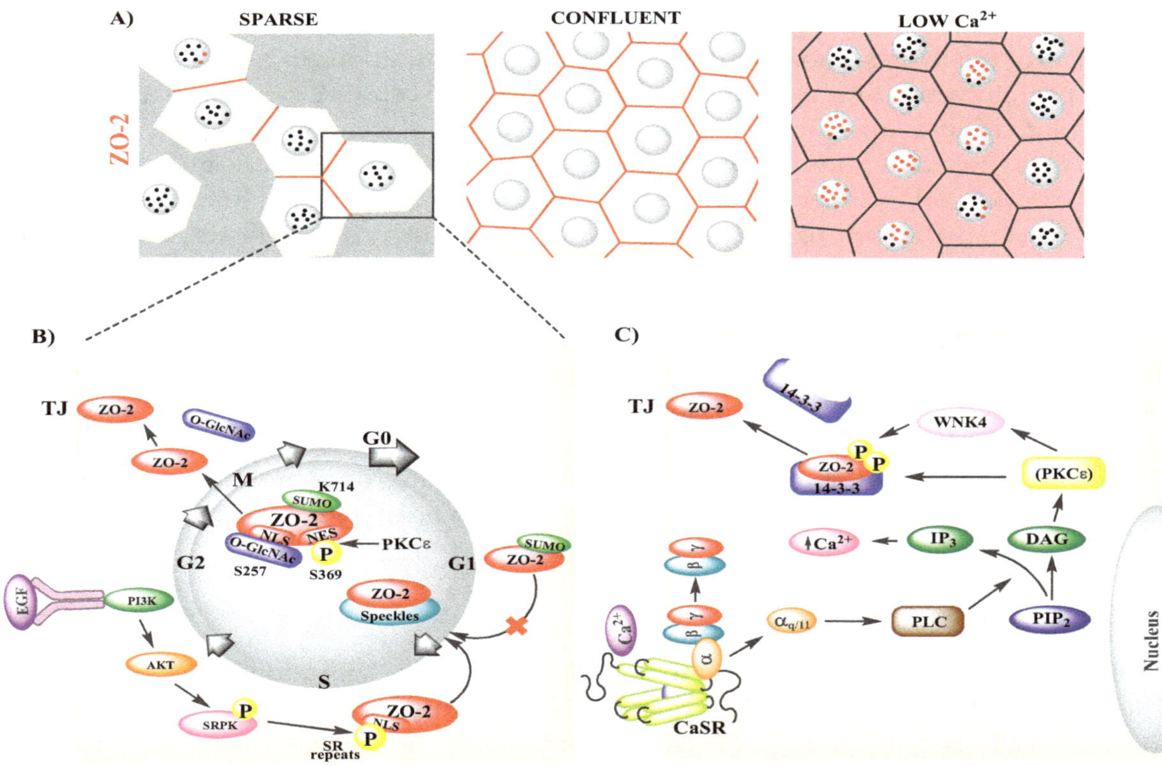

Figure 2. Intracellular movement of ZO-2. (**A**) In sparse cultures, ZO-2 is present in nuclear speckles (red dots) and TJs (red cell borders), whereas in confluent cells ZO-2 concentrates at TJs (red cell borders). In cells cultured with low Ca^{2+} (1–5 μM) ZO-2 distributes diffusely in the cytoplasm (pink cytoplasmic staining) and concentrates in nuclear speckles (red dots). (**B**) In sparse cultures, epidermal growth factor (EGF) induces AKT activation, which triggers ZO-2 phosphorylation of serine-arginine (SR) repeats by SR protein kinase (SRPK). Nuclear localization signals (NLSs) of ZO-2 and phosphorylated SR repeats induce the arrival of ZO-2 to nuclear speckles at late G1. ZO-2 exit from the nucleus requires the phosphorylation of Ser369 within a nuclear exportation signal (NES) by PKCε and the inactivation of bpNLS-2 by *O*-GlcNAc of Ser257, and is facilitated by SUMOylation of Lys714. (**C**) Extracellular Ca^{2+} activates the Ca^{2+}-sensing receptor (CaSR) present in the plasma membrane that signals through the $G\alpha_{q/11}$ subunit, activating PLC, which hydrolyzes PIP2 into DAG and IP3. DAG activates PKC ε, leading to ZO-2 phosphorylation by WNK4. This phosphorylation induces the disassembly of ZO-2 from 14-3-3 proteins in the cytoplasm and triggers the relocation of ZO-2 to the plasma membrane.

ZO-2 stability augments with TJ establishment and maturity. Thus, the half-life of ZO-2 in epithelial MDCK cells goes from 7 h in cells cultured in media with low Ca^{2+} (1–5 μM) that cannot form TJs, to 19.7 h in cultures with 1.8 mM Ca^{2+} [41] and from 8.6 h in sparse cultures to 19.1 h in confluent cultures [3].

In MDCK cells, treatment with L-mimosine, which selectively blocks cell cycle progression in the late G1 phase, revealed that ZO-2 enters the nucleus at the later stages of G1, while treatment with nocodazole, which arrests the cell cycle at the beginning of mitosis, showed that ZO-2 exits the nucleus during mitosis [42]. These observations explain why the nucleus is devoid of ZO-2 in confluent quiescent cells, while proliferating cells have ZO-2 at the nucleus (Figure 2B).

Nuclear importation of ZO-2 relies on nuclear localization signals (NLSs) located within the U2 region of the molecule, between PDZ1 and PDZ2 [43]. The first bipartite (bp) NLS of ZO-2 is conserved in several species, including human, dog, mouse, and chicken, and is functional in a reporter protein nuclear importation assay. The second bpNLS of the molecule, present in dog and mouse, is not active in this assay [44]. The functionality of the second bpNLS is hampered by the phosphorylation of three serine residues (S257, S259, and S261) present within the carboxyl segment of the signal, which neutralizes the positive charges of three contiguous arginine residues (Figure 1).

In addition, the U2 region of ZO-2 has 16, 14, 19, and 16 serine–arginine (SR) motifs in dog, human, mouse, and chicken, respectively. SR motifs are also present in transcription factors and pre-mRNA splicing factors and constitute a signal necessary and sufficient to target proteins for nuclear speckles. ZO-2 also has a docking motif for SR protein kinase 1 (SRPK1) in its U2 domain that specifically phosphorylates serine residues present in SR motifs. Accordingly, ZO-2 associates to SRPK1 and the serine phosphorylation of ZO-2 by SRPK1, which is triggered by the presence of EGF and the subsequent activation of PI3K and AKT, induces the translocation of ZO-2 into the nucleus and its distribution in speckles [44] (Figure 2B). The binding of the PDZ2 domain of ZO-2 to phosphatidylinositol 4,5-bisphosphate (PIP2) regulates ZO-2 recruitment to nuclear speckles, and ZO-2 silencing disrupts speckle morphology, suggesting that ZO-2 plays an active role in the formation and stabilization of nuclear speckles. Residues R302 and R347 in the PDZ2 of mouse ZO-2 are critical for PtdIns binding [45].

ZO-2 exits the nucleus as the monolayer acquires confluence through a process sensitive to leptomycin B, a drug that blocks the interaction between nuclear exportation signals (NES) and exportin1/CRM1 [39]. NES are leucine-rich sequences containing a characteristic spacing of leucine or other hydrophobic residues. Based on the archetypal NES consensus sequence of HIV-1, Rev, PKI-α, IκBα, and zyxin, four NES in ZO-2 canine sequences were identified, three in each of mouse and chicken, and none in human ZO-2. In canine ZO-2, two NES are located at the PDZ-2 domain and the other two at the GuK module (Figure 1). With reporter protein nuclear export assays, the functionality of three of these signals was confirmed, while that of NES-1 proved to be effective only after a phosphomimetic mutation was introduced in a serine residue (Ser369) within the signal that constituted a putative protein kinase C (PKC) phosphorylation site [46] (Figure 2B). Although each of the four NES present in ZO-2 can export a reporter protein like ovalbumin, in full-length ZO-2 the presence of the four intact NES is required to avoid nuclear accumulation of the protein [46]. Since Ser369 is phosphorylated by nPKCε, this phosphorylation constitutes a critical step for the exit of ZO-2 from the nucleus [47].

Two additional post-translational modifications facilitate the nuclear exportation of ZO-2. One is the *O*-linked β-*N*-acetylglucosamination of ZO-2 at Ser257. Since this residue is located within the second bpNLS of ZO-2, it is proposed that this modification inactivates the NLS and consequently facilitates the departure of ZO-2 from the nucleus [44]. The other modification is the SUMOylation at Lys730 of human ZO-2, as mutation of this site to arginine results in prolonged nuclear localization of ZO-2 in nuclear recruitment assays. In addition, a construct mimicking constitutive SUMOylation of ZO-2, preferentially localized in the cytoplasm, suggesting that SUMOylation in ZO-2 occurs in the nucleus and is a signal that inhibits entry to the nucleus and instead facilitates nuclear exportation [48] (Figure 2B).

Newly synthesized ZO-2 first appears at the nucleus of sparse cells and as the monolayer becomes confluent, it relocates from the nucleus to the plasma membrane [39]. Thus, at the earliest times after transfection, around 80% of ZO-2 transfected cells showed ZO-2 at the nucleus; this value diminished to 20% after 18 h. With the help of a nuclear microinjection assay, in which antibodies against ZO-2 were introduced into the nucleus, it was observed that newly synthesized and endogenous ZO-2 made a stopover at the nucleus before reaching the plasma membrane [47]. The nucleus seemed to serve as a cellular reservoir of ZO-2, as the amount of ZO-2 in this compartment was significantly higher than that which eventually reached the plasma membrane. In the latter, the amount of ZO-2 stayed relatively constant while the content of ZO-2 at the nucleus suffered drastic variations with time [47].

In contrast to sparse cultures, in confluent monolayers, blocking the nuclear exportation of ZO-2 with a PKCε permeable inhibitor peptide did not trigger the nuclear accumulation of ZO-2, indicating that in confluent cultures, ZO-2 travels directly to the plasma membrane without stopping at the nucleus [44].

In epithelial monolayers cultured in media with a low concentration of Ca^{2+}, ZO-2 appeared to be diffusely distributed in the cytoplasm and accumulated at the nucleus [39] (Figure 2A). If these cultures were transferred to normal Ca^{2+}-containing medium, TJs assembled and TER developed [49]. However, if the calcium-switch was done in ZO-2 silenced monolayers, a significant delay occurred in the arrival to the cell borders of ZO-1, occludin, and E-cadherin [17]. Integration of ZO-2 to the TJ is triggered by a signaling cascade that starts at the plasma membrane with the activation of the Ca^{2+}-sensing receptor (CaSR) [41], a G-protein coupled receptor [50]. CaSR signals through the $G\alpha_{q/11}$ subunit, which activates phospholipase C (PLC). PLC hydrolyzes PIP2 to diacyl glycerol (DAG) and inositol triphosphate (IP3). The latter induces the release of Ca^{2+} from the endoplasmic reticulum, while DAG leads to the activation of nPKCε. This kinase phosphorylates the kinase WNK4 that binds and phosphorylates ZO-2 in serine residues, triggering its disassembly from 14-3-3 and its relocation to TJs [41].

5. Interaction of ZO-2 with Nuclear Proteins

ZO-2 is present in the nuclear matrix interacting with lamin B1 and actin [51]. Nuclear ZO-2 concentrates at speckles, which are active places that coordinate transcription and splicing. Accordingly, nuclear ZO-2 co-localizes with the essential pre-mRNA splicing protein SC-35 [39], the speckle protein ZASP [52], and the C-terminal portion of scaffold attachment factor-B (SAF-B) [40]. The latter is involved in transcriptional repression (for review on SAFB1/SAFB2 see [53]), associates with E-boxes in DNA [54,55], recruits histone deacetylases allowing histones to wrap the DNA more tightly [55,56], and binds and inhibits SRPK1 [57] (Figure 3).

In addition, a proteomic analysis revealed the presence of ZO-2 in a complex derived from an immunoprecipitation (IP) of SRm160 from a HeLa cell nuclear extract. SRm160 is a serine/arginine repeat nuclear matrix protein that promotes splicing and stimulates 3′-end processing [58].

In HEK293 cells, liquid chromatography-tandem mass spectometry (LC-MS/MS) revealed the presence of ZO-2 in SIRT7 immunoisolates [59]. SIRT7 is a nucleoli enzyme that associates with RNA Pol I machinery and is required for rDNA transcription. This is noteworthy, since ZO-2 does not co-localize at the nucleoli with nucleolin [39]. LC-MS/MS has also shown the presence of ZO-2 in a soluble protein complex with: (a) Ubinuclein [60], a nuclear protein that enhances the stability and function of human histone H3.3 chaperone complex HUCA (HIRA/UBN1/CABIN1/ASF1a) [61] and that in confluent epithelial cells interacts with cingulin and ZO-1 at TJs [62]; (b) WDR36/UTP21 [60], a protein required for the biogenesis of 18S ribosomal RNA [63]; (c) ZMYM3 [60], a chromatin-interacting protein that promotes DNA repair by homologous recombination [64]; (d) the estrogen-related receptor α (ERRα) [60], a nuclear receptor that interacts with estrogen, activates reporters containing steroidogenesis factor 1 response elements [65], and modulates the activity of estrogen receptor α (ERα) in breast, uterus, and bone (for review see [65]); (e) MCM3, a protein involved in the initiation of eukaryotic genome replication [66]; and f) PUS7 [66], a pseudouridine synthase that enhances RNA stability [67].

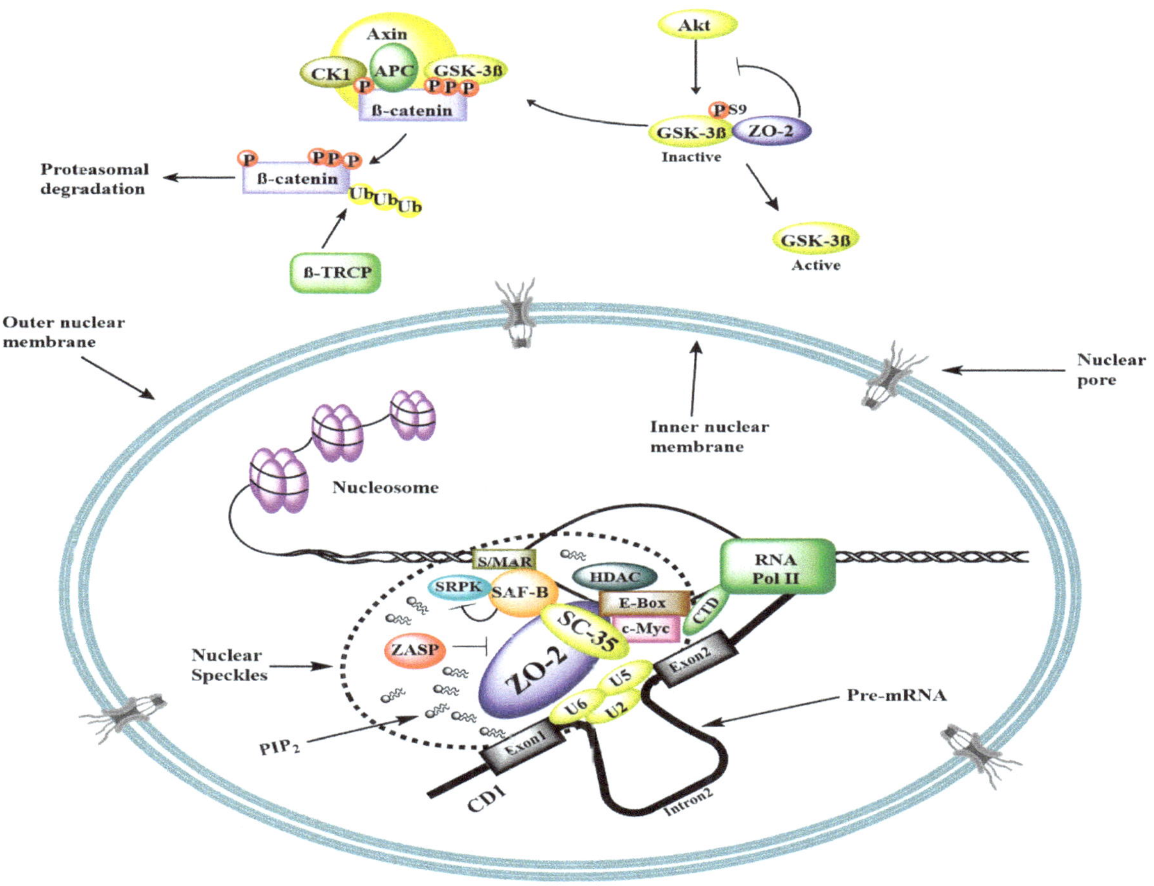

Figure 3. Schematic diagram of ZO-2 as a scaffold for transcription and splicing. ZO-2 expression blocks GSK-3β inhibitory phosphorylation at Ser[9]. GSK-3β phosphorylates β-catenin and targets it for ubiquitination and proteasomal degradation. Consequently, β-catenin entry into the nucleus and transcriptional activity is inhibited. At the nucleus, ZO-2 localizes in speckles associated to PIP2 and ZASP and inhibits transcription of CD1 through its interaction with an E-box in the promoter region, mediated by c-Myc transcription factor. In addition, ZO-2 is proposed to participate in mRNA splicing due to its association with the essential splicing protein SC-35 and the scaffold attachment factor SAF-B. S/MAR, scaffold/matrix attachment región; P, phosphate; Ub, ubiquitin; U2, U5, and U6, splicing ribonucleoproteins; PIP2, phosphatidylinositol (4,5)-bisphosphate.

Altogether, these observations suggest that nuclear ZO-2 forms part of a complex that couples chromatin structure to transcription and RNA processing.

6. ZO-2 as Repressor of Gene Transcription

Pull-down and gel shift assays revealed the interaction of ZO-2 with Fos, Jun, and C/EBP transcription factors. In addition, IP and immunolocalization experiments showed that ZO-2 associates with these transcription factors, both at the nucleus and the TJ region [68]. The role of ZO-2 in gene transcription was first studied in reporter gene assays with promoters regulated by AP-1 sites, revealing that ZO-2 downregulates these promoters in a dose-dependent manner [68]. Then, in reporter gene assays, ZO-2 was found to downregulate cyclin D1 (CD1) transcription in a dose-dependent manner, acting upon a region in the promoter that harbored an E-box. IP and chromatin IP (ChIP) assays further revealed that the transcription factor Myc and the histone deacetylase (HDAC) 1 bound to this E-box and formed a complex with ZO-2 to repress CD1 transcription. The over-expression of Myc increased ZO-2-mediated inhibition of CD1 transcription [55], whereas the speckle protein ZASP blocked this transcriptional inhibition [52] (Figure 3).

ZO-2 also represses gene transcription regulated by the Wnt signaling pathway, which leads to cell proliferation and un-differentiation. Reverse transcription quantitative PCR (RT-qPCR) showed that the endogenous expression of axin-2, a Wnt target gene, was repressed upon ZO-2 overexpression [42]. Moreover, with IPs of transfected and endogenous proteins, as well as proximity ligation assays, ZO-2 was found to bind to glycogen synthase kinase 3β (GSK-3β) [48], while ZO-2 overexpression decreased the amount of inactive GSK-3β phosphorylated at Ser9 [42] (Figure 3). This enzyme, which inhibits epithelial to mesenchymal transformation (EMT), phosphorylates slug transcription factor, inducing its subsequent degradation. Accordingly, an increase in GSK-3β phosphorylated at Ser9 was found in tumor cells associated with slug expression. This factor, which represses E-cadherin transcription, belongs to the snail family, and LC-MS/MS revealed that ZO-2 was present in slug IPs [69]. Likewise, affinity purification of epitope tagged proteins followed by MS (AP-MS) identified ZO-2 as an interacting partner of smug, another transcription factor of the snail family [70].

GSK-3β also phosphorylates β-catenin at Ser33, Ser37, and Thr41, leading to its ubiquitination and subsequent proteasomal degradation (Figure 3). Instead, when β-catenin is not degraded, it accumulates in the cytoplasm and is translocated into the nucleus where it forms a complex with T cell factor (TCF)/lymphoid enhancer family (LEF) transcription factor, which induces the transcription of a variety of target genes involved in cell cycle progression and EMT (for reviews see [71,72]). The effect of ZO-2 on the transcription of Wnt target genes was analyzed in two different Wnt reporter constructs, where luciferase expression was regulated either by artificial TCF/LEF binding sites or by the promoter of Siamois, a Wnt target gene. In both cases, dose-dependent repression was exerted by ZO-2. Moreover, the transcriptional activity regulated by artificial TCF/LEF binding sites increased in ZO-2 KD cells in comparison to parental cells, indicating that the absence of ZO-2 facilitated TCF-regulated transcriptional activity [73]. In agreement, in colon carcinoma cells, vitamin D3 promotes ZO-2 expression and inhibits the expression of a variety of β-catenin-TCF4 responsive genes [74].

The capacity of ZO-2 to repress gene transcription regulated by the Wnt pathway was studied in mice *in vivo*, where ZO-2 overexpression in the glomerulus was achieved by hydrodynamic transfection. Treatment with adriamycin, an anthracycline antibiotic, produced a nephrotic syndrome triggered by activation of Wnt signaling, whereas ZO-2 overexpression in these animals controlled proteinuria and podocyte effacement and favored urea and creatinine clearance through a process that inhibited the expression of snail and increased the amount of nephrin and phosphorylated β-catenin in the glomeruli. This indicates that ZO-2 protects against podocyte injury induced by aberrant Wnt signaling activation [75].

The ability of ZO-2 to inhibit gene transcription regulated by the Wnt pathway is controlled by the intracellular localization of ZO-2. Thus, this repressive activity was not observed when a constitutively SUMOylated ZO-2 construct that could not enter the nucleus was employed. Instead, using a ZO-2 SUMOylation mutant (K730R) that exhibited a strong delayed nuclear export increased ZO-2 repressive activity on promoters regulated by TCF/LEF binding. Accordingly, a ZO-2 construct with forced nuclear localization due to fused NLS completely inhibited transcriptional activity mediated by TCF/LEF, even in dense cultures [48].

To address if ZO-2 repression of this transcriptional activity requires the presence of β-catenin, a LEF chimera lacking the β-catenin binding site and instead containing the Herpes virus VP16 transactivation domain (ΔNLEF-1-VP16) was employed. ZO-2 exhibited transcriptional repression with this LEF chimera in reporter assays done either with a synthetic promoter or with the Siamois promoter, thus demonstrating that ZO-2 inhibits TCF/LEF mediated gene transcription in a β-catenin-independent manner [48]. Nevertheless, ZO-2 does form a complex with β-catenin. This was demonstrated in co-IP and proximity ligation assays employing transfected and endogenous proteins. Moreover, pull-down assays with recombinant ZO-2 and β-catenin proteins revealed that ZO-2/β-catenin interaction was direct. These observations hence suggest that ZO-2 repressed TCF/LEF mediated transcriptional activity in both a β-catenin-independent and -dependent manner [48].

Yes-associated protein (YAP), the main downstream effector of the mammalian Hippo pathway that regulates cell size, proliferation, and differentiation (for review see [76]), docks the ubiquitin ligase β-TrCP to the β-catenin destruction complex in cells where the Wnt signaling pathway is inactive, while Wnt ligands dislodge YAP from this complex and trigger the nuclear accumulation of YAP [77]. YAP Ser127 phosphorylation by the large tumor suppressor (LATS) kinase of the Hippo pathway creates a 14-3-3 binding site, which allows the cytoplasmic retention of YAP. However, YAP phosphorylation at Ser381 by LATS primes YAP for a subsequent CK1 phosphorylation that triggers YAP ubiquitination and proteasomal degradation [78]. Instead, the inactivation of the Hippo pathway allows the entry of the un-phosphorylated YAP into the nucleus, where YAP acts as a cofactor for transcriptional enhanced associate domain (TEAD) transcription factors that promote EMT (for review see [76]). MDCK ZO-2 KD cells exhibit a nuclear accumulation of YAP even when grown at high density and display an increased transcriptional activity of a reporter driven by artificial TEAD-binding sites or by the promoter of connective tissue growth factor (CTGF), a YAP target gene, while transfection of ZO-2 reduces promoter activity in both parental and ZO-2 KD cells. Moreover, the absence of ZO-2 increased the amount of CTGF mRNA determined by RT-qPCR, while the overexpression of ZO-2 reduced CTGF mRNA in parental cells. Altogether, these results indicate that ZO-2 repressed the transcriptional activity mediated by TEAD [73].

The involvement of ZO-2 in gene transcription was also studied in human vascular smooth muscle cells (VSMC). ZO-2 was barely observed in isolated control coronary arteries, but was highly expressed after dilation with a balloon catheter in the tunica media, which is composed of SMC, and in the intima, which is made up of endothelial cells. In contrast, the transcription factor stat1 was inversely regulated upon dilation injury. In primary cultures of human VSMC, ZO-2 silencing increased the nuclear and cytosolic expression of stat1 and its mRNA level. When these ZO-2-silenced VSMC cells were transfected with a luciferase reporter plasmid under the control of an interferon stimulated response element (ISRE), to which stat binds, an induction of transcription was observed in comparison to control VSMC. These results suggest that ZO-2 in VSMC repressed stat1 transcriptional activity. The latter in VSMC inhibited the proliferation rate, which was important because an increased proliferation of medial VSMC is one of the hallmarks of negative vascular remodeling [79].

7. ZO-2 and Cell Proliferation

ZO-2 over-expression inhibits cell proliferation blocking, the progression of cells through the G1/S boundary of the cell cycle. This effect is due to the amino terminal segment of ZO-2, which contains the three PDZ domains of ZO-2 [42]. ZO-2 has also been found to block cell proliferation induced by YAP [80].

In MDCK cells, ZO-2 overexpression reduces the level of CD1 mRNA [55], but this effect is not observed when the cultures are synchronized. In these cultures, the level of CD1 mRNA is higher upon arrest after incubation with 0.1% serum than in proliferating cells cultured with 10% serum [42]. ZO-2 also inhibits cell proliferation by augmenting CD1 degradation in the proteasome induced by GSK-3β phosphorylation. Thus, at the S phase, GSK-3β phosphorylates CD1 at Thr286, inducing its exit from the nucleus, ubiquitination, and subsequent degradation in the proteasome [81]. ZO-2 overexpression diminishes GSK-3β inhibitory phosphorylation at Ser-9 and triggers CD1 degradation [42]. Accordingly, LiCl, which inhibits GSK-3β [82,83], blocks the decrease in CD1 protein content induced by ZO-2 overexpression [42].

ZO-2 KD in preimplantation mouse embryos [13] and in monolayers of MDCK cells [73] has no effect on the number of blastocyst cells or proliferating cells, respectively. Instead, ZO-2 silencing in the spontaneously immortalized kidney cell line $mCCD_{d1}$, which displays highly differentiated properties of collecting duct principal cells, decreases cyclin D1 content and blocks cell cycle progression at G0/G1 [84]. This is noteworthy, because cells of the collecting duct show intense proliferation during embryogenesis but very low turnover in the adult kidney, suggesting that ZO-2 might play a regulatory role in this process.

In VSMC, ZO-2 silencing also inhibits cell proliferation, albeit through increased expression of stat1 transcription factor and activation of stat1 specific genes [79].

ZO-2 fused to a strong NLS of SV40, concentrates at the nucleus, increases cell proliferation, and triggers the expression of the M2-type isoenzyme of pyruvate kinase (PKM2) [85]. This enzyme, whose plasma concentration in cancer patients correlates with staging (for review see [86]), is involved in cancerous cell metabolism in a process known as the Warburg effect or aerobic glycolysis, where cancer cells take up high amounts of glucose and produce lactate even in the presence of oxygen. PKM2 catalyzes the last step of glycolysis, the conversion of phosphoenolpyruvate to pyruvate, with the concomitant production of ATP. Generation of ATP by PKM2, in contrast to mitochondrial respiration, is independent of oxygen and hence allows tumor cells to grow in hypoxic conditions. These results therefore suggest that the concentration of ZO-2 in the nuclei through the overexpression of PKM2 might confer a metabolic benefit to cancer cells.

8. ZO-2 in Apoptosis and Cell Degeneration

Apoptosis is a form of programmed cell death present in multicellular organisms. In contrast to necrosis, which is a traumatic cell death after acute cellular injury, apoptosis is a highly regulated and controlled process. Some factors, like Fas receptors and caspases, which are a family of cysteinyl-aspartate proteases that signal through a cascade, promote apoptosis, while proteins of the BCL-2 family inhibit apoptosis.

In human breast epithelial cells, upon apoptosis induction by staurosporine or anti-CD95/Fas antibody treatment, TJs are disrupted and ZO-2 is cleaved by caspases [87]. Caspase-mediated cell death requires cleavage of key proteins involved in cell survival, including PARP-1. This is an enzyme activated by DNA breaks that polyADP-ribosylates proteins. In response to mild DNA damage, PARylation favors DNA repair and cell survival, whereas severe DNA damage leads to apoptosis or necrosis by amplified PARP-1 activity, resulting in high donor nicotinamide adenine dinucleotide (NAD^+) consumption and depletion of ATP pools (for review see [88]). In TK6 lymphoblastoid cells, after induction of apoptosis with hydroquinone, a metabolite of benzene involved in benzene-induced leukemia, ZO-2 is polyADP-ribosylated via interaction with PARP-1 in the nucleus and its expression is upregulated [89].

ZO-2 overexpression does not augment the percentage of early or late apoptotic epithelial cells in culture [42]. Instead, a tandem inverted genomic duplication of the wild type *TJP2* gene in humans leads to progressive non-syndromic hearing loss, accompanied by a change in the expression of BCL2 family genes that favors apoptosis. Although this change in gene expression was studied in the lymphoblast cells of affected individuals, it suggests that ZO-2 overexpression leads to increased susceptibility to apoptosis in inner ear cells and that this mechanism triggers adult-onset hearing loss in this kindred [4].

Ubiquitination regulates protein stability. In particular, lysine 48-linked polyubiquitination chains target proteins for degradation in the proteasome. Protein ubiquitination is counter-regulated by deubiquitinating enzymes, among which the ubiquitin specific proteases (USP) constitute the largest family. Although USP53 has been suggested to be catalytically inactive due to the absence of a critical hystidine residue, mambo mice carrying a point mutation in the catalytic domain of USP53 exhibit rapid progressive hearing loss. USP53 is broadly expressed in the inner ear, where it interacts with ZO-1 and ZO-2. In mambo mice, biotin tracer assays indicated that TJs of the stria vascularis were not affected, while hair cell degeneration preceded the endocochlear potential decline. Moreover, the outer hair cells of these mice evaded degeneration in organ culture, indicating that the unfavorable extracellular conditions that these cells encountered *in vivo* in the mambo mice promoted their degeneration. These observations hence suggest that the inactivation of USP53 promoted the degradation of ZO-1/ZO-2, triggering a disturbance of cochlear homeostasis that resulted in hair cell degeneration and apoptosis [90]. Moreover, in humans, cholestasis and hearing loss was found to correlate with a truncating variant in USP53 [91].

YAP, through its PDZ binding motif, interacts with the first PDZ domain of ZO-2; this interaction facilitates the nuclear localization of YAP. Nuclear YAP exerts a pro-apoptotic function determined by the appearance of a 89 kDa PARP-1 fragment characteristic of caspase-3 and -7 cleavage. The amount of this PARP-1 fragment decreases when ZO-2 expression is silenced, suggesting that ZO-2 facilitates the apoptotic function of nuclear YAP [80,92].

9. ZO-2 Associates with the Cortical Actomyosin Ring and Regulates Cytoarchitecture

In epithelial cells, F-actin associated with the apical junctional complex appears as a thin band that surrounds the apical margins of the cells, also known as the cortical ring of actin. Myosin II staining follows that of actin, but is more punctate and more concentrated at the tricellular junctions [93].

ZO-2 directly interacts with actin, but is not a F-actin cross-linking protein [10]. ZO-2 also binds to 4.1 [94], a cytoskeleton protein that stabilizes spectrin–actin interactions and associates with ankyrin (for review see [95]).

ZO-2 at the cell border is associated with the integral TJ protein JAM-A and the submembranous protein afadin. PDZ-GEF1 binds to this complex and stimulates the activity of the GTPase Rap2c, which inhibits RhoA. Consequently, in JAM-A-deficient cells, the activity of RhoA and the phosphorylation of myosin light chains are enhanced [96]. Moreover, the lack of this complex, which inhibits RhoA signaling, explains the engrossment of the perijunctional ring of actomyosin and punctate distribution of myosin II, which are observed in MDCK cells without ZO-1 and ZO-2 [93].

However, it is noteworthy that, in ZO-1KO/ZO-2KD mammary Eph4 cells, AJs are organized in a fragmented manner containing E-cadherin and actin but lacking myosin-II. In these cells, transfection of either ZO-1 or ZO-2 restores myosin integration and the establishment of regular AJs through a process mediated by Rho activation [97]. The differences in myosin II organization observed in MDCK and Eph4 cells which do not express ZO-1/ZO-2 could be due to the cellular context, since, in contrast to MDCK cells [9], Eph4 cells do not express ZO-3 [98].

In MDCK cells lacking ZO-1 and ZO-2, the change in the appearance of actin and myosin is accompanied by expansion of the apical domain, cell border recruitment of phosphomyosin light chain and Rho kinase 1 (ROCK-I), and contraction of the actomyosin ring [93]. In accordance, atomic force microscopy revealed that the lack of ZO-1 and ZO-2 increased the tension of the apical domain due to elevated myosin II ATPase activity [99]. In addition, in MDCK cells with shRNAs targeting both ZO-1 and ZO-2, the normal convoluted pattern of cell–cell contact becomes much more linear, making the overall cell shape more polygonal [93]. This change does not happen in ZO-2 KD [100] and KO [101] cells, but is clearly observed in ZO-1 KD [100] and KO [101] cells, and can also be attained in control cells after treatment with blebbistatin, an inhibitor of myosin II [101]. Moreover, excessive expression of ZO-1 in ZO-1 KO cells induces an intensive contoured shape of cell–cell junctions. Taken together, these observations indicate that the change from tortuous to linear contacts present in the double ZO-1 and ZO-2 KD is due to the lack of ZO-1 and depends on an actomyosin-generated force.

Structured illumination microscopy of ZO-1 reveals two conformations of the protein, stretched and folded. The latter relies on the interaction between the extreme C-terminal segment and the PDZ3, SH3, U5, and GuK domains of ZO-1. The conformational state of ZO-1 depends on its hetero-dimerization with ZO-2 and the tension of the actomyosin cytoskeleton. Thus, upon ZO-2 depletion, ZO-1 is maintained in the stretched conformation only if the actomyosin tension is high, like when cells are grown on glass coverslips. However, ZO-1 acquires a folded conformation if ZO-2 is absent and actomyosin tension is reduced by treatment with blebbistatin, or in conditions of reduced substrate stiffness (e.g., 3D cultures in Matrigel). In its folded conformation, ZO-1 cannot associate with occludin or recruit the transcription factor zonulin to the membrane, hence, this condition promotes the expression of zonulin target genes like cyclin D1 and proliferating cell nuclear antigen (PCNA) [102]. These observations suggest that, in a condition of low tension due to low substrate stiffness, like the one encountered by epithelial tissues *in vivo* (for review see [103]), the presence of ZO-2 is needed to maintain ZO-1 in a stretched conformation capable of binding to integral TJ proteins and to transcription factors like zonulin, which regulates the expression of genes involved in cell proliferation and differentiation.

ZO-2 alters the cytoarchitecture of epithelial monolayers, as the absence or reduction of ZO-2 content widens the intercellular spaces and detaches regions of the monolayer from the substrate [17,18]. In addition, decreased expression of integrin β1 and claudin-7, which localizes along the basolateral membrane in association to integrin β1, is observed in ZO-2 KD cells. Lack of ZO-2 induces recruitment of vinculin to focal adhesions and proliferation of stress fibers associated with increased activation of RhoA and ROCK-II [18] (Figure 4). Likewise, treatment of scirrhous gastric cancer cells with transforming growth factor β (TGFβ), whose signaling is associated with the invasion of cancer cells, decreases ZO-2 expression and upregulates the activity of RhoA and myosin phosphorylation [104].

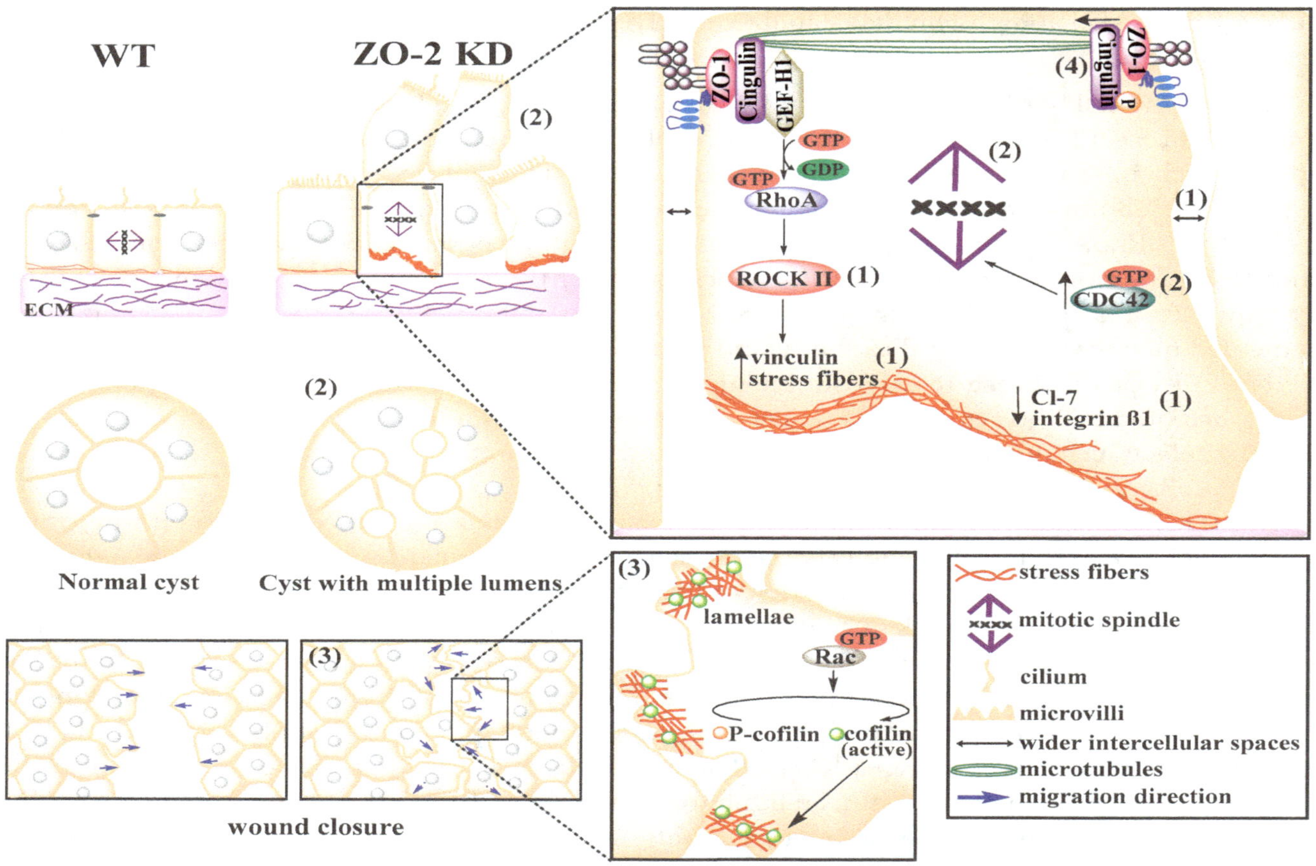

Figure 4. ZO-2 modulates cell size and tissue architecture. Absence of ZO-2 induces cell hypertrophy and development of an atypical cytoarchitecture characterized by: (**1**) Wider intercellular spaces and the detachment of cells from the substrate accompanied by a reduced expression of claudin-7 and integrin β1, an increased recruitment of vinculin to focal adhesions, and the proliferation of stress fibers induced by RhoA/ROCK-II activation; (**2**) an augmented activation of CDC42 that triggers the disorientation of the mitotic spindle and leads to the growth of cells on top of each other in 2D and 3D cultures and then to the formation of multiple lumens per cyst; (**3**) the activation of Rac and cofilin, which promote the formation of multiple lamellipodia, reduce directional persistence, and increase wound closure; and (**4**) the disappearance of a planar network of microtubules at the cell border.

ZO-2 KD cells also exhibit a profusion of lamellae mediated by an increase activity of Rac and cofilin. Rac promotes random cell migration in opposition to directional migration; accordingly, although ZO-2 KD cells move more than parental cells and close wounds more efficiently, they display reduced directional persistence [18] (Figure 4).

ZO-2 KD MDCK cells lose the microtubules that form a non-centrosomal network organized as a planar apical structure and display diminished phosphorylation of cingulin as well as a reduced interaction of cingulin with β-tubulin. ZO-2 favors the phosphorylation of cingulin by the adenosine monophosphate-activated protein kinase (AMPK), which allows microtubules/cingulin interactions at the apical junctional region. In addition, in ZO-2 KD monolayers, some cells grow on top of each other, while cysts of ZO-2 KD cells formed in the extracellular matrix Matrigel display multiple lumens instead of a single lumen per cyst. These aberrations might be due to the disorientation of the mitotic spindle, since in ZO-2 KD cells, the percentage of cells with mitotic spindles parallel to the substrate surface (0°–10°) decreases [18]. ZO-2 KD monolayers also display a reduced number of cilia [73] and more CDC42 activation [18], which controls spindle orientation [105] and centrosome polarization toward the direction of cell migration [106] (Figure 4). Taken together, these observations suggest that ZO-2 affects centrosome function.

The centrosome is a microtubule-organizing center composed of a centriole pair surrounded by pericentriolar material. In addition, centrioles template the formation of primary cilia in non-cycling cells. The mother centriole has sub-distal appendages needed for the organization of the cellular microtubular network and distal appendages involved in docking the centriole to the plasma membrane during ciliogenesis. Recently, using a proximity-dependent biotinylation assay to detect proteins interacting with centrosome and cilium proteins, ZO-2 was found to be associated with [107]: The sub-distal appendage proteins of the mother centriole CEP128, which regulates signaling at the primary cilium [108], and its associated protein cenexin [109] required for centrosome positioning in interphase cells, microtubule stability for proper spindle orientation during mitosis, and the formation of single lumens in cysts [110]; CEP135, a centriole assembly protein that regulates centriole duplication [111]; and centriolin, a protein of maternal centrioles required for the final stages of cytokinesis and entry into S phase [112].

10. ZO-2 as a Modulator of Cell Size

ZO-2 silencing in MDCK cells produces an increase in cell size also known as hypertrophy, where 67% of cells have a cell diameter of 55–60 nm instead of the 35–40 nm present in 73% of parental cells [73]. Hypertrophy has been associated with the activation of cyclin D without the concurrent upregulation of cyclin E, which triggers an increase in the physical growth of cells in early G1 and delayed entry into the S phase [113,114]. In ZO-2 KD cells this appears to be the case, as cyclin D1 content is increased, while entry into the S phase exhibits a delay [73].

In ZO-2 KD cells, hypertrophy is also due to inhibition of the Hippo pathway and activation of mTOR signaling [73]. The absence of ZO-2 induces the concentration of YAP at the nucleus and its transcriptional activity that promotes the expression of the catalytic pik3cb subunit of phosphatidylinositol 3-kinase (PI3K) and, through an miRNA-dependent process, blocks the translation of the phosphate and tensin homologue (PTEN). Consequently, the concentration of phosphatidylinositol(3,4,5)-trisphosphate (PIP3) increases and recruits PDK1 and AKT to the plasma membrane, triggering the phosphorylation and activation of AKT and the subsequent phosphorylation of mTORC1 and its downstream target, the ribosomal protein S6 kinase 1 (S6K1), which increases protein synthesis. Altogether, this leads to an increased protein/DNA ratio that produces hypertrophy [73], in agreement with previous observations in renal cells where hypertrophy appears to be a result of an increase in the protein synthesis/degradation rate [115,116]. Moreover, in renal compensatory hypertrophy induced by the removal of one kidney or uninephrectomy, the size of cells in the proximal tubules of the remaining kidney increased, while the expression of ZO-2 decreased and YAP accumulated in the nucleus. These results hence show that both *in vivo* and *in vitro* ZO-2 silencing correlates with cell hypertrophy.

11. ZO-2 as a Tumor Regulator Protein

Several lines of evidence suggest that ZO-2 is a tumor suppressor protein. First, ZO-2 is highly similar to Disc-Large (DLG) (amino segment with PDZ domains = 63% similarity, SH3 domain = 59% similarity, and GuK domain = 50% similarity), which acts as a tumor suppressor in the larva imaginal discs of Drosophila [6]. Second, some viral oncogenic proteins target ZO-2. Thus, the E4 region-encoded ORF1 (E4-ORF1), the primary oncogenic determinant of adenovirus type 9 that elicits mammary tumors in animals, associates through its PDZ binding motif with the first PDZ domain of ZO-2 and this interaction results in the aberrant sequestration of ZO-2 within the cytoplasm [117]. Likewise, E6 proteins from cancer-causing human papillomavirus (HPV)-16, -18, -31, -51, and -70 have a PDZ binding motif on their extreme carboxy termini that interacts with ZO-2. Instead, E6 proteins from HPV-66 or -40, which are, respectively, rarely or never associated with cancer, do not bind to ZO-2 [118]. E6/ZO-2 interaction appears to stabilize ZO-2, as ablation of E6 expression in HeLa cells reduces ZO-2 levels [118], whereas the expression of E6 from HPV-16 in the skin of transgenic female mice (K14E6) upregulated the expression of ZO-2 in the epidermis; and in monolayers of MDCK cells, transfection of E6 from HPV-16 blocked ZO-2 protein decay [119]. This is noteworthy because E6 proteins from high-risk HPV are known to target degradation of the cell–cell adhesion proteins Dlg [120], MAGI [120], MUPP1 [121], and Scribble [122]. E6 also triggers the mislocalization of ZO-2 as in MDCK cells transfected with E6, ZO-2 disappears from the cell borders and, in some cells, accumulates in the nucleus [119]. In this respect, in testicular carcinoma *in situ* and in bronchopulmonary cancers, ZO-2 disappears from the cell borders and stains the cytoplasm. This aberrant ZO-2 localization correlates with disruption of the blood–testis barrier [123] and the invasive properties of lung cancer cells *in vitro* [124].

c-Src, a non-receptor tyrosine kinase which is the cellular version of v-Src, the oncogene product of Rous sarcoma virus, is frequently overexpressed and active in various carcinomas. c-Src phosphorylates diverse cellular proteins involved in signaling pathways which lead to cancer progression. In MDCK cells, v-Src transfection opens the TJ, which is determined by a profound decrease in TER, and induces tyrosine phosphorylation of ZO-1 and ZO-2 [125]. Moreover, by proteomic analysis using the c-Src SH2 domain as the ligand, ZO-1 and ZO-2 were identified as binding proteins to both c-Src and C-terminal Src kinase (Csk). The latter phosphorylates and negatively regulates c-Src. Therefore ZO-1 and ZO-2 appear to function as scaffolding, regulating c-Src's transformation ability [125].

The *TJP2* gene has six alternatively spliced transcripts encoding various isoforms. In normal epithelia, expression was studied in two isoforms of ZO-2: ZO-2A and ZO-2C, which are respectively transcribed from the downstream promoter P_A and the upstream promoter P_C. The amino terminal region of ZO-2A has a 23-amino acid segment that is absent in ZO-2C [126,127]. In human pancreatic ductal adenocarcinoma and in breast ductal adenocarcinoma, ZO-2A isoform is absent, whereas in colon and acinar prostate adenocarcinomas, ZO-2 isoforms A and C are both expressed [128]. Loss of ZO-2A in pancreas ductal adenocarcinoma is not due to mutation, lack of transcription factors, or methylation of the immediate promoter, although methylation of the P_A promoter inactivates the promoter at the late stages of tumor development [126]. The sequence of promoter P_A reveals that ZO-2A transcription is regulated by numerous Sp1 sites [126].

Among *TJP2* transcripts, one with a longer exon which includes an AluYc sequence (*TJP2*-Alu transcript) was described. Alu proteins are retrotransposons that regulate gene expression patterns. *TJP2*-Alu transcript expression is not tissue specific, although its highest expression is in the testis. *TJP2*-Alu transcript is more highly expressed in normal tissues compared to tumors of the liver, uterus, and breast, however, colon tumor tissue displays a strong expression of *TJP2*-Alu transcript not present in normal tissue, making it a potential biomarker for colorectal cancer diagnosis [129].

The expression of ZO-2 is silenced in several carcinomas, including breast [128,130] and pancreas [127], as well as in a hypoxia-resistant cancer cell lines derived from a scirrhous gastric carcinoma [131]. In patients with incompletely enhancing glioblastoma multiforme (GBM), who have a longer survival rate, ZO-2 is expressed at higher levels than completely enhancing GBM, which is associated with shortened survival [132].

17β-estradiol (E2) and its receptors are risk factors for the initiation and progression of endocrine-related cancers, like those of the breast, prostate, ovarian, and endometrial cells (for reviews see [133,134]). In ovariectomized mice, E2 decreased the expression of ZO-2 in mouse epidermis and, although progesterone induced ZO-2 overexpression, it could not overturn ZO-2 silencing induced by E2 [119].

TGFβ1, which promotes the invasion of cancer cells, decreases the expression of ZO-2 in scirrhous gastric cancer cells [104]. Vascular endothelial growth factor-A (VEGF-A) in pancreas cancer cells increases cell motility and decreases ZO-2 expression [135]. Contrarily, vitamin D3, which promotes the differentiation of colon carcinoma cells, induces the expression of ZO-2 [74].

In summary, ZO-2 could be viewed as a tumor suppressor protein due to its structural similarity with known tumor suppressor proteins and considering that it is a target of oncogenic kinases, is silenced in various carcinomas, and exerts transcriptional repression on genes involved in cell proliferation and EMT. However, since oncogenic HPVs upregulate ZO-2 expression and forced nuclear accumulation of ZO-2 increases cell proliferation and confers metabolic benefit to cancer cells, we think ZO-2 might be considered as a tumor regulator rather than a tumor suppressor.

Author Contributions: L.G.M. wrote the manuscript. H.G.G, L.G.G and C.H.G searched for the information and organized it in different chapters. H.G.G made Figure 3 and Table 1. L.G.G made Figure 4 and C.H.G made Figures 1 and 2.

Acknowledgments: H.G.G.: L.G.G. and C.H.G. were recipients of doctoral fellowship from the Mexican council of Science and Technology (Conacyt) (282075, 340209 and 407499) and H.G.G. was also a recipient of a scholarship from Mexiquense Council of Science and Technology (Comecyt, 2018AD0009-11).

References

1. Gonzalez-Mariscal, L.; Quiros, M.; Diaz-Coranguez, M. ZO proteins and redox-dependent processes. *Antioxid Redox Signal.* **2011**, *15*, 1235–1253. [CrossRef] [PubMed]
2. Gonzalez-Mariscal, L.; Miranda, J.; Ortega-Olvera, J.M.; Gallego-Gutierrez, H.; Raya-Sandino, A.; Vargas-Sierra, O. Zonula Occludens Proteins in Cancer. *Curr. Pathobiol. Rep.* **2016**, *4*, 107–116. [CrossRef]
3. Gumbiner, B.; Lowenkopf, T.; Apatira, D. Identification of a 160-kDa polypeptide that binds to the tight junction protein ZO-1. *Proc. Natl. Acad. Sci. USA* **1991**, *88*, 3460–3464. [CrossRef] [PubMed]
4. Walsh, T.; Pierce, S.B.; Lenz, D.R.; Brownstein, Z.; Dagan-Rosenfeld, O.; Shahin, H.; Roeb, W.; McCarthy, S.; Nord, A.S.; Gordon, C.R.; et al. Genomic duplication and overexpression of TJP2/ZO-2 leads to altered expression of apoptosis genes in progressive nonsyndromic hearing loss DFNA51. *Am. J. Hum. Genet.* **2010**, *87*, 101–109. [CrossRef] [PubMed]
5. Itoh, M.; Morita, K.; Tsukita, S. Characterization of ZO-2 as a MAGUK family member associated with tight as well as adherens junctions with a binding affinity to occludin and alpha catenin. *J. Biol. Chem.* **1999**, *274*, 5981–5986. [CrossRef] [PubMed]
6. Jesaitis, L.A.; Goodenough, D.A. Molecular characterization and tissue distribution of ZO-2, a tight junction protein homologous to ZO-1 and the Drosophila discs-large tumor suppressor protein. *J. Cell Biol.* **1994**, *124*, 949–961. [CrossRef] [PubMed]
7. Jenkins, E.L.; Caputo, M.; Angelini, G.D.; Ghorbel, M.T. Chronic hypoxia down-regulates tight junction protein ZO-2 expression in children with cyanotic congenital heart defect. *Esc Heart Fail.* **2016**, *3*, 131–137. [CrossRef]
8. Gonzalez-Mariscal, L.; Miranda, J.; Raya-Sandino, A.; Dominguez-Calderon, A.; Cuellar-Perez, F. ZO-2, a tight junction protein involved in gene expression, proliferation, apoptosis, and cell size regulation. *Ann. New York Acad. Sci.* **2017**, *1397*, 35–53. [CrossRef]

9. Haskins, J.; Gu, L.; Wittchen, E.S.; Hibbard, J.; Stevenson, B.R. ZO-3, a novel member of the MAGUK protein family found at the tight junction, interacts with ZO-1 and occludin. *J. Cell Biol.* **1998**, *141*, 199–208. [CrossRef]
10. Wittchen, E.S.; Haskins, J.; Stevenson, B.R. Protein interactions at the tight junction. Actin has multiple binding partners, and ZO-1 forms independent complexes with ZO-2 and ZO-3. *J. Biol. Chem.* **1999**, *274*, 35179–35185. [CrossRef]
11. Wu, J.; Yang, Y.; Zhang, J.; Ji, P.; Du, W.; Jiang, P.; Xie, D.; Huang, H.; Wu, M.; Zhang, G.; et al. Domain-swapped dimerization of the second PDZ domain of ZO2 may provide a structural basis for the polymerization of claudins. *J. Biol. Chem.* **2007**, *282*, 35988–35999. [CrossRef] [PubMed]
12. Umeda, K.; Ikenouchi, J.; Katahira-Tayama, S.; Furuse, K.; Sasaki, H.; Nakayama, M.; Matsui, T.; Tsukita, S.; Furuse, M.; Tsukita, S. ZO-1 and ZO-2 independently determine where claudins are polymerized in tight-junction strand formation. *Cell* **2006**, *126*, 741–754. [CrossRef] [PubMed]
13. Sheth, B.; Nowak, R.L.; Anderson, R.; Kwong, W.Y.; Papenbrock, T.; Fleming, T.P. Tight junction protein ZO-2 expression and relative function of ZO-1 and ZO-2 during mouse blastocyst formation. *Exp. Cell Res.* **2008**, *314*, 3356–3368. [CrossRef] [PubMed]
14. Eckert, J.J.; Fleming, T.P. Tight junction biogenesis during early development. *Biochim. Et. Biophys. Acta.* **2008**, *1778*, 717–728. [CrossRef] [PubMed]
15. Xu, J.; Kausalya, P.J.; Phua, D.C.; Ali, S.M.; Hossain, Z.; Hunziker, W. Early embryonic lethality of mice lacking ZO-2, but Not ZO-3, reveals critical and nonredundant roles for individual zonula occludens proteins in mammalian development. *Mol. Cell Biol.* **2008**, *28*, 1669–1678. [CrossRef] [PubMed]
16. Xu, J.; Anuar, F.; Ali, S.M.; Ng, M.Y.; Phua, D.C.; Hunziker, W. Zona occludens-2 is critical for blood-testis barrier integrity and male fertility. *Mol. Biol. Cell* **2009**, *20*, 4268–4277. [CrossRef] [PubMed]
17. Hernandez, S.; Chavez Munguia, B.; Gonzalez-Mariscal, L. ZO-2 silencing in epithelial cells perturbs the gate and fence function of tight junctions and leads to an atypical monolayer architecture. *Exp. Cell Res.* **2007**, *313*, 1533–1547. [CrossRef] [PubMed]
18. Raya-Sandino, A.; Castillo-Kauil, A.; Dominguez-Calderon, A.; Alarcon, L.; Flores-Benitez, D.; Cuellar-Perez, F.; Lopez-Bayghen, B.; Chavez-Munguia, B.; Vazquez-Prado, J.; Gonzalez-Mariscal, L. Zonula occludens-2 regulates Rho proteins activity and the development of epithelial cytoarchitecture and barrier function. *Biochim. Biophys. Acta.* **2017**, *1864*, 1714–1733. [CrossRef]
19. Kim, M.A.; Kim, Y.R.; Sagong, B.; Cho, H.J.; Bae, J.W.; Kim, J.; Lee, J.; Park, H.J.; Choi, J.Y.; Lee, K.Y.; et al. Genetic analysis of genes related to tight junction function in the Korean population with non-syndromic hearing loss. *PLoS ONE* **2014**, *9*, e95646. [CrossRef]
20. Wang, H.Y.; Zhao, Y.L.; Liu, Q.; Yuan, H.; Gao, Y.; Lan, L.; Yu, L.; Wang, D.Y.; Guan, J.; Wang, Q.J. Identification of Two Disease-causing Genes TJP2 and GJB2 in a Chinese Family with Unconditional Autosomal Dominant Nonsyndromic Hereditary Hearing Impairment. *Chin. Med. J.* **2015**, *128*, 3345–3351. [CrossRef]
21. Carlton, V.E.; Harris, B.Z.; Puffenberger, E.G.; Batta, A.K.; Knisely, A.S.; Robinson, D.L.; Strauss, K.A.; Shneider, B.L.; Lim, W.A.; Salen, G.; et al. Complex inheritance of familial hypercholanemia with associated mutations in TJP2 and BAAT. *Nat. Genet.* **2003**, *34*, 91–96. [CrossRef]
22. Vitale, G.; Gitto, S.; Raimondi, F.; Mattiaccio, A.; Mantovani, V.; Vukotic, R.; D'Errico, A.; Seri, M.; Russell, R.B.; Andreone, P. Cryptogenic cholestasis in young and adults: ATP8B1, ABCB11, ABCB4, and TJP2 gene variants analysis by high-throughput sequencing. *J. Gastroenterol.* **2018**, *53*, 945–958. [CrossRef]
23. Sambrotta, M.; Strautnieks, S.; Papouli, E.; Rushton, P.; Clark, B.E.; Parry, D.A.; Logan, C.V.; Newbury, L.J.; Kamath, B.M.; Ling, S.; et al. Mutations in TJP2 cause progressive cholestatic liver disease. *Nat. Genet.* **2014**, *46*, 326–328. [CrossRef]
24. Ge, T.; Zhang, X.; Xiao, Y.; Wang, Y.; Zhang, T. Novel compound heterozygote mutations of TJP2 in a Chinese child with progressive cholestatic liver disease. *Bmc Med. Genet.* **2019**, *20*, 18. [CrossRef]
25. Zhou, S.; Hertel, P.M.; Finegold, M.J.; Wang, L.; Kerkar, N.; Wang, J.; Wong, L.J.; Plon, S.E.; Sambrotta, M.; Foskett, P.; et al. Hepatocellular carcinoma associated with tight-junction protein 2 deficiency. *Hepatology* **2015**, *62*, 1914–1916. [CrossRef]
26. Parsons, D.W.; Roy, A.; Yang, Y.; Wang, T.; Scollon, S.; Bergstrom, K.; Kerstein, R.A.; Gutierrez, S.; Petersen, A.K.; Bavle, A.; et al. Diagnostic Yield of Clinical Tumor and Germline Whole-Exome Sequencing for Children With Solid Tumors. *JAMA Oncol.* **2016**, *2*, 616–624. [CrossRef]
27. Vij, M.; Shanmugam, N.P.; Reddy, M.S.; Sankaranarayanan, S.; Rela, M. Paediatric hepatocellular carcinoma in tight junction protein 2 (TJP2) deficiency. *Virchows. Arch.* **2017**, *471*, 679–683. [CrossRef]

28. Dixon, P.H.; Sambrotta, M.; Chambers, J.; Taylor-Harris, P.; Syngelaki, A.; Nicolaides, K.; Knisely, A.S.; Thompson, R.J.; Williamson, C. An expanded role for heterozygous mutations of ABCB4, ABCB11, ATP8B1, ABCC2 and TJP2 in intrahepatic cholestasis of pregnancy. *Sci. Rep.* **2017**, *7*, 11823. [CrossRef]
29. Stanton, P.G. Regulation of the blood-testis barrier. *Semin. Cell Dev. Biol.* **2016**, *59*, 166–173. [CrossRef]
30. Ortega-Olvera, J.M.; Winkler, R.; Quitanilla-Vega, B.; Shibayama, M.; Chavez-Munguia, B.; Martin-Tapia, D.; Alarcon, L.; Gonzalez-Mariscal, L. The organophosphate pesticide methamidophos opens the blood-testis barrier and covalently binds to ZO-2 in mice. *Toxicol. Appl. Pharmacol.* **2018**, *360*, 257–272. [CrossRef]
31. Sambrotta, M.; Thompson, R.J. Mutations in TJP2, encoding zona occludens 2, and liver disease. *Tissue Barriers* **2015**, *3*, e1026537. [CrossRef]
32. Ben-Yosef, T.; Belyantseva, I.A.; Saunders, T.L.; Hughes, E.D.; Kawamoto, K.; Van Itallie, C.M.; Beyer, L.A.; Halsey, K.; Gardner, D.J.; Wilcox, E.R.; et al. Claudin 14 knockout mice, a model for autosomal recessive deafness DFNB29, are deaf due to cochlear hair cell degeneration. *Hum. Mol. Genet.* **2003**, *12*, 2049–2061. [CrossRef]
33. Kitajiri, S.; Katsuno, T.; Sasaki, H.; Ito, J.; Furuse, M.; Tsukita, S. Deafness in occludin-deficient mice with dislocation of tricellulin and progressive apoptosis of the hair cells. *Biol. Open* **2014**, *3*, 759–766. [CrossRef]
34. Kamitani, T.; Sakaguchi, H.; Tamura, A.; Miyashita, T.; Yamazaki, Y.; Tokumasu, R.; Inamoto, R.; Matsubara, A.; Mori, N.; Hisa, Y.; et al. Deletion of Tricellulin Causes Progressive Hearing Loss Associated with Degeneration of Cochlear Hair Cells. *Sci. Rep.* **2015**, *5*, 18402. [CrossRef]
35. Morozko, E.L.; Nishio, A.; Ingham, N.J.; Chandra, R.; Fitzgerald, T.; Martelletti, E.; Borck, G.; Wilson, E.; Riordan, G.P.; Wangemann, P.; et al. ILDR1 null mice, a model of human deafness DFNB42, show structural aberrations of tricellular tight junctions and degeneration of auditory hair cells. *Hum. Mol. Genet.* **2015**, *24*, 609–624. [CrossRef]
36. Nakano, Y.; Kim, S.H.; Kim, H.M.; Sanneman, J.D.; Zhang, Y.; Smith, R.J.; Marcus, D.C.; Wangemann, P.; Nessler, R.A.; Banfi, B. A claudin-9-based ion permeability barrier is essential for hearing. *Plos Genet.* **2009**, *5*, e1000610. [CrossRef]
37. Gow, A.; Davies, C.; Southwood, C.M.; Frolenkov, G.; Chrustowski, M.; Ng, L.; Yamauchi, D.; Marcus, D.C.; Kachar, B. Deafness in Claudin 11-null mice reveals the critical contribution of basal cell tight junctions to stria vascularis function. *J. Neurosci. Off. J. Soc. Neurosci.* **2004**, *24*, 7051–7062. [CrossRef]
38. Kitajiri, S.; Miyamoto, T.; Mineharu, A.; Sonoda, N.; Furuse, K.; Hata, M.; Sasaki, H.; Mori, Y.; Kubota, T.; Ito, J.; et al. Compartmentalization established by claudin-11-based tight junctions in stria vascularis is required for hearing through generation of endocochlear potential. *J. Cell Sci.* **2004**, *117*, 5087–5096. [CrossRef]
39. Islas, S.; Vega, J.; Ponce, L.; Gonzalez-Mariscal, L. Nuclear localization of the tight junction protein ZO-2 in epithelial cells. *Exp. Cell Res.* **2002**, *274*, 138–148. [CrossRef]
40. Traweger, A.; Fuchs, R.; Krizbai, I.A.; Weiger, T.M.; Bauer, H.C.; Bauer, H. The tight junction protein ZO-2 localizes to the nucleus and interacts with the heterogeneous nuclear ribonucleoprotein scaffold attachment factor-B. *J. Biol Chem.* **2003**, *278*, 2692–2700. [CrossRef]
41. Amaya, E.; Alarcon, L.; Martin-Tapia, D.; Cuellar-Perez, F.; Cano-Cortina, M.; Ortega-Olvera, J.M.; Cisneros, B.; Rodriguez, A.; Gamba, G.; Gonzalez-Mariscal, L. Activation of the Ca2+ sensing receptor and the PKC/WNK4 downstream signaling cascade induces arrival of ZO-2 to tight junctions and its separation from 14-3-3. *Mol. Biol. Cell* **2019**. In press. [CrossRef]
42. Tapia, R.; Huerta, M.; Islas, S.; Avila-Flores, A.; Lopez-Bayghen, E.; Weiske, J.; Huber, O.; Gonzalez-Mariscal, L. Zona occludens-2 inhibits cyclin D1 expression and cell proliferation and exhibits changes in localization along the cell cycle. *Mol. Biol. Cell* **2009**, *20*, 1102–1117. [CrossRef]
43. Lopez-Bayghen, E.; Jaramillo, B.E.; Huerta, M.; Betanzos, A.; Gonzalez-Mariscal, L. TJ Proteins That Make Round Trips to the Nucleus. In *Tight Junctions*; Gonzalez-Mariscal, L., Ed.; Springer US: Boston, MA, USA, 2006; pp. 76–100. [CrossRef]
44. Quiros, M.; Alarcon, L.; Ponce, A.; Giannakouros, T.; Gonzalez-Mariscal, L. The intracellular fate of zonula occludens 2 is regulated by the phosphorylation of SR repeats and the phosphorylation/O-GlcNAcylation of S257. *Mol. Biol. Cell* **2013**, *24*, 2528–2543. [CrossRef]
45. Meerschaert, K.; Tun, M.P.; Remue, E.; De Ganck, A.; Boucherie, C.; Vanloo, B.; Degeest, G.; Vandekerckhove, J.; Zimmermann, P.; Bhardwaj, N.; et al. The PDZ2 domain of zonula occludens-1 and -2 is a phosphoinositide binding domain. *Cell Mol. Life Sci.* **2009**, *66*, 3951–3966. [CrossRef]

46. Gonzalez-Mariscal, L.; Ponce, A.; Alarcon, L.; Jaramillo, B.E. The tight junction protein ZO-2 has several functional nuclear export signals. *Exp. Cell Res.* **2006**, *312*, 3323–3335. [CrossRef]
47. Chamorro, D.; Alarcon, L.; Ponce, A.; Tapia, R.; Gonzalez-Aguilar, H.; Robles-Flores, M.; Mejia-Castillo, T.; Segovia, J.; Bandala, Y.; Juaristi, E.; et al. Phosphorylation of zona occludens-2 by protein kinase C epsilon regulates its nuclear exportation. *Mol. Biol. Cell* **2009**, *20*, 4120–4129. [CrossRef]
48. Wetzel, F.; Mittag, S.; Cano-Cortina, M.; Wagner, T.; Kramer, O.H.; Niedenthal, R.; Gonzalez-Mariscal, L.; Huber, O. SUMOylation regulates the intracellular fate of ZO-2. *Cell. Mol. Life Sci. CMLS* **2016**, *74*, 373–392. [CrossRef]
49. Gonzalez-Mariscal, L.; Contreras, R.G.; Bolivar, J.J.; Ponce, A.; Chavez De Ramirez, B.; Cereijido, M. Role of calcium in tight junction formation between epithelial cells. *Am. J. Physiol.* **1990**, *259*, C978–C986. [CrossRef]
50. Jouret, F.; Wu, J.; Hull, M.; Rajendran, V.; Mayr, B.; Schofl, C.; Geibel, J.; Caplan, M.J. Activation of the Ca(2)+-sensing receptor induces deposition of tight junction components to the epithelial cell plasma membrane. *J. Cell Sci.* **2013**, *126*, 5132–5142. [CrossRef]
51. Jaramillo, B.E.; Ponce, A.; Moreno, J.; Betanzos, A.; Huerta, M.; Lopez-Bayghen, E.; Gonzalez-Mariscal, L. Characterization of the tight junction protein ZO-2 localized at the nucleus of epithelial cells. *Exp. Cell Res.* **2004**, *297*, 247–258. [CrossRef]
52. Lechuga, S.; Alarcon, L.; Solano, J.; Huerta, M.; Lopez-Bayghen, E.; Gonzalez-Mariscal, L. Identification of ZASP, a novel protein associated to Zona occludens-2. *Exp. Cell Res.* **2010**, *316*, 3124–3139. [CrossRef]
53. Hong, E.A.; Gautrey, H.L.; Elliott, D.J.; Tyson-Capper, A.J. SAFB1- and SAFB2-mediated transcriptional repression: Relevance to cancer. *Biochem. Soc. Trans.* **2012**, *40*, 826–830. [CrossRef]
54. Lin, J.; Xu, P.; LaVallee, P.; Hoidal, J.R. Identification of proteins binding to E-Box/Ku86 sites and function of the tumor suppressor SAFB1 in transcriptional regulation of the human xanthine oxidoreductase gene. *J. Biol. Chem.* **2008**, *283*, 29681–29689. [CrossRef]
55. Huerta, M.; Munoz, R.; Tapia, R.; Soto-Reyes, E.; Ramirez, L.; Recillas-Targa, F.; Gonzalez-Mariscal, L.; Lopez-Bayghen, E. Cyclin D1 is transcriptionally down-regulated by ZO-2 via an E box and the transcription factor c-Myc. *Mol. Biol. Cell* **2007**, *18*, 4826–4836. [CrossRef]
56. Townson, S.M.; Kang, K.; Lee, A.V.; Oesterreich, S. Structure-function analysis of the estrogen receptor alpha corepressor scaffold attachment factor-B1: Identification of a potent transcriptional repression domain. *J. Biol. Chem.* **2004**, *279*, 26074–26081. [CrossRef]
57. Nikolakaki, E.; Kohen, R.; Hartmann, A.M.; Stamm, S.; Georgatsou, E.; Giannakouros, T. Cloning and characterization of an alternatively spliced form of SR protein kinase 1 that interacts specifically with scaffold attachment factor-B. *J. Biol. Chem.* **2001**, *276*, 40175–40182. [CrossRef]
58. McCracken, S.; Longman, D.; Marcon, E.; Moens, P.; Downey, M.; Nickerson, J.A.; Jessberger, R.; Wilde, A.; Caceres, J.F.; Emili, A.; et al. Proteomic analysis of SRm160-containing complexes reveals a conserved association with cohesin. *J. Biol. Chem.* **2005**, *280*, 42227–42236. [CrossRef]
59. Tsai, Y.C.; Greco, T.M.; Boonmee, A.; Miteva, Y.; Cristea, I.M. Functional proteomics establishes the interaction of SIRT7 with chromatin remodeling complexes and expands its role in regulation of RNA polymerase I transcription. *Mol. Cell. Proteom. MCP* **2012**, *11*, 60–76. [CrossRef]
60. Havugimana, P.C.; Hart, G.T.; Nepusz, T.; Yang, H.; Turinsky, A.L.; Li, Z.; Wang, P.I.; Boutz, D.R.; Fong, V.; Phanse, S.; et al. A census of human soluble protein complexes. *Cell* **2012**, *150*, 1068–1081. [CrossRef]
61. Tang, Y.; Puri, A.; Ricketts, M.D.; Rai, T.S.; Hoffmann, J.; Hoi, E.; Adams, P.D.; Schultz, D.C.; Marmorstein, R. Identification of an ubinuclein 1 region required for stability and function of the human HIRA/UBN1/CABIN1/ASF1a histone H3.3 chaperone complex. *Biochemistry* **2012**, *51*, 2366–2377. [CrossRef]
62. Lupo, J.; Conti, A.; Sueur, C.; Coly, P.A.; Coute, Y.; Hunziker, W.; Burmeister, W.P.; Germi, R.; Manet, E.; Gruffat, H.; et al. Identification of new interacting partners of the shuttling protein ubinuclein (Ubn-1). *Exp. Cell Res.* **2012**, *318*, 509–520. [CrossRef]
63. Bernstein, K.A.; Gallagher, J.E.; Mitchell, B.M.; Granneman, S.; Baserga, S.J. The small-subunit processome is a ribosome assembly intermediate. *Eukaryot Cell* **2004**, *3*, 1619–1626. [CrossRef]
64. Leung, J.W.; Makharashvili, N.; Agarwal, P.; Chiu, L.Y.; Pourpre, R.; Cammarata, M.B.; Cannon, J.R.; Sherker, A.; Durocher, D.; Brodbelt, J.S.; et al. ZMYM3 regulates BRCA1 localization at damaged chromatin to promote DNA repair. *Genes Dev.* **2017**, *31*, 260–274. [CrossRef]

65. Bonnelye, E.; Vanacker, J.M.; Dittmar, T.; Begue, A.; Desbiens, X.; Denhardt, D.T.; Aubin, J.E.; Laudet, V.; Fournier, B. The ERR-1 orphan receptor is a transcriptional activator expressed during bone development. *Mol. Endocrinol.* **1997**, *11*, 905–916. [CrossRef]
66. Wan, C.; Borgeson, B.; Phanse, S.; Tu, F.; Drew, K.; Clark, G.; Xiong, X.; Kagan, O.; Kwan, J.; Bezginov, A.; et al. Panorama of ancient metazoan macromolecular complexes. *Nature* **2015**, *525*, 339–344. [CrossRef]
67. Shaheen, R.; Tasak, M.; Maddirevula, S.; Abdel-Salam, G.M.H.; Sayed, I.S.M.; Alazami, A.M.; Al-Sheddi, T.; Alobeid, E.; Phizicky, E.M.; Alkuraya, F.S. PUS7 mutations impair pseudouridylation in humans and cause intellectual disability and microcephaly. *Hum. Genet.* **2019**, *138*, 231–239. [CrossRef]
68. Betanzos, A.; Huerta, M.; Lopez-Bayghen, E.; Azuara, E.; Amerena, J.; Gonzalez-Mariscal, L. The tight junction protein ZO-2 associates with Jun, Fos and C/EBP transcription factors in epithelial cells. *Exp. Cell Res.* **2004**, *292*, 51–66. [CrossRef]
69. Kao, S.H.; Wang, W.L.; Chen, C.Y.; Chang, Y.L.; Wu, Y.Y.; Wang, Y.T.; Wang, S.P.; Nesvizhskii, A.I.; Chen, Y.J.; Hong, T.M.; et al. GSK3beta controls epithelial-mesenchymal transition and tumor metastasis by CHIP-mediated degradation of Slug. *Oncogene* **2014**, *33*, 3172–3182. [CrossRef]
70. Huttlin, E.L.; Ting, L.; Bruckner, R.J.; Gebreab, F.; Gygi, M.P.; Szpyt, J.; Tam, S.; Zarraga, G.; Colby, G.; Baltier, K.; et al. The BioPlex Network: A Systematic Exploration of the Human Interactome. *Cell* **2015**, *162*, 425–440. [CrossRef]
71. Clevers, H.; Nusse, R. Wnt/beta-catenin signaling and disease. *Cell* **2012**, *149*, 1192–1205. [CrossRef]
72. Nusse, R.; Clevers, H. Wnt/beta-Catenin Signaling, Disease, and Emerging Therapeutic Modalities. *Cell* **2017**, *169*, 985–999. [CrossRef]
73. Dominguez-Calderon, A.; Avila-Flores, A.; Ponce, A.; Lopez-Bayghen, E.; Calderon-Salinas, J.V.; Luis Reyes, J.; Chavez-Munguia, B.; Segovia, J.; Angulo, C.; Ramirez, L.; et al. ZO-2 silencing induces renal hypertrophy through a cell cycle mechanism and the activation of YAP and the mTOR pathway. *Mol. Biol. Cell* **2016**, *27*, 1581–1595. [CrossRef]
74. Palmer, H.G.; Gonzalez-Sancho, J.M.; Espada, J.; Berciano, M.T.; Puig, I.; Baulida, J.; Quintanilla, M.; Cano, A.; de Herreros, A.G.; Lafarga, M.; et al. Vitamin D(3) promotes the differentiation of colon carcinoma cells by the induction of E-cadherin and the inhibition of beta-catenin signaling. *J. Cell Biol.* **2001**, *154*, 369–387. [CrossRef]
75. Bautista-Garcia, P.; Reyes, J.L.; Martin, D.; Namorado, M.C.; Chavez-Munguia, B.; Soria-Castro, E.; Huber, O.; Gonzalez-Mariscal, L. Zona occludens-2 protects against podocyte dysfunction induced by ADR in mice. *Am. J. Physiol. Ren. Physiol.* **2013**, *304*, F77–F87. [CrossRef]
76. Meng, Z.; Moroishi, T.; Guan, K.L. Mechanisms of Hippo pathway regulation. *Genes Dev.* **2016**, *30*, 1–17. [CrossRef]
77. Azzolin, L.; Panciera, T.; Soligo, S.; Enzo, E.; Bicciato, S.; Dupont, S.; Bresolin, S.; Frasson, C.; Basso, G.; Guzzardo, V.; et al. YAP/TAZ incorporation in the beta-catenin destruction complex orchestrates the Wnt response. *Cell* **2014**, *158*, 157–170. [CrossRef]
78. Zhao, B.; Li, L.; Tumaneng, K.; Wang, C.Y.; Guan, K.L. A coordinated phosphorylation by Lats and CK1 regulates YAP stability through SCF(beta-TRCP). *Genes Dev.* **2010**, *24*, 72–85. [CrossRef]
79. Kusch, A.; Tkachuk, S.; Tkachuk, N.; Patecki, M.; Park, J.K.; Dietz, R.; Haller, H.; Dumler, I. The tight junction protein ZO-2 mediates proliferation of vascular smooth muscle cells via regulation of Stat1. *Cardiovasc. Res.* **2009**, *83*, 115–122. [CrossRef]
80. Oka, T.; Remue, E.; Meerschaert, K.; Vanloo, B.; Boucherie, C.; Gfeller, D.; Bader, G.D.; Sidhu, S.S.; Vandekerckhove, J.; Gettemans, J.; et al. Functional complexes between YAP2 and ZO-2 are PDZ domain-dependent, and regulate YAP2 nuclear localization and signalling. *Biochem J.* **2010**, *432*, 461–472. [CrossRef]
81. Diehl, J.A.; Cheng, M.; Roussel, M.F.; Sherr, C.J. Glycogen synthase kinase-3beta regulates cyclin D1 proteolysis and subcellular localization. *Genes Dev.* **1998**, *12*, 3499–3511. [CrossRef]
82. Klein, P.S.; Melton, D.A. A molecular mechanism for the effect of lithium on development. *Proc. Natl. Acad. Sci. USA* **1996**, *93*, 8455–8459. [CrossRef]
83. Hedgepeth, C.M.; Conrad, L.J.; Zhang, J.; Huang, H.C.; Lee, V.M.; Klein, P.S. Activation of the Wnt signaling pathway: A molecular mechanism for lithium action. *Dev. Biol.* **1997**, *185*, 82–91. [CrossRef]
84. Qiao, X.; Roth, I.; Feraille, E.; Hasler, U. Different effects of ZO-1, ZO-2 and ZO-3 silencing on kidney collecting duct principal cell proliferation and adhesion. *Cell Cycle* **2014**, *13*, 3059–3075. [CrossRef]

85. Traweger, A.; Lehner, C.; Farkas, A.; Krizbai, I.A.; Tempfer, H.; Klement, E.; Guenther, B.; Bauer, H.C.; Bauer, H. Nuclear Zonula occludens-2 alters gene expression and junctional stability in epithelial and endothelial cells. *Differ. Res. Biol. Divers.* **2008**, *76*, 99–106. [CrossRef]
86. Iqbal, M.A.; Gupta, V.; Gopinath, P.; Mazurek, S.; Bamezai, R.N. Pyruvate kinase M2 and cancer: An updated assessment. *Febs Lett.* **2014**, *588*, 2685–2692. [CrossRef]
87. Bojarski, C.; Weiske, J.; Schoneberg, T.; Schroder, W.; Mankertz, J.; Schulzke, J.D.; Florian, P.; Fromm, M.; Tauber, R.; Huber, O. The specific fates of tight junction proteins in apoptotic epithelial cells. *J. Cell Sci.* **2004**, *117*, 2097–2107. [CrossRef]
88. Chaitanya, G.V.; Steven, A.J.; Babu, P.P. PARP-1 cleavage fragments: Signatures of cell-death proteases in neurodegeneration. *Cell Commun Signal.* **2010**, *8*, 31. [CrossRef]
89. Liu, J.; Yuan, Q.; Ling, X.; Tan, Q.; Liang, H.; Chen, J.; Lin, L.; Xiao, Y.; Chen, W.; Liu, L.; et al. PARP1 may be involved in hydroquinoneinduced apoptosis by poly ADPribosylation of ZO2. *Mol. Med. Rep.* **2017**, *16*, 8076–8084. [CrossRef]
90. Kazmierczak, M.; Harris, S.L.; Kazmierczak, P.; Shah, P.; Starovoytov, V.; Ohlemiller, K.K.; Schwander, M. Progressive Hearing Loss in Mice Carrying a Mutation in Usp53. *J. Neurosci. : Off. J. Soc. Neurosci.* **2015**, *35*, 15582–15598. [CrossRef]
91. Maddirevula, S.; Alhebbi, H.; Alqahtani, A.; Algoufi, T.; Alsaif, H.S.; Ibrahim, N.; Abdulwahab, F.; Barr, M.; Alzaidan, H.; Almehaideb, A.; et al. Identification of novel loci for pediatric cholestatic liver disease defined by KIF12, PPM1F, USP53, LSR, and WDR83OS pathogenic variants. *Genet. Med. : Off. J. Am. Coll. Med. Genet.* **2019**, *21*, 1164–1172. [CrossRef]
92. Oka, T.; Schmitt, A.P.; Sudol, M. Opposing roles of angiomotin-like-1 and zona occludens-2 on pro-apoptotic function of YAP. *Oncogene* **2012**, *31*, 128–134. [CrossRef]
93. Fanning, A.S.; Van Itallie, C.M.; Anderson, J.M. Zonula occludens-1 and -2 regulate apical cell structure and the zonula adherens cytoskeleton in polarized epithelia. *Mol. Biol. Cell* **2012**, *23*, 577–590. [CrossRef]
94. Mattagajasingh, S.N.; Huang, S.C.; Hartenstein, J.S.; Benz, E.J., Jr. Characterization of the interaction between protein 4.1R and ZO-2. A possible link between the tight junction and the actin cytoskeleton. *J. Biol Chem* **2000**, *275*, 30573–30585. [CrossRef]
95. Baines, A.J.; Lu, H.C.; Bennett, P.M. The Protein 4.1 family: Hub proteins in animals for organizing membrane proteins. *Biochim. Et Biophys. Acta* **2014**, *1838*, 605–619. [CrossRef]
96. Monteiro, A.C.; Sumagin, R.; Rankin, C.R.; Leoni, G.; Mina, M.J.; Reiter, D.M.; Stehle, T.; Dermody, T.S.; Schaefer, S.A.; Hall, R.A.; et al. JAM-A associates with ZO-2, afadin, and PDZ-GEF1 to activate Rap2c and regulate epithelial barrier function. *Mol. Biol. Cell* **2013**, *24*, 2849–2860. [CrossRef]
97. Yamazaki, Y.; Umeda, K.; Wada, M.; Nada, S.; Okada, M.; Tsukita, S.; Tsukita, S. ZO-1- and ZO-2-dependent integration of myosin-2 to epithelial zonula adherens. *Mol. Biol. Cell* **2008**, *19*, 3801–3811. [CrossRef]
98. Umeda, K.; Matsui, T.; Nakayama, M.; Furuse, K.; Sasaki, H.; Furuse, M.; Tsukita, S. Establishment and characterization of cultured epithelial cells lacking expression of ZO-1. *J. Biol Chem* **2004**, *279*, 44785–44794. [CrossRef]
99. Cartagena-Rivera, A.X.; Van Itallie, C.M.; Anderson, J.M.; Chadwick, R.S. Apical surface supracellular mechanical properties in polarized epithelium using noninvasive acoustic force spectroscopy. *Nat. Commun.* **2017**, *8*, 1030. [CrossRef]
100. Van Itallie, C.M.; Fanning, A.S.; Bridges, A.; Anderson, J.M. ZO-1 stabilizes the tight junction solute barrier through coupling to the perijunctional cytoskeleton. *Mol. Biol. Cell* **2009**, *20*, 3930–3940. [CrossRef]
101. Tokuda, S.; Higashi, T.; Furuse, M. ZO-1 knockout by TALEN-mediated gene targeting in MDCK cells: Involvement of ZO-1 in the regulation of cytoskeleton and cell shape. *PLoS ONE* **2014**, *9*, e104994. [CrossRef]
102. Spadaro, D.; Le, S.; Laroche, T.; Mean, I.; Jond, L.; Yan, J.; Citi, S. Tension-Dependent Stretching Activates ZO-1 to Control the Junctional Localization of Its Interactors. *Curr. Biol. : Cb* **2017**, *27*, 3783–3795.e8. [CrossRef]
103. Handorf, A.M.; Zhou, Y.; Halanski, M.A.; Li, W.J. Tissue stiffness dictates development, homeostasis, and disease progression. *Organogenesis* **2015**, *11*, 1–15. [CrossRef]
104. Shinto, O.; Yashiro, M.; Kawajiri, H.; Shimizu, K.; Shimizu, T.; Miwa, A.; Hirakawa, K. Inhibitory effect of a TGFbeta receptor type-I inhibitor, Ki26894, on invasiveness of scirrhous gastric cancer cells. *Br. J. Cancer* **2010**, *102*, 844–851. [CrossRef]
105. Jaffe, A.B.; Kaji, N.; Durgan, J.; Hall, A. Cdc42 controls spindle orientation to position the apical surface during epithelial morphogenesis. *J. Cell Biol.* **2008**, *183*, 625–633. [CrossRef]

106. Etienne-Manneville, S.; Hall, A. Integrin-mediated activation of Cdc42 controls cell polarity in migrating astrocytes through PKCzeta. *Cell* **2001**, *106*, 489–498. [CrossRef]
107. Gupta, G.D.; Coyaud, E.; Goncalves, J.; Mojarad, B.A.; Liu, Y.; Wu, Q.; Gheiratmand, L.; Comartin, D.; Tkach, J.M.; Cheung, S.W.; et al. A Dynamic Protein Interaction Landscape of the Human Centrosome-Cilium Interface. *Cell* **2015**, *163*, 1484–1499. [CrossRef]
108. Monnich, M.; Borgeskov, L.; Breslin, L.; Jakobsen, L.; Rogowski, M.; Doganli, C.; Schroder, J.M.; Mogensen, J.B.; Blinkenkjaer, L.; Harder, L.M.; et al. CEP128 Localizes to the Subdistal Appendages of the Mother Centriole and Regulates TGF-beta/BMP Signaling at the Primary Cilium. *Cell Rep.* **2018**, *22*, 2584–2592. [CrossRef]
109. Kashihara, H.; Chiba, S.; Kanno, S.I.; Suzuki, K.; Yano, T.; Tsukita, S. Cep128 associates with Odf2 to form the subdistal appendage of the centriole. *Genes Cells : Devoted Mol. Cell. Mech.* **2019**, *24*, 231–243. [CrossRef]
110. Hung, H.F.; Hehnly, H.; Doxsey, S. The Mother Centriole Appendage Protein Cenexin Modulates Lumen Formation through Spindle Orientation. *Curr. Biol. : Cb* **2016**, *26*, 793–801. [CrossRef]
111. Ganapathi Sankaran, D.; Stemm-Wolf, A.J.; Pearson, C.G. CEP135 isoform dysregulation promotes centrosome amplification in breast cancer cells. *Mol. Biol. Cell* **2019**, *30*, 1230–1244. [CrossRef]
112. Gromley, A.; Jurczyk, A.; Sillibourne, J.; Halilovic, E.; Mogensen, M.; Groisman, I.; Blomberg, M.; Doxsey, S. A novel human protein of the maternal centriole is required for the final stages of cytokinesis and entry into S phase. *J. Cell Biol.* **2003**, *161*, 535–545. [CrossRef]
113. Liu, B.; Preisig, P. TGF-beta1-mediated hypertrophy involves inhibiting pRB phosphorylation by blocking activation of cyclin E kinase. *Am. J. Physiol.* **1999**, *277*, F186–F194.
114. Liu, B.; Preisig, P.A. Compensatory renal hypertrophy is mediated by a cell cycle-dependent mechanism. *Kidney Int.* **2002**, *62*, 1650–1658. [CrossRef]
115. Jurkovitz, C.T.; England, B.K.; Ebb, R.G.; Mitch, W.E. Influence of ammonia and pH on protein and amino acid metabolism in LLC-PK1 cells. *Kidney Int.* **1992**, *42*, 595–601. [CrossRef]
116. Ling, H.; Vamvakas, S.; Gekle, M.; Schaefer, L.; Teschner, M.; Schaefer, R.M.; Heidland, A. Role of lysosomal cathepsin activities in cell hypertrophy induced by NH4Cl in cultured renal proximal tubule cells. *J. Am. Soc. Nephrol. : Jasn* **1996**, *7*, 73–80.
117. Glaunsinger, B.A.; Weiss, R.S.; Lee, S.S.; Javier, R. Link of the unique oncogenic properties of adenovirus type 9 E4-ORF1 to a select interaction with the candidate tumor suppressor protein ZO-2. *Embo J.* **2001**, *20*, 5578–5586. [CrossRef]
118. Thomas, M.; Myers, M.P.; Massimi, P.; Guarnaccia, C.; Banks, L. Analysis of Multiple HPV E6 PDZ Interactions Defines Type-Specific PDZ Fingerprints That Predict Oncogenic Potential. *Plos Pathog.* **2016**, *12*, e1005766. [CrossRef]
119. Hernandez-Monge, J.; Garay, E.; Raya-Sandino, A.; Vargas-Sierra, O.; Diaz-Chavez, J.; Popoca-Cuaya, M.; Lambert, P.F.; Gonzalez-Mariscal, L.; Gariglio, P. Papillomavirus E6 oncoprotein up-regulates occludin and ZO-2 expression in ovariectomized mice epidermis. *Exp. Cell Res.* **2013**, *319*, 2588–2603. [CrossRef]
120. Grm, H.S.; Banks, L. Degradation of hDlg and MAGIs by human papillomavirus E6 is E6-AP-independent. *J. Gen. Virol.* **2004**, *85*, 2815–2819. [CrossRef]
121. Lee, S.S.; Glaunsinger, B.; Mantovani, F.; Banks, L.; Javier, R.T. Multi-PDZ domain protein MUPP1 is a cellular target for both adenovirus E4-ORF1 and high-risk papillomavirus type 18 E6 oncoproteins. *J. Virol.* **2000**, *74*, 9680–9693. [CrossRef]
122. Nakagawa, S.; Huibregtse, J.M. Human scribble (Vartul) is targeted for ubiquitin-mediated degradation by the high-risk papillomavirus E6 proteins and the E6AP ubiquitin-protein ligase. *Mol. Cell. Biol.* **2000**, *20*, 8244–8253. [CrossRef]
123. Fink, C.; Weigel, R.; Hembes, T.; Lauke-Wettwer, H.; Kliesch, S.; Bergmann, M.; Brehm, R.H. Altered expression of ZO-1 and ZO-2 in Sertoli cells and loss of blood-testis barrier integrity in testicular carcinoma in situ. *Neoplasia* **2006**, *8*, 1019–1027. [CrossRef]
124. Luczka, E.; Syne, L.; Nawrocki-Raby, B.; Kileztky, C.; Hunziker, W.; Birembaut, P.; Gilles, C.; Polette, M. Regulation of membrane-type 1 matrix metalloproteinase expression by zonula occludens-2 in human lung cancer cells. *Clin. Exp. Metastasis* **2013**. [CrossRef]
125. Saito, K.; Enya, K.; Oneyama, C.; Hikita, T.; Okada, M. Proteomic identification of ZO-1/2 as a novel scaffold for Src/Csk regulatory circuit. *Biochem. Biophys. Res. Commun.* **2008**, *366*, 969–975. [CrossRef]

126. Chlenski, A.; Ketels, K.V.; Engeriser, J.L.; Talamonti, M.S.; Tsao, M.S.; Koutnikova, H.; Oyasu, R.; Scarpelli, D.G. zo-2 gene alternative promoters in normal and neoplastic human pancreatic duct cells. *Int J. Cancer* **1999**, *83*, 349–358. [CrossRef]
127. Chlenski, A.; Ketels, K.V.; Tsao, M.S.; Talamonti, M.S.; Anderson, M.R.; Oyasu, R.; Scarpelli, D.G. Tight junction protein ZO-2 is differentially expressed in normal pancreatic ducts compared to human pancreatic adenocarcinoma. *Int. J. Cancer* **1999**, *82*, 137–144. [CrossRef]
128. Chlenski, A.; Ketels, K.V.; Korovaitseva, G.I.; Talamonti, M.S.; Oyasu, R.; Scarpelli, D.G. Organization and expression of the human zo-2 gene (tjp-2) in normal and neoplastic tissues. *Biochim. Et Biophys. Acta* **2000**, *1493*, 319–324. [CrossRef]
129. Kim, Y.J.; Jung, Y.D.; Kim, T.O.; Kim, H.S. Alu-related transcript of TJP2 gene as a marker for colorectal cancer. *Gene* **2013**, *524*, 268–274. [CrossRef]
130. Tokes, A.M.; Szasz, A.M.; Juhasz, E.; Schaff, Z.; Harsanyi, L.; Molnar, I.A.; Baranyai, Z.; Besznyak, I., Jr.; Zarand, A.; Salamon, F.; et al. Expression of tight junction molecules in breast carcinomas analysed by array PCR and immunohistochemistry. *Pathol. Oncol. Res.* **2012**, *18*, 593–606. [CrossRef]
131. Kato, Y.; Yashiro, M.; Noda, S.; Tendo, M.; Kashiwagi, S.; Doi, Y.; Nishii, T.; Matsuoka, J.; Fuyuhiro, Y.; Shinto, O.; et al. Establishment and characterization of a new hypoxia-resistant cancer cell line, OCUM-12/Hypo, derived from a scirrhous gastric carcinoma. *Br. J. Cancer* **2010**, *102*, 898–907. [CrossRef]
132. Pope, W.B.; Chen, J.H.; Dong, J.; Carlson, M.R.; Perlina, A.; Cloughesy, T.F.; Liau, L.M.; Mischel, P.S.; Nghiemphu, P.; Lai, A.; et al. Relationship between gene expression and enhancement in glioblastoma multiforme: Exploratory DNA microarray analysis. *Radiology* **2008**, *249*, 268–277. [CrossRef]
133. Deroo, B.J.; Korach, K.S. Estrogen receptors and human disease. *J. Clin. Investig.* **2006**, *116*, 561–570. [CrossRef]
134. Acconcia, F.; Marino, M. The Effects of 17beta-estradiol in Cancer are Mediated by Estrogen Receptor Signaling at the Plasma Membrane. *Front. Physiol.* **2011**, *2*, 30. [CrossRef]
135. Doi, Y.; Yashiro, M.; Yamada, N.; Amano, R.; Noda, S.; Hirakawa, K. VEGF-A/VEGFR-2 signaling plays an important role for the motility of pancreas cancer cells. *Ann. Surg. Oncol.* **2012**, *19*, 2733–2743. [CrossRef]

Claudin-19 is Regulated by Extracellular Osmolality in Rat Kidney Inner Medullary Collecting Duct Cells

Annalisa Ziemens [1], Svenja R. Sonntag [1,2], Vera C. Wulfmeyer [1,3], Bayram Edemir [4], Markus Bleich [1] and Nina Himmerkus [1,*]

[1] Institute of Physiology, Christian-Albrecht-University Kiel, Hermann-Rodewald-Str. 5, 24118 Kiel, Germany
[2] Department of Ophthalmology, University of Lübeck, Ratzeburger Allee 160, 23538 Lübeck, Germany
[3] Department of Nephrology and Hypertension, Hannover Medical School, Carl-Neuberg-Str. 1, 30625 Hannover, Germany
[4] Department of Internal Medicine IV, Hematology and Oncology, University Hospital Halle, Ernst-Grube-Str. 40, 06120 Halle (Saale), Germany
[*] Correspondence: n.himmerkus@physiologie.uni-kiel.de.

Abstract: The inner medullary collecting duct (IMCD) is subject to severe changes in ambient osmolality and must either allow water transport or be able to seal the lumen against a very high osmotic pressure. We postulate that the tight junction protein claudin-19 is expressed in IMCD and that it takes part in epithelial adaptation to changing osmolality at different functional states. Presence of claudin-19 in rat IMCD was investigated by Western blotting and immunofluorescence. Primary cell culture of rat IMCD cells on permeable filter supports was performed under different osmotic culture conditions and after stimulation by antidiuretic hormone (AVP). Electrogenic transepithelial transport properties were measured in Ussing chambers. IMCD cells cultivated at 300 mosm/kg showed high transepithelial resistance, a cation selective paracellular pathway and claudin-19 was mainly located in the tight junction. Treatment by AVP increased cation selectivity but did not alter transepithelial resistance or claudin-19 subcellular localization. In contrast, IMCD cells cultivated at 900 mosm/kg had low transepithelial resistance, anion selectivity, and claudin-19 was relocated from the tight junctions to intracellular vesicles. The data shows osmolality-dependent transformation of IMCD epithelium from tight and sodium-transporting to leaky, with claudin-19 expression in the tight junction associated to tightness and cation selectivity under low osmolality.

Keywords: barrier function; paracellular permeability; antidiuretic hormone

---

1. Introduction

The inner medullary collecting duct (IMCD) is the last segment of the nephron. It is embedded in the unique structure of the inner medulla together with descending and ascending thin limbs of Henle's loop and the accompanying vasa recta, which are specialized capillaries [1,2]. Maintenance of body water homeostasis under extreme variations in water intake is achieved under hormonal regulation, especially AVP signaling. Thereby, urine osmolality can vary between 50 mosm/kg and 1500 mosm/kg in humans. Of high importance is the interstitial osmolality provided by the main osmolytes, NaCl and urea, and the cortico-medullary concentration gradient established by the loop of Henle [3,4] countercurrent concentration system. During antidiuresis, interstitial and intraluminal osmolality is high and the transepithelial gradient is levelled by the water flux from the lumen to the interstitium through water channel insertion into the cell membrane. Under water diuresis, interstitial osmolality is lower; however, high NaCl concentration is conserved and mainly urea concentration is changing with diuretic state [5,6]. Therefore, the now transcellularly water tight collecting duct faces a high transepithelial osmotic gradient. The interstitial accumulation of urea during antidiuresis is tightly controlled by antidiuretic hormone (AVP) but also by other factors (e.g., endothelin, PGE2) [4]. Hypertonicity and urea are thereby cell stressors and cellular osmo-homeostasis has to be controlled and adapted under these changing conditions [7,8]. Osmo-protective responses include accumulation of compatible organic osmolytes, abundance of heat shock proteins, and control of transmembrane water flux by water channels. Ionic and osmotic gradients under both antidiuresis or water diuresis between the luminal and the interstitial side of the epithelium do not only challenge the transcellular barrier, but also the paracellular route. The paracellular cleft between epithelial cells is specifically sealed by the tight junction (TJ). This complex is composed of a variety of different proteins fulfilling a broad spectrum of tasks in addition to barrier formation and provision of permeability, including anchoring the TJ proteins to the cytoskeleton and cell signaling. The proteins responsible for tightness and permeability are claudins, a family comprising more than 25 different members to date (for review, see [9–11]). From a functional perspective, IMCD paracellular properties can vary with respect to water permeability and ion selectivity. Under diuretic conditions, in the absence of AVP, the collecting duct is sealed against water permeation [12]. In antidiuresis, water channels provide a high transcellular water permeability. Early studies in the 1990s by Flamion et al. suggested a potential additional paracellular route for water in IMCD under strong antidiuresis [13]. So far, different claudins have been described for collecting ducts (claudin-3, -4, -7, -8, -10, -19) [14]. Especially, claudin-4 and -8 have been implicated in paracellular selectivity and ion transport [15]. Claudin-19 was described first in 2004 [16] and in the kidney, it is mainly expressed in the thick ascending limb of Henle's Loop (TAL) [17–19]. Mutations in humans lead to familial hypomagnesemia with hypercalciuria and nephrocalcinosis (FHHNC, OMIM 610036, [17]) which also can be caused by mutations in claudin-16 [20,21]. Both TAL claudins are required for functional TAL Ca^{2+} and Mg^{2+} transport [19,22]. In 2006, Lee et al. described a different expression pattern identified by antibody staining, showing claudin-19 in the distal tubule (i.e., mainly TAL) and the collecting duct in humans and rats [23]. In RNA-seq analysis of microdissected rat kidney tubule segments, claudin-19 mRNA has been found in more upstream collecting duct segments but hardly any in IMCD [24]. By immunofluorescence staining, our group repetitively encountered claudin-19 expression in segments more distal to the TAL and decided to investigate its expression and function in the inner medulla.

We hypothesize that claudin-19 is part of the IMCD TJ, that IMCD cells adapt to osmolality or AVP stimulation, and that claudin-19 takes part in this adaptation. We used the well-established model of primary cultivated inner medullary collecting duct cells grown on permeable supports (IMCDs) [25,26]. Cells were cultivated under hormone stimulation and different osmotic challenges. Under cell culture conditions with proliferating cells not yet confluent and therefore not yet forming a complete epithelial layer, naturally occurring highly variable osmotic conditions are not easy to copy. We focused on long term (days) basolateral changes, omitting transepithelial gradients. Plasma-isotonicity represented cortical conditions and two increasing medullary osmolalities represented either the medullary axis or different diuretic states. Under low osmolality, claudin-19 localized to the TJ of IMCD, most likely fulfilling the sealing function and being responsible for high transepithelial resistance.

2. Results

2.1. *Claudin19 Is Expressed in Inner Medullary Collecting Duct*

We investigated claudin-19 expression using WB analysis in both mouse and rat tissue. Along the nephron of the mouse (Figure 1A), claudin-19 protein was absent in glomeruli and proximal tubule (PT). In TAL and all segments further downstream, claudin-19 was expressed and protein abundance was highest in TAL, medullary CD (mCD), and in papilla tissue. Qualitatively similar results were obtained along the rat nephron and here we could specifically detect claudin-19 also in dissected and isolated IMCD (Figure 1B). We performed claudin-19 immunofluorescence staining in mouse and rat inner medulla (Figure 1C,D). Claudin-19 was expressed in both, IMCD and thin limbs. In IMCD, claudin-19 was located mostly in the TJ, i.e., line- or dot-shaped fluorescence along the apical cell cleft but also extending to the lateral cell borders (Figure 1D). In addition, some intracellular vesicular staining was visible at the basal cell pole of the IMCD cells. Costaining with the intercalated cell anion exchanger AE1 confirmed IMCD localization. Thin limb TJs showed the typical meandering pattern (Figure 1C [27]). Next, we investigated claudin-19 protein in rat kidney tissue in a more quantitative approach and tried to extract membrane proteins first using the mild nonionic detergent triton-x-100 (Triton), followed by a harsher extraction of the remaining membrane proteins by incubation with the anionic surfactant sodium dodecyl sulfate (SDS). As shown in Figure 1E, inner medullary claudin-19 is mainly already solubilized by Triton extraction, whereas outer medullary claudin-19 (i.e., mainly TAL) only solubilized in the presence of SDS, indicating a different TJ structure and organization.

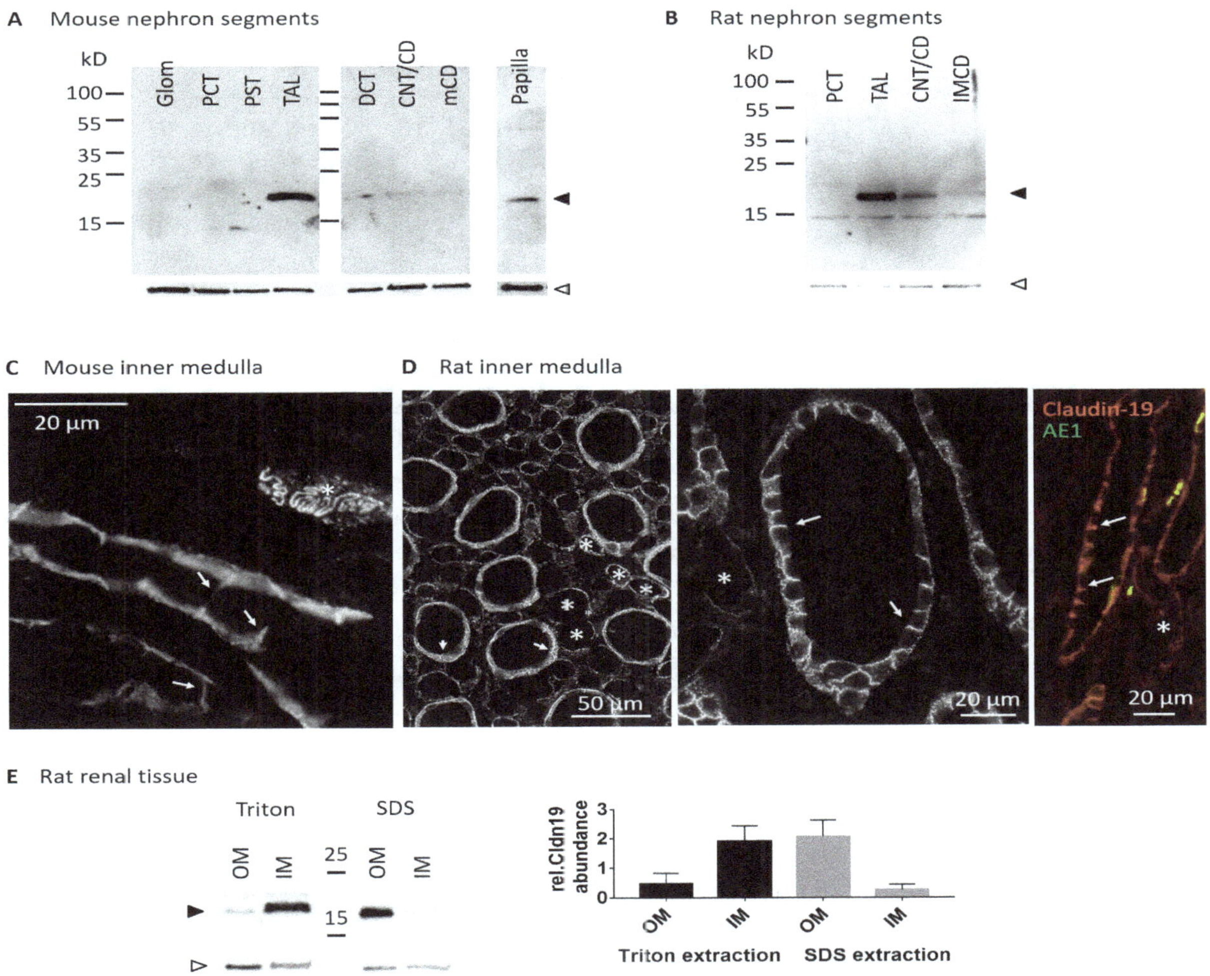

Figure 1. Claudin-19 in rodent kidney tissue. (**A**) Western blot of mouse kidney segments: glomeruli (Glom), proximal convoluted tubule (PCT), proximal straight tubule (PST), thick ascending limb (TAL), distal convoluted tubule (DCT), connecting tubule/cortical CD (CNT/CD), medullary CD (mCD), and papilla. Upper part: claudin-19 (black arrowhead), lower part: β-actin (white arrowhead). (**B**) Western blot of rat kidney segments: proximal convoluted tubule (PCT), thick ascending limb (TAL), connecting tubule and cortical CD (CNT/CD) and inner medullary collecting duct (IMCD). Claudin-19 and β-actin indicated as in A. Immunofluorescence of claudin-19 in mouse (**C**) and rat (**D**) inner medulla, arrows indicate claudin-19 TJ expression in IMCD, asterisks indicate thin limbs. AE1 (green) costaining in claudin-19 (red) positive IMCD in rat inner medulla. (**E**) Western blot of rat kidney tissue: Triton and SDS solubilized protein extracts of Cortex, outer medulla (OM) and inner medulla (IM), claudin-19 and β-actin indicated as in A. Summary of $n = 2$ independent extraction experiments showing Triton solubilization mainly in IM and SDS solubilization mainly in OM.

2.2. AVP Signaling Is Preserved in IMCD Cells at Different Osmotic Culture Conditions

In a first set of experiments, we tested the IMCD culture cells for basic physiological function. Therefore, IMCD cells were cultured under 300 mosm/kg (300-IMCD) and 600 mosm/kg (600-IMCD) conditions for five days and finally treated for 24 h with AVP. As shown in Figure 2 for 300-IMCD, subcellular localization of AQP2 was assessed by immunostaining. Although some AQP2 still remained in intracellular vesicles, AQP2 membrane staining markedly increased, revealing intact AVP signaling also after several days of cell culture. The same results were obtained for 600-IMCD (data not shown).

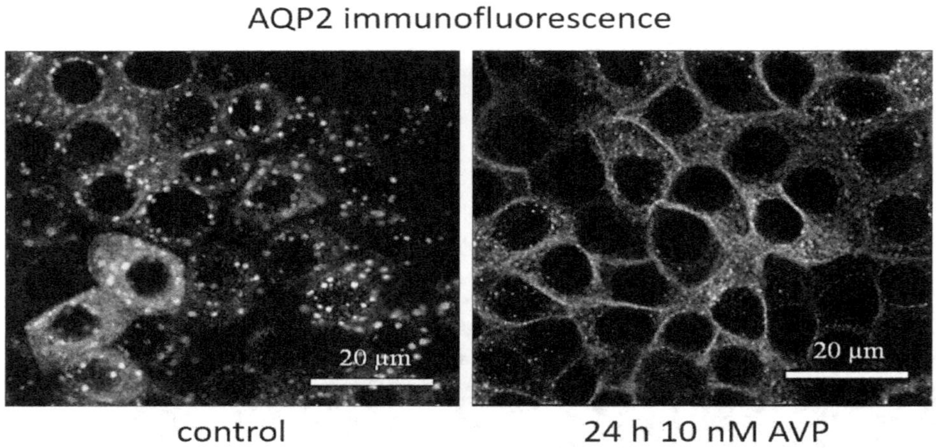

Figure 2. Effect of antidiuretic hormone (AVP) on subcellular AQP2 localization in 300-IMCD. AVP stimulation induced AQP2 insertion into the plasma membrane.

2.3. AVP Treatment Changes IMCD Cell Electrophysiological Properties but not Claudin-19 Localization

In two independent experimental series, 300-IMCD (Figure 3) or 600-IMCD (Figure 4) were cultivated for 24 h in the absence (control) or presence of 10 nmol/L AVP prior to the electrophysiological measurements. Under 300 mosm/kg conditions, mimicking strong water-diuresis or cortical osmolality, IMCD cells showed low, but consistently negative transepithelial voltage (V_{te}) and a transepithelial resistance (R_{te}) in the range of 310 Ωcm^2. These properties were not changed by AVP treatment. Accordingly, equivalent short-circuit current I'_{sc} also remained unaltered by AVP treatment (Figure 3A). Paracellular permeability properties were tested in the continuous presence of 50 µmol/L amiloride (V_{te} in the presence of amiloride was virtually abolished as shown in insert in Figure 3A) by applying an iso-osmotic NaCl concentration gradient with low NaCl (30 mmol/L) at the luminal side. In Figure 3B, two representative original chart recordings are shown to illustrate the development of a lumen-positive diffusion potential (DP) after application of the iso-osmotic NaCl gradient, indicating preferred diffusion of Na^+ ions towards the lumen. After AVP treatment, DP increased, resulting in higher cation selectivity as summarized in Figure 3C. To test whether claudin-19 subcellular localization was altered by AVP stimulation, we compared filters of both groups for the relative distribution of claudin-19 within the cells. Under control as well as under AVP treatment, almost all claudin-19 staining was localized in thin lines representing membrane, i.e., TJ staining (Figure 3D,E). There was no difference in the tight junction scores (2.70 ± 0.06 and 2.77 ± 0.05), respectively.

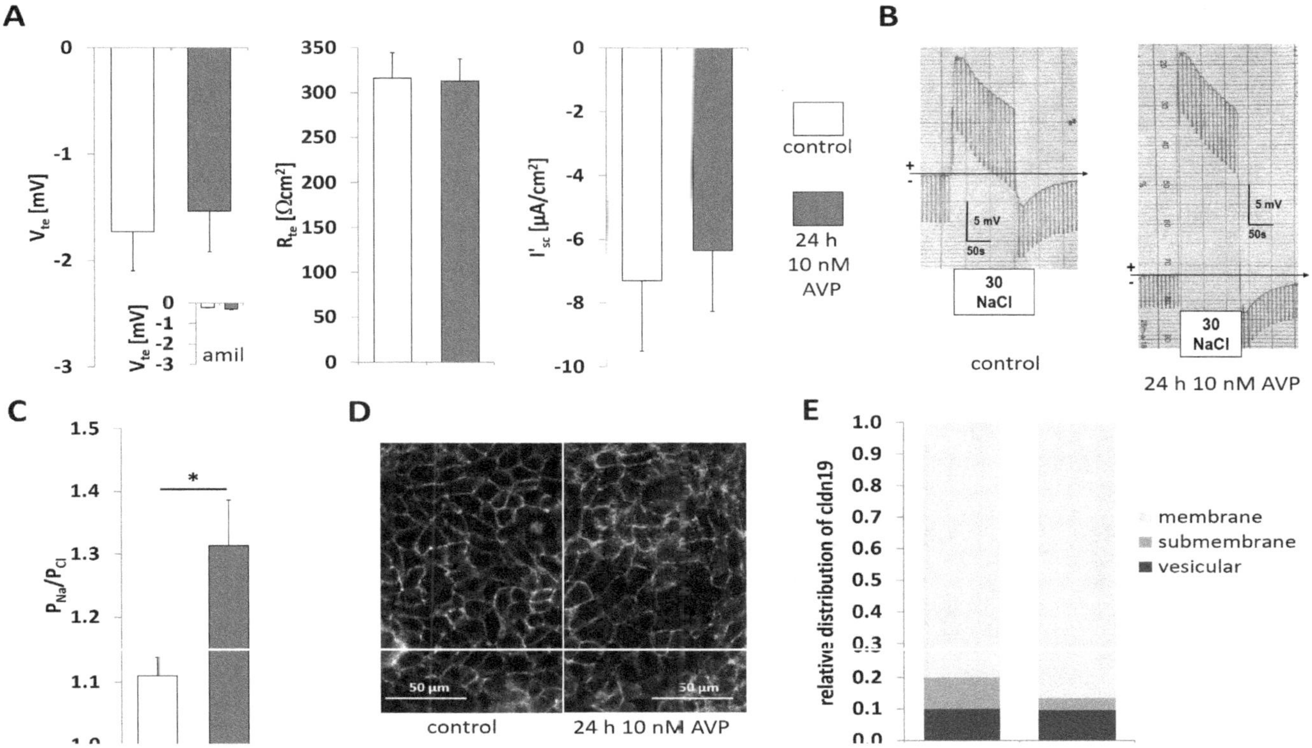

Figure 3. 300-IMCD under AVP stimulation (10 nmol/L). (**A**) Electrophysiological properties with transepithelial voltage V_{te}, transepithelial resistance R_{te}, and equivalent short-circuit current I'_{sc}, with or without 24 h prestimulation with AVP. Insert in V_{te} panel: V_{te} in the presence of 50 µM amiloride. (**B**) Original experiments showing 30 mmol/L NaCl diffusion potentials, (**C**) summarized permeability ratio PNa/PCl. Data are means ± SEM, $n = 13{,}13$; * $p < 0.05$. (**D**) Immunofluorescence of claudin-19 in 300-IMCD with or without 24 h prestimulation with AVP, (**E**) summarized subcellular localization of claudin-19 in the categories of membrane TJ staining, submembrane and intracellular vesicular staining, $n = 13$, 13.

The same parameters were also measured under 600 mosm/kg (Figure 4) culture conditions, which is a more physiological osmolar condition for inner medulla. Osmolality was adjusted by adding urea as well as NaCl. In comparison to 300-IMCD (Figure 3), V_{te} and R_{te} of 600-IMCDs was lower. In addition, V_{te} and I'_{sc} became less negative under AVP treatment, whereas R_{te} increased (Figure 4A). DP was measured after luminal exchange to iso-osmotic 50 mmol/L NaCl to generate a NaCl concentration gradient comparable to the previous experiments in 300-IMCD. In absolute contrast to 300-IMCD (Figure 3B,C), we now observed lumen negative diffusion voltage (Figure 4B), indicating pronounced anion selectivity (Figure 4C). There was no significant effect of AVP on paracellular selectivity as shown in original recordings (Figure 4B) and summary (Figure 4C). Claudin-19 subcellular distribution was not different between control situation und AVP treatment (Figure 4D,E; tight junction score 1.76 ± 0.07 and 1.93 ± 0.03, respectively). However, claudin-19 subcellular distribution was again markedly changed by osmolality, with a shift to submembrane and even more intracellular vesicular representation at 600-IMCD.

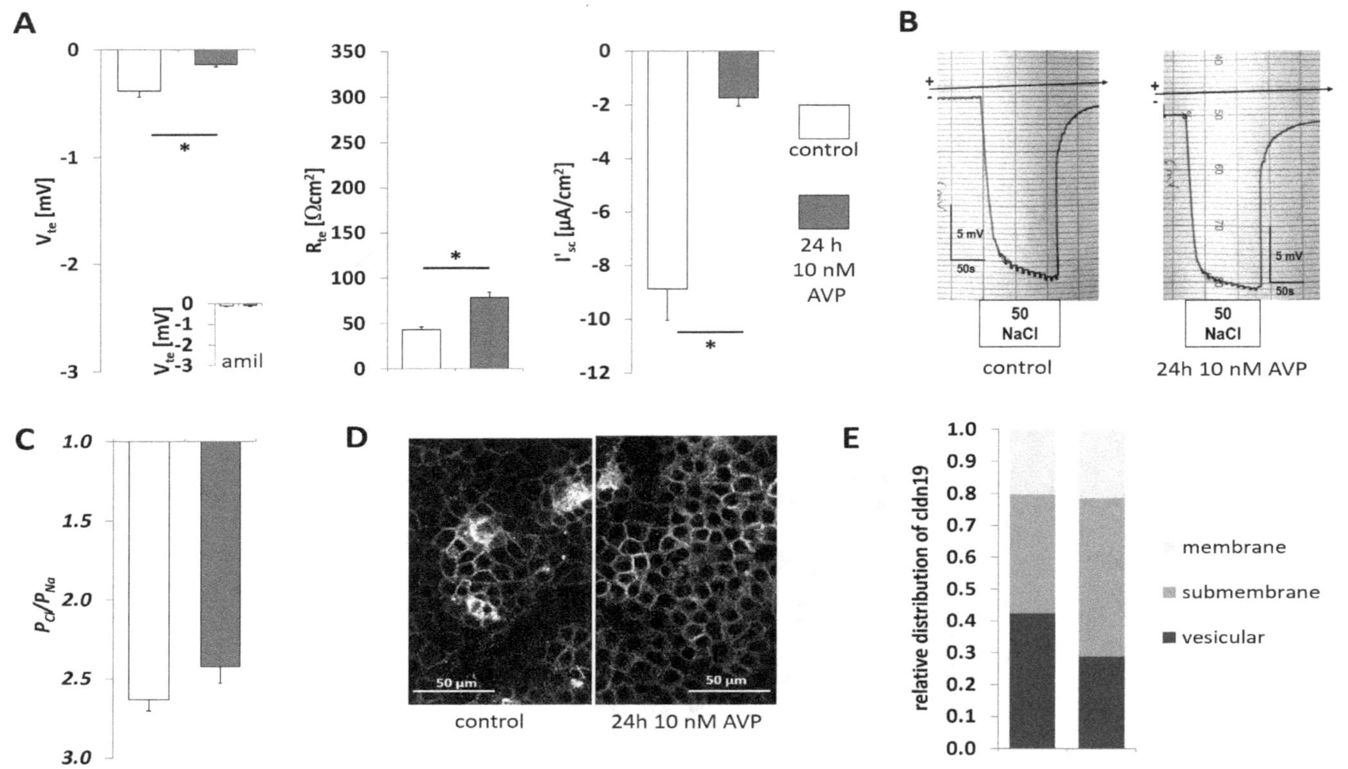

Figure 4. 600-IMCD under AVP stimulation (10 nmol/L). (**A**) Electrophysiological properties with transepithelial voltage V_{te}, transepithelial resistance R_{te}, and equivalent short-circuit current I'_{sc}, with or without 24 h prestimulation with AVP. Insert in V_{te} panel: V_{te} in the presence of 50 µM amiloride. (**B**) Original experiments showing 50 mmol/L NaCl diffusion potentials, (**C**) summarized permeability ratio P_{Cl}/P_{Na}. Data are means ± SEM, $n = 12{,}12$; * $p < 0.05$. (**D**) Immunofluorescence of claudin-19 in 600-IMCD, (**E**) summarized subcellular localization of claudin-19 in the categories of membrane TJ staining, submembrane and intracellular vesicular staining, $n = 8, 7$.

2.4. *Differences in Extracellular Osmolality Determine IMCD-Cell Trans- and Paracellular Properties*

The observation that osmolality strongly changed the transepithelial properties of IMCD-cells prompted us to further examine this strikingly strong effect of ambient osmolality on claudin-19 function. Therefore, we performed a new independent set of experiments cultivating IMCD cells in parallel under 300 mosm/kg, 600 mosm/kg, and 900 mosm/kg (900-IMCD) to mimic stronger antidiuresis. Transepithelial resistance measured daily under culture condition (TER) was monitored and cells reached peak resistance values at day 6 under all three osmolality regimens (Figure 5). Interestingly, the average resistance varied considerably between the three groups. Whereas cells grown under plasma isotonic conditions (300-IMCD) reached TER values of around 550 Ωcm^2, TER of 600-IMCDs was five times and TER of 900-IMCDs even 10 times lower.

Subsequently, filters were measured in Ussing Chamber experiments at day 5. Figure 6 shows electrophysiological properties in more detail. Only 300-IMCDs showed noteworthy lumen negative V_{te}, which disappeared under high osmolality. The resistance measurements during cell culture (TER, Figure 5) were corroborated. 300-IMCD displayed ca. 400 Ωcm^2, 600-IMCDs ca. 70 Ωcm^2, and 900-IMCD only ca. 27 Ωcm^2 (Figure 6A). Consistently, only 300-IMCD showed a lumen negative transcellular transport current I'_{sc} in this set of experiments. Paracellular selectivity was then assessed using comparable transepithelial NaCl concentration gradients adjusted to medium NaCl concentrations, respectively (Table 1, Figure 6B). Whereas 300-IMCD consistently showed a lumen positive DP, indicating cation selectivity (left original recording in Figure 6B), and all 900-IMCD a lumen negative DP, indicating anion selectivity (right original recording in Figure 6B), 600-IMCD displayed either lumen positive or lumen negative DP. Ion permeability ratios for Na^+ and Cl^- are summarized in Figure 6C. 300-IMCDs were cation selective and 900-IMCDs anion selective, 600-IMCD showing a moderate anion selectivity close to unity. Absolute permeability was calculated from these ratios and R_{te}. As the effect of osmolality on R_{te} of IMCD was much more pronounced than the changes in ion selectivity, the effect of higher osmolality was an increase for P_{Na} and even more for P_{Cl}. Figure 6D depicts the calculated permeabilities for Na^+ and Cl^- with an up to 25-fold increase in P_{Cl} comparing 300-IMCD and 900-IMCD.

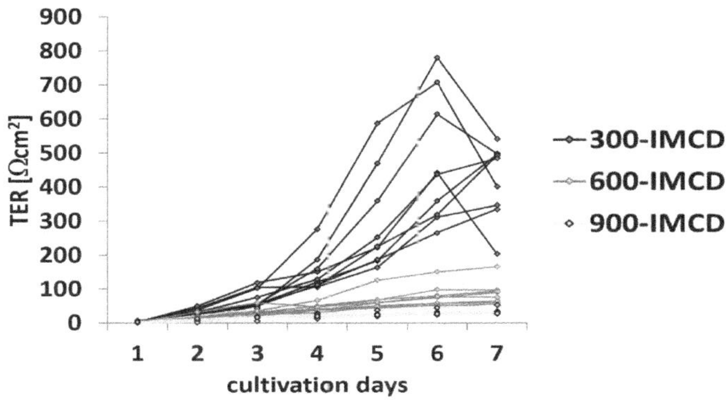

Figure 5. Time course of transepithelial resistance (TER) under culture conditions after seeding for nine different preparations with the filters of 300-IMCD, 600-IMC, and 900-IMCD, respectively.

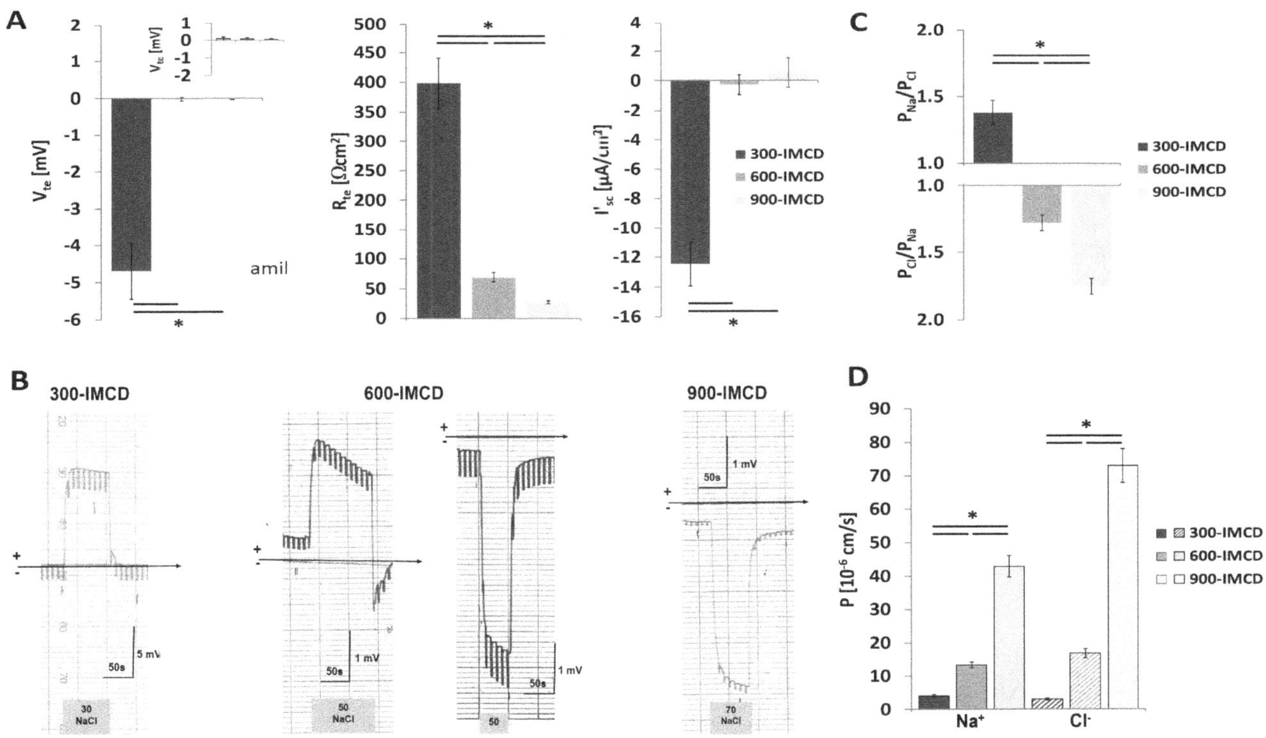

Figure 6. Comparison between 300-IMCD, 600-IMCD, and 900-IMCD after five days of cultivation. (**A**) Electrophysiological properties with transepithelial voltage V_{te}, transepithelial resistance R_{te} and equivalent short-circuit current I'_{sc}. Insert in Vte panel: Vte in the presence of 50 µM amiloride. (**B**) Original experiments showing 30 mmol/L NaCl (300-IMCD), 50 mmol/L (600-IMCD), and 70 mmol/L (900-IMCD) diffusion potentials. (**C**) Summarized permeability ratio PNa/PCl and PCl/PNa, respectively. (**D**) Calculated Na^+ and Cl^- permeabilities. Data are means ± SEM, $n = 24, 39, 46$; * $p < 0.05$.

Table 1. Composition of experimental solutions used in Ussing chamber experiments. Concentrations are given in mmol/L. pH was adjusted to pH 7.4.

	Control Solution			Low NaCl Solution		
	'300'	'600'	'900'	'300'	'600'	'900'
NaCl	145	245	345	30	50	70
KH_2PO_4	0.4	0.4	0.4	0.4	0.4	0.4
K_2HPO_4	1.6	1.6	1.6	1.6	1.6	1.6
$MgCl_2$	1	1	1	1	1	1
Ca-gluconate	1.3	1.3	1.3	1.3	1.3	1.3
glucose	5	5	5	5	5	5
urea	-	100	200	-	100	200
mannitol	-	-	-	230	390	550

2.5. *Effect of the Different Osmolality on IMCD Transport Properties*

300-IMCD showed a lumen negative I'_{sc} which was almost completely abolished by luminal amiloride application, indicating functionally expressed ENaC (Figure 7A). 600-IMCD showed basically no amiloride-dependent I'_{sc} but, astonishingly, 900-IMCD became slightly lumen positive after amiloride application. Sodium transport was verified by the measurement of changes in sodium concentration under cell culture conditions within 24 h. Luminal and basolateral cell culture medium was collected and compared directly before the Ussing chamber experiments. Only 300-IMCD showed sodium reabsorption decreasing the luminal sodium concentration (Figure 7B). 600-IMCD and 900-IMCD did not achieve a significant difference in luminal versus basolateral sodium concentration. To investigate whether osmolality affected the overall tightness of the epithelium for larger molecules, we applied 40 kDa FITC–dextran to one compartment of the cell culture and measured the fluorescence in the trans-compartment. In comparison to the empty cell culture filters, confluent cell layers showed hardly any FITC–dextran permeability confirming intact cell layers. Only 900-IMCD showed a slightly higher 40 kD FITC–dextran permeability (Figure 7C), still minute in comparison to empty filters.

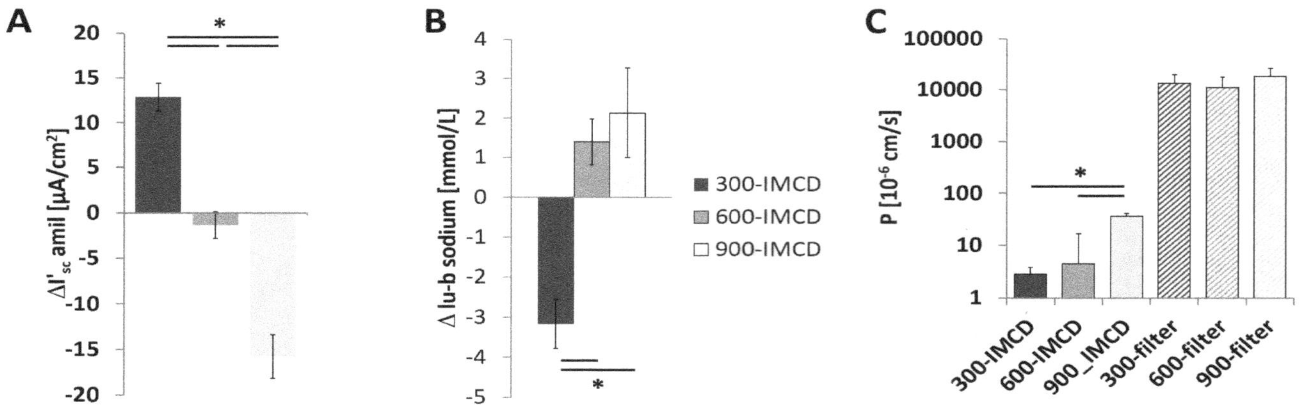

Figure 7. Transepithelial transport parameters of 300-IMCD, 600-IMCD, and 900-IMCD after five days' cultivation. (**A**) Amiloride-sensitive equivalent short-circuit current $\Delta I'_{sc}$. (**B**) Na^+ concentration difference between luminal and basolateral cultivation medium, (**C**) summarized 40 kD FITC–dextran permeability. Data are means ± SEM, IMCD filters: n = 10,11,11; empty filters: n = 3, 3, 3; * $p < 0.05$.

2.6. *Differences in Extracellular Osmolality Change Claudin-19 Subcellular Localization*

Finally, we compared the subcellular localization of claudin-19 in the different IMCD cultures and corroborated the findings already implicated by the AVP experiments. Whereas in 300-IMCD most claudin-19 was localized to the tight junction and appeared as a sharp line (Figure 8A), the percentage of submembrane and intracellular vesicular staining increased at higher cell culture medium osmolality (Figure 8B,C). Representative immunofluorescence stainings for claudin-19 are shown in (Figure 8A–C) and the observations are summarized as relative distribution between membrane, submembrane, and vesicular compartments (Figure 8D). 300-IMCD showed nearly 70% TJ staining (TJ score 2.69 ± 0.10), 600-IMCD expression pattern was intermediate (TJ score 1.80 ± 0.11), and in 900-IMCD, more than 90% of claudin-19 was distributed to small intracellular vesicular-like structures (TJ score 1.13 ± 0.08).

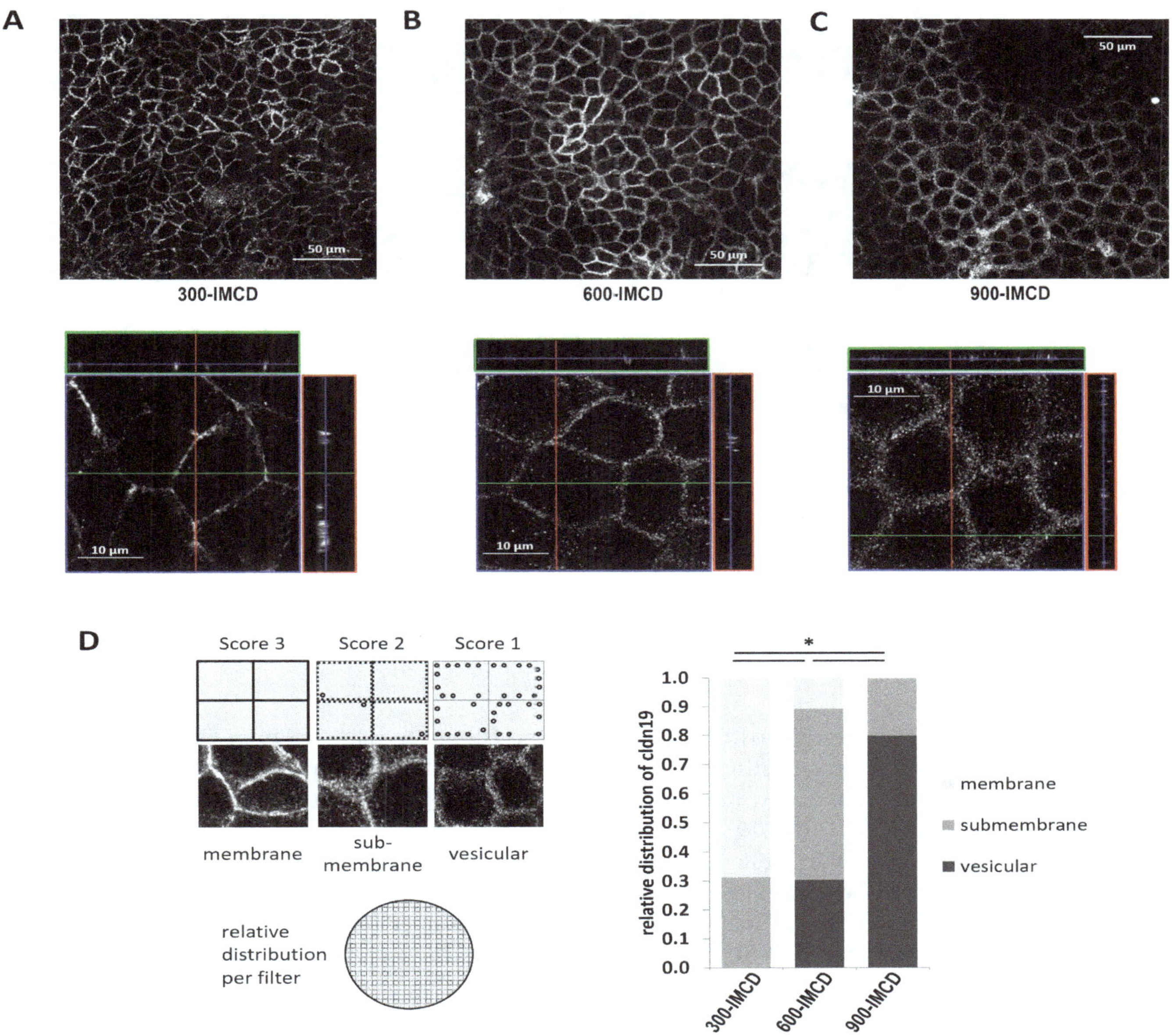

Figure 8. Immunofluorescence of claudin-19 (**A**) in 300-IMCD, (**B**) in 600-IMCD, and (**C**) in 900-IMCD, shown in an xy-overview (upper panel) and a representative z-stack (lower panel). (**D**) Summarized subcellular localization of claudin-19 in the categories of membrane TJ staining, submembrane and intracellular vesicular staining (schematic drawing), n = 12, 14, 12.

2.7. Further Observations

This study mainly focused on the expression and localization of claudin-19 under different osmotic cell culture conditions. However, in comparison to the changing subcellular localization of claudin-19, which we investigated in detail, we could not find a similar pattern in the expression of claudin-8 (Figure 9A). In addition, the TJ associated scaffolding protein ZO-1 localized nicely to the tight junction belt of IMCD under all conditions (Figure 9B), independent of osmolality.

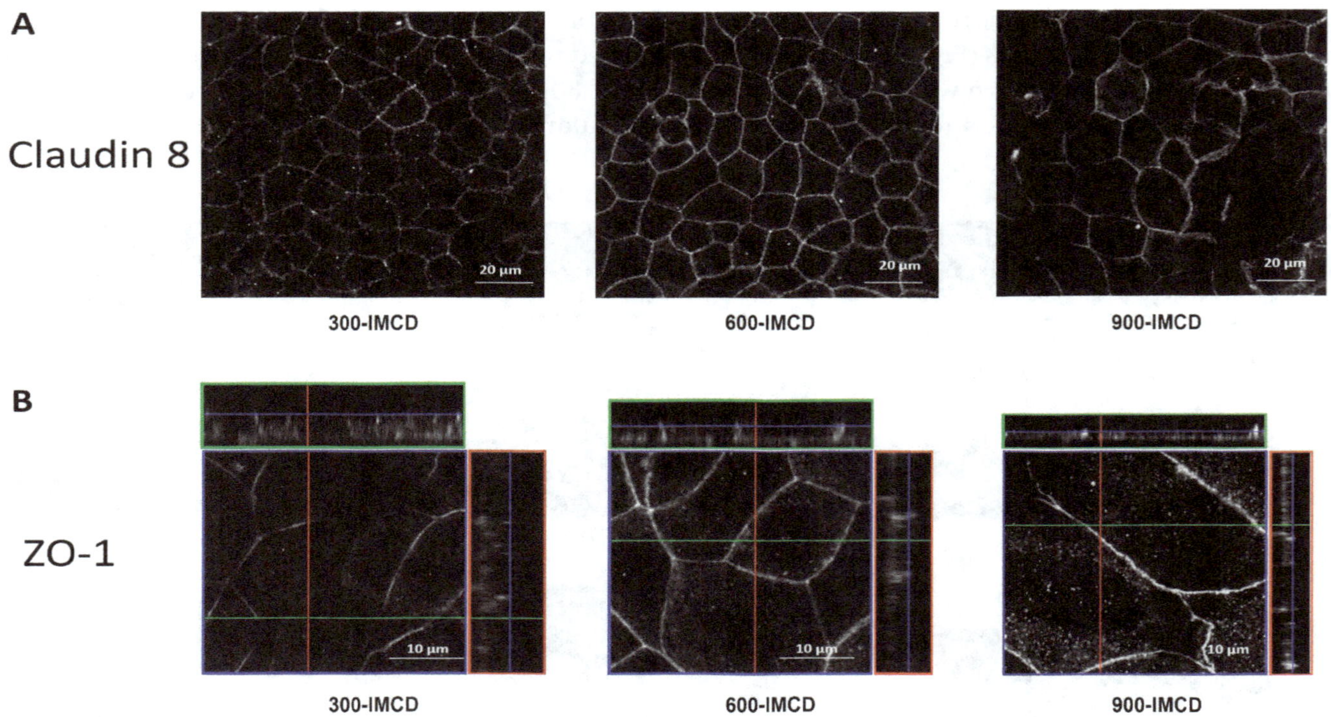

Figure 9. Immunofluorescence of claudin-8 in xy-overview in 300-IMCD, 600-IMCD, and 900-IMCD (**A**). Z-stacks of immunofluorescence of ZO-1 in 300-IMCD, 600-IMCD, and 900-IMCD (**B**).

Furthermore, the z-stacks in Figure 8; Figure 9 show that osmotic cell culture conditions also influenced cell height and size. Especially 900-IMCD, although covering the total area of the filter, were rather flat and more expanded, suggesting a reduced number of available cells.

3. Discussion

We and others have investigated the expression and function of a variety of claudins along the nephron [28,29]; however, there are still several claudins in this large protein family which are not fully characterized with respect to their expression and functional role. Refinement of Western Blot analysis and immunofluorescence of claudin-19 protein in isolated renal tubules showed substantial expression in medullary and inner medullary collecting ducts. As these tubules are exposed to varying interstitial osmolality depending on the diuretic state and at the same time have to serve different needs with respect to permeability and barrier function, we hypothesized that claudin-19 is involved in medullary and inner medullary collecting duct function and might be under the control of antidiuretic hormone and osmolality.

We used primary IMCD cells cultivated under hormonal and osmotic challenges to investigate their electrophysiological transport properties as well as subcellular localization of claudin-19 under these conditions. AVP partially influenced electrophysiological properties and did this in an osmolality depended manner: 300 mosm/kg—increase in cation selectivity (Figure 3), 600 mosm/kg—decrease in transcellular transport and increase in R_{te} (Figure 4). Claudin-19 subcellular localization, however, remained unaffected by the hormone. More strikingly, low osmolality increased R_{te} substantially and translocated claudin-19 to the tight junction, whereas high osmolality resulted in rather low R_{te} and removal of claudin-19 from the TJ to intracellular vesicles (Figures 6–8). In addition, paracellular selectivity changed from cation to anion selectivity with increasing extracellular osmolality.

Although the data are shown in comparison to isotonic conditions, we consider 600 mosm/kg as "physiological" osmolality for our experiments, a tonicity more attributed to the outer medulla [3] and the more outer parts of inner medulla. This is in line with former publications [25,30] and takes the cone-like structure of the inner medulla into account. However, towards the tip of the papilla, even higher osmolalities are to be expected in rodents as rats, which excrete urine with osmolality values between 1500 and 2500 mosm/kg [27,31]. The electrophysiological findings of our primary cell culture under hyperosmotic conditions (600-IMCD and 900-IMCD) reproduced the findings from native, isolated perfused papillary collecting ducts rather well, with a slightly lumen positive V_{te} (< 1 mV) and R_{te} of about 148 Ωcm^2 [32]. We could corroborate the presence of claudin-19 protein in IMCD in native tissue (Figure 1) as well as in primary cell culture (Figures 3, 4 and 8). This is in accordance with the finding of Lee et al. [23] but did not find its way into the recently published expression profiles along the nephron where claudin-19 so far is not included in the collecting duct [28,29]. Our data is also in line with the transcriptome, which is negative for the tubule segments upstream the TAL but positive for the TAL and downstream, including IMCD. The lower metabolism of cells in the medulla might explain why protein abundance is high in comparison to the low mRNA levels found for claudin-19 in the IMCD transcriptome [24].

Acute effects of changes in osmolality on paracellular properties have been investigated mainly in established renal epithelia cell lines. In MDCK cells, mimicking proximal tubules, small steps of osmolality increase (10–20 mosm/kg by addition of glucose or mannitol) were tested to imitate diabetic nephropathy with conflicting results. After 24–72 h hyperosmotic challenge using glucose, R_{te} decreased, while cation selectivity increased (increase in claudin-2 expression and reduction in claudin-1 and -3; [33]). In contrast, mannitol increased R_{te} by decreasing claudin-2 expression and increasing claudin-1 within the same time-frame [34]. Collecting ducts face more pronounced changes in osmolality due to the unique urinary concentration mechanisms [4]. A lot of studies have been performed in IMCD3 cells, collecting duct cells originally derived from mouse inner medulla, challenging them acutely by increasing tonicity and looking for cellular adaptations within hours or the first days. Fast changes to higher osmolality triggered apoptotic events [35]; a combination of NaCl and urea was better tolerated [36]. Altogether, IMCD3 adapted well to higher osmolality if changes were induced stepwise [37]. In primary IMCD cells, changes in tonicity also lead to adaptation, partially to apoptosis and to vast changes in gene expression [8,26]. In our experiments, we observed tonicity-dependent changes in cell shape, with expanded and flat cells under high osmolality. This might be attributed to the higher susceptibility of the cells to undergo apoptosis. Still the epithelial monolayer was intact. In this study, we did not further investigate apoptosis, but we confined our experiments on physiological transport properties and tight junction protein expression. Adaptation described for IMCD3 also included changes in the TJ proteins [38,39]. Claudin-4 was thereby under the control of the scaffolding multi-PDZ domain protein 1 (MUPP-1) and showed higher expression in IMCD3 cells adapted to high osmolalities. In native murine papilla, claudin-4 expression was upregulated after water deprivation of mice [39]. In addition, we could show that ascending thin limbs reacted to water load by increasing the cation preference of their paracellular pathway [27]. Otherwise, not much is known about inner medullary adaptation of the paracellular pathway to changing environments. For this study, we focused our investigation mainly on claudin-19 in long-term adaptation experiments and on functional states.

Plasma isotonic conditions are most probably an environment rarely faced by IMCD cells in native inner medulla. One explanation for the changes seen here, especially functional ENaC expression (Figure 7A) and directed Na^+ transport (Figure 7B), could be that the cells undergo a transition to cortical collecting duct properties; a kind of 'corticalization'. All three ENaC subunits are expressed in all parts of the connecting tubule and the collecting duct [40]. However, without aldosterone stimulation, functional ENaC in the luminal membrane plays a crucial role mainly in the connecting tubule [41] and even in the cortical collecting duct of mice, lumen negative V_{te} as an indicator of the presence of ENaC is not measured [42] in the absence of aldosterone. As already mentioned, IMCDs did not show negative V_{te} [32]. Accordingly, water homeostasis has been implicated in the regulation of ENaC. Crambert et al. could show that water deprivation in mice increased ENaC expression in cortical, but not outer medullary CDs, and in a collecting cell line, hyperosmolality sufficed to reduce ENaC expression [43]. On the other side, low osmolality at the luminal side, but not necessarily at the basolateral side, would be indicative for water diuresis, which would require a sealing against water flux and the assimilation of luminal and basolateral osmolalities [12].

How could known claudin-19 properties support collecting duct function? Under isotonic conditions, claudin-19 appears to be localized in the TJ (Figures 3 and 8). In TAL, claudin-19 confers Ca^{2+} and Mg^{2+} permeability to the tight junction, together with claudin-16 [22,44]. If claudin-19 alone is expressed in MDCK cells, it seems to act as cation barrier [18], whereas expression in a per-se anion selective background (porcine kidney cells, LLC-PK1), claudin-19 reduces anion selectivity [45]. Altogether, translocation of claudin-19 to the tight junction might be responsible for the increase in R_{te} and the accompanying cation selectivity. So far, no other direct interaction partners of claudin-19, in analogy to claudin-16/19 interaction, have been described. Obviously, the TJ structure and accessibility to detergents was remarkably different between the outer and the inner medulla (Figure 1). The TAL presents the main tubular component of the inner stripe of outer medulla (we excluded the outer stripe for a more secure exclusion of cortex), whereas in the inner medulla, the collecting duct is probably the main component by cell mass. Hence, outer medulla claudin-19 originates mainly from TAL and inner medulla claudin-19 mainly from IMCD, both showing completely different dissolving profiles. It has been proposed that the tight junction complex can be part of or be associated with cholesterol-enriched "lipid rafts" of plasma membranes. These lipid rafts consist of cholesterol and sphingomyelin to form a tightly packed membrane lipid region [46]. The extractability of, e.g., claudin-4, is strongly dependent on the detergent used for solubilization [47]. Triton solubility increases with lipid raft disruption. Sugibayashi et al. showed in MDCK cells that this disruption increased solubility to Triton for claudin-4, -5, and occludin, but not for claudin-1, -2, and -3, indicating that in this cell system, claudin-4 and claudin-5 were organized in lipid rafts, whereas claudin-1, -2, and -3 were not [48]. In analogy, we would propose that claudin-19 could be part of a lipid raft (TAL) or not (CD) depending on the structure and composition of the respective TJ.

Increasing osmolalities in the cultivation medium dramatically decreased R_{te} and increased ion permeability, especially for Cl^-. High osmolalities on both sides of the IMCD epithelium mimic conditions of strong antidiuresis, also independent of the original stimulus by AVP [49]. The gradual removal of claudin-19 into intracellular compartments and vesicular structures (Figure 8) could partially explain the reduction in R_{te}. However, literature and our own first findings emphasize the complexity of the regulation and composition of the paracellular pathway. A parallel increase in claudin-4 as described by Lanaspa et al. [39] could contribute in addition, especially with regard to the increase in anion selectivity. Claudin-8, a direct interaction partner of claudin-4, however, does not show regulation, at least on the level of subcellular localization (Figure 9). Under high osmotic conditions on both sides of the epithelium, osmotic pressure across the epithelium is low and, potentially, tricellular sealing, an additional route for water, ions, and macromolecules, could be less stringent [50]. One indicator for this could be the slightly increasing permeability for the high-molecular 40 kD FITC dextran marker in 600-IMCD and especially 900-IMCD. Other claudins (e.g., claudin-7) and expressional or post-translational regulations of TJ properties have to be considered in future studies to finally complete the picture.

Although AVP is the main controlling hormone of renal water handling [4,49], we could corroborate that local osmolality changes were able to regulate IMCD transport properties and claudin expression independently, similar to the findings of Lanaspa and coworkers [38,39]. Whereas in these studies TJ proteins have been described as targets for regulation, intriguingly, it has also been proposed that the TJ itself could serve as sensor for osmolality and tonicity changes, thereby becoming the origin of regulation (reviewed in [51]). In MDCK cells, acute changes in transepithelial osmotic gradients dramatically changed paracellular selectivity and claudin localization as well as cytoskeleton remodeling [52]. Besides these changes by cell culture osmolality, AVP itself showed effects on paracellular properties as it increased cation selectivity in 300-IMCD. However, these changes were independent of claudin-19, or at least not dependent on subcellular localization of the claudin. We speculate from findings in medullary TAL that mechanisms like claudin-19 phosphorylation could be part of the effect of AVP.

In TAL, we could show that AVP directly increased cation selectivity within minutes, without changing claudin-10b localization, indicating that phosphorylation rather than recruitment to the membrane may be involved [53].

In conclusion, we propose that claudin-19 is part of the collecting duct TJ complex. Its presence in the TJ is under the control of ambient interstitial tonicity. Under iso-osmotic conditions, it is responsible for a high electrical transepithelial resistance and contributes to the tonicity-induced changes in paracellular ion selectivity.

4. Materials and Methods

4.1. Animals

All experiments were performed in accordance with the German law on animal protection and approved by the local authorities (animal ethics protocol number V312-72241.121-2). C57 Bl6/J mice, Wistar or Sprague-Dawley rats were housed under a 12 h light cycle with free access to water and chow.

4.2. Cell Culture

Cell culture medium of 300 mosm/kg was used as basic medium. It consisted of high glucose DMEM (PAA Laboratories, Coelbe, Germany) enriched with 1% Ultroser G (CytoGen, Wetzlar, Germany), 1% nonessential amino acids (PAA Laboratories), 1% L-Glutamine (PAA Laboratories; 200 mmol/L), and 1% penicillin/streptomycin (PAA Laboratories). For 600 mosm/kg medium, 100 mmol/L NaCl and 100 mmol/L urea were added; for 900 mosm/kg medium, 200 mmol/L NaCl and 200 mmol/L urea were added, respectively.

IMCD cells for primary cell cultures were isolated as described previously [25,26]. Briefly, 200–300 g Wistar or Sprague-Dawley rats (n = 20) of both sexes were sacrificed. Under sterile conditions, the inner medulla of both kidneys was dissected, transferred to enzyme solution (0.2% hyaluronidase type I (Sigma Aldrich, St. Louis, MO, USA) and 0.2% collagenase type II (Biochrom AG, Berlin, Germany) in PBS and digested for 90 min in a thermoshaker (37 °C, 850 rpm). The suspension of cells and tubular fragments was washed in PBS twice by centrifugation und resuspension in 600 mosm/kg medium. Cells were seeded on collagen IV-coated (mouse collagen IV, BD Biosciences, Heidleberg, Germany; 0.2 µg/cm) Costar Transwell Permeable supports (Sigma Aldrich, St. Louis, USA); 0.33 cm^2) at densities of 350,000 cells/cm^2 or 600,000 cells/cm^2. Cells were cultivated at 37 °C and 8% CO_2. After a 24 h settlement period in 600 mosm/kg medium, cells were transferred to 300, 600, or 900 mosm/kg medium, respectively, and cultivation lasted 5–7 days in total. Medium was changed and transepithelial resistance was monitored daily (EVOM2µ, World Precision Instruments). All filters were checked for confluency after the experimental procedure by visual control for complete epithelial coverage of the filter. Filters with gaps in confluency were excluded from analysis.

4.3. Immunohistochemistry

After Ussing chamber measurements, the filters were fixed in 0.4% PFA in PBS at 4 °C overnight. In addition, cryosections (3 µm) of perfusion fixed kidneys of Sprague-Dawley rats and of C57Bl/6 J mice [54] were heated in 0.3% triton-x-100 in PBS (PBST-T) for antigen retrieval. Filters as well as kidney slices were washed extensively in PBS-T and blocked with 1% bovine serum albumin in PBS-T. Filters and cryosections were incubated with primary antibody in 1% BSA overnight at 4°C. The claudin-19 antibody, a kind gift of J. Hou (Department of Internal Medicine, Washington University Renal Division, USA), has been successfully used in a Claudin-19 KD animal model [21]) and was used in a dilution of 1:300, Antibodies used in addition: rabbit anti-aquaporin-, 1:100 (Alomone labs, Jerusalem, Israel), rabbit anti-claudin-8 and rabbit anti-ZO-1 (both Thermo Fisher, Waltham, USA), guinea pig anti-AE1 (kind gift of C. Wagner, University of Zürich, Institute of Physiology, Zürich, Switzerland). After three additional washing steps (PBS-T), filters were incubated with the respective secondary antibody (1:400; AlexaFluo 488, donkey anti-rabbit, AlexaFluo 633, goat anti-rabbit; AlexaFluo 488 goat anti-guinea pig, Thermo Fisher) for 60 min at room temperature. After final extensive washing (PBS-T), filters and cryosections were embedded in Moviol-Dabco. Pictures were taken either confocally (LSM 510, Axiovert 200 M; Zeiss, Jena, Germany) or with Apotome2 (AX 10, Zeiss). Filters were analyzed for the subcellular staining pattern of claudin-19 in three categories as ordinal scale reflecting the functional expression of claudin-10 (see scheme in Figure 8: score 3 for membrane associated tight junction staining (clear lines), score 2 for submembrane staining (diffuse scattering lines), and score 1 for vesicular staining (dotted intracellular staining). For each filter the percentage of the different categories was given with 5–10% precision and each filter was attributed then an overall TJ score between 1 (all claudin-19 in intracellular vesicles) and 3 (all claudin-19 in TJ).

4.4. Western Blotting

Enzymatic and enzymatically assisted dissection was used to generate sorted tubular samples [27,55]. After flank incision under deep anesthesia the left kidney of Sprague-Dawley rats was perfused with enzyme solution: 2 mg/mL collagenase Type II in incubation solution (in mmol/L: 140 NaCl, 0.4 KH_2PO_4, 1.6 K_2HPO_4, 1 $MgSO_4$, 10 sodium acetate, 1 α-ketoglutarate, 1.3 calcium gluconate, 5 glycine, containing 48 mg/L trypsin inhibitor and 25 mg/L DNase I, pH 7.4, at 37 °C). Mice kidneys were similarly perfused via the abdominal vein [45]. Tubular segments from 6–8 week C57 Bl/6J or Sprague-Dawley rats were sorted after digestion as described in [45,55] or dissected manually from thin transversal kidney slices at 4 °C in sorting solution (incubation solution supplemented with 0.5 mg/L albumin). For Western blotting, approximately 20–50 tubules per segment were sorted and stored in 20–25 µL 1× Laemmli solution at −80 °C prior to use. For membrane protein solubilization, rat kidneys (6–8 week) were cut in transversal slices at 4 °C and under visual control the papilla (inner medulla, IM) and outer medulla (OM) were prepared, the later with a safety margin to cortex. Therefore, OM mainly comprises the inner stripe. Tissues were snap frozen in liquid nitrogen and stored at −80 °C. After thawing tissues were homogenized with a glass teflon homogenizer in Triton buffer (in mmol/L: 20 Tris–HCl, 50 NaCl, 2 EDTA, 1% Triton-x-100, pH 7.6, protease inhibitor cocktail (Roche, Basel, Switzerland) and incubated for 2 h on ice. After 30 min centrifugation at 21,000× *g* and 4 °C, the supernatant was transferred to a new reaction tube, a sample was kept for determination of protein content (NanoDrop, Thermo Scientific, Waltham, MA, USA) and the tube stored at −20 °C after addition of Laemmli concentrate. The remaining pellet was resuspended in SDS buffer (in mmol/L: 20 Tris–HCl, 50 NaCl, 2 EDTA, 1% SDS, pH 7.6, protease inhibitor cocktail (Roche)), incubated for one additional hour at 4 °C, and then processed as described for the Triton samples. After thawing and denaturation, samples were separated in 15% polyacrylamide minigels and electrophoretically transferred to nitrocellulose membranes. Membranes were blocked for 60 min with 5% BSA in PBS–Tween (0.1% Tween in PBS) and incubated with primary antibody (anti-claudin-19, 1:2000) at 4 °C overnight. Membranes were further incubated for 2 h at RT with the secondary antibodies (Dianova, goat anti-rabbit-HRP) and imaged using a ChemiDoc Imaging system (Biorad, Herkules, USA). Procedure was repeated with anti-β-actin for loading control.

4.5. Electrophysiology

IMCD cells grown on filter supports were measured in a modified Ussing Chamber (EP Devices, Leuven, Belgium). Measurements were performed under the osmolality corresponding to the respective cultivation conditions and experimental solutions are enlisted in Table 1. After a short equilibration period, luminal amiloride (50 µmol/L) was applied to inhibit ENaC driven transepithelial Na^+ transport. In the presence of luminal amiloride, the transepithelial voltage was close to zero in all cases. Next, diffusion potentials were obtained by consecutively replacing the luminal solution by low-NaCl solution and application of a 1:5 luminal to basolateral NaCl concentration gradient (iso-osmotic; 30, 50, and 70 mmol/L NaCl, respectively, see Table 1). All obtained potentials were corrected for their respective calculated liquid junction potentials [56]. Permeability ratio P_{Na}/P_{Cl} was calculated using the Goldman–Hodgkin–Katz flux equation. Absolute permeability of Na^+ was obtained using the simplified Kimizuka–Koketsu equation [54,57]. Absolute permeability for Cl^- was calculated from the respective permeability ratio and P_{Na}.

4.6. FITC Dextran Permeability Measurement

The 40 kDa FITC–dextran conjugate ((Sigma Aldrich, St. Louis, USA) stock solution was dialyzed (8000–10,000 MWCO) to remove free FITC. Under cell culture conditions, 26 μM FITC dextran was added to the luminal compartment of nine filters per osmolality condition. Fluorescence in the basolateral compartment was measured at time points 0 and 30 min in a plate reader (Tecan, Männedorf, Switzerland). Calibration curves were conducted to calculate FITC–dextran concentration. FITC–dextran flow rate was given as change in concentration over time in the chamber volume. Permeability coefficient (P) was calculated relating the flow rate to the respective filter area and the initial concentration difference as driving force. Permeability of empty filters under the different osmolality conditions was determined accordingly.

4.7. Electrolyte Measurement

Luminal and basolateral cell culture supernatants of the last 24 h of cultivation were collected, Na^+ and K^+ concentrations measured by flame photometry (EFOX 5053, Eppendorf, Hamburg, Germany), and basolateral to luminal differences in electrolyte concentrations were calculated.

4.8. Statistics

Electrophysiological data are presented as means ± SEM. Two groups were tested by unpaired *t*-test or Mann–Whitney test. Three groups were tested with one-way ANOVA followed by Tukey post-hoc testing or Kruskal–Wallis test and Dunn's post-hoc test. All statistical analyses were performed using GraphPad Prism including the test for normal distribution (D'Agostino & Pearson normality test). Immunofluorescence subcellular staining data are presented as a stacked-column graph, giving the respective category as a percentage. Filter-scores were tested with nonparametric Mann–Whitney *U*-Test or Kruskal–Wallis test and Dunn's post-hoc test.

Author Contributions: Conceptualization, A.Z., S.R.S., N.H. and B.E.; Methodology, A.Z., S.R.S., M.B.; Software, N.H. and S.R.S; Validation, N.H. and M.B.; Formal Analysis, A.Z., S.R.S. and N.H.; Investigation, A.Z., S.R.S., V.C.W. and N.H.; Resources, M.B.; Data Curation, N.H., A.Z., S.R.S. and V.C.W.; Writing—Original Draft Preparation, A.Z., S.R.S. and N.H.; Writing—Review & Editing, V.C.W., B.E. and M.B.; Visualization, A.Z., S.R.S. and N.H.; Supervision, M.B. and N.H.; Project Administration, N.H., M.B.; Funding Acquisition, M.B.

Acknowledgments: We greatly appreciate the expert technical assistance from T. Stegman and R. Lingg as well as the support by the animal facility of the Christian-Albrechts-University Kiel. We thank Jianghui Hou, Division of Nephrology, Washington University School of Medicine, St. Louis, USA, for providing the claudin-19 antibody. Funding was provided by the Christian-Albrechts-University Kiel.

Abbreviations

IMCD	Inner medullary collecting duct
AVP	Arginine vasopressin
PG2	Prostaglandin E2
TJ	Tight junction
TAL	Thick ascending limb
PT	Proximal tubule
PCT	Proximal convoluted tubule

PST	Proximal straight
DCT	Distal convoluted tubule
CNT/CD	Connecting tubule and cortical collecting duct
mCD	Medullary collecting duct (outer stripe of outer medulla)
SDS	sodium dodecyl sulfate
AQP2	Aquaporin 2
V_{te}	Transepithelial voltage
R_{te}	Transepithelial resistance
I'_{sc}	Equivalent short circuit current
TER	Transepithelial resistance (cell culture)
DP	Diffusion potential

References

1. Dantzler, W.H.; Pannabecker, T.L.; Layton, A.T.; Layton, H.E. Urine concentrating mechanism in the inner medulla of the mammalian kidney: Role of three-dimensional architecture. *Acta Physiol.* **2011**, *202*, 361–378. [CrossRef] [PubMed]
2. Pannabecker, T.L. Comparative physiology and architecture associated with the mammalian urine concentrating mechanism: Role of inner medullary water and urea transport pathways in the rodent medulla. *Am. J. Physiol. Regul. Integr. Comp. Physiol.* **2013**, *304*, R488–R503. [CrossRef] [PubMed]
3. Sands, J.M.; Layton, H.E. The physiology of urinary concentration: An update. *Semin. Nephrol.* **2009**, *29*, 178–195. [CrossRef] [PubMed]
4. Fenton, R.A.; Knepper, M.A. Mouse models and the urinary concentrating mechanism in the new millennium. *Physiol. Rev.* **2007**, *87*, 1083–1112. [CrossRef] [PubMed]
5. Sheen, M.R.; Kim, J.-A.; Lim, S.W.; Jung, J.-Y.; Han, K.-H.; Jeon, U.S.; Park, S.-H.; Kim, J.; Kwon, H.M. Interstitial tonicity controls TonEBP expression in the renal medulla. *Kidney Int.* **2009**, *75*, 518–525. [CrossRef] [PubMed]
6. Maril, N.; Margalit, R.; Mispelter, J.; Degani, H. Sodium magnetic resonance imaging of diuresis: Spatial and kinetic response. *Magn. Reson. Med.* **2005**, *53*, 545–552. [CrossRef] [PubMed]
7. Berl, T. How Do Kidney Cells Adapt to Survive in Hypertonic Inner Medulla? *Trans. Am. Clin. Climatol. Assoc.* **2009**, *120*, 389. [PubMed]
8. Schulze Blasum, B.; Schröter, R.; Neugebauer, U.; Hofschröer, V.; Pavenstädt, H.; Ciarimboli, G.; Schlatter, E.; Edemir, B. The kidney-specific expression of genes can be modulated by the extracellular osmolality. *FASEB J.* **2016**, *30*, 3588–3597. [CrossRef] [PubMed]
9. Günzel, D.; Yu, A.S.L. Claudins and the modulation of tight junction permeability. *Physiol. Rev.* **2013**, *93*, 525–569. [CrossRef]
10. Günzel, D.; Fromm, M. Claudins and other tight junction proteins. *Compr. Physiol.* **2012**, *2*, 1819–1852.
11. González-Mariscal, L.; Tapia, R.; Chamorro, D. Crosstalk of tight junction components with signaling pathways. *Biochim. Biophys. Acta* **2008**, *1778*, 729–756. [CrossRef] [PubMed]
12. Gong, Y.; Himmerkus, N.; Sunq, A.; Milatz, S.; Merkel, C.; Bleich, M.; Hou, J. ILDR1 is important for paracellular water transport and urine concentration mechanism. *Proc. Natl. Acad. Sci. USA* **2017**, *114*, 5271–5276. [CrossRef] [PubMed]
13. Flamion, B.; Spring, K.R.; Abramow, M. Adaptation of inner medullary collecting duct to dehydration involves a paracellular pathway. *Am. J. Physiol.* **1995**, *268*, F53–F63. [CrossRef] [PubMed]
14. Muto, S. Physiological roles of claudins in kidney tubule paracellular transport. *Am. J. Physiol. Ren. Physiol.* **2017**, *312*, F9–F24. [CrossRef] [PubMed]
15. Hou, J.; Renigunta, A.; Yang, J.; Waldegger, S. Claudin-4 forms paracellular chloride channel in the kidney and requires claudin-8 for tight junction localization. *Proc. Natl. Acad. Sci. USA* **2010**, *107*, 18010–18015. [CrossRef] [PubMed]
16. Luk, J.M.; Tong, M.-K.; Mok, B.W.; Tam, P.-C.; Yeung, W.S.B.; Lee, K.-F. Sp1 site is crucial for the mouse claudin-19 gene expression in the kidney cells. *FEBS Lett.* **2004**, *578*, 251–256. [CrossRef] [PubMed]

17. Konrad, M.; Schaller, A.; Seelow, D.; Pandey, A.V.; Waldegger, S.; Lesslauer, A.; Vitzthum, H.; Suzuki, Y.; Luk, J.M.; Becker, C.; et al. Mutations in the tight-junction gene claudin 19 (CLDN19) are associated with renal magnesium wasting, renal failure, and severe ocular involvement. *Am. J. Hum. Genet.* **2006**, *79*, 949–957. [CrossRef] [PubMed]
18. Angelow, S.; El-Husseini, R.; Kanzawa, S.A.; Yu, A.S.L. Renal localization and function of the tight junction protein, claudin-19. *Am. J. Physiol. Ren. Physiol.* **2007**, *293*, F166–F177. [CrossRef] [PubMed]
19. Hou, J.; Renigunta, A.; Konrad, M.; Gomes, A.S.; Schneeberger, E.E.; Paul, D.L.; Waldegger, S.; Goodenough, D.A. Claudin-16 and claudin-19 interact and form a cation-selective tight junction complex. *J. Clin. Investig.* **2008**, *118*, 619–628. [CrossRef]
20. Weber, S.; Hoffmann, K.; Jeck, N.; Saar, K.; Boeswald, M.; Kuwertz-Broeking, E.; Meij, I.I.C.; Knoers, N.V.A.M.; Cochat, P.; Suláková, T.; et al. Familial hypomagnesaemia with hypercalciuria and nephrocalcinosis maps to chromosome 3q27 and is associated with mutations in the PCLN-1 gene. *Eur. J. Hum. Genet.* **2000**, *8*, 414–422. [CrossRef]
21. Hou, J.; Renigunta, A.; Gomes, A.S.; Hou, M.; Paul, D.L.; Waldegger, S.; Goodenough, D.A. Claudin-16 and claudin-19 interaction is required for their assembly into tight junctions and for renal reabsorption of magnesium. *Proc. Natl. Acad. Sci. USA* **2009**, *106*, 15350–15355. [CrossRef] [PubMed]
22. Günzel, D.; Haisch, L.; Pfaffenbach, S.; Krug, S.M.; Milatz, S.; Amasheh, S.; Hunziker, W.; Müller, D. Claudin function in the thick ascending limb of Henle's loop. *Ann. N. Y. Acad. Sci.* **2009**, *1165*, 152–162. [CrossRef] [PubMed]
23. Lee, N.P.Y.; Tong, M.K.; Leung, P.P.; Chan, V.W.; Leung, S.; Tam, P.-C.; Chan, K.-W.; Lee, K.-F.; Yeung, W.S.B.; Luk, J.M. Kidney claudin-19: Localization in distal tubules and collecting ducts and dysregulation in polycystic renal disease. *FEBS Lett.* **2006**, *580*, 923–931. [CrossRef] [PubMed]
24. Lee, J.W.; Chou, C.-L.; Knepper, M.A. Deep Sequencing in Microdissected Renal Tubules Identifies Nephron Segment-Specific Transcriptomes. *J. Am. Soc. Nephrol.* **2015**, *26*, 2669–2677. [CrossRef] [PubMed]
25. Maric, K.; Oksche, A.; Rosenthal, W. Aquaporin-2 expression in primary cultured rat inner medullary collecting duct cells. *Am. J. Physiol.* **1998**, *275*, F796–F801. [CrossRef] [PubMed]
26. Schenk, L.K.; Rinschen, M.M.; Klokkers, J.; Kurian, S.M.; Neugebauer, U.; Salomon, D.R.; Pavenstaedt, H.; Schlatter, E.; Edemir, B. Cyclosporin-A Induced Toxicity in Rat Renal Collecting Duct Cells: Interference with Enhanced Hypertonicity Induced Apoptosis. *Cell. Physiol. Biochem.* **2010**, *26*, 887–900. [CrossRef]
27. Sonntag, S.R.; Ziemens, A.; Wulfmeyer, V.C.; Milatz, S.; Bleich, M.; Himmerkus, N. Diuretic state affects ascending thin limb tight junctions. *Am. J. Physiol. Renal Physiol.* **2018**, *314*, F190–F195. [CrossRef]
28. Gong, Y.; Hou, J. Claudins in barrier and transport function-the kidney. *Pflug. Arch.* **2017**, *469*, 105–113. [CrossRef]
29. Yu, A.S.L. Claudins and the kidney. *J. Am. Soc. Nephrol.* **2015**, *26*, 11–19. [CrossRef]
30. Ruhfus, B.; Bauernschmitt, H.G.; Kinne, R.K. Properties of a polarized primary culture from rat renal inner medullary collecting duct (IMCD) cells. *Cell. Dev. Biol. Anim.* **1998**, *34*, 227–231. [CrossRef]
31. Cil, O.; Ertunc, M.; Onur, R. The diuretic effect of urea analog dimethylthiourea in female Wistar rats. *Hum. Exp. Toxicol.* **2012**, *31*, 1050–1055. [CrossRef]
32. Husted, R.F.; Hayashi, M.; Stokes, J.B. Characteristics of papillary collecting duct cells in primary culture. *Am. J. Physiol.* **1988**, *255*, F1160–F1169. [CrossRef] [PubMed]
33. Mongelli-Sabino, B.M.; Canuto, L.P.; Collares-Buzato, C.B. Acute and chronic exposure to high levels of glucose modulates tight junction-associated epithelial barrier function in a renal tubular cell line. *Life Sci.* **2017**, *188*, 149–157. [CrossRef] [PubMed]
34. Canuto, L.P.; Collares-Buzato, C.B. Increased osmolality enhances the tight junction-mediated barrier function in a cultured renal epithelial cell line. *Cell Biol. Int.* **2019**, *43*, 73–82. [CrossRef] [PubMed]
35. Michea, L.; Ferguson, D.R.; Peters, E.M.; Andrews, P.M.; Kirby, M.R.; Burg, M.B. Cell cycle delay and apoptosis are induced by high salt and urea in renal medullary cells. *Am. J. Physiol. Ren. Physiol.* **2000**, *278*, F209–F218. [CrossRef] [PubMed]
36. Santos, B.C.; Chevaile, A.; Hébert, M.J.; Zagajeski, J.; Gullans, S.R. A combination of NaCl and urea enhances survival of IMCD cells to hyperosmolality. *Am. J. Physiol.* **1998**, *274*, F1167–F1173. [CrossRef] [PubMed]
37. Capasso, J.M.; Rivard, C.J.; Berl, T. Long-term adaptation of renal cells to hypertonicity: Role of MAP kinases and Na-K-ATPase. *Am. J. Physiol. Ren. Physiol.* **2001**, *280*, F768–F776. [CrossRef] [PubMed]

38. Lanaspa, M.A.; Almeida, N.E.; Andres-Hernando, A.; Rivard, C.J.; Capasso, J.M.; Berl, T. The tight junction protein, MUPP1, is up-regulated by hypertonicity and is important in the osmotic stress response in kidney cells. *Proc. Natl. Acad. Sci. USA* **2007**, *104*, 13672–13677. [CrossRef] [PubMed]
39. Lanaspa, M.A.; Andres-Hernando, A.; Rivard, C.J.; Dai, Y.; Berl, T. Hypertonic stress increases claudin-4 expression and tight junction integrity in association with MUPP1 in IMCD3 cells. *Proc. Natl. Acad. Sci. USA* **2008**, *105*, 15797–15802. [CrossRef]
40. Hager, H.; Kwon, T.H.; Vinnikova, A.K.; Masilamani, S.; Brooks, H.L.; Frøkiaer, J.; Knepper, M.A.; Nielsen, S. Immunocytochemical and immunoelectron microscopic localization of alpha-, beta-, and gamma-ENaC in rat kidney. *Am. J. Physiol. Ren. Physiol.* **2001**, *280*, F1093–F1106. [CrossRef]
41. Ronzaud, C.; Loffing, J.; Bleich, M.; Gretz, N.; Gröne, H.-J.; Schütz, G.; Berger, S. Impairment of sodium balance in mice deficient in renal principal cell mineralocorticoid receptor. *J. Am. Soc. Nephrol.* **2007**, *18*, 1679–1687. [CrossRef] [PubMed]
42. Leviel, F.; Hübner, C.A.; Houillier, P.; Morla, L.; El Moghrabi, S.; Brideau, G.; Hatim, H.; Parker, M.D.; Kurth, I.; Kougioumtzes, A.; et al. The Na^+-dependent chloride-bicarbonate exchanger SLC4A8 mediates an electroneutral Na^+ reabsorption process in the renal cortical collecting ducts of mice. *J. Clin. Investig.* **2010**, *120*, 1627–1635. [CrossRef] [PubMed]
43. Crambert, G.; Ernandez, T.; Lamouroux, C.; Roth, I.; Dizin, E.; Martin, P.-Y.; Féraille, E.; Hasler, U. Epithelial sodium channel abundance is decreased by an unfolded protein response induced by hyperosmolality. *Physiol. Rep.* **2014**, *2*, e12169. [CrossRef] [PubMed]
44. Milatz, S.; Himmerkus, N.; Wulfmeyer, V.C.; Drewell, H.; Mutig, K.; Hou, J.; Breiderhoff, T.; Müller, D.; Fromm, M.; Bleich, M.; et al. Mosaic expression of claudins in thick ascending limbs of Henle results in spatial separation of paracellular Na^+ and Mg^{2+} transport. *Proc. Natl. Acad. Sci. USA* **2017**, *114*, E219–E227. [CrossRef] [PubMed]
45. Gong, Y.; Renigunta, V.; Himmerkus, N.; Zhang, J.; Renigunta, A.; Bleich, M.; Hou, J. Claudin-14 regulates renal Ca(+)(+) transport in response to CaSR signalling via a novel microRNA pathway. *EMBO J.* **2012**, *31*, 1999–2012. [CrossRef] [PubMed]
46. Nusrat, A.; Parkos, C.A.; Verkade, P.; Foley, C.S.; Liang, T.W.; Innis-Whitehouse, W.; Eastburn, K.K.; Madara, J.L. Tight junctions are membrane microdomains. *J. Cell Sci.* **2000**, *113*, 1771–1781. [PubMed]
47. Mitic, L.L. Expression, solubilization, and biochemical characterization of the tight junction transmembrane protein claudin-4. *Protein Sci.* **2003**, *12*, 218–227. [CrossRef] [PubMed]
48. Sugibayashi, K.; Onuki, Y.; Takayama, K. Displacement of tight junction proteins from detergent-resistant membrane domains by treatment with sodium caprate. *Eur. J. Pharm. Sci.* **2009**, *36*, 246–253. [CrossRef]
49. Wilson, J.L.L.; Miranda, C.A.; Knepper, M.A. Vasopressin and the regulation of aquaporin-2. *Clin. Exp. Nephrol.* **2013**, *17*, 751–764. [CrossRef]
50. Krug, S.M.; Amasheh, S.; Richter, J.F.; Milatz, S.; Günzel, D.; Westphal, J.K.; Huber, O.; Schulzke, J.D.; Fromm, M. Tricellulin forms a barrier to macromolecules in tricellular tight junctions without affecting ion permeability. *Mol. Biol. Cell* **2009**, *20*, 3713–3724. [CrossRef]
51. Tokuda, S.; Yu, A.S.L. Regulation of Epithelial Cell Functions by the Osmolality and Hydrostatic Pressure Gradients: A Possible Role of the Tight Junction as a Sensor. *Int. J. Mol. Sci.* **2019**, *20*, 3513. [CrossRef] [PubMed]
52. Tokuda, S.; Hirai, T.; Furuse, M. Effects of osmolality on paracellular transport in MDCK II cells. *PLoS ONE* **2016**, *11*, 1–21. [CrossRef] [PubMed]
53. Himmerkus, N.; Plain, A.; Marques, R.D.; Sonntag, S.R.; Paliege, A.; Leipziger, J.; Bleich, M. AVP dynamically increases paracellular Na^+ permeability and transcellular NaCl transport in the medullary thick ascending limb of Henle's loop. *Pflug. Arch.* **2017**, *469*, 149–158. [CrossRef] [PubMed]
54. Plain, A.; Wulfmeyer, V.C.; Milatz, S.; Klietz, A.; Hou, J.; Bleich, M.; Himmerkus, N. Corticomedullary difference in the effects of dietary Ca^{2+} on tight junction properties in thick ascending limbs of Henle's loop. *Pflug. Arch.* **2016**, *468*, 293–303. [CrossRef] [PubMed]
55. Pohl, M.; Kaminski, H.; Castrop, H.; Bader, M.; Himmerkus, N.; Bleich, M.; Bachmann, S.; Theilig, F. Intrarenal

renin angiotensin system revisited: Role of megalin-dependent endocytosis along the proximal nephron. *J. Biol. Chem.* **2010**, *285*, 41935–41946. [CrossRef] [PubMed]

56. Barry, P.H.; Lynch, J.W. Liquid junction potentials and small cell effects in patch-clamp analysis. *J. Membr. Biol.* **1991**, *121*, 101–117. [CrossRef]
57. Kimizuka, H.; Koketsu, K. Ion transport through cell membrane. *J. Theor. Biol.* **1964**, *6*, 290–305. [CrossRef]

8

The Integral Role of Tight Junction Proteins in the Repair of Injured Intestinal Epithelium

Zachary M. Slifer and Anthony T. Blikslager *

Department of Clinical Sciences, College of Veterinary Medicine, North Carolina State University, Raleigh, NC 27607, USA; zmslifer@ncsu.edu

* Correspondence: Anthony_Blikslager@ncsu.edu

Abstract: The intestinal epithelial monolayer forms a transcellular and paracellular barrier that separates luminal contents from the interstitium. The paracellular barrier consists of a highly organized complex of intercellular junctions that is primarily regulated by apical tight junction proteins and tight junction-associated proteins. This homeostatic barrier can be lost through a multitude of injurious events that cause the disruption of the tight junction complex. Acute repair after injury leading to the reestablishment of the tight junction barrier is crucial for the return of both barrier function as well as other cellular functions, including water regulation and nutrient absorption. This review provides an overview of the tight junction complex components and how they link to other plasmalemmal proteins, such as ion channels and transporters, to induce tight junction closure during repair of acute injury. Understanding the components of interepithelial tight junctions and the mechanisms of tight junction regulation after injury is crucial for developing future therapeutic targets for patients experiencing dysregulated intestinal permeability.

Keywords: barrier function; tight junction; repair; occludin; claudin; NHE2; ClC-2

---

1. Intestinal Epithelium as a Selective Barrier

The intestine is lined with a monolayer of columnar epithelium that maintains two critical functions: (1) selectively filtering luminal contents, including nutrients, water and electrolytes, to allow for their translocation into the circulation and (2) forming a barrier to prevent the translocation of luminal toxins, commensal or pathogenic microorganisms, and foreign antigens into the circulation [1]. Under homeostatic conditions, these functions are regulated by both transcellular and paracellular pathways, the latter of which are primarily maintained by apical tight junction proteins through paracellular pore and leak permeability pathways [2,3]. The paracellular pathway is associated with the charge and size selective transport of materials through the space between intestinal epithelial cells.

Intestinal barrier homeostasis is disrupted through tight junction protein dysregulation, which occurs via a variety of injurious events, including microbial degradation and bacterial toxin exposure, exposure to cytotoxic agents, exposure to pro-inflammatory cytokines such as IFNγ and TNFα, intestinal autoimmune disease such as Celiac disease, and intestinal ischemia [4–9]. The loss of tight junction integrity results in the formation of a third pathway, known as the high-capacity and nonselective unrestricted permeability pathway, which can allow for the unrestricted movement of microorganisms and large proteins across the paracellular space [10]. An inability to rapidly repair the tight junctions in order to restore epithelial barrier function is detrimental to the patient, as it can result in various pathologies, including sepsis and multiple organ dysfunction [11,12]. Therefore, understanding factors that can regulate the tight junction complex during repair of injured intestinal epithelium is crucial for developing future therapeutic targets.

2. Tight Junction Protein Structure

Tight junctions are made up of a number of protein elements, including transmembrane claudins (total of 27 mammalian claudins), as well as myelin and lymphocyte (MAL) and related proteins for vesicle trafficking and membrane link (MARVEL) [13–18]. MARVEL domain-containing proteins are a component of a larger group of tight junction-associated MARVEL proteins (TAMPS) that include transmembrane proteins such as occludin and tricellulin [18–20]. Other tight junction-associated transmembrane molecules include junctional adhesion molecules (JAM-1, -2, and -3) that can regulate the formation of tight junctions and migration of neutrophils [21–26]. Additionally, intracellular scaffold proteins such as zonula occludens (ZO) -1, -2, and -3 play an integral role in tight junction protein assembly and link tight junction transmembrane proteins with the actin cytoskeleton [27–30].

Intestinal claudins exist in two different classes: sealing claudins and pore-forming claudins [31]. Increased membrane expression of sealing proteins results in a 'tighter' epithelial barrier, further restricting the movement of luminal contents through the paracellular space. Sealing tight junction proteins include claudins-1, -3, -4, -5, -8, -11, -14, 18, and -19 [17]. Alternatively, increased membrane expression of pore-forming proteins (including claudins-2, -10a/-10b, -15, -16, and -17) reduces the selectivity for luminal contents that can pass between epithelial cells, thereby increasing paracellular permeability [17]. Each pore-forming claudin has an ionic charge specificity for cations or anions as well as ionic size selectivity, thereby increasing the permeability for ions based on claudin-specific permeability characteristics. An interesting interaction between the two classes of claudins has been observed with the displacement of pore-forming claudins by sealing claudins. For example, in a claudin-8 transfected Madin–Darby canine kidney II (MDCK II) cell line in which claudin-8 expression occurred in the absence of doxycycline, claudin-8 displacement of claudin-2 was visualized upon immunofluorescent imaging [32,33]. Specifically, claudin-8 served to replace claudin-2 in tight junction strands in this model, which reduced the number of functional paracellular cation pores [32].

Tight junction protein expression in the intestinal tract is tissue- and age-specific. For example, claudin-2 is normally expressed in the human small intestine but is reported to be expressed only in the colonic crypt of fetal humans and absent in the adult colon under homeostatic conditions [34]. Overall, claudin-2 generally exhibits higher expression in leaky epithelial tissues, such as colonic tissues from a subset of patients with inflammatory bowel disease [35]. Additionally, its protein expression was detected throughout the crypt–villus axis of human small intestines but was only detected in undifferentiated crypt cells of human colonic tissue [36]. Other claudins follow suit regarding crypt–luminal axis expression with some pore-forming claudins (-2, -10, -13, -15) being restricted to the intestinal crypt base in murine tissue while other sealing claudins (-3, -4, -7, -8) are expressed in luminal epithelial cells [37–40].

Special Functions of Select Tight Junction Proteins

As discussed throughout the remainder of this review, regulation of tight junction proteins is a vital component of epithelial barrier repair after injury. However, specific tight junction proteins can have additional special functions that are species and age dependent. In the case of claudin-4, there is an age-dependent disparity between cellular localization of tight junction proteins [41]. Intestinal porcine claudin-4 is localized to the apical surface of porcine jejunal enterocytes within the first two days of age and only localizes to the lateral surface between adjacent epithelial cells beyond two days of age [41]. This age-specificity of claudin-4 localization in piglet jejunum may be due to immunological naivety that newborn piglets experience. Piglets must be able to acquire and absorb immunoglobulins and other immune-related molecules, including cytokines and antimicrobial peptides, from colostrum within the first day of life to prevent death via bacterial sepsis [42]. It is reported that this age-specific, apical surface localization of claudin-4 occurs in jejunal enterocytes at the same period of time in which the vital immune macromolecules are absorbed into the bloodstream [41]. Therefore, this special function of claudin-4 localization is necessary to allow for the proper uptake of immune molecules by the piglet, and subsequent relocalization to the lateral surface may aid in sealing of the paracellular space between adjacent epithelial cells once the immune constituents are absorbed.

Another tight junction protein, tricellulin, serves a special function in the tight junction barrier where more than two epithelial cells meet. Tight junction strands between two adjacent epithelial cells typically associated laterally to pair with another tight junction strand, forming bicellular tight junctions between the two cells [43]. However, these bicellular tight junctions are not continuous at tricellular epithelial cell contacts and have therefore been described as tricellular tight junction proteins. While traditional tight junction proteins, such as occludin and claudins, are found in both bicellular and tricellular tight junctions, tricellulin is concentrated to the tricellular tight junction and its knockdown in the EpH4 cell line of immortalized mouse mammary gland epithelium resulted in altered organization of bicellular tight junction proteins [20]. Additionally, tricellulin has been shown to not affect the permeability for ions while forming a barrier to macromolecules in tricellulin-transfected MDCK II cells overexpressing tricellulin in the tricellular tight junctions [44]. Overall, it is crucial to consider all components and special functions of tight junction proteins when studying tight junction structure. Furthermore, the understanding of these tight junction special functions may be crucial to restoring barrier function following injury.

3. Acute Mechanisms of Repair in Injured Intestinal Epithelium

When the monolayer of intestinal epithelium is injured, such as that which occurs during ischemia/reperfusion injury or exposure to pathogenic microbes such as rotavirus [45,46], detachment of the epithelium from the basement membrane and separation of adjacent epithelial cells from one another due to dysregulation and loss of tight junctional proteins occurs. Furthermore, the loss of polarity-establishing tight junctional proteins results in the loss of cell polarity, which abolishes apical and basolateral positioning of localized molecules such as ion channels/transporters, resulting in their mislocalization [47]. When homeostatic positioning of ion channels and transporters is lost, this can subsequently lead to the dysregulation of a multitude of cellular functions including water absorption/secretion, intracellular and organelle pH, and nutrient absorption.

Once the cause of intestinal injury is resolved, such as restoration of blood flow in ischemic injury, rapid mechanisms of intestinal mucosal repair take place in a well-orchestrated series of reparative events. Initially, small intestinal villi contract via the contraction of myofibroblasts adjacent to the epithelial basement membrane and centrally along the central lacteal. Villus contraction is characterized histologically by a quantitatively diminished villus height [48] and occurs in response to mediators such as PGE_2 [49]. Villus contraction results in reduction of the denuded surface area that remains to be covered by epithelial cells. Simultaneously, restitution of epithelial cells shouldering the site of injury occurs to cover the denuded area [50]. These cells depolarize to disassemble microvilli, allowing for subsequent lamellipodia-driven movement via actin–myosin treadmilling, while maintaining transient attachment to the basement membrane through integrins [11]. Although the underlying intestinal layers may not appear exposed to luminal contents since the mucosa is no longer denuded, the unrestricted permeability pathway via poorly formed tight junctions allows for microorganisms and macromolecules to cross the epithelial barrier. In order for the tight junction barrier and cell polarity to be restored, tight junction proteins internalized during injury, such as the endocytosis of occludin that accompanies anoxic injury in Caco-2 cells, must be reinserted back into the membrane via recycling endosomes [51–53]. Ultimately, following these acute repair mechanisms, crypt cells can proliferate and differentiate to restore the proper number of epithelial cells to the monolayer in order to regain full homeostatic function.

4. Regulation of Tight Junctions via Ion Channels/Transporters

Closure of the tight junction after acute intestinal injury is paramount in restoring barrier function and returning to homeostatic functioning. Tight junction proteins can be regulated by many factors, including cytokines, growth factors, and nutrients. For example, transport of glucose by SGLT1 has been shown to result in the physiological opening of tight junctions in an NHE3-dependent mechanism [54]. Alternatively, ion channel/transporters, including proteins from the Na^+/H^+ exchanger (NHE) family as well as chloride channel protein 2 (ClC-2) have also been shown to regulate tight junction proteins, specifically after intestinal ischemic injury [55–57]. This review will examine the reparative role of these transport proteins specifically related to restoration of junctions.

4.1. NHE2 and Intestinal Repair

The gastrointestinal epithelium is home to many ion transporters that are collectively responsible for regulating homeostatic cell functions, including the regulation of nutrient absorption, cytosolic and organelle pH, water absorption and secretion, and cell volume [58]. One major family of ion transporters in the human GI tract is the *SLC9* gene family, also known as the NHE family. NHE isoforms belonging to the *SLC9A* gene subgroup (*SLC9A1-9*) can be either plasmalemmal or intracellular, depending on the isoform and tissue location within the gut [59]. Additionally, the Na^+/H^+ exchanger 5 (NHE5) is the only isoform for which expression has not been shown in the gastrointestinal tract [60]. These proteins are responsible for the electroneutral antiport of Na^+ into intestinal epithelium in exchange for H^+ secreted from the cell to maintain cellular pH and volume.

An additional function of NHEs that continues to be explored is the link between NHEs and the tight junction. One mechanism that links NHEs to the tight junction is through binding to the actin cytoskeleton. Specifically, Na^+/H^+ exchanger 3 (NHE3) has been shown to bind directly to the actin cytoskeleton and indirectly through various binding partners, including ezrin [61,62]. The ezrin protein is known to link the plasma membrane to the cytoskeleton in its active, phosphorylated conformation through binding to actin with its C-terminal region [63–65]. This interaction with the cytoskeleton has been shown to regulate plasma membrane tension, which is involved in motility and endocytosis [66]. Ezrin links the cytoskeleton to the plasma membrane through binding of its N-terminal region to either membrane lipids or cytoplasmic regions of transmembrane proteins, including NHE3 [66,67]. Thus, by linking transmembrane proteins such as NHE3 to the cytoskeleton, there is an indirect link between transmembrane proteins and tight junction proteins.

Of the NHE isoforms that have been described in the gut, Na^+/H^+ exchanger 2 (NHE2) is one of the least described NHEs in regards to its homeostatic and pathophysiologic functionality. However, NHE2 has been linked to paracellular barrier function and tight junction regulation during the recovery of injured intestinal epithelium [55,56]. In both porcine and murine models of intestinal ischemic injury, NHE2, rather than NHE1 or NHE3, appears to be the primary NHE responsible for regulating tight junction proteins during the recovery of ischemia-injured intestines [55,56]. During ex vivo recovery of porcine intestinal ischemia, selective pharmacologic inhibition of NHE2 enhanced epithelial recovery, as evidenced by significant elevations in transepithelial electrical resistance (TER) while inhibition of NHE1 or NHE3 did not elicit a recovery response [56]. In the same study, this NHE2-specific inhibitory effect on recovery was independent of epithelial restitution, and NHE2 was shown to co-immunoprecipitate with ezrin/radixin/moesin (ERM)-binding phosphoprotein 50 (EBP50), also known as NHE regulatory factor 1 (NHERF1), in ischemia-injured porcine ileum. These data suggest that NHE2 regulates restoration of the tight junction barrier during recovery of intestinal ischemia and is potentially linked to the actin cytoskeleton through binding partners (Figure 1). Although NHE2 is also implicated in the in vivo recovery of murine intestinal ischemia, the genetic knockout of NHE2 in the murine model has the inverse effect when compared to pharmacologic inhibition of NHE2 in the porcine model of intestinal ischemia [55]. Specifically, NHE2 null mice exhibit increased blood-to-lumen ^3H-mannitol flux at 1.5 and 3 hours after ischemic injury as well as a change in localization of occludin and claudin-1 from the membrane to the cytosol when compared to wild-type mice [55]. Additionally, epithelial restitution after intestinal ischemia was unaffected by the absence of NHE2 in this model. It is important to note that pharmacologic inhibition or genetic knockout of NHE2 may affect intracellular pH (pH_i) since NHEs are known to contribute to pH_i changes, and these potential pH_i changes can affect charge selectivity of the paracellular pathway [68,69]. However, further studies will be required to determine if NHE-mediated changes in pH_i are linked to alterations in the tight junction. Together, this information suggests that NHE2 regulates acute recovery after intestinal ischemic injury in a tight junction-dependent manner, whereas its absence delays restoration of tight junction barrier function.

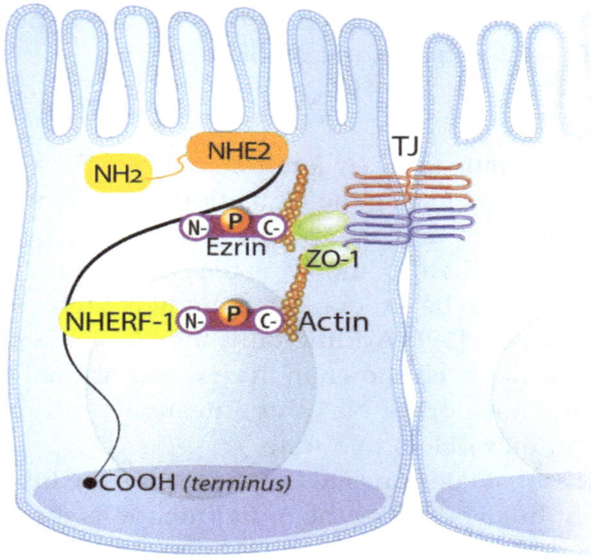

Figure 1. Schematic representation of the potential link of NHE2 to the actin cytoskeleton and subsequently the tight junction through binding partners. The primary candidate protein linking NHE2 to the actin cytoskeleton is phosphorylated ezrin. Based on information known about NHE3 and data from NHE2 in vivo studies, NHE2 may bind directly to ezrin or indirectly through additional binding partners, including NHERF1/EBP50.

4.2. ClC-2 and Intestinal Repair

Chloride secretion from intestinal epithelium into the lumen is crucial for homeostatic water absorption/secretion via maintaining an osmotic balance with luminal accumulation of both chloride and sodium ions. This subsequently allows for proper mucosal hydration of the epithelial layer, which protects the lumen as food passes through the intestine [70]. The primary protein responsible for chloride transport into the intestinal lumen is the apically located cystic fibrosis transmembrane receptor (CFTR) [71]. However, another contributor to transepithelial chloride transport within intestinal epithelium is the voltage-gated ClC-2 protein, one of nine mammalian proteins belonging to the chloride channel (ClC) protein family [72]. ClC-2 has been shown to localize in the plasma membrane at tight junction complexes within mouse intestinal epithelium [73] or has plasmalemmal basolateral localization within guinea pig colons [74], suggesting species- or tissue-specific localization of ClC-2.

In addition to its role in transepithelial chloride transport, ClC-2 has been shown to regulate intestinal tight junction barrier function in various injury models. After porcine intestinal ischemic injury, stimulation of ClC-2 with the ClC-2 agonist lubiprostone during ex vivo recovery on Ussing chambers resulted in marked increases in TER and reduced mucosal-to-serosal mannitol flux [57]. Contrasting the effect of ClC-2 stimulation with lubiprostone, the genetic absence of ClC-2 in a murine model of intestinal ischemia resulted in significant increases in blood-to-lumen mannitol clearance while also reducing expression of membrane-bound occludin and claudin-1 after up to 3 hours of in vivo recovery [75]. In this murine model, occludin co-localized with ClC-2 after co-immunoprecipitation studies, and its localization to the tight junction region was diffuse in ClC-2 null mice after up to 3 hours of recovery [75]. Additionally, in a murine model of dextran sulfate sodium (DSS)-induced colitis, the absence of ClC-2 increased disease severity, as measured through significant losses in body weight and significant increases in disease activity index [76]. ClC-2 null mice treated with DSS also demonstrated significantly increased expression of claudin-2 and reduced occludin expression in the same study. Interestingly, a recent in vitro study established Caco-2 cells overexpressing ClC-2 (Caco-2^{ClCN2}), and this ClC-2 overexpression resulted in a decrease of the pore-forming claudin-2 protein while maintaining claudin-1 and claudin-4 protein levels to that of control cells [77]. As an aside, although cell volume and pH_i is partially regulated by ClC-2 and thus the genetic knockout of ClC-2 can affect these intracellular factors, studies will be needed to determine if ClC-2-mediated changes in these intracellular factors have an effect on the tight junction [78,79]. Based on these studies, there appears to be a mechanistic link between ClC-2 and the regulation of membrane claudin expression, but further studies will need to be carried out to determine how ClC-2 plays a role in claudin expression patterns. Nonetheless, current data suggest the critical role of ClC-2 in barrier function during recovery from epithelial injury while also reinforcing the link between ClC-2 and the tight junction barrier.

The link between ClC-2 and the tight junction was initially shown to exist through intracellular caveolar trafficking of occludin via interaction with both caveolin-1 and the small GTPase Rab5 in a cell line derived from human intestinal Caco-2 cells (Figure 2) [80]. This connection between ClC-2, occludin, and caveolin-1 was further supported in vivo with a model of DSS-induced colitis. ClC-2 null mice treated with DSS had significantly displaced occludin/caveolin-1 densitometry readings toward high-density, detergent-soluble fractions of sucrose density gradient-based fractions when compared to wild-type mice treated with DSS [76]. These data suggest that after DSS-induced colitis, occludin and caveolin-1 are strongly associated in the cytosol of mice lacking ClC-2 but not in mice normally expressing ClC-2. In tandem, overexpression of ClC-2 in Caco-2^{ClCN2} cells was reported to not only exhibit enhanced tight junction barrier function through significant increases in TER and reductions in apical-to-basal inulin flux, but this ClC-2 overexpression further connected ClC-2 to caveolin-1 and caveolar trafficking of occludin [77]. Specifically, ClC-2 overexpression in Caco-2^{ClCN2} cells exhibited both significantly increased occludin protein and reduced endocytosis of occludin when compared to control cells while simultaneously diminishing both caveolin-1 protein and caveolae assembly [77]. Furthermore, this study reported that selective inhibition of ClC-2 lead to both reduced occludin protein and increased caveolin-1 protein. Taken together, there is strong evidence from both in vitro and in vivo models that links ClC-2 to the tight junction protein occludin and its regulation by caveolar trafficking. Based on the presented evidence, it is believed that ClC-2 facilitates the shuttling of endocytosed tight junction proteins back to the apical–lateral membrane to repair injured tight junctions. However, further mechanistic studies are required out to determine the precise mechanisms of these events.

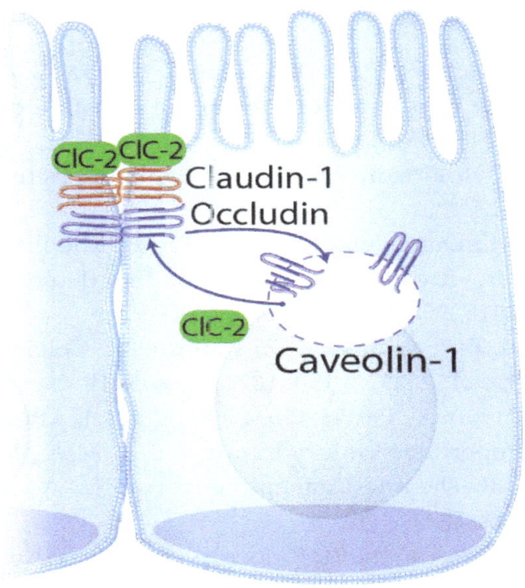

Figure 2. Schematic representation of the link between ClC-2 and caveolar trafficking of occludin. This schematic is a slight modification from a previously published figure [81] to more closely associate ClC-2 to both the tight junction complex and to caveolin-1-associated endocytosis and recycling of tight junction proteins such as occludin. Note that the representation of ClC-2 at the tight junction complex is not exclusively apical in localization, which leaves room for ClC-2 to be more closely associated with occludin in the membrane.

5. Conclusions

Injury of intestinal epithelium affects both the epithelial cells and the junctional structures that link them. A great deal of attention has been focused on mechanisms of epithelial restitution, but a lesser level of attention has been paid to the reassembly of tight junctions within repairing epithelium. This intriguing process appears to be intimately associated with ion channels, which in the case of NHE2 and ClC-2, is facilitated by a close association with tight junction regulatory proteins. With ClC-2 in particular, the mechanism of ion channel-facilitated tight junction reassembly has been linked to endosomal recycling of tight junction proteins, with evidence of restoration of the positioning of tight junction integral membrane proteins during the reparative process, and increased membrane expression of sealing claudins with cellular over-expression of ClC-2. However, how precisely ion channels interact with structures such as endosomes, and how this facilitates insertion of sealing tight junction proteins at the repairing tight junction will require further study. Nonetheless, it does appear that ion channels such as NHE2 and ClC-2 have a greater cellular function than ion transport alone. It is conceivable that the transport of select ions accompanies a structural change that sets off a series of signaling events associated with tight junction reassembly, but this will require additional mechanistic work. Ultimately, further studies to uncover the relationship between ion channels and reassembly of tight junctions has the potential to lead to novel therapeutic targets for patients with increased intestinal paracellular permeability.

References

1. Groschwitz, K.R.; Hogan, S.P. Intestinal barrier function: Molecular regulation and disease pathogenesis. *J. Allergy Clin. Immunol.* **2009**, *124*, 3–22. [CrossRef] [PubMed]
2. Turner, J.R. Intestinal mucosal barrier function in health and disease. *Nat. Rev. Immunol.* **2009**, *9*, 799–809. [CrossRef] [PubMed]
3. Liang, G.H.; Weber, C.R. Molecular aspects of tight junction barrier function. *Curr. Opin. Pharmacol.* **2014**, *19*, 84–89. [CrossRef] [PubMed]
4. König, J.; Wells, J.; Cani, P.D.; García-Ródenas, C.L.; Macdonald, T.; Mercenier, A.; Whyte, J.; Troost, F.; Brummer, R.J. Human Intestinal Barrier Function in Health and Disease. *Clin. Transl. Gastroenterol.* **2016**, *7*, e196. [CrossRef] [PubMed]
5. El Asmar, R.; Panigrahi, P.; Bamford, P.; Berti, I.; Not, T.; Coppa, G.V.; Catassi, C.; Fasano, A. Host-dependent zonulin secretion causes the impairment of the small intestine barrier function after bacterial exposure. *Gastroenterology* **2002**, *123*, 1607–1615. [CrossRef]
6. Al-Sadi, R. Mechanism of cytokine modulation of epithelial tight junction barrier. *Front. Biosci.* **2009**, *14*, 2765–2778. [CrossRef]
7. Broekaert, N.; Devreese, M.; Demeyere, K.; Berthiller, F.; Michlmayr, H.; Varga, E.; Adam, G.; Meyer, E.; Croubels, S. Comparative in vitro cytotoxicity of modified deoxynivalenol on porcine intestinal epithelial cells. *Food Chem. Toxicol.* **2016**, *95*, 103–109. [CrossRef]
8. Schumann, M.; Siegmund, B.; Schulzke, J.D.; Fromm, M. Celiac Disease: Role of the Epithelial Barrier. *Cell. Mol. Gastroenterol. Hepatol.* **2017**, *3*, 150–162. [CrossRef]
9. Li, Q.; Zhang, Q.; Wang, C.; Liu, X.; Qu, L.; Gu, L.; Li, N.; Li, J. Altered distribution of tight junction proteins after intestinal ischaemia/reperfusion injury in rats. *J. Cell. Mol. Med.* **2009**, *13*, 4061–4076. [CrossRef]
10. Odenwald, M.A.; Turner, J.R. The intestinal epithelial barrier: A therapeutic target? *Nat. Rev. Gastroenterol. Hepatol.* **2017**, *14*, 9–21. [CrossRef]
11. Blikslager, A.T.; Moeser, A.J.; Gookin, J.L.; Jones, S.L.; Odle, J. Restoration of Barrier Function in Injured Intestinal Mucosa. *Physiol. Rev.* **2007**, *87*, 545–564. [CrossRef] [PubMed]
12. Gonzalez, L.M.; Moeser, A.J.; Blikslager, A.T. Animal models of ischemia-reperfusion-induced intestinal injury: Progress and promise for translational research. *Am. J. Physiol. Gastrointest Liver Physiol.* **2015**, *308*, G63–G75. [CrossRef] [PubMed]
13. Furuse, M. Claudin-1 and -2: Novel Integral Membrane Proteins Localizing at Tight Junctions with No Sequence Similarity to Occludin. *J. Cell Biol.* **1998**, *141*, 1539–1550. [CrossRef] [PubMed]
14. Tsukita, S.; Tanaka, H.; Tamura, A. The Claudins: From Tight Junctions to Biological Systems. *Trends Biochem.*

Sci. **2019**, *44*, 141–152. [CrossRef]
15. Chiba, H.; Osanai, M.; Murata, M.; Kojima, T.; Sawada, N. Transmembrane proteins of tight junctions. *Biochim. Biophys. Acta BBA Biomembr.* **2008**, *1778*, 588–600. [CrossRef]
16. Günzel, D.; Fromm, M. Claudins and Other Tight Junction Proteins. *Compr. Physiol.* **2012**, *2*, 1819–1852.
17. Günzel, D.; Yu, A.S.L. Claudins and the Modulation of Tight Junction Permeability. *Physiol. Rev.* **2013**, *93*, 525–569. [CrossRef]
18. Van Itallie, C.M.; Anderson, J.M. Phosphorylation of tight junction transmembrane proteins: Many sites, much to do. *Tissue Barriers* **2018**, *6*, e1382671. [CrossRef]
19. Furuse, M. Occludin: A novel integral membrane protein localizing at tight junctions. *J. Cell Biol.* **1993**, *123*, 1777–1788. [CrossRef]
20. Ikenouchi, J.; Furuse, M.; Furuse, K.; Sasaki, H.; Tsukita, S.; Tsukita, S. Tricellulin constitutes a novel barrier at tricellular contacts of epithelial cells. *J. Cell Biol.* **2005**, *171*, 939–945. [CrossRef]
21. Martin-Padura, I.; Lostaglio, S.; Schneemann, M.; Williams, L.; Romano, M.; Fruscella, P.; Panzeri, C.; Stoppacciaro, A.; Ruco, L.; Villa, A.; et al. Junctional Adhesion Molecule, a Novel Member of the Immunoglobulin Superfamily That Distributes at Intercellular Junctions and Modulates Monocyte Transmigration. *J. Cell Biol.* **1998**, *142*, 117–127. [CrossRef] [PubMed]
22. Palmeri, D. Vascular Endothelial Junction-associated Molecule, a Novel Member of the Immunoglobulin Superfamily, Is Localized to Intercellular Boundaries of Endothelial Cells. *J. Biol. Chem.* **2000**, *275*, 19139–19145. [CrossRef] [PubMed]
23. Liang, T.W.; Chiu, H.H.; Gurney, A.; Sidle, A.; Tumas, D.B.; Schow, P.; Foster, J.; Klassen, T.; Dennis, K.; Demarco, R.A.; et al. Vascular endothelial-junctional adhesion molecule (VE-JAM)/JAM 2 interacts with T, NK, and dendritic cells through JAM 3. *J. Immunol.* **2002**, *168*, 1618–1626. [CrossRef] [PubMed]
24. Ebnet, K.; Aurrand-Lions, M.; Kiefer, F.; Butz, S.; Zander, K.; Zu Brickwedde, M.K.M.; Kuhn, A.; Suzuki, A.; Imhof, B.A.; Vestweber, D. The junctional adhesion molecule (JAM) family members JAM-2 and JAM-3 associate with the cell polarity protein PAR-3: A possible role for JAMs in endothelial cell polarity. *J. Cell Sci.* **2003**, *116*, 3879–3891. [CrossRef] [PubMed]
25. Garrido-Urbani, S.; Bradfield, P.F.; Imhof, B.A. Tight junction dynamics: The role of junctional adhesion molecules (JAMs). *Cell Tissue Res.* **2014**, *355*, 701–715. [CrossRef] [PubMed]
26. Zemans, R.L.; Colgan, S.P.; Downey, G.P. Transepithelial migration of neutrophils: Mechanisms and implications for acute lung injury. *Am. J. Respir. Cell Mol. Biol.* **2009**, *40*, 519–535. [CrossRef] [PubMed]
27. Stevenson, B.R.; Siliciano, J.D.; Mooseker, M.S.; Goodenough, D.A. Identification of ZO-1: A high molecular weight polypeptide associated with the tight junction (zonula occludens) in a variety of epithelia. *J. Cell Biol.* **1986**, *103*, 755–766. [CrossRef]
28. Gumbiner, B.; Lowenkopf, T.; Apatira, D. Identification of a 160-kDa polypeptide that binds to the tight junction protein ZO-1. *Proc. Natl. Acad. Sci. USA* **1991**, *88*, 3460–3464. [CrossRef]
29. Jesaitis, L.A. Molecular characterization and tissue distribution of ZO-2, a tight junction protein homologous to ZO-1 and the Drosophila discs-large tumor suppressor protein. *J. Cell Biol.* **1994**, *124*, 949–961. [CrossRef]
30. Balda, M.S. Assembly of the tight junction: The role of diacylglycerol. *J. Cell Biol.* **1993**, *123*, 293–302. [CrossRef]
31. Shen, L.; Weber, C.R.; Raleigh, D.R.; Yu, D.; Turner, J.R. Tight junction pore and leak pathways: A dynamic duo. *Annu. Rev. Physiol.* **2011**, *73*, 283–309. [CrossRef] [PubMed]
32. Yu, A.S.L.; Yin, H.; Havrilla, C.M.; Gao, L.; Morrow, J.D.; Porter, N.A.; Enck, A.H.; Lencer, W.I.; Schneeberger, E.E. Claudin-8 Expression in Madin-Darby Canine Kidney Cells Augments the Paracellular Barrier to Cation Permeation. *J. Biol. Chem.* **2003**, *278*, 17350–17359. [CrossRef]
33. France, M.M.; Turner, J.R. The mucosal barrier at a glance. *J. Cell Sci.* **2017**, *130*, 307–314. [CrossRef]
34. Luettig, J.; Rosenthal, R.; Barmeyer, C.; Schulzke, J. Claudin-2 as a mediator of leaky gut barrier during intestinal inflammation. *Tissue Barriers* **2015**, *3*, e977176. [CrossRef] [PubMed]
35. Zeissig, S.; Burgel, N.; Gunzel, D.; Richter, J.; Mankertz, J.; Wahnschaffe, U.; Kroesen, A.J.; Zeitz, M.; Fromm, M.; Schulzke, J.D. Changes in expression and distribution of claudin 2, 5 and 8 lead to discontinuous tight junctions and barrier dysfunction in active Crohn's disease. *Gut* **2007**, *56*, 61–72. [CrossRef] [PubMed]
36. Escaffit, F.; Boudreau, F.; Beaulieu, J.F. Differential expression of claudin-2 along the human intestine: Implication of GATA-4 in the maintenance of claudin-2 in differentiating cells. *J. Cell. Physiol.* **2005**, *203*, 15–26. [CrossRef] [PubMed]

37. Garcia-Hernandez, V.; Quiros, M.; Nusrat, A. Intestinal epithelial claudins: Expression and regulation in homeostasis and inflammation. *Ann. N. Y. Acad. Sci.* **2017**, *1397*, 66–79. [CrossRef] [PubMed]
38. Rahner, C.; Mitic, L.L.; Anderson, J.M. Heterogeneity in expression and subcellular localization of claudins 2, 3, 4, and 5 in the rat liver, pancreas, and gut. *Gastroenterology* **2001**, *120*, 411–422. [CrossRef]
39. Fujita, H.; Chiba, H.; Yokozaki, H.; Sakai, N.; Sugimoto, K.; Wada, T.; Kojima, T.; Yamashita, T.; Sawada, N. Differential Expression and Subcellular Localization of Claudin-7, -8, -12, -13, and -15 Along the Mouse Intestine. *J. Histochem. Cytochem.* **2006**, *54*, 933–944. [CrossRef]
40. Holmes, J.L.; Van Itallie, C.M.; Rasmussen, J.E.; Anderson, J.M. Claudin profiling in the mouse during postnatal intestinal development and along the gastrointestinal tract reveals complex expression patterns. *Gene Expr. Patterns* **2006**, *6*, 581–588. [CrossRef]
41. Pasternak, J.A.; Kent-Dennis, C.; Van Kessel, A.G.; Wilson, H.L. Claudin-4 Undergoes Age-Dependent Change in Cellular Localization on Pig Jejunal Villous Epithelial Cells, Independent of Bacterial Colonization. *Mediat. Inflamm.* **2015**, *2015*, 1–14. [CrossRef] [PubMed]
42. Butler, J.E.; Lager, K.M.; Splichal, I.; Francis, D.; Kacskovics, I.; Sinkora, M.; Wertz, N.; Sun, J.; Zhao, Y.; Brown, W.R.; et al. The piglet as a model for B cell and immune system development. *Vet. Immunol. Immunopathol.* **2009**, *128*, 147–170. [CrossRef] [PubMed]
43. Tsukita, S.; Furuse, M.; Itoh, M. Multifunctional strands in tight junctions. *Nat. Rev. Mol. Cell Biol.* **2001**, *2*, 285–293. [CrossRef] [PubMed]
44. Krug, S.M.; Amasheh, S.; Richter, J.F.; Milatz, S.; Günzel, D.; Westphal, J.K.; Huber, O.; Schulzke, J.D.; Fromm, M. Tricellulin Forms a Barrier to Macromolecules in Tricellular Tight Junctions without Affecting Ion Permeability. *Mol. Biol. Cell* **2009**, *20*, 3713–3724. [CrossRef]
45. Ciarlet, M.; Conner, M.E.; Finegold, M.J.; Estes, M.K. Group A Rotavirus Infection and Age-Dependent Diarrheal Disease in Rats: A New Animal Model To Study the Pathophysiology of Rotavirus Infection. *J. Virol.* **2002**, *76*, 41–57. [CrossRef]
46. Estes, M.K.; Kang, G.; Zeng, C.Q.; Crawford, S.E.; Ciarlet, M. Pathogenesis of Rotavirus Gastroenteritis. *Novartis Found. Symp.* **2001**, *238*, 82–96; discussion 96–100.
47. Klunder, L.J.; Faber, K.N.; Dijkstra, G.; Van Ijzendoorn, S.C. Mechanisms of Cell Polarity–Controlled Epithelial Homeostasis and Immunity in the Intestine. *Cold Spring Harb. Perspect. Biol.* **2017**, *9*, a027888. [CrossRef]
48. Moore, R.; Carlson, S.; Madara, J.L. Villus contraction aids repair of intestinal epithelium after injury. *Am. J. Physiol. Liver Physiol.* **1989**, *257*, G274–G283. [CrossRef]
49. Gookin, J.L.; Galanko, J.A.; Blikslager, A.T.; Argenzio, R.A. PG-mediated closure of paracellular pathway and not restitution is the primary determinant of barrier recovery in acutely injured porcine ileum. *Am. J. Physiol. Liver Physiol.* **2003**, *285*, G967–G979. [CrossRef]
50. Dignass, A.U. Mechanisms and Modulation of Intestinal Epithelial Repair. *Inflamm. Bowel Dis.* **2001**, *7*, 68–77. [CrossRef]
51. Utech, M.; Mennigen, R.; Bruewer, M. Endocytosis and recycling of tight junction proteins in inflammation. *J. Biomed. Biotechnol.* **2010**, *2010*, 484987. [CrossRef] [PubMed]
52. Stamatovic, S.M.; Johnson, A.M.; Sladojevic, N.; Keep, R.F.; Andjelkovic, A.V. Endocytosis of tight junction proteins and the regulation of degradation and recycling. *Ann. N. Y. Acad. Sci.* **2017**, *1397*, 54–65. [CrossRef] [PubMed]
53. Jin, Y.; Blikslager, A.T. Myosin light chain kinase mediates intestinal barrier dysfunction via occludin endocytosis during anoxia/reoxygenation injury. *Am. J. Physiol. Physiol.* **2016**, *311*, C996–C1004. [CrossRef] [PubMed]
54. Turner, J.R.; Black, E.D.; Ward, J.; Tse, C.M.; Uchwat, F.A.; Alli, H.A.; Donowitz, M.; Madara, J.L.; Angle, J.M. Transepithelial resistance can be regulated by the intestinal brush-border Na(+)/H(+) exchanger NHE3. *Am. J. Physiol. Cell Physiol.* **2000**, *279*, C1918–C1924. [CrossRef]
55. Moeser, A.J.; Nighot, P.K.; Ryan, K.A.; Simpson, J.E.; Clarke, L.L.; Blikslager, A.T. Mice lacking the Na+/H+ exchanger 2 have impaired recovery of intestinal barrier function. *Am. J. Physiol. Gastrointest Liver Physiol.* **2008**, *295*, G791–G797. [CrossRef]
56. Moeser, A.J.; Nighot, P.K.; Ryan, K.A.; Wooten, J.G.; Blikslager, A.T. Prostaglandin-mediated inhibition of Na+/H+ exchanger isoform 2 stimulates recovery of barrier function in ischemia-injured intestine. *Am. J. Physiol. Gastrointest Liver Physiol.* **2006**, *291*, G885–G894. [CrossRef]
57. Moeser, A.J.; Nighot, P.K.; Engelke, K.J.; Ueno, R.; Blikslager, A.T. Recovery of mucosal barrier function in ischemic porcine ileum and colon is stimulated by a novel agonist of the ClC-2 chloride channel, lubiprostone.

Am. J. Physiol. Liver Physiol. **2007**, *292*, G647–G656. [CrossRef]
58. Feranchak, A.P. Ion channels in digestive health and disease. *J. Pediatr. Gastroenterol. Nutr.* **2003**, *37*, 230–241. [CrossRef]
59. Donowitz, M.; Ming Tse, C.; Fuster, D. SLC9/NHE gene family, a plasma membrane and organellar family of Na(+)/H(−) exchangers. *Mol. Asp. Med.* **2013**, *34*, 236–251. [CrossRef]
60. Gurney, M.A.; Laubitz, D.; Ghishan, F.K.; Kiela, P.R. Pathophysiology of Intestinal Na(+)/H(+) exchange. *Cell Mol. Gastroenterol. Hepatol.* **2017**, *3*, 27–40. [CrossRef]
61. Kurashima, K.; D'Souza, S.; Szaszi, K.; Ramjeesingh, R.; Orlowski, J.; Grinstein, S. The apical Na(+)/H(+) exchanger isoform NHE3 is regulated by the actin cytoskeleton. *J. Biol. Chem.* **1999**, *274*, 29843–29849. [CrossRef] [PubMed]
62. Donowitz, M.; Li, X. Regulatory Binding Partners and Complexes of NHE3. *Physiol. Rev.* **2007**, *87*, 825–872. [CrossRef] [PubMed]
63. Yao, X.; Cheng, L.; Forte, J.G. Biochemical Characterization of Ezrin-Actin Interaction. *J. Biol. Chem.* **1996**, *271*, 7224–7229. [CrossRef] [PubMed]
64. Vaheri, A.; Carpén, O.; Heiska, L.; Helander, T.S.; Jääskeläinen, J.; Majander-Nordenswan, P.; Sainio, M.; Timonen, T.; Turunen, O. The ezrin protein family: Membrane-cytoskeleton interactions and disease associations. *Curr. Opin. Cell Biol.* **1997**, *9*, 659–666. [CrossRef]
65. Bosk, S.; Braunger, J.A.; Gerke, V.; Steinem, C. Activation of F-Actin Binding Capacity of Ezrin: Synergism of PIP2 Interaction and Phosphorylation. *Biophys. J.* **2011**, *100*, 1708–1717. [CrossRef] [PubMed]
66. Brückner, B.R.; Pietuch, A.; Nehls, S.; Rother, J.; Janshoff, A. Ezrin is a Major Regulator of Membrane Tension in Epithelial Cells. *Sci. Rep.* **2015**, *5*, 14700. [CrossRef]
67. Hayashi, H.; Tamura, A.; Krishnan, D.; Tsukita, S.; Suzuki, Y.; Kocinsky, H.S.; Aronson, P.S.; Orlowski, J.; Grinstein, S.; Alexander, R.T. Ezrin is Required for the Functional Regulation of the Epithelial Sodium Proton Exchanger, NHE3. *PLoS ONE* **2012**, *8*, e55623.
68. Praetorius, J.; Andreasen, D.; Jensen, B.L.; Ainsworth, M.A.; Friis, U.G.; Johansen, T. NHE1, NHE2, and NHE3 contribute to regulation of intracellular pH in murine duodenal epithelial cells. *Am. J. Physiol. Liver Physiol.* **2000**, *278*, G197–G206. [CrossRef]
69. Schneeberger, E.E.; Lynch, R.D. Tight junctions. Their structure, composition, and function. *Circ. Res.* **1984**, *55*, 723–733. [CrossRef]
70. Barrett, K.E.; Keely, S.J. Chloride Secretion by the Intestinal Epithelium: Molecular Basis and Regulatory Aspects. *Annu. Rev. Physiol.* **2000**, *62*, 535–572. [CrossRef] [PubMed]
71. Frizzell, R.A.; Hanrahan, J.W. Physiology of Epithelial Chloride and Fluid Secretion. *Cold Spring Harb. Perspect. Med.* **2012**, *2*, a009563. [CrossRef] [PubMed]
72. Poroca, D.R.; Pelis, R.M.; Chappe, V.M. ClC Channels and Transporters: Structure, Physiological Functions, and Implications in Human Chloride Channelopathies. *Front. Pharmacol.* **2017**, *8*, 151. [CrossRef] [PubMed]
73. Gyomorey, K.; Yeger, H.; Ackerley, C.; Garami, E.; Bear, C.E. Expression of the chloride channel ClC-2 in the murine small intestine epithelium. *Am. J. Physiol. Physiol.* **2000**, *279*, C1787–C1794. [CrossRef] [PubMed]
74. Catalan, M.; Niemeyer, M.I.; Cid, L.; Sepúlveda, F.V. Basolateral ClC-2 chloride channels in surface colon epithelium: Regulation by a direct effect of intracellular chloride. *Gastroenterology* **2004**, *126*, 1104–1114. [CrossRef] [PubMed]
75. Nighot, P.K.; Blikslager, A.T. ClC-2 regulates mucosal barrier function associated with structural changes to the villus and epithelial tight junction. *Am. J. Physiol. Liver Physiol.* **2010**, *299*, G449–G456. [CrossRef] [PubMed]
76. Nighot, P.; Young, K.; Nighot, M.; Rawat, M.; Sung, E.J.; Maharshak, N.; Plevy, S.E.; Ma, T.; Blikslager, A. Chloride channel ClC-2 is a key factor in the development of DSS-induced murine colitis. *Inflamm. Bowel Dis.* **2013**, *19*, 2867–2877. [CrossRef] [PubMed]
77. Nighot, P.K.; Leung, L.; Ma, T.Y. Chloride channel ClC- 2 enhances intestinal epithelial tight junction barrier function via regulation of caveolin-1 and caveolar trafficking of occludin. *Exp. Cell Res.* **2017**, *352*, 113–122. [CrossRef]
78. Roman, R.M.; Smith, R.L.; Feranchak, A.P.; Clayton, G.H.; Doctor, R.B.; Fitz, J.G. ClC-2 chloride channels contribute to HTC cell volume homeostasis. *Am. J. Physiol. Liver Physiol.* **2001**, *280*, G344–G353. [CrossRef]
79. Nighot, M.P.; Nighot, P.K.; Ma, T.Y.; Malinowska, D.H.; Shull, G.E.; Cuppoletti, J.; Blikslager, A.T. Genetic Ablation of the ClC-2 Cl- Channel Disrupts Mouse Gastric Parietal Cell Acid Secretion. *PLoS ONE* **2015**, *10*, e0138174. [CrossRef]

80. Nighot, P.K.; Blikslager, A.T. Chloride channel ClC-2 modulates tight junction barrier function via intracellular trafficking of occludin. *Am. J. Physiol. Physiol.* **2012**, *302*, C178–C187. [CrossRef]
81. Jin, Y.; Blikslager, A.T. ClC-2 regulation of intestinal barrier function: Translation of basic science to therapeutic target. *Tissue Barriers* **2015**, *3*, e1105906. [CrossRef] [PubMed]

Caprate Modulates Intestinal Barrier Function in Porcine Peyer's Patch Follicle-Associated Epithelium

Judith Radloff [1,2], Valeria Cornelius [1], Alexander G. Markov [3] and Salah Amasheh [1,*]

1 Institute of Veterinary Physiology, Department of Veterinary Medicine, Freie Universität Berlin, 14163 Berlin, Germany; judith.radloff@vetmeduni.ac.at (J.R.); valeria.cornelius@fu-berlin.de (V.C.)
2 Department for Biomedical Sciences, University of Veterinary Medicine, 1210 Vienna, Austria
3 Department of General Physiology, St. Petersburg State University, Universitetskaya nab. 7/9, 199034 St. Petersburg, Russia; markov_51@mail.ru
* Correspondence: salah.amasheh@fu-berlin.de

Abstract: Background: Many food components influence intestinal epithelial barrier properties and might therefore also affect susceptibility to the development of food allergies. Such allergies are triggered by increased antibody production initiated in Peyer's patches (PP). Usually, the presentation of antigens in the lumen of the gut to the immune cells of the PP is strongly regulated by the follicle-associated epithelium (FAE) that covers the PP. As the food component caprate has been shown to impede barrier properties in villous epithelium, we hypothesized that caprate also affects the barrier function of the PP FAE, thereby possibly contributing a risk factor for the development of food allergies. Methods: In this study, we have focused on the effects of caprate on the barrier function of PP, employing in vitro and ex vivo experimental setups to investigate functional and molecular barrier properties. Incubation with caprate induced an increase of transepithelial resistance, and a marked increase of permeability for the paracellular marker fluorescein in porcine PP to 180% of control values. These effects are in accordance with changes in the expression levels of the barrier-forming tight junction proteins tricellulin and claudin-5. Conclusions: This barrier-affecting mechanism could be involved in the initial steps of a food allergy, since it might trigger unregulated contact of the gut lumen with antigens.

Keywords: tissue barrier; tight junction; claudins; tricellulin

---

1. Introduction

The gastrointestinal mucosa is one of the largest surface areas of the body and constantly interacts with possibly harmful substances or non-hazardous food components. The intestinal epithelial cells and the associated immune cells of the gut-associated-lymphocytic tissue (GALT) need to be able to distinguish the category of a molecule.

Peyer's patches (PP) represent the dominating part of the GALT. They are structured into a germinal center that consists of macrophages, lymphocytes, and dendritic cells (DC) and a covering epithelial cell layer, namely the follicle-associated epithelium (FAE) [1]. The FAE of rat and pig has recently been characterized in detail, giving a greater insight into the barrier properties directly above the PP [2,3]. Generally, the FAE has a more porous basal membrane and contains fewer goblet cells, resulting in a thinner mucus layer, than the villous epithelium (VE) [4,5]. The FAE also includes microfold cells (M-cells), which are specialized for antigen translocation to the underlying cells of the PP [4]. The presentation of antigens can also be initiated via DC that reach through the epithelial cell layer into the gut lumen and directly sample bacteria. During this process, the barrier integrity is preserved, as DC also express tight junction (TJ) proteins [6]. Overall, the barrier of the FAE can be classified as tight when compared with the surrounding leaky VE, as sealing claudins such as claudin-1, -3, -4, -5, and -8 are predominantly expressed in PP samples of various species [2,3,7].

Numerous food components have been shown to influence barrier integrity in the VE, including the medium chain fatty acids laurate and caprate [8–10], the amino acids glutamine and glycine [11,12], and plant components quercetin and berberine [13,14], all which lead to a change of transepithelial resistance (TER) and paracellular permeability. In turn, this can enhance or limit the presentation of common food components to the immune cells of the PP, triggering or impeding the excessive formation of immunoglobulin E (IgE) [15]. The consequence is commonly known as food allergy.

Food allergies are defined as an adverse health effect arising from a specific immune response that occurs reproducibly on exposure to a given food [16]. With food allergies being more prevalent than ever, the need for an effective treatment or prevention mechanism is obvious [17]. This remains difficult for as long as the precise causes remain unidentified.

One of these causes might involve the barrier-weakening effect of the medium chain fatty acid caprate. Caprate has been shown to decrease the TER reversibly, leading to an increase of paracellular permeability for fluorescein and various fluorescein isothyocyanate (FITC)-dextrans (FD4, FD10) in the intestinal cell line HT29-B6 [8]. This can be attributed to the reversible decrease of the sealing TJ proteins claudin-5 and tricellulin, an effect opening the paracellular pathway [8]. Because of its reversible barrier-weakening properties, caprate is used as an absorption enhancer, improving membrane permeation for pharmaceutically active agents [18]. Since caprate is also contained in common foods, such as milk and coconut oil, it is widely consumed by most people, albeit in low concentrations [19].

Despite the numerous studies on the effect of caprate on intestinal barrier function [8,20,21], the effect of caprate on the FAE has not yet been investigated. Therefore, we have focused on the barrier-influencing properties of caprate on the FAE of the porcine PP and on the consequences with regard to gastrointestinal immunology.

2. Results

2.1. Cell Culture Study

Initial studies on the effect of caprate on the porcine intestinal barrier function were carried out by employing the Intestinal Porcine Epithelial Cell line-J2 (IPEC-J2) and 5 mM caprate.

Incubation of IPEC-J2 cells with 5 mM caprate led to a rapid and consistent decrease of barrier function (from 5.85 ± 0.38 kΩ to 3.68 ± 0.59 kΩ, ** $p < 0.01$, $n = 20$) when compared with the control group (not significant (n. s.), $n = 20$; Figure 1). This decrease was evident during the whole timespan of the experiment (120 min). After a recovery period of 24–48 h, TER values of the caprate group were comparable with control levels (4.78 ± 0.63 kΩ to 5.05 ± 0.67 kΩ, $n = 20$).

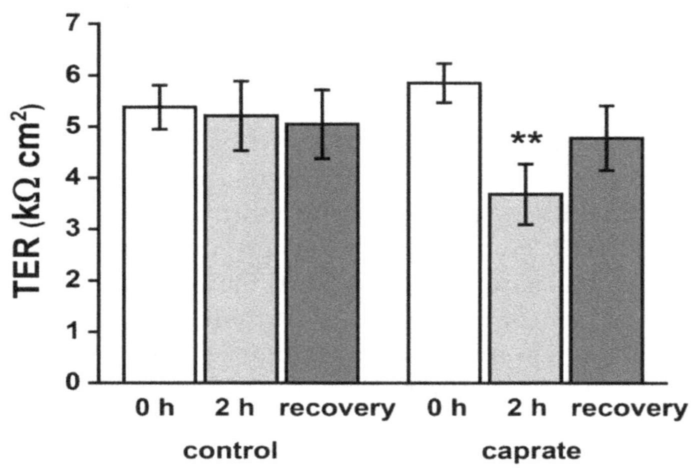

Figure 1. Transepithelial electrical resistance (TER) of IPEC-J2 cell monolayers. Incubation with 5 mM caprate leads to a reversible decrease of TER (** $p < 0.01$, $n = 20$).

2.2. Ussing Chamber Experiments

The effect of caprate on PP tissue taken from adult pigs was analyzed by the Ussing chamber technique. Tissue was incubated with 5 mM caprate while TER values and paracellular permeability measurements for sodium fluorescein were carried out.

In contrast to the IPEC-J2 experiments, the TER of PP increased significantly after incubation of the tissue with caprate (240 min; 92.97 ± 4.94% to 116.73 ± 7.55%, * $p < 0.05$, $n = 11–12$) (Figure 2A). However, the permeability measurements of sodium fluorescein revealed a marked increase to 180.24 ± 29.52% (* $p < 0.05$, $n = 8$; Figure 2B).

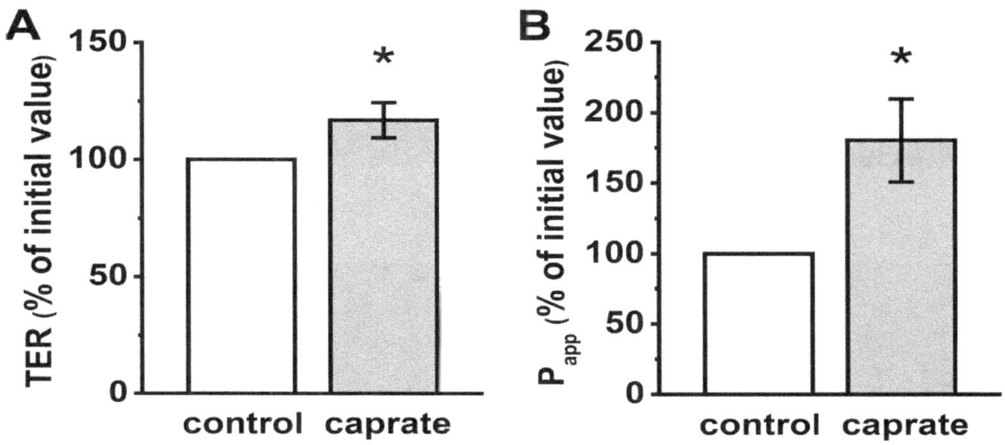

Figure 2. (**A**) Transepithelial electrical resistance (TER) and (**B**) apparent permeability (P_{app}) of the paracellular flux marker fluorescein in porcine Peyer's patches (PP). 5 mM caprate lead to markedly higher TER values (* $p < 0.05$, $n = 11–12$), and increased permeability for fluorescein ($n = 8$, * $p < 0.05$).

2.3. Immunoblotting

To account for the detected functional differences in TER, the expression levels of sealing TJ proteins were analyzed; the immunoblots revealed major changes (Figure 3). Whereas claudin-3 only showed a positive trend in expression levels (162.7 ± 31.23%, $p = 0.064$, $n = 8$), claudin-5 was significantly increased (198.82 ± 40.11%, $p < 0.05$, $n = 8$). Tricellulin on the other hand was significantly reduced (66.64 ± 13.69%, $p < 0.05$, $n = 6$). Sealing claudin-1 in PP tissue (139.32 ± 24.62%, $n = 8$) was not affected by caprate incubation.

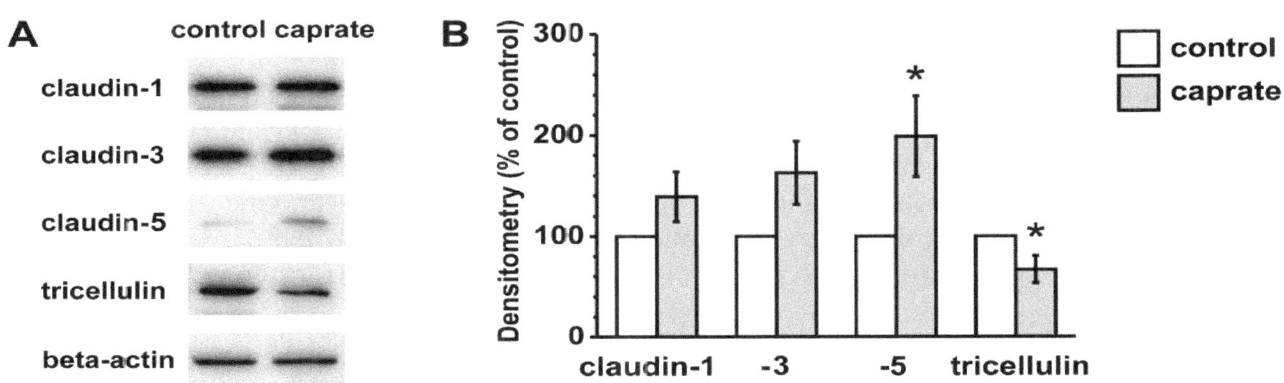

Figure 3. (**A**) Immunoblots and (**B**) densitometry of TJ proteins in PP tissue (control versus caprate). Densitometric analysis of TJ protein signals after caprate incubation reveals the lower expression of tricellulin, whereas stronger expression of claudin-5 can be detected ($n = 6–8$, * $p < 0.05$).

2.4. *Immunohistochemistry*

To analyze the location of single TJ proteins, immunoflourescent staining of tissue specimens was performed following the Ussing chamber experiments. Immunohistochemistry of these self-same tissues revealed specific signals for TJ proteins in all tested PP tissue specimens, as shown for claudin-3 and claudin-5 (Figure 4).

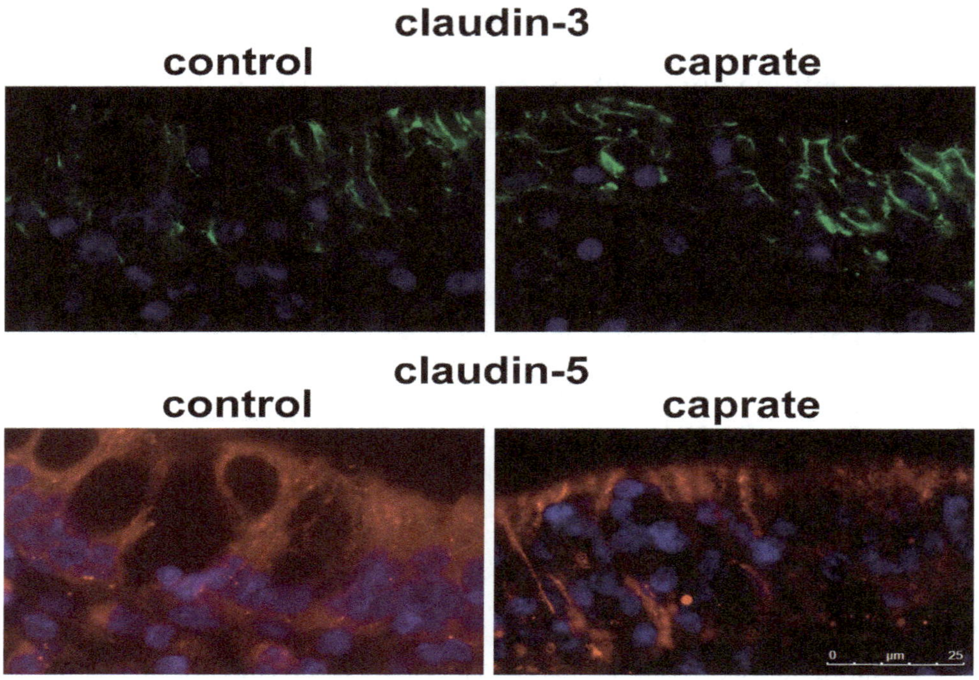

Figure 4. Immunohistochemistry. TJ proteins claudin-3 and claudin-5 are detectable as paracellular signals in the surface epithelium in PP tissue (bar: 25 µm).

3. Discussion

3.1. *Suitability of a Porcine Model for Human Gastrointestinal Pathophysiology*

Porcine models exhibit certain benefits when exploring molecular mechanisms of intestinal pathophysiological conditions. In 2007, Humphray et al. showed that 60% of the porcine and human genome sequence match [22]. The morphology and anatomy of the human gastrointestinal tract is comparable with that of a pig [23]. Moreover, the physiology of digestive processes and the subsequent absorption of nutrients are also similar [24,25]. Last but not least, the pig is the only commonly used laboratory animal that has comparable nutritional needs with humans, on account of it being classified as an omnivore [26].

Since its origin in 1989, the IPEC-J2 cell line has been extensively used for studies on the gastrointestinal barrier and the effects of nutrients. It represents one of the best cell lines of non-human origin for intestinal research because of its non-transformed properties [27,28]. Taking all of these aspects into account, we chose porcine intestinal epithelial tissue specimens and the IPEC-J2 cell line as promising models for studying the effects of caprate on barrier function with regard to immunologically relevant mechanisms, which might also provide a perspective as a model for porcine PP FAE.

3.2. *Caprate in Modern Foods*

Coconut oil is one of the main contributors of caprate in Western nutrition. Since it is currently regarded as a superfood, the intake of coconut oil, and therefore caprate, has increased tremendously. The health benefits of coconut oil include weight loss, cardioprotective effects, and improved immunity and memory [29]. The milk of various mammalian species is the other main source of caprate in current nutrition [19,30]. Existing data on the health benefits of milk include preventive effects against diabetes mellitus, cancer, hypertension, and obesity [31].

Even without the specific ingestion of caprate as an absorption enhancer, the daily intake of caprate can be high because of the recommended consumption of 2.5–3 servings of dairy products [32]. Digestion effects can be expectedly rather limited, as the medium chain fatty acid absorbed is non-modified, even devoid pancreatic enzymes [33]. Moreover, medium-chain triglycerides are degraded into fatty acids and the glycerol backbone, a mechanism adding to luminal caprate concentration. After absorption, the medium chain fatty acids are transferred into the portal vein for transport to the liver for hepatic metabolism [34]. However, prior to absorption, luminal caprate effects might even add up with effects of other bioactive compounds, e.g., laurate, which then might reach barrier-affecting concentrations earlier [9].

3.3. *Effects of Caprate*

Since foods containing caprate are consumed regularly, the effects of the fatty acid need to be brought into focus. Currently, caprate is used as an absorption enhancer to improve the bioavailability of pharmaceutically active agents [18]. A human in vitro cell culture experiment, employing the colonic HT-29/B6 cell line, has revealed the cellular mechanisms involving the selective removal of claudin-5 and tricellulin from the TJ [8]. Recent in vitro experiments have provided evidence of the antibacterial effect of caprate in porcine epithelial cells when challenged with *Salmonella enteritidis* or an enteropathogenic *Escherichia coli* strain [35]. Furthermore, caprate has also been shown to reduce body fat and therefore, body weight [36].

Data concerning the intestinal barrier function obtained in this study has shown that the effect of caprate can also occur in other species. Significantly lower TER values were recorded for the porcine IPEC-J2 cell line after caprate incubation (5 mM). This effect was reversible in both species (compare with [8] for human cells). However, transferring this functional result to the ex vivo setup using the Ussing chamber technique was not possible, as the TER markedly increased in PP tissue. This apparent discrepancy might reflect a limited suitability of the plain IPEC-J2 model for modeling the PP FAE. However, it might still provide a suitable matrix for, e.g., co-culturing experiments approaching a functional FAE model, in vitro.

Permeability measurements in PP tissue revealed a higher permeability for fluorescein, which can be compared with glucose in size (330 Da). A higher permeability for fluorescein, but also for FITC-dextran (4k, 10k), was observed after incubation with 10 mM caprate in HT-29/B6 cells [8].

Increased paracellular permeability might give rise to unregulated contact between the immune cells of the PP and food components. The most common foods that provoke allergic reactions are nuts, eggs, milk, soy, wheat, and shell fish [37]. The uptake of antigens within the PP is strictly regulated, as it is usually undertaken by specific binding to receptors, through specialized epithelial cells, the M-cells, or via the extending dendrites of DC belonging to the underlying PP [4,6,38]. An impairment of the intestinal barrier function would interrupt these controlled processes.

3.4. *Caprate Effects on Tight Junction Protein Expression*

The study has shown that claudin-5 and tricellulin have been removed from bi- and tricellular TJ through caprate incubation [8]. TJ are apicolateral epithelial cell membrane contacts determining paracellular permeability [39–42]. Claudin-5 is classically regarded as a sealing claudin, since it increases barrier properties by effects on TER [43,44].

Tricellulin was the first barrier-forming TJ protein to be found in tricellular TJ, and supression of tricellulin leads to the disruption of bicellular and tricellular TJ [45]. Overexpression of tricellulin increases barrier function and decreases the permeability of ions and larger solutes [46]. As caprate induces a reversible removal of claudin-5 and tricellulin from the TJ, functional results under caprate incubation can easily be attributed to structural changes of TJ formation.

The immunoblots after Ussing chamber experiments in our study also revealed changes in claudin-5 and tricellulin. Quantitative analysis of PP tissue samples showed an increased expression level of claudin-5 and decreased levels of tricellulin. These results suggest the differential sensitivity of the FAE towards the effects of caprate when compared with the intestinal epithelial monolayers that were used in the previously performed initial incubation experiment [8]. This is understandable when we keep the main task of the PP in mind. A structural change in TJ formation prevents the accidental passage of antigen to the underlying PP. Our immunohistological assessment has shown the paracellular localization of claudin-3 and claudin-5. Although a decrease of claudin-5 signals could not be observed in the current study, the overall increase of its expression level might have contributed to a compensation, in accordance with the observed increase of TER. This apparent inconsistency might also be explained by the different contribution of single tight junction proteins to transepithelial resistance and permeability of flux markers, as tricellulin is affecting the paracellular passage of molecules [46], whereas claudin-5 primarily affects the passage of ions [44]. However, our results indicate that it is possible to functionally uncouple resistance and solute permeability by affecting different tight junction proteins in parallel, which includes the possibility that an increased expression of a sealing protein such as claudin-5 does not necessarily provide an expected primary functional effect.

However, further studies, particularly in humans, are required to demonstrate that caprate may affect the susceptibility to the development of food allergies.

4. Materials and Methods

4.1. Cell Culturing and Experiments

The porcine jejunal cell line IPEC-J2 is derived from the jejunum of unsuckled piglets and is commonly regarded as an in vitro model for the analysis of the intestinal barrier function in both pigs and humans [27]. For routine cultivation, Dulbecco's Modified Eagle Medium (MEM)/ Nutrient F 12 Ham's (Biochrom, Berlin, Germany) was used as the medium basis, with 10% fetal calf serum, 1% penicillin-streptomycin, 5 ng /mL epithelial growth factor (Biochrom, Berlin, Germany) and insulin, transferrin and selenium (Life Technologies, Carlsbad, CA, USA) as supplements. Cells were passaged twice a week and fed every other day.

For experiments, cells (10^5) were seeded on cell culture inserts with a pore size of 0.45 μm (Millicell, Darmstadt, Germany). The seeded inserts were placed into multiwell plates and cultivated for approximately 14 days until a functional barrier had formed. This was assessed via measurements of TER employing a chopstick electrode and an Epithelial Volt/Ohm Meter (EVOM) measuring device (World Precision Instruments, Hertfordshire, United Kingdom). Only filters that showed TER values above 2 kΩ were chosen for experimentation. Passages ranging from 9 to 15 were used in this experimental setup. TER values were measured initially in plain culture medium. Subsequently, the culture media was replaced with medium A (plain culture medium) or medium B (culture medium with 5 mM caprate). During the incubation period of 120 min, TER values were measured every 20 min, after which the medium was replaced with plain culture medium. Following a recovery period of 24–48 h, TER was measured again. The cells were kept at 37 °C during the recovery period.

4.2. Tissue Samples

Samples of the distal small intestine of 12 adult pigs were taken immediately after slaughter at a commercial slaughterhouse (Salzbrunn, Germany). PP tissue was transported in cold (4–6 °C) buffer solution containing (in mmol·L-1): Na^+ (149.9), Cl^- (128.8), K^+ (5), Ca^{2+} (1.2), Mg^{2+} (1.2), HCO_3^- (25), $H_2PO_4^{2-}$ (0.6), HPO_4^- (1.2), and D-glucose (10). The pH was set to 7.4 before transport.

4.3. Ussing Chamber Experiments

Tissue samples were mounted in a common Ussing chamber setup (Mussler Scientific Instruments, Aachen, Germany), and data recording commenced after 45 min of preincubation with fluorescein, but only once electrophysiological values were stable. The incubation experiment itself was carried out under voltage clamp conditions. The experimental buffer contained (in mmol·L-1): Na^+ (149.9), Cl^- (128.8), K^+ (5), Ca^{2+} (1.2), Mg^{2+} (1.2), HCO_3^- (25), $H_2PO_4^{2-}$ (0.6), HPO_4^- (1.2), and D-glucose (10). It was constantly gassed with 95% O_2 and 5% CO_2, resulting in a pH of 7.4. After 30 min, 5 mM caprate was added to the apical side. TER was recorded continuously throughout the overall incubation time of 4 h. Electrophysiological data are expressed in relation to the initial value before the addition of caprate.

4.4. Permeability Measurements

Once the tissue had been mounted in the Ussing chamber, fluorescein (100 μmol·L^{-1}) was added to the apical side of each chamber. During the incubation experiment, permeability samples were taken every 60 min from the basolateral side. The volume that was taken was replaced with fresh buffer solution. Samples were measured photometrically at 514 nm by employing the EnSpire Multimode Plate Reader (Perkin Elmer, Waltham, MA, USA). Paracellular flux and permeability were then calculated by using the following equations.

$$c = (c_{t-1} \times V_S + c_t \times V_K)/V_K \quad (1)$$

$$J = (c_t - c_{t-1}) \times V_K/\Delta t \times A \quad (2)$$

$$P_{app} = J/c_a. \quad (3)$$

In Equation (1), the measured concentration of fluorescein in the basolateral sample has to be corrected for the dilution that takes effect because of the replacement of the experimental buffer, with ct being the measured concentration, c_{t-1} the measured concentration of the previous period, V_S the sample volume, and V_K the volume of the Ussing chamber. Equation (2) gives the paracellular flux (J) for fluorescein, with Δt being the duration of the measurement period, and A being the area within the Ussing chamber. Finally, the paracellular permeability can be calculated by using Equation (3), with c_a being the fluorescein concentration in the apical side of the Ussing chamber. Results were expressed as a fraction of the control values.

4.5. Protein Extraction and Quantification

Samples for protein extraction were taken out of the Ussing chamber and directly placed into liquid nitrogen. These samples were homogenized in a Tris-buffer containing in mmol·L -1: Tris (10), NaCl (150), Triton X-100 (0.5), SDS (0.1), and enzymatic protease inhibitors (Complete, Boehringer, Mannheim, Germany). After homogenization, samples were centrifuged for 1 min at 13,000 rpm (Eppendorf centrifuge 5418, Eppendorf AG, Hamburg, Germany). The supernatant was then cooled on ice for 30 min and retrieved after a second centrifugation step for 15 min at 15,000 g at 4 °C (sigma 3-30ks, Sigma-Aldrich, Munich, Germany).

Protein quantification of PP tissues was carried out by using Bradford reagent (Bio-Rad Laboratories GmbH, Munich, Germany) as instructed in the leaflet provided, and the EnSpire Multimode Plate Reader (Perkin Elmer, Waltham, MA, USA) was used for detection.

4.6. Immunoblotting and Densitometry

Protein (4.1 μg) and Laemmli buffer (Bio-Rad Laboratories GmbH, Munich, Germany) were mixed and loaded onto a 12.5% polyacrylamide gel. Electrophoresis was allowed to run for 5 min at 200 mV, followed by 60 min at 100 mV. Afterwards, the samples were transferred onto a PVDF membrane, which was blocked for 60 min in 5% milk (in Tris-buffered saline with 0.1% Tween 20) and later incubated with the antibodies for claudin-1, claudin-3, claudin-5, tricellulin, or beta-actin (Life Technologies, Carlsbad, CA, USA) overnight at 4 °C. Antibodies were used as recommended by the manufacturer (1 μg/mL). To enable visual detection, peroxidase-conjugated goat anti-mouse or anti-rabbit antibodies (Cell Signaling Technology, Danvers, MA, USA) were employed to bind the primary antibodies. Signals were then visualized by employing the chemiluminescence detection system of Clarity Western ECL Blotting Substrate (Bio-Rad Laboratories GmbH, Munich, Germany) and a luminescence imager (ChemiDoc MP, Munich, Germany).

The density of individual immunoblotting bands was analyzed by using the imager-associated software (ImageLab, BioRad, Hercules, CA, USA). The measured values were divided by the coherent beta-actin band signal. Finally, the signal for the caprate challenge was expressed in relation to the control values for VE and PP, which were set to 100%.

4.7. Immunohistochemistry

Tissue samples were placed into 4% formaldehyde (Roti-Histofix, Karlsruhe, Germany) for 24 h, dehydrated by using an increasing alcohol gradient, transferred to xylol, and embedded in paraffin. Paraffin blocks were cut into 3-μm-thick sections on a Leica RM 2245 microtome (Leica Microsystems Heidelberg, Germany). Before staining, the sections on the slides were warmed for 30 min at 60 °C (incubator), deparaffinized in xylol, and taken through a decreasing alcohol gradient. Depending on the protein, EDTA or citrate buffer was used for heat-induced antigen retrieval (45 min). Following the permeabilization of the samples with Triton X-100 (Carl Roth, Karlsruhe, Germany), sections were incubated with 5% goat serum in PBS and subsequently with antibodies raised against claudin-3 and -5 (1:100, Life Technologies, Carlsbad, CA, USA) for 60 min at 37 °C. After a washing step in blocking solution, samples were incubated with goat anti-mouse and -rabbit Alexa Fluor-488, and -594 (1:1000, Life Technologies, Carlsbad, CA, USA) and 4′,6-diamidino-2-phenylindole (DAPI, 1:5000) for 60 min at 37 °C. Sections were then mounted with ProTags Mount Fluor (Biocyc, Luckenwalde, Germany). Analyses and imaging of these sections was carried out by employing a Leica microscope of the DMI 6000 series (Leica Microsystems, Heidelberg, Germany).

4.8. Statistical Analysis

Results are always expressed as the mean ± standard error of the mean, with n being the number of animals used. For in vitro experiments, n represents the number of experiments carried out. When indicated, relative values were used for statistical evaluation.

Data from the cell culture experiments were analyzed by employing the paired Student's t-test, whereas the results of the Ussing chamber experiments were evaluated with the unpaired Student's t-test. Significant difference was considered at $p < 0.05$.

5. Conclusions

The results of this study show an increased permeability in the PP and a reorganization of the TJ structure after caprate incubation. Since caprate is a prevalent food component, these data imply a constant increasing effect on the paracellular permeability of the PP FAE. This might also accelerate the development of allergies, since more molecules are presented to the immune cells of the PP.

Author Contributions: J.R. and V.C. performed the experiments, J.R., A.G.M. and S.A. discussed the data and wrote the manuscript.

Acknowledgments: We thank Martin Grunau, Uwe Tietjen, and Evgeny Falchuk for excellent technical assistance.

Abbreviations

Cldn	Claudin
DC	Dendritic cells
FAE	Follicle-associated epithelium
GALT	Gut-associated lymphocytic tissue

IgE Immunoglobulin E
M-cell Microfold cell
PP Peyer's patch
TER Transepithelial electrical resistance
VE Villous epithelium

References

1. Neutra, M.R.; Mantis, N.J.; Kraehenbuhl, J.P. Collaboration of epithelial cells with organized mucosal lymphoid tissues. *Nat. Immunol.* **2001**, *2*, 1004–1009. [CrossRef] [PubMed]
2. Markov, A.G.; Falchuk, E.F.; Kruglova, N.M.; Radloff, J.; Amasheh, S. Claudin expression in follicle-associated epithelium of rat Peyer's patches defines a major restriction of the paracellular pathway. *Acta Physiol.* **2016**, *216*, 112–119. [CrossRef] [PubMed]
3. Radloff, J.; Falchuk, E.L.; Markov, A.G.; Amasheh, S. Molecular Characterization of Barrier Properties in Follicle-Associated Epithelium of Porcine Peyer's Patches Reveals Major Sealing Function of Claudin-4. *Front. Physiol.* **2017**, *8*, 579. [CrossRef] [PubMed]
4. Owen, R.L.; Jones, A.L. Epithelial cell specialization within human Peyer's patches: An ultrastructural study of intestinal lymphoid follicles. *Gastroenterology* **1974**, *66*, 189–203. [CrossRef]
5. Takeuchi, T.; Gonda, T. Distribution of the pores of epithelial basement membrane in the rat small intestine. *J. Vet. Med. Sci.* **2004**, *66*, 695–700. [CrossRef] [PubMed]
6. Rescigno, M.; Urbano, M.; Valzasina, B.; Francolini, M.; Rotta, G.; Bonasio, R.; Granucci, F.; Kraehenbuhl, J.P.; Ricciardi-Castagnoli, P. Dendritic cells express tight junction proteins and penetrate gut epithelial monolayers to sample bacteria. *Nat. Immunol.* **2001**, *2*, 361–367. [CrossRef]
7. Tamagawa, H.; Takahashi, I.; Furuse, M.; Yoshitake-Kitano, Y.; Tsukita, S.; Ito, T.; Matsuda, H.; Kiyono, H. Characteristics of claudin expression in follicle-associated epithelium of Peyer's patches: Preferential localization of claudin-4 at the apex of the dome region. *Lab. Invest.* **2003**, *83*, 1045–1053. [CrossRef]
8. Krug, S.M.; Amasheh, M.; Dittmann, I.; Christoffel, I.; Fromm, M.; Amasheh, S. Sodium caprate as an enhancer of macromolecule permeation across tricellular tight junctions of intestinal cells. *Biomaterials* **2013**, *34*, 275–282. [CrossRef]
9. Dittmann, I.; Amasheh, M.; Krug, S.M.; Markov, A.G.; Fromm, M.; Amasheh, S. Laurate permeates the paracellular pathway for small molecules in the intestinal epithelial cell model HT-29/B6 via opening the tight junctions by reversible relocation of claudin-5. *Pharm. Res.* **2014**, *31*, 2539–2548. [CrossRef]
10. Radloff, J.; Zakrzewski, S.S.; Pieper, R.; Markov, A.G.; Amasheh, S. Porcine milk induces a strengthening of barrier function in porcine jejunal epithelium in vitro. *Ann. N. Y. Acad. Sci.* **2017**, *1397*, 110–118. [CrossRef]
11. Li, W.; Sun, K.; Ji, Y.; Wu, Z.; Wang, W.; Dai, Z.; Wu, G. Glycine regulates expression and distribution of claudin-7 and ZO-3 proteins in intestinal porcine epithelial cells. *J. Nutr.* **2016**, *146*, 964–969. [CrossRef] [PubMed]
12. Wang, B.; Wu, Z.; Ji, Y.; Sun, K.; Dai, Z.; Wu, G. L-Glutamine enhances tight junction integrity by activating CaMK kinase 2-AMP-activated protein kinase signaling in intestinal porcine epithelial cells. *J. Nutr.* **2016**, *146*, 501–508. [CrossRef]
13. Amasheh, M.; Schlichter, S.; Amasheh, S.; Mankertz, J.; Zeitz, M.; Fromm, M.; Schulzke, J.D. Quercetin enhances epithelial barrier function and increases claudin-4 expression in Caco-2 cells. *J. Nutr.* **2008**, *138*, 1067–1073. [CrossRef]
14. Amasheh, M.; Fromm, A.; Krug, S.M.; Amasheh, S.; Andres, S.; Zeitz, M.; Fromm, M.; Schulzke, J.D. TNFalpha-induced and berberine-antagonized tight junction barrier impairment via tyrosine kinase, Akt and NFkappaB signaling. *J. Cell Sci.* **2010**, *123*, 4145–4155. [CrossRef] [PubMed]
15. Berin, M.C. Pathogenesis of IgE-mediated food allergy. *Clin. Exp. Allergy* **2015**, *45*, 1483–1496. [CrossRef] [PubMed]
16. Boyce, J.A.; Assa'ad, A.; Burks, A.W.; Jones, S.M.; Sampson, H.A.; Wood, R.A.; Plaut, M.; Cooper, S.F.; Fenton, M.J.; Arshad, S.H.; et al. Guidelines for the Diagnosis and Management of Food Allergy in the United States: Summary of the NIAID-Sponsored Expert Panel Report. *J. Allergy Clin. Immunol.* **2010**, *126*, 1105–1118. [CrossRef] [PubMed]
17. Tang, M.L.; Mullins, R.J. Food allergy: Is prevalence increasing? *Intern. Med. J.* **2017**, *47*, 256–261. [CrossRef]
18. Lindmark, T.; Söderholm, J.D.; Olaison, G.; Alvan, G.; Ocklind, G.; Artursson, P. Mechanism of Absorption

Enhancement in Humans After Rectal Administration of Ampicillin in Suppositories Containing Sodium Caprate. *Pharm. Res.* **1997**, *14*, 930–935. [CrossRef] [PubMed]

19. Beare-Rogers, J.L.; Dieffenbacher, A.; Holm, J.V. Lexicon of lipid nutrition (IUPAC Technical Report). *Pure. Appl. Chem.* **2001**, *73*, 685–744. [CrossRef]
20. Tomita, M.; Hayashi, M.; Awazu, S. Absorption-enhancing mechanism of sodium caprate and decanoylcarnitine in Caco-2 cells. *J. Pharmacol. Exp. Ther.* **1995**, *272*, 739–743. [CrossRef] [PubMed]
21. Lv, X.Y.; Li, J.; Zhang, M.; Wang, C.M.; Fan, Z.; Wang, C.Y.; Chen, L. Enhancement of Sodium Caprate on Intestine Absorption and Antidiabetic Action of Berberine. *AAPS PharmSciTech.* **2010**, *11*, 372–382. [CrossRef]
22. Humphray, S.J.; Scott, C.E.; Clark, R.; Marron, B.; Bender, C.; Camm, N.; Davis, J.; Jenks, A.; Noon, A.; Patel, M.; et al. A high utility integrated map of the pig genome. *Genome Biol.* **2007**, *8*, R139. [CrossRef] [PubMed]
23. Kararli, T.T. Comparison of the gastrointestinal anatomy, physiology, and biochemistry of humans and commonly used laboratory animals. *Biopharm. Drug. Dispos.* **1995**, *16*, 351–380. [CrossRef] [PubMed]
24. Rowan, A.M.; Moughan, P.J.; Wilson, M.N.; Maher, K.; Tasman-Jones, C. Comparison of the ileal and faecal digestibility of dietary amino acids in adult humans and evaluation of the pig as a model animal for digestion studies in man. *Br. J. Nutr.* **1994**, *71*, 29–42. [CrossRef] [PubMed]
25. Patterson, J.K.; Lei, X.G.; Miller, D.D. The Pig as an Experimental Model for Elucidating the Mechanisms Governing Dietary Influence on Mineral Absorption. *Exp. Biol. Med.* **2008**, *233*, 651–664. [CrossRef] [PubMed]
26. Miller, E.R.; Ullrey, D.E. The pig as a model for human nutrition. *Annu. Rev. Nutr.* **1987**, *7*, 361–382. [CrossRef] [PubMed]
27. Schierack, P.; Nordhoff, M.; Pollmann, M.; Weyrauch, K.D.; Amasheh, S.; Lodemann, U.; Jores, J.; Tachu, B.; Kleta, S.; Blikslager, A.; et al. Characterization of a porcine intestinal epithelial cell line for in vitro studies of microbial pathogenesis in swine. *Histochem. Cell. Biol.* **2006**, *125*, 293–305. [CrossRef] [PubMed]
28. Vergauwen, H. The IPEC-J2 Cell Line. In *The Impact of Food Bioactives on Health: In Vitro and Ex Vivo Models*; Springer: Cham, Switzerland, 2015. [CrossRef]
29. Sankararaman, S.; Sferra, T.J. Are We Going Nuts on Coconut Oil? *Curr. Nutr. Rep.* **2018**, *7*, 107–115. [CrossRef] [PubMed]
30. Davies, D.T.; Holt, C.; Christie, W.W. The composition of milk. In *Biochemistry of Lactation*; Mepham, T.B., Ed.; Elsevier Science Publishers: New York, NY, USA, 1983; pp. 71–117.
31. Prentice, A.M. Dairy products in global public health. *Am. J. Clin. Nutr.* **2014**, *99*, 1212–1216. [CrossRef]
32. Guillot, E.; Lemarchal, P.; Dhorne, T.; Rerat, A. Intestinal absorption of medium chain fatty acids: In vivo studies in pigs devoid of exocrine pancreatic secretion. *Br. J. Nutr.* **1994**, *72*, 545–553. [CrossRef]
33. You, Y.Q.; Ling, P.R.; Qu, J.Z.; Bistrian, B.R. Effects of medium-chain triglycerides, long-chain triglycerides, or 2-monododecanoin on fatty acid composition in the portal vein, intestinal lymph, and systemic circulation in rats. *JPEN J. Parenter. Enteral. Nutr.* **2008**, *32*, 169–175. [CrossRef]
34. Quann, E.E.; Fulgoni, V.L., III; Auestad, N. Consuming the daily recommended amounts of dairy products would reduce the prevalence of inadequate micronutrient intakes in the United States: Diet modeling study based on NHANES 2007–2010. *Nutr. J.* **2015**, *14*, 90. [CrossRef]
35. Martínez-Vallespín, B.; Vahjen, W.; Zentek, J. Effects of medium chain fatty acids on the structure and immune response of IPEC-J2 cells. *Cytotechnology* **2016**, *68*, 1925–1936. [CrossRef]
36. Rego Costa, A.; Rosado, E.L.; Soares-Mota, M. Influence of the dietary intake of medium chain triglycerides on body composition, energy expenditure and satiety: A systematic review. *Nutr. Hosp.* **2012**, *27*, 103–108. [CrossRef]
37. Husain, Z.; Schwartz, R.A. Food allergy update: More than a peanut of a problem. *Int. J. Dermatol.* **2013**, *52*, 286–294. [CrossRef]
38. Fujita, K.; Katahira, J.; Horiguchi, Y.; Sonoda, N.; Furuse, M.; Tskuita, S. Clostridium perfringens enterotoxin binds to the second extracellular loop of claudin-3, a tight junction membrane protein. *FEBS Lett.* **2000**, *476*, 258–261. [CrossRef]
39. Farquhar, M.G.; Palade, G.E. Junctional complexes in various epithelia. *J. Cell Biol.* **1963**, *17*, 375–412. [CrossRef]
40. Martinez-Palomo, A.; Erlij, D.; Bracho, H. Localization of permeability barriers in the frog skin epithelium. *J. Cell. Biol.* **1971** *50*, 277–287. [CrossRef]

41. Günzel, D.; Fromm, M. Claudins and other tight junction proteins. *Compr. Physiol.* **2012**, *2*, 1819–1852. [CrossRef]
42. Markov, A.G.; Veshnyakova, A.; Fromm, M.; Amasheh, M.; Amasheh, S. Segmental expression of claudin proteins correlates with tight junction barrier properties in rat intestine. *J. Comp. Physiol. B* **2010**, *180*, 591–598. [CrossRef]
43. Wen, H.; Watry, D.D.; Marcondes, M.C.; Fox, H.S. Selective decrease in paracellular conductance of tight junctions: Role of the first extracellular domain of claudin-5. *Mol. Cell. Biol.* **2004**, *24*, 8408–8417. [CrossRef] [PubMed]
44. Amasheh, S.; Schmidt, T.; Mahn, M.; Florian, P.; Mankertz, J.; Tavalali, S.; Gitter, A.H.; Schulzke, J.D.; Fromm, M. Contribution of claudin-5 to barrier properties in tight junctions of epithelial cells. *Cell Tissue Res.* **2005**, *321*, 89–96. [CrossRef] [PubMed]
45. Ikenouchi, J.; Furuse, M.; Furuse, K.; Sasaki, K.; Tsukita, S.; Tsukita, S. Tricellulin constitutes a novel barrier at tricellular contacts of epithelial cells. *J. Cell Biol.* **2005**, *171*, 939–945. [CrossRef] [PubMed]
46. Krug, S.M.; Amasheh, S.; Richter, J.F.; Milatz, S.; Günzel, D.; Westphal, J.K.; Huber, O.; Schulzke, J.D.; Fromm, M. Tricellulin forms a barrier to macromolecules in tricellular tight junctions without affecting ion permeability. *Mol. Biol. Cell.* **2009**, *20*, 3713–3724. [CrossRef] [PubMed]

10

Claudin-1: A Double-Edged Sword in Cancer

Ajaz A. Bhat [1], Najeeb Syed [1], Lubna Therachiyil [2,3], Sabah Nisar [1], Sheema Hashem [1], Muzafar A. Macha [4,5], Santosh K. Yadav [1], Roopesh Krishnankutty [2], Shanmugakonar Muralitharan [6], Hamda Al-Naemi [6], Puneet Bagga [7], Ravinder Reddy [7], Punita Dhawan [5], Anthony Akobeng [8], Shahab Uddin [2], Michael P. Frenneaux [9], Wael El-Rifai [10] and Mohammad Haris [1,6,*]

1. Division of Translational Medicine, Research Branch, Sidra Medicine, Doha 26999, Qatar; abhat@sidra.org (A.A.B.); nsyed@sidra.org (N.S.); snisar1@sidra.org (S.N.); shashem@sidra.org (S.H.); syadav@sidra.org (S.K.Y.)
2. Translational Research Institute, Academic Health System, Hamad Medical Corporation, Doha 3050, Qatar; LTherachiyil@hamad.qa (L.T.); RKrishnankutty@hamad.qa (R.K.); SKhan34@hamad.qa (S.U.)
3. Department of Pharmaceutical Sciences, College of Pharmacy, QU Health, Qatar University, Doha 2713, Qatar
4. Department of Biotechnology, Central University of Kashmir, Ganderbal, Jammu and Kashmir 191201, India; muzafar.macha1@gmail.com
5. Department of Biochemistry and Molecular Biology, University of Nebraska Medical Center, Omaha, NE 68198, USA; punita.dhawan@unmc.edu
6. Laboratory Animal Research Center, Qatar University, Doha 2713, Qatar; smurli28@qu.edu.qa (S.M.); halnaemi@qu.edu.qa (H.A.-N.)
7. Center for Magnetic Resonance and Optical Imaging, Department of Radiology, Perelman School of Medicine at the University of Pennsylvania, Philadelphia, PA 19104, USA; puneetb@pennmedicine.upenn.edu (P.B.); krr@pennmedicine.upenn.edu (R.R.)
8. Department of Pediatric Gastroenterology, Sidra Medicine, Doha 26999, Qatar; aakobeng@sidra.org
9. Academic Health System, Hamad Medical Corporation, Doha 3050, Qatar; mfrenneaux@hamad.qa
10. Department of Surgery, University of Miami Miller School of Medicine, Miami, FL 33136, USA; welrifai@med.miami.edu

* Correspondence: mharis@sidra.org.

Abstract: Claudins, a group of membrane proteins involved in the formation of tight junctions, are mainly found in endothelial or epithelial cells. These proteins have attracted much attention in recent years and have been implicated and studied in a multitude of diseases. Claudins not only regulate paracellular transepithelial/transendothelial transport but are also critical for cell growth and differentiation. Not only tissue-specific but the differential expression in malignant tumors is also the focus of claudin-related research. In addition to up- or down-regulation, claudin proteins also undergo delocalization, which plays a vital role in tumor invasion and aggressiveness. Claudin (CLDN)-1 is the most-studied claudin in cancers and to date, its role as either a tumor promoter or suppressor (or both) is not established. In some cancers, lower expression of CLDN-1 is shown to be associated with cancer progression and invasion, while in others, loss of CLDN-1 improves the patient survival. Another topic of discussion regarding the significance of CLDN-1 is its localization (nuclear or cytoplasmic vs perijunctional) in diseased states. This article reviews the evidence regarding CLDN-1 in cancers either as a tumor promoter or suppressor from the literature and we also review the literature regarding the pattern of CLDN-1 distribution in different cancers, focusing on whether this localization is associated with tumor aggressiveness. Furthermore, we utilized expression data from The Cancer Genome Atlas (TCGA) to investigate the association between CLDN-1 expression and overall survival (OS) in different cancer types. We also used TCGA data to compare CLDN-1 expression in normal and tumor tissues. Additionally, a pathway interaction analysis was performed to investigate the interaction of CLDN-1 with other proteins and as a future therapeutic target.

Keywords: claudin 1; tight junctions; tumor; metastasis; epithelial to mesenchymal transition

1. Introduction

Claudins and occludin, a group of cell junctional proteins, serve as the backbone of the tight junctions. Claudin family members perform dual roles; some have barrier activities, while others mediate the permeability of small molecules and ions. In addition to the localization pattern, the differential expression of claudins between normal and tumor tissue has drawn attention to these proteins as potential prime candidates for future cancer therapy. Another hot topic of discussion is the tumor-promoter or tumor-suppressor role of claudins. This opens a wide area of research in elucidating how the tissue-specific expression of claudins and their interaction with other molecules in the cell may result in these two opposing effects. Both defective tight junctions and the absence of tight junctions have shown to be associated with the development and progression of certain cancers. In this article, we include a brief introduction of tight junctions, the structure of claudins, and their role in various cancers. We also perform the bioinformatics analysis on TCGA data to supplement the literature review.

1.1. Tight Junctions

Adjacent epithelial cells are sealed into an epithelial barrier by the most apical intercellular junctions called tight junctions. Tight junctions, as a network of continuous strands, separate the plasma membrane into apical and basolateral domains [1,2]. Tight junctions between adjacent cells associate to form paired strands imparting mechanical strength to the cells [3,4] and serve as barriers to control the movement of small molecules and ions across the paracellular space [5–7]. Apart from their mechanical strength, maintaining polarity and paracellular movement, tight junction proteins are able to recruit signaling proteins for various cellular processes [4]. Alterations in the structure and function of tight junctions result in a multitude of diseases, especially adenocarcinoma of various organs [8–10]. The failure of tight junctiontight junctions or tight junction proteins is one of the many key factors that contribute to the progression of cancer, but this is not a universal phenomenon as there can be many other direct or indirect factors that contribute to the development of cancer. The second reason that loss of tight junctions or tight junction proteins is not a universal phenomenon to the development of cancers is that besides the epithelial cancers, there are also non-epithelial tumors such as small subset of laryngeal neoplasms [11], angiomas, lipomas and neuromas [12] which do not display failure of tight junctions but other contributing factors come into play.

1.2. Claudins

Claudins are integral to the structure and function of tight junctions with four membrane-spanning regions, which include two extracellular loops, N- and C-terminal cytoplasmic domains. The extracellular loops are highly conserved, and the C-terminal domain is important for localizing the claudins to tight junctions (Figure 1). Being a part of a multigene family, there are about 27 members of claudins that are unique in their tissue-specific expression and their molecular weight ranges from 20–34 kDa [3]. Claudins play an important role in regulating transepithelial permeability by regulating the epithelium's paracellular permeability to small molecules and ions [5,7,13]. Post-translational modifications such as phosphorylation alter the paracellular functions of claudins, which in turn modulate diverse signal mechanisms [14–17].

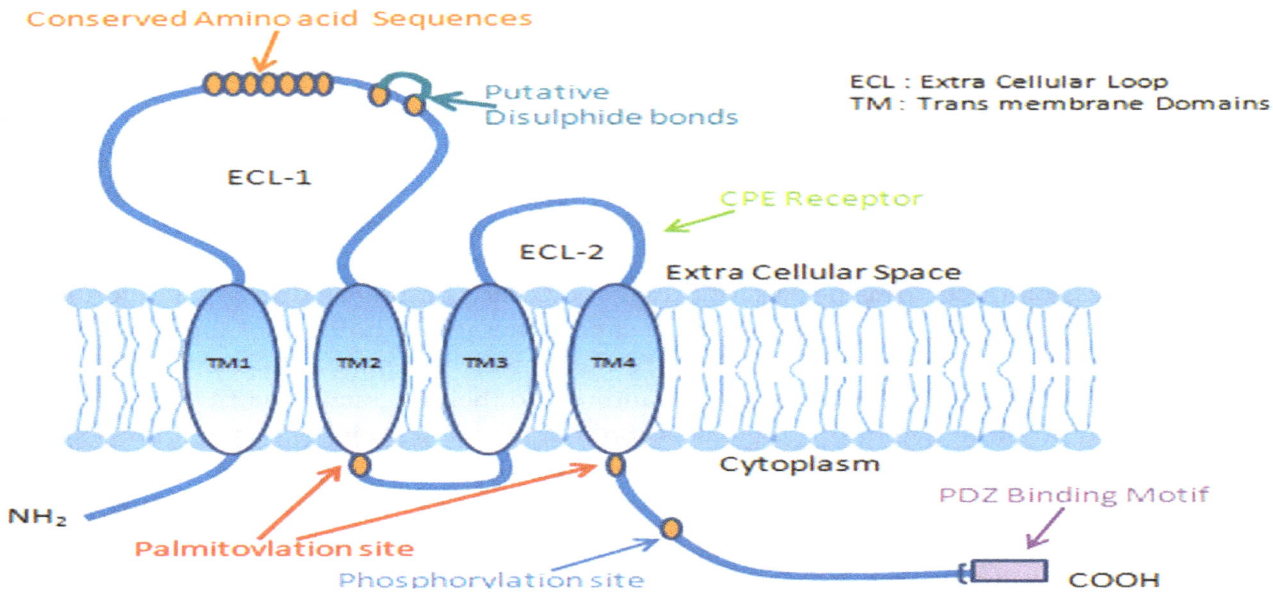

Figure 1. A schematic presentation of tight junction complex involving claudins and other major components. Claudins contain four transmembrane domains (TMD-1, TMD-2, TMD-3, and TMD-4) and two extracellular (ECL) loops. The PDZ-binding domain of the –COOH terminal of claudin undergoes post-transcriptional modification and has been implicated in signal transduction.

1.3. Claudins and Cancer

One of the important factors in cellular transformation and tumorigenesis is the loss of cell-to-cell adhesion [1]. Accordingly, the claudin family of proteins is significantly involved in the progression and growth of several cancers [9,10,18]. Tumor progression is characterized by migration, invasion, and metastasis of cancer cells. Claudins are believed to play a significant role in these processes as their loss contributes to the loss of cell junctions in a tissue-dependent manner [18,19]. Claudins have been also reported to play a vital role in the epithelial–mesenchymal transition (EMT) (Figure 2), a process that favors the spread of carcinomas, generation of cancer stem cells (CSCs) or tumor-initiating cells (TICs), and chemo-resistance [20–23]. The loss of claudins in epithelial cells results in disrupted tight junction function responsible for impaired cell polarity and epithelial integrity [6,7]. Several studies have reported the mislocalization and altered expression of claudins in various cancers [19,24]. The CLDN-1 and CLDN-7 members of the claudin family are primarily found to be downregulated in several invasive cancers including breast, esophageal, and prostate cancers [9,19,25–27]. However, in contrast, overexpression of CLDN-1 has been observed in colon, nasopharyngeal, ovarian and oral squamous cell cancers [9,10], while CLDN-3 and -4 are highly overexpressed in ovarian cancer and upregulated in breast, gastric, pancreatic, prostate and uterine cancers [28–30]. Human carcinomas such as those of the breast, liver, ovary, prostate, colon, liver and stomach are found to exhibit altered expression of claudins [19]. The expression and localization patterns of some of the claudins serve as an important prognostic predictor in many cancers [30,31]. The consensus of whether claudin expression increases or decreases during tumorigenesis is still a debatable topic and open to more research.

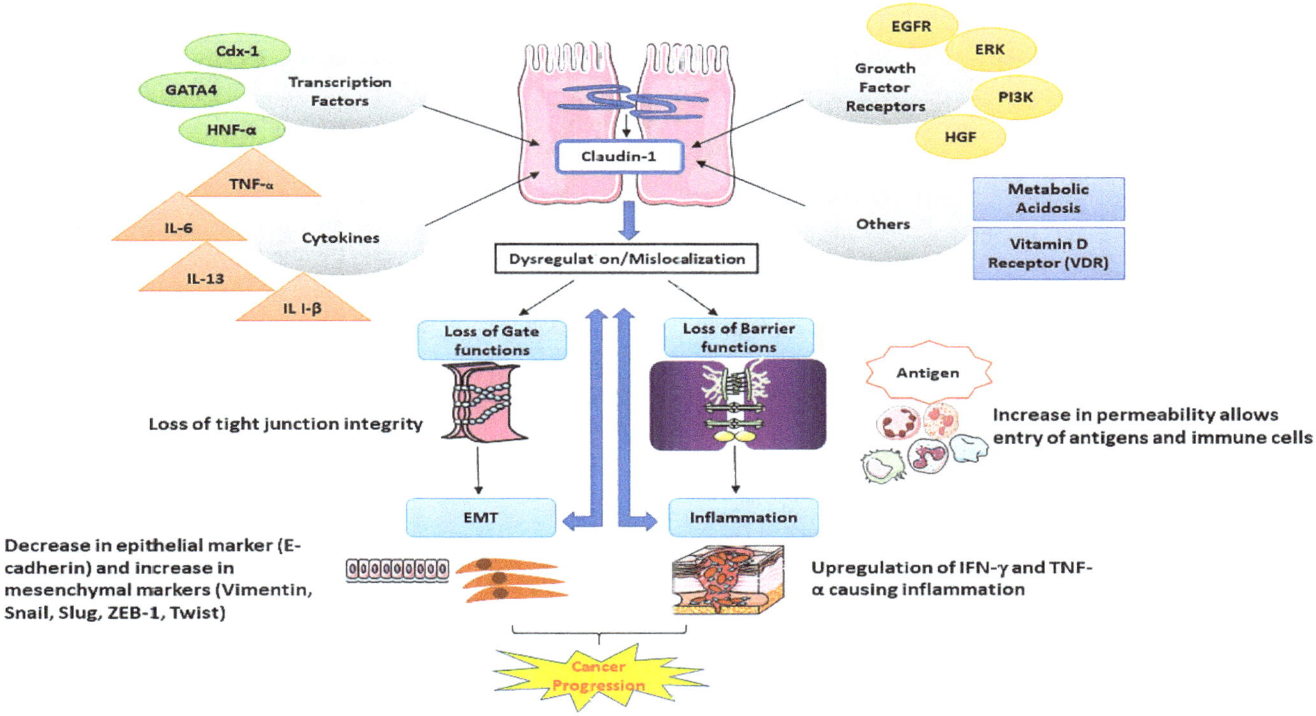

Figure 2. Schematic model of role and regulation of CLDN-1 in a normal or diseased state. In normal physiological conditions, CLDN-1 expression/ integrity is regulated by transcription factors, growth factors and cytokines, which in turn maintain the normal gate function and barrier function of tight junctions. Dysregulation of CLDN-1 expression can result in the compromise of membrane barrier functions and gate functions, which subsequently can lead to the upregulation of the expression of pro-inflammatory markers such as IFN-γ and TNF-α. In cancer, the loss of CLDN-1 facilitates the malignant transformation of cancer cells and epithelial-mesenchymal transition (EMT).

2. Claudin-1 and Cancer; Tumor Promoter or Suppressor

CLDN-1 is a membrane protein that, along with occludin and other claudins form the backbone of the tight junctions and is essential for epithelial barrier functions [32]. It was the first member of the claudin family to be identified with a molecular weight of 22 kDa and is strongly expressed in the intestine, spleen, brain, liver, kidney, and testis [19,33]. Studies have shown the direct involvement of CLDN-1 in the development and progression of several cancers, such as colon cancers [34], oral squamous cell carcinomas [35], breast cancers [36], melanomas [37,38], and in many other cancers as discussed in this review. In some cancers, CLDN-1 has the opposite role where the decreased expression of CLDN-1 is associated with cancer progression, invasion and development of the metastatic phenotype [37,39]. The expression of CLDN-1 in different types of cancer and cancer subtypes is summarized in Table 1. Based on the literature, CLDN-1 is one of the most deregulated claudins in human cancer and can function as a tumor promoter or suppressor depending on the type of cancer (Figure 3) (Table 2). The role of CLDN-1 as a tumor promoter is mostly through its effect on the invasion or motility of cancer cells. Considering the importance of claudins in cancer, targeting claudin expression appears to have promise in the treatment of cancer. The specific role of CLDN-1 in various cancers is discussed in the following sections.

Table 1. Expression of CLDN-1 in different types of cancer.

Type of Cancer	Subtypes	Expression of CLDN-1	References
Breast Cancer	Luminal A	Downregulated	[40]
	Luminal B	Downregulated	[40]
	Triple negative/Basal like	Upregulated	[40]
	HER2 enriched	Downregulated	[41]
	Claudin-low	Downregulated	[41]
Thyroid Cancer	Papillary Thyroid Cancer	Upregulated	[42]
	Follicular Thyroid Cancer	Upregulated	[43]
Colorectal Cancer	Ulcerative Colitis associated Colorectal Cancer	Upregulated	[44]
	Sporadic Colorectal Cancer	Upregulated	[34]
Gastric Adenocarcinoma	-	Upregulated	[45]
Head and Neck Squamous Cell Carcinoma	-	Upregulated	[46]
Hypopharyngeal Squamous Cell Carcinoma	-	Upregulated	[47]
Hepatocellular Carcinoma	-	Downregulated	[48,49]
Lung Adenocarcinoma	-	Downregulated	[50]
Pancreatic Ductal Carcinoma	-	Upregulated	[51]
Epithelial Ovarian Cancer	-	Upregulated	[52,53]
Oral Squamous Cell Carcinoma	-	Upregulated	[35]
Melanoma	-	Upregulated	[38]
Prostate adenocarcinoma	-	Downregulated	[54,55]

"-" no subtypes.

Table 2. Role of CLDN-1 in different cancers.

Cancer Type	Activity	Findings	References
Melanoma	Tumor Promoter	Cytoplasmic expression of CLDN-1 contributes to the migratory capacity of melanoma cells	[56]
Oral Squamous Cell Carcinoma	Tumor Promoter	CLDN-1 enhances the invasive activity of OSC-4 and NOS-2 cell lines by activation of MT1-MMP and MMP-2	[35]
Prostate Cancer	Tumor Suppressor	Loss of CLDN-1 associated with progression of Prostate cancer	[54]
Lung Cancer	Tumor Suppressor	Knockdown of CLDN-1 increased invasive and metastatic activity of lung adenocarcinoma cells	[57]
Breast Cancer	Tumor Promoter in ER-Subtypes	Increases cell migration and also exhibits an anti-apoptotic effect	[40]
	Tumor Suppressor in ER+ Subtypes	Acts as a suppressor of mammary epithelial proliferation Increases apoptosis of breast cancer cells	
Thyroid Cancer	Tumor Promoter	High expression of CLDN-1 found in follicular thyroid carcinoma (FTC-133) and Papillary Thyroid Carcinoma cells	[42,43]
Ovarian Cancer	Tumor Promoter	High expression of CLDN-1 correlated with shorter overall survival in ovarian carcinoma effusions	[58]
Colon Cancer	Tumor Promoter	High CLDN-1 expression in colon carcinoma and metastasis CLDN-1 upregulates the repressor ZEB-1 to reduce expression of E-cadherin in colon cancer cells	[34,59]
Gastric Cancer	Tumor Promoter	High expression of CLDN-1 in gastric cancer associated with poor survival	[60]
Hypopharyngeal Squamous Cell Carcinoma	Tumor Promoter	High expression of CLDN-1 associated with lymph node metastasis and degree of tumor differentiation	[61]
Hepatocellular Carcinoma	Tumor Promoter	CLDN-1 promoted epithelial-mesenchymal transition (EMT) in HCC cells by overexpression of mesenchymal markers (N-cadherin and vimentin)	[48,62,63]
Pancreatic Cancer	Tumor Promoter	TNF-α upregulated CLDN-1 expression, leading to increased proliferation of pancreatic cancer cells	[64]

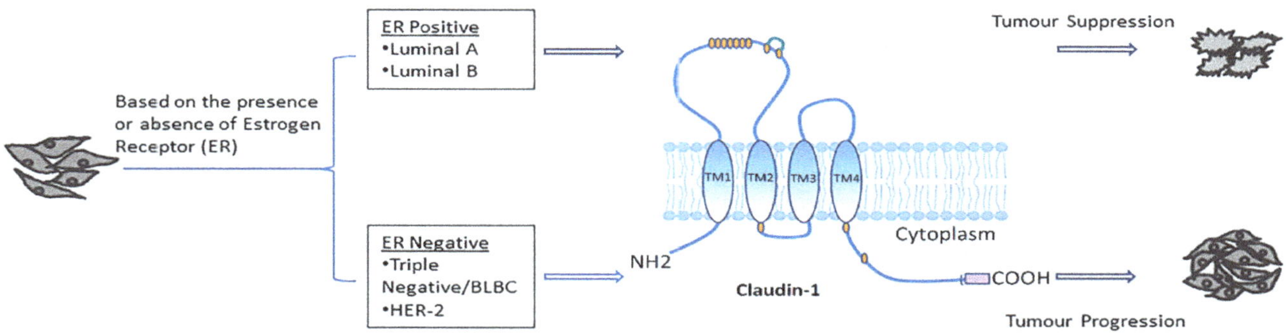

Figure 3. CLDN-1 expression in different subtypes of breast cancer as characterized by the presence or absence of estrogen receptor (ER). Luminal A, and Luminal B subtypes of human invasive breast cancer (ER-positive) exhibit low levels of CLDN-1, which suggests the suppressor role of CLDN-1 in these tumors. However, aggressive forms (ER-negative) exhibit overall high levels of CLDN-1 expression, which signifies CLDN-1 role as a tumor promoter.

2.1. Claudin-1 and Breast Cancer

Breast cancer is the second major cause of death in women, and its heterogeneous molecular nature is a significant obstacle in treatment planning [65]. It has several subtypes, such as human epidermal growth factor receptor 2 (HER2), triple-negative or basal-like, Luminal A and Luminal B type depending on the presence or absence of several hormone receptors like HER2, estrogen, and progesterone [66–68]. Recently, another subtype of breast cancer known as the claudin-low subtype has been reported [67]. Each subtype of breast cancer exhibits unique prognostic features and different molecular markers [69].

The *CLDN-1* gene has been found to be upregulated during the early involution of the mammary gland [70]. The differential expression of CLDN-1 observed in different cancers outlines the complexity of the potential role that it plays in the cancer process. The CLDN-1 expression level in breast cancer differs depending on the cancer subtypes [71]. Studies have shown a correlation between increased malignancy, invasiveness and recurrence of breast cancer with total or partial loss of CLDN-1 expression [36,70]. In most of the invasive human breast cancers such as ER+ luminal A and luminal B, CLDN-1 expression is found to be downregulated, while an increased expression and cytoplasmic delocalization of CLDN-1 has been observed in some of the aggressive ER- basal-like breast cancer (BLBC) subtypes [40,72,73]. CLDN-1 is also found to be downregulated in HER2 enriched and claudin low breast cancer subtypes [41]. CLDN-1 acts as a tumor suppressor in ER+ and as a tumor promoter in ER- cancer subtypes [25]. In hereditary and sporadic breast cancer, CLDN-1 is found to be involved in tumorigenesis by suppressing the proliferation of mammary epithelial cells [74]. Further, CLDN-1 overexpression in MDA-MB 361 breast cancer cells resulted in increased apoptosis [75,76]. While one study reported that the activation of CLDN-1 was repressed by the binding of E-cadherin to CLDN-1 promoter [77], knockdown of CLDN-1 has been found to be associated with decreased cell migration and induction of EMT in breast cancer cells [76]. Another study showed a unique pattern of expression for CLDN-1 in ER-ve and ER+ve tumors. The authors showed that the protein expressions of CLDN-1 were significantly higher in the basal-like subtype of breast cancers (ER-ve, Her-2-ve, EGFR+ve, CK5/6+ve, a subtype largely linked to poor outcome [40]. CLDN-1 expression has also been observed in a small percentage of invasive human breast cancers that exhibit different pathological lesions leading to complexity in CLDN-1 expression [78]. CLDN-1 also possesses tumor-promoting effects by increasing cell migration and by exhibiting anti-apoptotic effects in some breast cancer cell lines like MCF-7 [76,79].

Several proteins interact with CLDN-1 to fuel the progression of breast cancer, including the following: Ephrin B1, ESCRT, CD9 and EpCAM [80–83]. CLDN-1 mediates the tyrosine phosphorylation of Ephrin B1, a transmembrane protein, in a receptor independent manner which provides the evidence that ephrin-B1 inhibits the formation of the tight cell–cell adhesion in a wide variety of epithelial and cancer cells regardless of the existence of cognate Eph receptors [80]. Endosomal sorting complexes required for transport (ESCRT) machinery are a set of proteins present in the cytosol that are involved in the maintenance of cell polarity and the regulation of membrane-bound proteins [81]. When the function of ESCRT is inhibited, CLDN-1 accumulates in the cytoplasm causing the tight junctions to disassemble and lose cell polarity [25]. The loss of ESCRT function is also linked with increased proliferation and less stable tissue structure in the cancer cells. CLDN-1 is also found to interact with CD9, a transmembrane protein that regulates cell migration, proliferation, differentiation and fusion [82]. CD9 prevents the association between CLDN-1 and tight junctions that could cause the progression of the tumor. The subcellular co-localization of CLDN-1 and CD9 supports their interaction, and this was confirmed in many cell clines including different human breast cancer cell lines [82]. EpCAM (also known as epithelial cell adhesion molecule), another surface transmembrane glycoprotein known to be expressed in some invasive carcinomas is involved in cell proliferation and metastasis and has been shown to protect CLDN-1 from degradation. [83]. This could be a cause for the cytoplasmic accumulation of CLDN-1 in some breast cancer cell lines [76,83]. Several transcript variants for CLDN-1 were found in human invasive breast cancer as a result of splicing and mis splicing events suggesting that through alternative splicing CLDN-1 is downregulated in invasive type of breast cancers [72].

2.2. Claudin-1 and Thyroid Cancer

Thyroid cancer is the most commonly occurring endocrine malignancy [84,85]. A study by Nemeth et al. performed independent microarray expression analyses of two types of thyroid carcinomas, namely papillary thyroid cancer (PTC) and follicular thyroid cancer (FTC) [42]. The study showed that high expression of CLDN-1 is specific for the regional lymph node metastasis associated with PTC [42] and found increased expression of *CLDN-1* gene in PTC [86,87]. Sobel et al. reported high levels of CLDN-1 expression in serous papillary endometrial carcinoma [88]. Another study about the role of CLDN-1 in follicular-cell derived thyroid carcinoma cell lines (FTC-133 and FTC-238) found higher expression of CLDN-1 in the nuclei of FTC-238 cells as compared to the FTC-133 cells [43]. The same study demonstrated the increased pathogenic character of FTC-133 cells by RASV12 transfection was associated with high expression of CLDN-1 and enhanced cell proliferation and migration [43]. Conversely, the downregulation of CLDN-1 by siRNA caused decreased cell invasion and migration accompanied by decreased phospho-PKC expression in the FTC-238 cells, suggesting that the aggressiveness of follicular thyroid carcinoma associated with high CLDN-1 expression can be influenced by PKC activity [43]. Another study described the reduced expression of CLDN-1 in follicular carcinomas vs adenomas, specifically in the poorly-differentiated and undifferentiated types of human thyroid carcinomas [89]. The expression of CLDN-1 was significantly different between malignant and benign thyroid neoplasms, and between follicular and papillary carcinomas [90]. Similarly, papillary carcinomas showed significantly higher positive CLDN-1 expression. While negative CLDN-1 expression was observed in the tissue samples of normal thyroid and solitary-follicular-patterned-nodules [91].

2.3. Claudin-1 and Colorectal Cancer

Colorectal cancer (CRC) is the fourth leading cause of cancer-related deaths and the third most frequently diagnosed malignancy worldwide [92]. Increased expression of CLDN-1 is associated with the progression and metastasis of colon carcinoma [34,93]. In mouse xenograft studies, tumor growth and metastasis is regulated by genetic modulation of CLDN-1 [94]. The nucleus and cytoplasm of colon carcinoma cells and metastatic lesions showed intensified CLDN-1 expression [34]. Many studies suggest that the genes encoding tight junction proteins (TGPs) in CRC are differentially expressed and involved in the process of invasion and cellular transformation [95]. Several studies reported up-regulation of CLDN-1 in CRC [34,44,96,97]. A similar study showed that CLDN-1 overexpression induced a highly invasive and metastatic potential in CRC cells [34]. Noncancerous cells with normal CLDN-1 expression were found to form a monolayer, whereas cells that overexpressed CLDN-1 grew as aggregates. CLDN-1 regulates cellular morphology and behavior in the colonic epithelium [34,44]. The possible involvement of CLDN-1 in the tumorigenesis of ulcerative colitis (UC)-associated CRC has also been demonstrated [44]. Another study demonstrated that the prognostic factor for CRC is the independent expression of CLDN-1 [98]. Delocalization of CLDN-1 from the membrane to cytoplasm and nuclei of cancer cells supports cancer growth and malignancy [34]. In colon cancer cells, CLDN-1 decreases the expression of E-cadherin by upregulating ZEB-1 repressor resulting in invasion and reduction of anoikis [59].

The level of CLDN-1 mRNA was found to be higher in the distal site of the colon as compared to the proximal site and demonstrated significant effects on xenografted tumors growth in athymic mice by changes in the expression of CLDN-1, showing its role in CRC tumorigenesis [34]. Both mRNA and protein levels of CLDN-1 were found to be upregulated in sporadic human CRC compared to the normal mucosa [34]. Dhawan and colleagues reported that T84 cell transfection with CLDN -1 resulted in aggregation and multilayer formation in transfected T84 cells as compared to the T84 parent cells. The interactions between claudin family members are both homophilic and heterophilic and are considered to play a significant role in the progression of CRC and several other cancers [99]. The progression of colon cancer has been linked with the dysregulation of the CLDN-1 expression causing disorganization of the tight junction fibrils leading to increased paracellular permeability [100]. Increased potential for invasion and metastasis has been demonstrated in xenografts that express CLDN-1 [101]. CLDN-1 positively correlates with CRC cell proliferation and influences the growth and evolution of the tumor. Its expression was also found to be associated with accelerated serrated lesions of CRC and was related to anoikis resistance and cellular dis-cohesion [101]. Moreover, serrated polyps with over-expressed CLDN-1 were found to have a higher potential for the development and progression into higher-grade lesions [101].

Activation of the Wnt signaling pathway is strongly implicated in the development of colorectal cancer [102]. Wnt signaling is activated by the loss of the adenomatous polyposis coli (APC) protein or by the activation of β-catenin mutations [103,104]. CLDN-1, one of the target genes in the Wnt signaling pathway, has two β-catenin binding sites (TCF/LEF) in its promoter region for the activation of the transcription process [105,106]. A study showed that the expression of CLDN-1 was found to be elevated in the intestinal adenomas of the APC in mice as compared to normal tissue [94]. Additionally, high expressions of CLDN-1 are seen in the dysplastic areas of the colon in patients with chronic inflammatory disease [94]. CLDN-1 was overexpressed in metastatic colorectal cancer (mCRC) samples as compared to normal mucosa with differential expression in other CRC subtypes. Consensus molecular subtype CMS2, transit-amplifying and C5 subtypes of the mCRC exhibited higher expression of CLDN-1 [107].

2.4. Claudin-1 and Gastric Cancer

CLDN-1 is highly expressed in gastric cancers [108,109]. High expression of CLDN-1 was reported in intestinal type gastric cancer that correlated with lymph node metastasis, advanced TNM (classification of malignant tumors) stage, recruitment, and activation of MMP-2 and MMP-9, which are all responsible for enhanced cell invasion and metastasis [60,109]. The invasion of gastric adenocarcinoma cells is associated with the levels of CLDN-1 expression as CLDN-1 is found to be upregulated in gastric carcinoma and participates in the metastatic behavior of these cancer cells [45].

One study demonstrated that the localization and correlation of CLDN-1 expression are linked with anoikis resistance in gastric cancer through mediating membrane β-catenin expression and by inducing cell aggregation and inhibiting apoptosis cascade [110]. The authors also observed that the levels of CLDN-1 expression in gastric cancer tissues decreased from well to moderate to poorly differentiated tumors, suggesting that reduced CLDN-1 expression is an adverse prognostic factor predicting a lower survival rate [110]. However, another study showed that in comparison to CLDN-4, the expression of CLDN-1 was higher in well-to-moderately differentiated gastric adenocarcinomas [111].

2.5. Claudin-1 and Hypopharyngeal Squamous Cell Carcinoma

Head and neck squamous cell carcinoma (HNSCC) is the sixth most frequent tumor worldwide [112]. Tissue microarray and immunohistochemistry assays of surgical samples suggested that CLDN-1 expression is increased in squamous cell cancer [46,113]. It has been shown previously that CLDN-1 induces the generation of tumor lymphatic vessels and increases the lymph node metastasis [47]. Additionally, a study demonstrated that CLDN-1 expression in squamous cancers differs in an organ-specific manner [113]. CLDN-1 was found to be upregulated in hypopharyngeal squamous cell carcinoma (HSCC). This study showed a positive association of CLDN-1 expression with the degree of tumor differentiation and lymph node metastasis [61].

2.6. Claudin-1 and Hepatocellular Carcinoma

Hepatocellular carcinoma (HCC) is the third major cause of death due to cancer and the fifth most common cancer malignancy worldwide [114]. Studies have reported the involvement of CLDN-1in the tumorigenesis and metastasis of HCC [48,63]. HCC cell line experiments demonstrated the role of CLDN-1 in the process of cancer cell invasion [115]. Primary HCC samples were found to be positive for CLDN-1, suggesting its significant role in the formation of metastasis and hepato-carcinogenesis [115]. Another study demonstrated the involvement of CLDN-1 in the epithelial to mesenchymal transition in HCC and hepato-carcinogenesis [62]. One study established that overexpression of CLDN-1 induces MMP-2 in SNU-354, -423 and -449 HCC cells resulting in increased invasion and migration of the cancer cells compared to the normal liver cells and other CLDN-1 expressing HCC cells such as SNU-398 and SNU-475 [63]. However in another study reduced expression of CLDN-1 was reported to be a marker for a poor prognosis in HCC [49], and a further study showed that reduced expression of CLDN-1 reinforced the invasive and cancer stem cell (CSC) like properties of HCC cell lines (Huh7 and Hep3B) in vitro, while the forced expression of CLDN-1 diminished the CSC-like properties of HCC cells [116].

2.7. Claudin-1 and Lung Adenocarcinoma

Lung cancer is one of the leading causes of death worldwide. Several studies have shown that CLDN-1 has a significant role in the pathogenesis of lung cancer [117–119]. In lung cancer, CLDN-1 acts as a cancer invasion/metastasis suppressor [57]. CLDN-1 was found to be associated with increased expressions of cancer metastasis suppressors such as connective tissue growth factor (CTGF), thrombospondin 1 (THBS1), deleted in liver cancer 1 (DLC1), occludin (OCLN), zona occludens 1 (ZO-1) and reduced expressions of cancer metastasis enhancers such as secreted phosphoprotein 1 (SPP1), cut-like homeobox 1 (CUTL1), transforming growth factor-alpha (TGF-α), solute carrier family 2 (facilitated glucose transporter) member 3 (SLC2A3) and placental growth factor (PGF) in lung adenocarcinoma [57]. For patients with lung adenocarcinoma, CLDN-1 is a potential drug treatment target and a useful predictor of prognosis. Studies showed that the invasive ability of HOP62 lung adenocarcinoma cells is increased by knockdown of endogenous expression of CLDN-1 [57]. Immunohistochemistry and RT-PCR analysis showed that CLDN-1 is either reduced or undetected in adenocarcinomas [57]. The authors demonstrated that CLDN-1 overexpression inhibited adenocarcinoma cell dissociation in wound-healing time-lapse images [57].

Inflammatory mediators, such as TNFα, plays a significant role in the process of tumorigenesis [120]. Studies have shown that CLDN-1 is involved in the mediation of inflammatory responses initiated by TNFα in different cancers [64,121,122]. An experiment in human lung cancer cell lines, observed that TNFα induced the expression of CLDN-1, and knockdown of CLDN-1 blocked 75% of TNFα-induced gene expression. In CL1-5 lung cancer cells, cell migration activity was inhibited by over-expression of CLDN-1 and restored by CLDN-1 knockdown in addition to cell invasion ability. The above findings signify a signal mediator role of CLDN-1 in TNFα induced gene expression and cell migration [57]. One study demonstrated that CLDN-1 expression correlated with Ras and epidermal-growth-factor-receptor (EGFR) expression suggesting the involvement of the latter two signaling pathways in the regulation of CLDN-1 in lung adenocarcinoma [119]. The results of this study demonstrated an association between CLDN-1 and Ras/EFGR in the development of lung cancer and the combination of both has strong clinical significance [119]. Surprisingly, as compared to the previous studies that showed that overexpression of CLDN-1 suppressed metastatic abilities of lung adenocarcinoma cells [50,57], this study showed that patients with positive expressions of both CLDN-1 and Ras/EGFR were found to have poor prognosis as compared to CLDN-1(+) Ras/EGFR(−), CLDN-1(−) Ras/EGFR(+), and CLDN-1(−) and Ras/EGFR(−) patients [119]. Another study found that enhanced cell migration by tumor necrosis factor and a similar morphology like fibroblast was found to be reduced by small CLDN-1 interfering RNA in the cells of lung cancer [121].

2.8. *Claudin-1 and Pancreatic Cancer*

Pancreatic cancer is the fourth major cause of deaths caused by cancer worldwide, with a strong capacity for metastasis and recurrence [123]. In pancreatic cancer (PC), increased expression of CLDN-1 was found to be associated with epithelial-mesenchymal transition. CLDN-1 is expressed by ductal pancreatic adenocarcinomas as well as intra-ductal papillary mucinous pancreatic tumors. One study demonstrated that 58% positive CLDN-1 immunostaining in ductal pancreatic adenocarcinomas and intraductal papillary pancreatic tumors [51]. Another study showed that through activation of mitogen-activated protein kinase 2 (MEK2), CLDN-1 was involved in cell dissociation of PC cells [124]. A further study observed the role of CLDN-1 in the progression of human PC using the PANC-1 cell line [64]. Increased expression of PARP [poly-(ADP-ribose) polymerase], an apoptosis marker, and decreased PANC-1 proliferation was observed after treatment with TNF-α [64]. Increased proliferation in PANC-1 cells was observed after treatment with TNF-α and CLDN-1 siRNA against CLDN-1, suggesting the cells were resistant to TNF-α-induced apoptosis when transfected with CLDN-1 siRNA. These findings clearly demonstrated that the CLDN-1 expression plays a role in the proliferation of PC cells [64].

2.9. *Claudin-1 and Ovarian Cancer*

Ovarian cancer affected a significant number of women worldwide and is the seventh most frequent cause of deaths due to cancer in women [125]. The upregulation of the *CLDN-1* gene is found to be associated with ovarian cancer [52]. Studies have shown that overexpression of CLDN-1 caused reduced cell differentiation and a high invasive growth rate [126]. The role of CLDN-1 has been widely studied in two different types of ovarian cancers, namely, ovarian serous and ovarian endometroid carcinoma [127]. The expression of CLDN-1 was shown to be negatively regulated by microRNA-155 (miR-155) which results in reduced proliferation and invasion of human ovarian cancer-initiating cells [52]. Reports have shown that elevated level of CLDN-11, 4, and 7 promotes the growth of both benign and malignant epithelial ovarian cancers [52]. Extensive studies conducted and analyzed for the association of CLDN-1 with survival and anatomical site showed 85% elevation of CLDN-1 expression [58]. Recently, the level of CLDN-1 expression was also investigated in borderline tumors of the ovary (BOT) [53]. A significantly higher level of CLDN-1 expression was associated with the peritoneal implants and micropapillary patterns that are specifically seen only in serous BOT [53].

2.10. *Claudin-1 and Oral Squamous Cell Carcinoma*

Almost 90% of all oral carcinomas are oral squamous cell carcinomas (OSCC) [128]. It has been shown previously that the invasive activity of OSCC cells is enhanced by CLDN-1 through activation of MMP-1 and 2, resulting in increased cleavage of Laminin-5 γ2 chains. The authors of the study further revealed elevated expression of CLDN-1 in OSC-4 and NOS-2 cell lines which are highly invasive [129]. One study demonstrated the association of high CLDN-1 expression with aggressive histopathologic features such as perineural and vascular invasion and suggested that CLDN-1 might be directly or indirectly involved in the progression of OSCC [130]. Another study found that the absence of CLDN-1 was associated with poorly differentiated tumors [131]. Immunohistochemical analysis revealed that the presence of CLDN-1 in the invasive front of tumor islands was associated with neck mode metastasis. The results obtained from this study further suggested that the expression of CLDN-1 is linked with the recurrence of OSCC [132].

2.11. Claudin-1 and Melanoma

Melanoma, which arises from melanocytes, causes 75% of deaths related to skin cancers [133]. CLDN-1 was found to be upregulated in melanoma tissues [134]. In melanoma, CLDN-1 is abnormally/aberrantly expressed in the cytoplasm of malignant cells and not in the cell membrane. This may be related to its influence on protein kinase C (PKC) activity [38]. PKC activation caused an increase in transcription and protein expression of CLDN-1 and thus, cell motility [38]. When melanoma cells transfected with CLDN-1, it increased the secretions of matrix metalloproteinase- 2 (MMP-2) reflecting its contribution to the cell invasion process. The data from French, et al. supported the hypothesis that the invasive capacity of melanoma cells is increased by cytoplasmic expression of CLDN-1 and not by the elevated nuclear expression of CLDN-1 [56]. In melanoma patients with brain metastases, the expression of CLDN-1 was downregulated, and the introduction of CLDN-1 retrovirus reduced the tumor aggressiveness and tumor migration ability and diminished micro-metastasis in the brain. This shows that reduction in CLDN-1 supports tumor progression and metastasis and that CLDN-1 can be used as a prognostic predictor for melanoma patients with increased risk of brain metastasis [135].

2.12. Claudin-1 and Prostate Cancer

Prostate cancer is the second most diagnosed malignancy and fifth-most leading cause of cancer deaths in men [136]. The Gleason grading score system is the most commonly used method to evaluate the aggressiveness of prostate cancers, whereas, the changes in the glandular architecture indicate the tumor grades [54]. The typical glandular architecture is supported by cellular polarity and cell-to-cell contact, and thus the alterations and dysregulations of proteins mediating normal cellular connection may impact the histology and Gleason grade [54]. It is suggested that the loss of tight junction protein, CLDN-1, is associated with cancer invasion, progression and the transformation into metastatic phenotype in prostate cancers. A study reported that the lower expression of CLDN-1 correlated with higher prostate-specific antigen in prostate cancer [54]. Further studies are needed to thoroughly investigate the association between prognostic factors and claudins expression in prostate cancer.

3. Claudin-1 and Tight Junction Barrier Function

CLDN-1 is an integral membrane protein that in conjunction with other claudins forms the tight junctions and together plays an essential role in epithelial barrier functions. CLDN-1 has a significant role in epithelial differentiation and loss of CLDN-1 can impair the functioning of tight junctions [137]. Numerous studies have reported the involvement of CLDN-1 in transepithelial electrical resistance (TER) [138,139] and paracellular permeability [140] showing its importance in tight junction barrier functions. The study showed that CLDN-1 increased the TER and reduced paracellular flux in Madin-Darby Canine Kidney (MCDK) cells [138]. While another study reported that induced CLDN-1-myc in MDCK cells resulted in the formation of aberrant tight junction strands independently without the participation of ZO-1 and occludin [139]. CLDN-1 controls the flux of solutes by localizing at tight junctions and modulating the paracellular permeability. Any defect in the expression of CLDN-1 can result in tight junction dysfunction causing increased paracellular permeability leading to various pathologies such as in Neonatal ichthyosis-sclerosing cholangitis (NISCH) syndrome [140]. This study showed that the silencing of CLDN-1 leads to increased hepatic paracellular permeability [140]. Mostly, an increase in the tight junction proteins should lead to an increase in the tight junction integrity but this is not a universal phenomenon as one study reported that high expression of CLDN-1 resulted in decreased TER and increased permeability causing loss of barrier function in intestinal epithelial cells (IEC-18) treated with TNF-α [141]. The authors explained the reason for this contradictory finding is that the loss of barrier function was due to the reduced expression of occludin protein and not CLDN-1 [141]. Besides, the other study reported that high expression of CLDN-1 resulted in blood–brain-barrier (BBB) leakiness during post-stroke recovery and targeting of CLDN-1 by a CLDN-1 peptide improved the permeability of brain endothelial barrier [142]. So, these studies suggest that the upregulation of CLDN-1 cannot necessarily be universal to the increased barrier function and there might be other contributing factors that regulate these functions, and also we should not rule out the tissue specific expression of claudins as other possibility to their dichotomous roles.

4. Gene Expression, Survival and Pathway Interaction Analysis of claudin-1 across different Cancers

Survival Analysis was performed using online survival analysis tools to assess the influence of CLDN-1 expression on survival in different types of cancer using the Cancer Genome Atlas (TCGA) datasets. This analysis was to investigate whether an alteration in gene expression correlates with poor survival or with tumor recurrence. The data showed that the gene expression of CLDN-1 did not affect survival significantly in cancers like breast invasive carcinoma (BRCA), cervical squamous cell carcinoma and endocervical adenocarcinoma (CESC) and pancreatic adenocarcinoma (PAAD) ($p > 0.05$), but did significantly associate with survival for cancers like thyroid carcinoma (THCA), adrenocortical carcinoma (ACC), rectum adenocarcinoma (READ) (* $p < 0.05$) (Figure 4). All survival analysis was performed using online survival analysis tools.

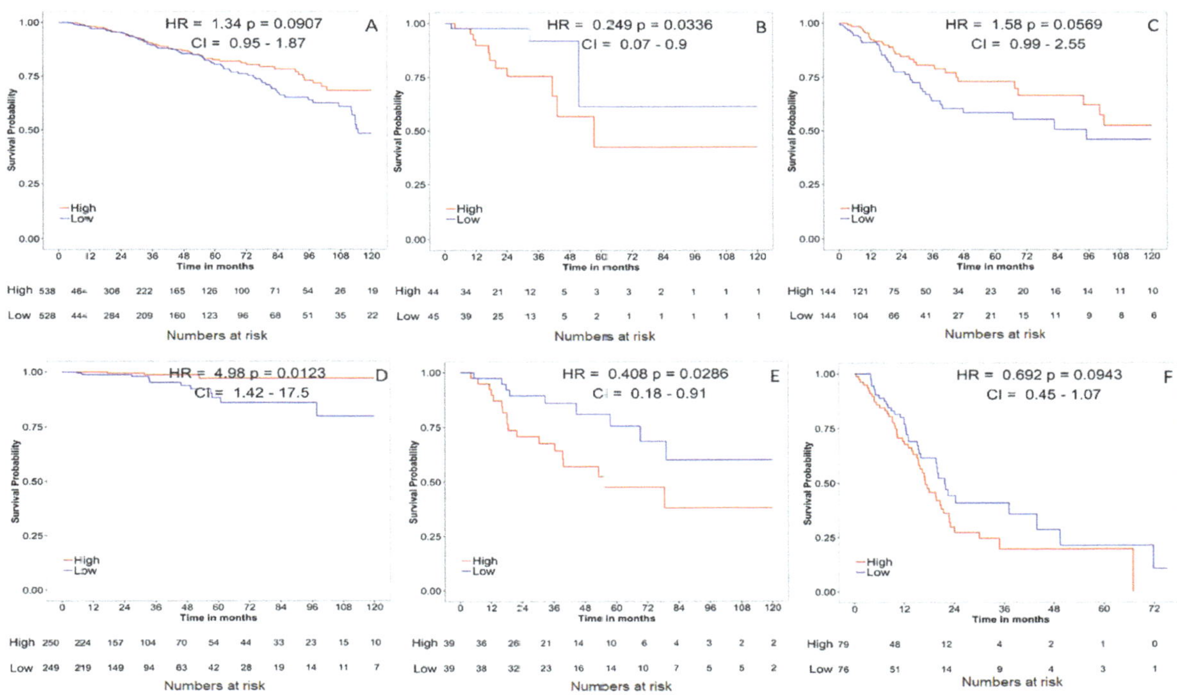

Figure 4. Survival analysis of CLDN-1 in various cancers. The red line denotes higher expression, and the blue line indicates lower expression. (**A**) Breast invasive carcinoma (BRCA); (**B**) rectum adenocarcinoma (READ); (**C**) cervical squamous cell carcinoma and endocervical adenocarcinoma (CESC); (**D**) thyroid carcinoma (THCA); (**E**) adrenocortical carcinoma (ACC); (**F**) pancreatic adenocarcinoma (PAAD).

We further compared the expression of CLDN-1 between tumor and normal tissues for ACC, CESC, PAAD, READ, THCA and BRCA from TCGA datasets. We found that most of the cancers showed significant expression differences of CLDN-1 between tumor and normal type. We observed that for most of the cancers, tumor tissues have higher expression than normal tissues but for cancer like ACC, CLDN-1 expression in normal tissue was found to be greater than tumor tissues (Figure 5). We did not find any significant difference in CLDN-1 expression between tumor and normal tissue of BRCA.

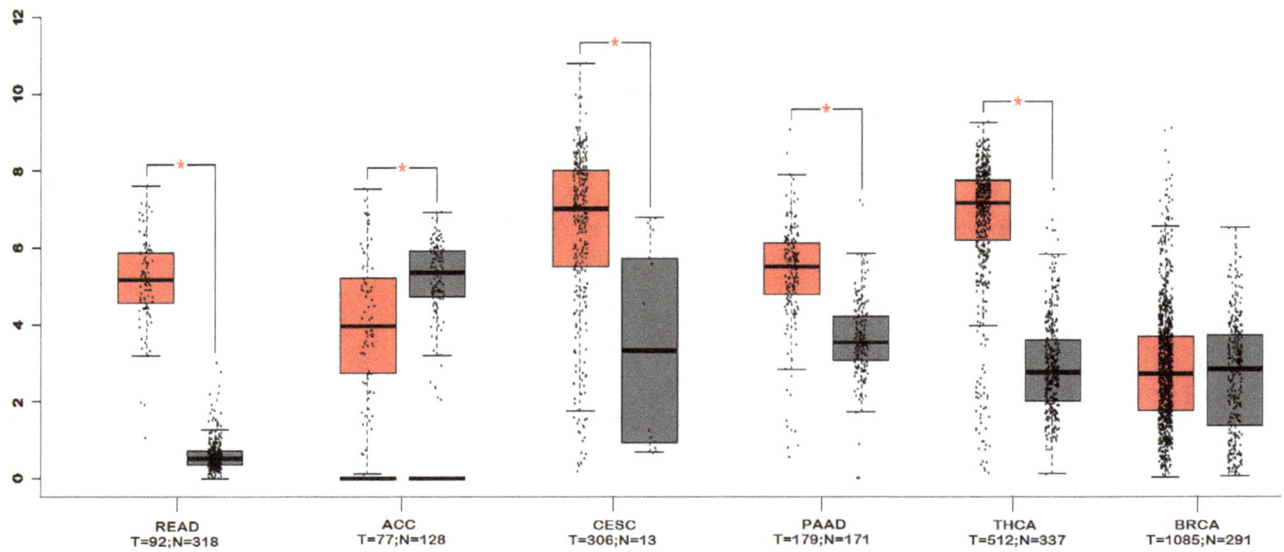

Figure 5. Boxplot showing the distribution of CLDN-1 expression in tumors and normal tissues for different types of cancers. Significant differences are shown with an asterisk (*). Boxplots were generated using GEPIA1 webserver and p-value < 0.01 was considered as significant.

To determine the interaction of CLDN-1 with other genes, we performed gene interaction analysis using the Gene MANIA prediction server. We found that CLDN-1 significantly interacted with several key genes that play an important role in normal cell physiology. Any disturbance in CLDN-1 expression or its partners may result in the manifestation of various diseases including cancers. The interaction of CLDN-1 with other key molecules can be individually analyzed in different cancers. This may highlight the key pathways which can be therapeutically targeted to suppress cancer growth or metastasis (Figure 6).

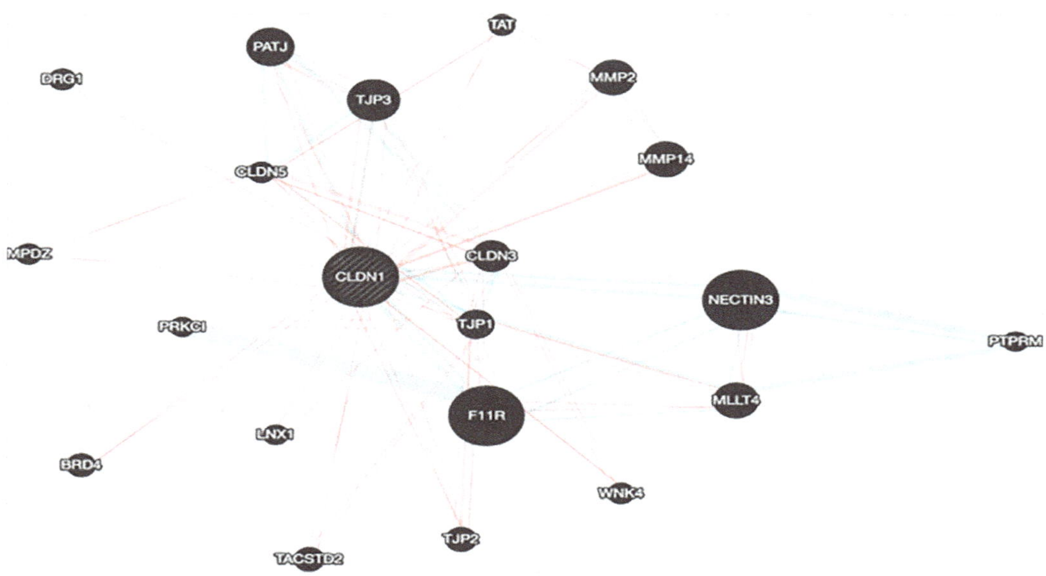

Figure 6. CLDN-1 interaction network using the Gene MANIA prediction server.

5. Claudin-1 as a Drug Target

The involvement of CLDN-1 in various pathological conditions has provided new perceptions into drug development targeting CLDN-1. The approach of targeting CLDN-1 either by monoclonal antibodies or chimeric antibodies has great potential but needs more research to reach the level of clinical trials. The initial studies have laid an important foundation towards the new strategies that could be employed and further modified towards the potential usefulness of CLDN-1 as a therapeutic target. The localization of CLDN-1 as a transmembrane protein makes it a perfect target for the enhanced drug absorption for preventing infection and treating cancer. One of the studies observed that the human hepatocytes treated with mouse anti-CLDN-1 monoclonal antibodies (mAbs), showed improved drug absorption and prevented hepatitis C virus (HCV) infection [143]. A human-mouse chimeric CLDN-1 mAb (clone 3A2) demonstrated cellular cytotoxicity against CLDN-1 expressing cancer cells [144]. The other aspect of claudins that is being exploited for therapeutic targeting is their role in regulating paracellular permeability in different tissues. *Clostridium perfringens* enterotoxin (cCPE) binds with claudin through its claudin binding domain and inhibits the claudin function. It was observed that blocking CLDN-1 with cCPE variants in the Huh7.5 hepatoma cell line inhibited infection of Huh7.5 cells with HCV in a dose-dependent manner and this also opened the epidermal barrier in the reconstructed human epidermis [145]. To eliminate the possible limitation facing CLDN-1 targeted therapies due to genotype-dependent escape via CLDN-6 and CLDN-9 and to improve anti-HCV activity, humanized anti- CLDN-1 monoclonal antibody (mAb) could be an alternative. One study developed functional mAB against extracellular domains of CLDN-1 and found that these antibodies have a very high affinity for intact CLDN-1, efficiently inhibited HCV infections both in vitro and in vivo, further demonstrating that anti-CLDN1 mAbs could be useful in inhibiting HCV infections [146]. In a very recent study, CLDN-1 was successfully targeted with anti-CLDN1 near-infrared fluorophore to track the colorectal cancer cells, and it may provide a novel way for fluorescence-guided surgery of tumor [147].

The main concern with claudin-targeted therapies is the presence of claudins in both normal epithelial cells and cancer cells that makes the targeting difficult. However, it has been observed that claudins are localized at the tight junctions in normal tissues, while in malignant tissues, there is a dysregulation of claudins localization from the tight junctions to the cell surface [24,148]. Claudins with aberrant localization in malignant tumors can be recognized by utilizing the C-terminal claudin-binding domain of cCPE fused with protein synthesis inhibitory factor (C-CPE-PSIF), causing less cytotoxicity to normal cells and a study has shown how CLDN-4 can be used as a target for tumor therapy by fusion of cCPE with PSIF. The results of the study showed that C-CPE-PSIF was cytotoxic to cells with undeveloped tight junctions (preconfluent cultures of Caco-2) and was not cytotoxic to cells with developed tight junctions (postconfluent cultures of Caco-2) [149]. Several other studies have used CLDN-4 as a target for tumor therapy and showed the accumulation of anti-CLDN-4 mAbs specifically in the tumors and reduced the growth of human colorectal and gastric tumors in mice [150]. One study detected CLDN-4 upregulation non-invasively in mice pancreatic ductal adenocarcinoma xenografts by using MRI and ^{18}FDG-PET [151] (for detailed reviews see [152]). In the future, a similar approach can be applied to CLDN-1 by preparation of CLDN-1 targeting molecule and can be tested for cytotoxicity to normal cells.

6. Claudins and Autosomal Recessive Disorders

As explained in the previous sections, the claudin family of proteins is an integral part of tight junctions that determine paracellular selectivity and permeability to small ions by acting as pores or barriers in polarized epithelia. We also discussed how overexpression or reduction of claudins could both promote and limit cancer progression, revealing complex dichotomous roles for claudins depending on cellular context. Besides the fact that the abnormal or deregulated expression of claudins has been associated with different human diseases like cancer, there are also other human disorders such as autosomal recessive disorders that have been reported due to clearly defined mutations in the corresponding claudin genes. Such disorders are mostly observed in skin, liver, kidney, the inner ear, and the eye. The first evidence that showed a mutation in the claudin family of tight junction proteins causes human disorders were from the group of Lifton [153]. In their study, they reported that the mutations in the human gene, paracellin-1 (PCLN-1)/CLDN-16 causes an autosomal recessive disorder called Familial hypomagnesemia with hypercalcinuria and nephrocalcinosis (FHHNC) characterized with renal Mg^{2+} and Ca^{+} wasting. Later the same group revealed additional evidence that loss of function mutations in paracellin-1 PCLN-1/CLDN-16, are causative of FHHNC [154]. PCLN-1 is related to the claudin family of tight junction proteins and is in tight junctions of the thick ascending limb of Henle (TAL). CLDN-16 is a cattle ortholog of PCLN-1 with ~ 90% sequence homology, and PCLN-1/CLDN16 mutations have been shown to be strongly associated with bovine chronic interstitial nephritis with diffuse zonal fibrosis (CINF) [155]. Although both renal disorders FHHN and CINF are caused by PCLN-1/CLDN16 mutations, but the clinical features of both diseases are quite different which may be due to specific mutations/deletions in the same gene or through species specificity. Since the first report [153], several other tight junction disorders have been shown to cause human diseases including mutations in claudin proteins such as CLDN-1 [156], CLDN-9 [157], CLDN-10 [158], CLDN-14 [159], CLDN-16 [160], CLDN-19 [161] (for detailed reviews see [162]).

7. Conclusions

It is clear from the literature that among all the tight junction proteins, the claudin family of proteins is particularly important in regulating normal cell physiology. Among the claudin family, CLDN-1 is the most extensively studied protein and has been shown to be involved directly or indirectly in the development and progression of cancer, and also has a suppressive role in some cancers. CLDN-1 acts alone or in combination with other molecules to exert its tumor promoting or suppressing effect. Likewise, the shuttling of CLDN-1 between the cell membrane, cytoplasm and nucleus is a deciding factor in the development and progression of cancers. Another important aspect is the involvement of CLDN-1 in many signaling pathways, especially in Wnt and Notch signaling. The association of CLDN-1 with patient survival or recurrence in many cancers suggests its importance as a prognostic marker and as a potential therapeutic target. Also, pathway interaction analysis revealed CLDN-1 interacting partners, which can be further explored as potential drug targets. Based on the complexity of the topic, there is no one statement we can make for CLDN-1 role in cancer or barrier function since it is more intricate (Claudins are upregulated or downregulated in cancer and may or may not play a role in barrier function). In other words, universal statements concerning CLDN-1 and cancer or CLDN-1 and barrier function are dangerous oversimplifications.

Author Contributions: Writing—original draft preparation, A.A.B., L.T., S.N., S.U., M.H.; writing—review and editing, S.H., M.A.M., P.D., W.E.-R., M.P.F., N.S., R.K.; visualization, S.M., H.A.-N., A.A.; supervision, A.A.B., M.H., W.E.-R., S.U., P.B., R.R.; project administration, A.A.B., N.S., S.K.Y., M.H., S.U.; funding acquisition, A.A.B., M.H. All authors have read and agreed to the published version of the manuscript.

Acknowledgments: The authors would like to express their gratitude to Vineeta Tanwar (Research Scientist, Ohio State University, Ohio, Columbus, USA) for help in English editing and valuable suggestions to improve the quality of the manuscript. The publication of this article was funded by the Qatar National Library.

Abbreviations

CLDN-1	Claudin-1 protein
CLDN-1	Claudin-1 gene
EMT	Epithelial-mesenchymal transition
CSCs	Cancer stem cells
HER2	Human epidermal growth factor receptor 2
BLBC	Basal-like breast cancer
ER	Estrogen receptor
ESCRT	Endosomal sorting complexes required for transport

EpCAM	Epithelial cell adhesion molecule
PTC	Papillary thyroid cancer
FTC	Follicular thyroid cancer
CRC	Colorectal cancer
mCRC	Metastatic colorectal cancer
UC	Ulcerative colitis
APC	Adenomatous polyposis coli
HNSCC	Head and neck squamous cell carcinoma
HSCC	Hypopharyngeal squamous cell carcinoma
HCC	Hepatocellular carcinoma
CTGF	Connective tissue growth factor
THBS1	Thrombospondin 1
DLC1	Deleted in liver cancer 1
OCLN	Occludin
ZO-1	Zona occludins 1
SPP1	Secreted phosphoprotein 1
CUTL1	Cut-like homeobox 1
TGF-α	Transforming growth factor-alpha
SLC2A3	Solute carrier family 2 (facilitated glucose transporter) member 3
PGF	Placental growth factor
EGFR	Epidermal growth factor
PC	Pancreatic cancer
MEK2	Mitogen activated protein kinase 2
miR-155	microRNA-155
BOT	Borderline tumors of the ovary
OSCC	Oral squamous cell carcinoma
MMP	Matrix metalloproteinase
BRCA	Breast invasive carcinoma
READ	Rectum adenocarcinoma
CESC	Cervical squamous cell carcinoma
THCA	Thyroid carcinoma
ACC	Adrenocortical carcinoma
PAAD	Pancreatic adenocarcinoma
mAbs	Monoclonal antibodies
cCPE	*Clostridium perfringens enterotoxin*
HCV	Hepatitis C virus
TCGA	The cancer genome atlas

References

1. Tsukita, S.; Yamazaki, Y.; Katsuno, T.; Tamura, A.; Tsukita, S. Tight junction-based epithelial microenvironment and cell proliferation. *Oncogene* **2008**, *27*, 6930–6938. [CrossRef]
2. Itoh, M.; Bissell, M.J. The organization of tight junctions in epithelia: Implications for mammary gland biology and breast tumorigenesis. *J. Mammary Gland Biol. Neoplasia* **2003**, *8*, 449–462. [CrossRef] [PubMed]
3. Tsukita, S.; Furuse, M.; Itoh, M. Multifunctional strands in tight junctions. *Nat. Rev. Mol. Cell Biol.* **2001**, *2*, 285–293. [CrossRef] [PubMed]
4. Krause, G.; Winkler, L.; Mueller, S.L.; Haseloff, R.F.; Piontek, J.; Blasig, I.E. Structure and function of claudins. *Biochim. Biophys. Acta* **2008**, *1778*, 631–645. [CrossRef] [PubMed]
5. Tsukita, S.; Furuse, M. Claudin-based barrier in simple and stratified cellular sheets. *Curr. Opin. Cell Biol.* **2002**, *14*, 531–536. [CrossRef]
6. Anderson, J.M.; Van Itallie, C.M. Physiology and function of the tight junction. *Cold Spring Harb. Perspect. Biol.* **2009**, *1*, a002584. [CrossRef]
7. Kirschner, N.; Rosenthal, R.; Furuse, M.; Moll, I.; Fromm, M.; Brandner, J.M. Contribution of tight junction proteins to ion, macromolecule, and water barrier in keratinocytes. *J. Investig. Derm.* **2013**, *133*, 1161–1169. [CrossRef]

8. Soler, A.P.; Miller, R.D.; Laughlin, K.V.; Carp, N.Z.; Klurfeld, D.M.; Mullin, J.M. Increased tight junctional permeability is associated with the development of colon cancer. *Carcinogenesis* **1999**, *20*, 1425–1431. [CrossRef]
9. Singh, A.B.; Sharma, A.; Dhawan, P. Claudin family of proteins and cancer: An overview. *J. Oncol.* **2010**, *2010*, 541957. [CrossRef]
10. Kwon, M.J. Emerging roles of claudins in human cancer. *Int. J. Mol. Sci.* **2013**, *14*, 18148–18180. [CrossRef]
11. Saraydaroglu, O.; Narter, S.; Ozsen, M.; Coskun, H. Non-epithelial tumors of the larynx: Case series of 12 years. *Eur. Arch. Oto-Rhino-Laryngol.* **2019**, *276*, 2843–2847. [CrossRef] [PubMed]
12. Reinert, S. Principles of Management of Neoplastic Salivary Gland Disease. *Maxillofacial Surgery*, 3rd ed. 2017, pp. 657–667. Available online: https://www.sciencedirect.com/book/9780702060564/maxillofacial-surgery (accessed on 13 January 2020).
13. Amasheh, S.; Milatz, S.; Krug, S.M.; Markov, A.G.; Gunzel, D.; Amasheh, M.; Fromm, M. Tight junction proteins as channel formers and barrier builders. *Ann. N. Y. Acad. Sci.* **2009**, *1165*, 211–219. [CrossRef] [PubMed]
14. Ikari, A.; Ito, M.; Okude, C.; Sawada, H.; Harada, H.; Degawa, M.; Sakai, H.; Takahashi, T.; Sugatani, J.; Miwa, M. Claudin-16 is directly phosphorylated by protein kinase a independently of a vasodilator-stimulated phosphoprotein-mediated pathway. *J. Cell Physiol.* **2008**, *214*, 221–229. [CrossRef] [PubMed]
15. Fujibe, M.; Chiba, H.; Kojima, T.; Soma, T.; Wada, T.; Yamashita, T.; Sawada, N. Thr203 of claudin-1, a putative phosphorylation site for map kinase, is required to promote the barrier function of tight junctions. *Exp. Cell Res.* **2004**, *295*, 36–47. [CrossRef] [PubMed]
16. Akizuki, R.; Shimobaba, S.; Matsunaga, T.; Endo, S.; Ikari, A. Claudin-5, -7, and -18 suppress proliferation mediated by inhibition of phosphorylation of akt in human lung squamous cell carcinoma. *Biochim. Biophys Acta* **2017**, *1864*, 293–302. [CrossRef] [PubMed]
17. Findley, M.K.; Koval, M. Regulation and roles for claudin-family tight junction proteins. *Iubmb. Life* **2009**, *61*, 431–437. [CrossRef]
18. Lal-Nag, M.; Morin, P.J. The claudins. *Genome Biol.* **2009**, *10*, 235. [CrossRef]
19. Hewitt, K.J.; Agarwal, R.; Morin, P.J. The claudin gene family: Expression in normal and neoplastic tissues. *BMC Cancer* **2006**, *6*, 186. [CrossRef]
20. Pope, J.L.; Bhat, A.A.; Sharma, A.; Ahmad, R.; Krishnan, M.; Washington, M.K.; Beauchamp, R.D.; Singh, A.B.; Dhawan, P. Claudin-1 regulates intestinal epithelial homeostasis through the modulation of notch-signalling. *Gut* **2014**, *63*, 622–634. [CrossRef]
21. Singh, A.B.; Uppada, S.B.; Dhawan, P. Claudin proteins, outside-in signaling, and carcinogenesis. *Pflug. Arch.* **2017**, *469*, 69–75. [CrossRef]
22. Turksen, K. Claudins and cancer stem cells. *Stem Cell Rev.* **2011**, *7*, 797–798. [CrossRef] [PubMed]
23. Zhou, B.; Flodby, P.; Luo, J.; Castillo, D.R.; Liu, Y.; Yu, F.X.; McConnell, A.; Varghese, B.; Li, G.; Chimge, N.O.; et al. Claudin-18-mediated yap activity regulates lung stem and progenitor cell homeostasis and tumorigenesis. *J. Clin. Investig.* **2018**, *128*, 970–984. [CrossRef] [PubMed]
24. Morin, P.J. Claudin proteins in human cancer: Promising new targets for diagnosis and therapy. *Cancer Res.* **2005**, *65*, 9603–9606. [CrossRef] [PubMed]
25. Zhou, B.; Moodie, A.; Blanchard, A.A.; Leygue, E.; Myal, Y. Claudin 1 in breast cancer: New insights. *J. Clin. Med.* **2015**, *4*, 1960–1976. [CrossRef]
26. Usami, Y.; Chiba, H.; Nakayama, F.; Ueda, J.; Matsuda, Y.; Sawada, N.; Komori, T.; Ito, A.; Yokozaki, H. Reduced expression of claudin-7 correlates with invasion and metastasis in squamous cell carcinoma of the esophagus. *Hum. Pathol.* **2006**, *37*, 569–577. [CrossRef]
27. Miyamoto, K.; Kusumi, T.; Sato, F.; Kawasaki, H.; Shibata, S.; Ohashi, M.; Hakamada, K.; Sasaki, M.; Kijima, H. Decreased expression of claudin-1 is correlated with recurrence status in esophageal squamous cell carcinoma. *Biomed. Res.* **2008**, *29*, 71–76. [CrossRef]
28. Honda, H.; Pazin, M.J.; Ji, H.; Wernyj, R.P.; Morin, P.J. Crucial roles of sp1 and epigenetic modifications in the regulation of the cldn4 promoter in ovarian cancer cells. *J. Biol. Chem.* **2006**, *281*, 21433–21444. [CrossRef]
29. Honda, H.; Pazin, M.J.; D'Souza, T.; Ji, H.; Morin, P.J. Regulation of the cldn3 gene in ovarian cancer cells. *Cancer Biol.* **2007** *6*, 1733–1742. [CrossRef]
30. Sheehan, G.M.; Kallakury, B.V.; Sheehan, C.E.; Fisher, H.A.; Kaufman, R.P., Jr.; Ross, J.S. Loss of claudins-1 and -7 and expression of claudins-3 and -4 correlate with prognostic variables in prostatic adenocarcinomas. *Hum. Pathol.* **2007**, *38*, 564–569. [CrossRef]

31. Resnick, M.B.; Konkin, T.; Routhier, J.; Sabo, E.; Pricolo, V.E. Claudin-1 is a strong prognostic indicator in stage ii colonic cancer: A tissue microarray study. *Mod. Pathol.* **2005**, *18*, 511–518. [CrossRef]
32. Furuse, M.; Hata, M.; Furuse, K.; Yoshida, Y.; Haratake, A.; Sugitani, Y.; Noda, T.; Kubo, A.; Tsukita, S. Claudin-based tight junctions are crucial for the mammalian epidermal barrier: A lesson from claudin-1-deficient mice. *J. Cell Biol.* **2002**, *156*, 1099–1111. [CrossRef] [PubMed]
33. Morita, K.; Furuse, M.; Fujimoto, K.; Tsukita, S. Claudin multigene family encoding four-transmembrane domain protein components of tight junction strands. *Proc. Natl Acad Sci USA* **1999**, *96*, 511–516. [CrossRef] [PubMed]
34. Dhawan, P.; Singh, A.B.; Deane, N.G.; No, Y.; Shiou, S.R.; Schmidt, C.; Neff, J.; Washington, M.K.; Beauchamp, R.D. Claudin-1 regulates cellular transformation and metastatic behavior in colon cancer. *J. Clin. Investig.* **2005**, *115*, 1765–1776. [CrossRef]
35. Oku, N.; Sasabe, E.; Ueta, E.; Yamamoto, T.; Osaki, T. Tight junction protein claudin-1 enhances the invasive activity of oral squamous cell carcinoma cells by promoting cleavage of laminin-5 gamma2 chain via matrix metalloproteinase (mmp)-2 and membrane-type mmp-1. *Cancer Res.* **2006**, *66*, 5251–5257. [CrossRef]
36. Tokes, A.M.; Kulka, J.; Paku, S.; Szik, A.; Paska, C.; Novak, P.K.; Szilak, L.; Kiss, A.; Bogi, K.; Schaff, Z. Claudin-1, -3 and -4 proteins and mrna expression in benign and malignant breast lesions: A research study. *Breast Cancer Res.* **2005**, *7*, R296–R305. [CrossRef] [PubMed]
37. Cohn, M.L.; Goncharuk, V.N.; Diwan, A.H.; Zhang, P.S.; Shen, S.S.; Prieto, V.G. Loss of claudin-1 expression in tumor-associated vessels correlates with acquisition of metastatic phenotype in melanocytic neoplasms. *J. Cutan. Pathol.* **2005**, *32*, 533–536. [CrossRef]
38. Leotlela, P.D.; Wade, M.S.; Duray, P.H.; Rhode, M.J.; Brown, H.F.; Rosenthal, D.T.; Dissanayake, S.K.; Earley, R.; Indig, F.E.; Nickoloff, B.J.; et al. Claudin-1 overexpression in melanoma is regulated by pkc and contributes to melanoma cell motility. *Oncogene* **2007**, *26*, 3846–3856. [CrossRef]
39. Morohashi, S.; Kusumi, T.; Sato, F.; Odagiri, H.; Chiba, H.; Yoshihara, S.; Hakamada, K.; Sasaki, M.; Kijima, H. Decreased expression of claudin-1 correlates with recurrence status in breast cancer. *Int. J. Mol. Med.* **2007**, *20*, 139–143. [CrossRef]
40. Blanchard, A.A.; Skliris, G.P.; Watson, P.H.; Murphy, L.C.; Penner, C.; Tomes, L.; Young, T.L.; Leygue, E.; Myal, Y. Claudins 1, 3, and 4 protein expression in er negative breast cancer correlates with markers of the basal phenotype. *Virchows Arch.* **2009**, *454*, 647–656. [CrossRef]
41. Lu, S.; Singh, K.; Mangray, S.; Tavares, R.; Noble, L.; Resnick, M.B.; Yakirevich, E. Claudin expression in high-grade invasive ductal carcinoma of the breast: Correlation with the molecular subtype. *Mod. Pathol.* **2013**, *26*, 485–495. [CrossRef]
42. Nemeth, J.; Nemeth, Z.; Tatrai, P.; Peter, I.; Somoracz, A.; Szasz, A.M.; Kiss, A.; Schaff, Z. High expression of claudin-1 protein in papillary thyroid tumor and its regional lymph node metastasis. *Pathol. Oncol. Res.* **2010**, *16*, 19–27. [CrossRef] [PubMed]
43. Zwanziger, D.; Badziong, J.; Ting, S.; Moeller, L.; Schmid, K.; Siebolts, U.; Wickenhauser, C.; Dralle, H.; Fuhrer, D. The impact of claudin-1 on follicular thyroid carcinoma aggressiveness. *Endocr. Relat. Cancer* **2015**, *22*, 819–830. [CrossRef] [PubMed]
44. Kinugasa, T.; Akagi, Y.; Yoshida, T.; Ryu, Y.; Shiratuchi, I.; Ishibashi, N.; Shirouzu, K. Increased claudin-1 protein expression contributes to tumorigenesis in ulcerative colitis-associated colorectal cancer. *Anticancer Res.* **2010**, *30*, 3181–3186. [PubMed]
45. Wu, Y.-L.; Zhang, S.; Wang, G.-R.; Chen, Y.-P. Expression transformation of claudin-1 in the process of gastric adenocarcinoma invasion. *World J. Gastroenterol.* **2008**, *14*, 4943–4948. [CrossRef]
46. Nelhubel, G.A.; Karoly, B.; Szabo, B.; Lotz, G.; Kiss, A.; Tovari, J.; Kenessey, I. The prognostic role of claudins in head and neck squamous cell carcinomas. *Pathol. Oncol. Res.* **2014**, *20*, 99–106. [CrossRef]
47. Li, W.J.; Zhang, Z.L.; Yu, X.M.; Cai, X.L.; Pan, X.L.; Yang, X.Y. Expression of claudin-1 and its relationship with lymphatic microvessel generation in hypopharyngeal squamous cell carcinoma. *Genet. Mol. Res.* **2015**, *14*, 11814–11826. [CrossRef]
48. Holczbauer, A.; Gyongyosi, B.; Lotz, G.; Szijarto, A.; Kupcsulik, P.; Schaff, Z.; Kiss, A. Distinct claudin expression profiles of hepatocellular carcinoma and metastatic colorectal and pancreatic carcinomas. *J. Histochem. Cytochem.* **2013** *61*, 294–305. [CrossRef]
49. Higashi, Y.; Suzuki, S.; Sakaguchi, T.; Nakamura, T.; Baba, S.; Reinecker, H.C.; Nakamura, S.; Konno, H. Loss of claudin-1 expression correlates with malignancy of hepatocellular carcinoma. *J. Surg. Res.* **2007**, *139*, 68–76. [CrossRef]

50. Paschoud, S.; Bongiovanni, M.; Pache, J.-C.; Citi, S. Claudin-1 and claudin-5 expression patterns differentiate lung squamous cell carcinomas from adenocarcinomas. *Mod. Pathol.* **2007**, *20*, 947–954. [CrossRef]
51. Tsukahara, M.; Nagai, H.; Kamiakito, T.; Kawata, H.; Takayashiki, N.; Saito, K.; Tanaka, A. Distinct expression patterns of claudin-1 and claudin-4 in intraductal papillary–mucinous tumors of the pancreas. *Pathol. Int.* **2005**, *55*, 63–69. [CrossRef]
52. English, D.P.; Santin, A.D. Claudins overexpression in ovarian cancer: Potential targets for clostridium perfringens enterotoxin (cpe) based diagnosis and therapy. *Int. J. Mol. Sci.* **2013**, *14*, 10412–10437. [CrossRef] [PubMed]
53. El-Balat, A.; Schmeil, I.; Gasimli, K.; Sänger, N.; Karn, T.; Ahr, A.; Becker, S.; Arsenic, R.; Holtrich, U.; Engels, K. Claudin-1 is linked to presence of implants and micropapillary pattern in serous borderline epithelial tumours of the ovary. *J. Clin. Pathol.* **2018**, *71*, 1060. [CrossRef]
54. Seo, K.W.; Kwon, Y.K.; Kim, B.H.; Kim, C.I.; Chang, H.S.; Choe, M.S.; Park, C.H. Correlation between claudins expression and prognostic factors in prostate cancer. *Korean J. Urol.* **2010**, *51*, 239–244. [CrossRef] [PubMed]
55. Väre, P.; Loikkanen, I.; Hirvikoski, P.; Vaarala, M.; Soini, Y. Low claudin expression is associated with high gleason grade in prostate adenocarcinoma. *Oncol. Rep.* **2008**, *19*, 25–31. [CrossRef] [PubMed]
56. French, A.D.; Fiori, J.L.; Camilli, T.C.; Leotlela, P.D.; O'Connell, M.P.; Frank, B.P.; Subaran, S.; Indig, F.E.; Taub, D.D.; Weeraratna, A.T. Pkc and pka phosphorylation affect the subcellular localization of claudin-1 in melanoma cells. *Int. J. Med. Sci.* **2009**, *6*, 93–101. [CrossRef] [PubMed]
57. Chao, Y.-C.; Pan, S.-H.; Yang, S.-C.; Yu, S.-L.; Che, T.-F.; Lin, C.-W.; Tsai, M.-S.; Chang, G.-C.; Wu, C.-H.; Wu, Y.-Y.; et al. Claudin-1 is a metastasis suppressor and correlates with clinical outcome in lung adenocarcinoma. *Am. J. Respir. Crit. Care Med.* **2009**, *179*, 123–133. [CrossRef] [PubMed]
58. Kleinberg, L.; Holth, A.; Trope, C.G.; Reich, R.; Davidson, B. Claudin upregulation in ovarian carcinoma effusions is associated with poor survival. *Hum. Pathol.* **2008**, *39*, 747–757. [CrossRef] [PubMed]
59. Singh, A.B.; Sharma, A.; Smith, J.J.; Krishnan, M.; Chen, X.; Eschrich, S.; Washington, M.K.; Yeatman, T.J.; Beauchamp, R.D.; Dhawan, P. Claudin-1 up-regulates the repressor zeb-1 to inhibit e-cadherin expression in colon cancer cells. *Gastroenterology* **2011**, *141*, 2140–2153. [CrossRef]
60. Eftang, L.L.; Esbensen, Y.; Tannaes, T.M.; Blom, G.P.; Bukholm, I.R.; Bukholm, G. Up-regulation of cldn1 in gastric cancer is correlated with reduced survival. *BMC Cancer* **2013**, *13*, 586. [CrossRef]
61. Li, W.; Dong, Q.; Li, L.; Zhang, Z.; Cai, X.; Pan, X. Prognostic significance of claudin-1 and cyclin b1 protein expression in patients with hypopharyngeal squamous cell carcinoma. *Oncol. Lett.* **2016**, *11*, 2995–3002. [CrossRef]
62. Suh, Y.; Yoon, C.H.; Kim, R.K.; Lim, E.J.; Oh, Y.S.; Hwang, S.G.; An, S.; Yoon, G.; Gye, M.C.; Yi, J.M.; et al. Claudin-1 induces epithelial–mesenchymal transition through activation of the c-abl-erk signaling pathway in human liver cells. *Oncogene* **2013**, *32*, 4873–4882. [CrossRef] [PubMed]
63. Stebbing, J.; Filipovic, A.; Giamas, G. Claudin-1 as a promoter of emt in hepatocellular carcinoma. *Oncogene* **2013**, *32*, 4871–4872. [CrossRef] [PubMed]
64. Kondo, J.; Sato, F.; Kusumi, T.; Liu, Y.; Motonari, O.; Sato, T.; Kijima, H. Claudin-1 expression is induced by tumor necrosis factor-alpha in human pancreatic cancer cells. *Int. J. Mol. Med.* **2008**, *22*, 645–649. [PubMed]
65. Cronin, K.A.; Lake, A.J.; Scott, S.; Sherman, R.L.; Noone, A.M.; Howlader, N.; Henley, S.J.; Anderson, R.N.; Firth, A.U.; Ma, J.; et al. Annual report to the nation on the status of cancer, part i: National cancer statistics. *Cancer* **2018**, *124*, 2785–2800. [CrossRef] [PubMed]
66. Sinn, H.P.; Kreipe, H. A brief overview of the who classification of breast tumors, 4th edition, focusing on issues and updates from the 3rd edition. *Breast Care (Basel)* **2013**, *8*, 149–154. [CrossRef]
67. Dai, X.; Li, T.; Bai, Z.; Yang, Y.; Liu, X.; Zhan, J.; Shi, B. Breast cancer intrinsic subtype classification, clinical use and future trends. *Am. J. Cancer Res.* **2015**, *5*, 2929–2943.
68. Yeo, S.K.; Guan, J.L. Breast cancer: Multiple subtypes within a tumor? *Trends Cancer* **2017**, *3*, 753–760. [CrossRef]
69. Dai, X.; Xiang, L.; Li, T.; Bai, Z. Cancer hallmarks, biomarkers and breast cancer molecular subtypes. *J. Cancer* **2016**, *7*, 1281–1294. [CrossRef]
70. Blanchard, A.; Shiu, R.; Booth, S.; Sorensen, G.; DeCorby, N.; Nistor, A.; Wong, P.; Leygue, E.; Myal, Y. Gene expression profiling of early involuting mammary gland reveals novel genes potentially relevant to human breast cancer. *Front. Biosci.* **2007** *12*, 2221–2232. [CrossRef]
71. Ricardo, S.; Gerhard, R.; Cameselle-Teijeiro, J.F.; Schmitt, F.; Paredes, J. Claudin expression in breast cancer: High or low, what to expect? *Histol. Histopathol.* **2012**, *27*, 1283–1295.

72. Blanchard, A.A.; Zelinski, T.; Xie, J.; Cooper, S.; Penner, C.; Leygue, E.; Myal, Y. Identification of claudin 1 transcript variants in human invasive breast cancer. *PLoS ONE* **2016**, *11*, e0163387. [CrossRef] [PubMed]
73. Swisshelm, K.; Machl, A.; Planitzer, S.; Robertson, R.; Kubbies, M.; Hosier, S. Semp1, a senescence-associated cdna isolated from human mammary epithelial cells, is a member of an epithelial membrane protein superfamily. *Gene* **1999**, *226*, 285–295. [CrossRef]
74. Kramer, F.; White, K.; Kubbies, M.; Swisshelm, K.; Weber, B.H. Genomic organization of claudin-1 and its assessment in hereditary and sporadic breast cancer. *Hum. Genet.* **2000**, *107*, 249–256. [CrossRef] [PubMed]
75. Hoevel, T.; Macek, R.; Swisshelm, K.; Kubbies, M. Reexpression of the tj protein cldn1 induces apoptosis in breast tumor spheroids. *Int. J. Cancer* **2004**, *108*, 374–383. [CrossRef]
76. Zhou, B.; Blanchard, A.; Wang, N.; Ma, X.; Han, J.; Schroedter, I.; Leygue, E.; Myal, Y. Claudin 1 promotes migration and increases sensitivity to tamoxifen and anticancer drugs in luminal-like human breast cancer cells mcf7. *Cancer Investig.* **2015**, 429–439. [CrossRef]
77. Martinez-Estrada, O.M.; Culleres, A.; Soriano, F.X.; Peinado, H.; Bolos, V.; Martinez, F.O.; Reina, M.; Cano, A.; Fabre, M.; Vilaro, S. The transcription factors slug and snail act as repressors of claudin-1 expression in epithelial cells. *Biochem. J.* **2006**, *394*, 449–457. [CrossRef]
78. Myal, Y.; Leygue, E.; Blanchard, A.A. Claudin 1 in breast tumorigenesis: Revelation of a possible novel "claudin high" subset of breast cancers. *J. Biomed. Biotechnol.* **2010**, *2010*, 956897. [CrossRef]
79. Akasaka, H.; Sato, F.; Morohashi, S.; Wu, Y.; Liu, Y.; Kondo, J.; Odagiri, H.; Hakamada, K.; Kijima, H. Anti-apoptotic effect of claudin-1 in tamoxifen-treated human breast cancer mcf-7 cells. *BMC Cancer* **2010**, *10*, 548. [CrossRef]
80. Tanaka, M.; Kamata, R.; Sakai, R. Phosphorylation of ephrin-b1 via the interaction with claudin following cell-cell contact formation. *Embo J.* **2005**, *24*, 3700–3711. [CrossRef]
81. Schmidt, O.; Teis, D. The escrt machinery. *Curr. Biol.* **2012**, *22*, R116–R120. [CrossRef]
82. Kovalenko, O.V.; Yang, X.H.; Hemler, M.E. A novel cysteine cross-linking method reveals a direct association between claudin-1 and tetraspanin cd9. *Mol. Cell Proteom.* **2007**, *6*, 1855–1867. [CrossRef] [PubMed]
83. Wu, C.J.; Mannan, P.; Lu, M.; Udey, M.C. Epithelial cell adhesion molecule (epcam) regulates claudin dynamics and tight junctions. *J. Biol. Chem.* **2013**, *288*, 12253–12268. [CrossRef] [PubMed]
84. Kondo, T.; Ezzat, S.; Asa, S.L. Pathogenetic mechanisms in thyroid follicular-cell neoplasia. *Nat. Rev. Cancer* **2006**, *6*, 292–306. [CrossRef] [PubMed]
85. Kitahara, C.M.; Sosa, J.A. The changing incidence of thyroid cancer. *Nat. Rev. Endocrinol.* **2016**, *12*, 646–653. [CrossRef]
86. Hucz, J.; Kowalska, M.; Jarzab, M.; Wiench, M. Gene expression of metalloproteinase 11, claudin 1 and selected adhesion related genes in papillary thyroid cancer. *Endokrynol. Pol.* **2006**, *57*, 18–25.
87. Fluge, O.; Bruland, O.; Akslen, L.A.; Lillehaug, J.R.; Varhaug, J.E. Gene expression in poorly differentiated papillary thyroid carcinomas. *Thyroid* **2006**, *16*, 161–175. [CrossRef]
88. Sobel, G.; Nemeth, J.; Kiss, A.; Lotz, G.; Szabo, I.; Udvarhelyi, N.; Schaff, Z.; Paska, C. Claudin 1 differentiates endometrioid and serous papillary endometrial adenocarcinoma. *Gynecol. Oncol.* **2006**, *103*, 591–598. [CrossRef]
89. Tzelepi, V.N.; Tsamandas, A.C.; Vlotinou, H.D.; Vagianos, C.E.; Scopa, C.D. Tight junctions in thyroid carcinogenesis: Diverse expression of claudin-1, claudin-4, claudin-7 and occludin in thyroid neoplasms. *Mod. Pathol.* **2008**, *21*, 22–30. [CrossRef]
90. Süren, D.; Yildirim, M.; Sayiner, A.; Alikanoğlu, A.S.; Atalay, I.; Gündüz, U.R.; Kaya, V.; Gündüz, Ş.; Oruç, M.T.; Sezer, C. Expression of claudin 1, 4 and 7 in thyroid neoplasms. *Oncol. Lett.* **2017**, *13*, 3722–3726. [CrossRef]
91. Abd El Atti, R.M.; Shash, L.S. Potential diagnostic utility of cd56 and claudin-1 in papillary thyroid carcinoma and solitary follicular thyroid nodules. *J. Egypt. Natl. Cancer Inst.* **2012**, *24*, 175–184. [CrossRef]
92. Arnold, M.; Sierra, M.S.; Laversanne, M.; Soerjomataram, I.; Jemal, A.; Bray, F. Global patterns and trends in colorectal cancer incidence and mortality. *Gut* **2017**, *66*, 683. [CrossRef] [PubMed]
93. Ouban, A. Claudin-1 role in colon cancer: An update and a review. *Histol. Histopathol.* **2018**, 11980.
94. Pope, J.L.; Ahmad, R.; Bhat, A.A.; Washington, M.K.; Singh, A.B.; Dhawan, P. Claudin-1 overexpression in intestinal epithelial cells enhances susceptibility to adenamatous polyposis coli-mediated colon tumorigenesis. *Mol. Cancer* **2014**, *13*, 167. [CrossRef] [PubMed]
95. Grone, J.; Weber, B.; Staub, E.; Heinze, M.; Klaman, I.; Pilarsky, C.; Hermann, K.; Castanos-Velez, E.; Ropcke, S.; Mann, B.; et al. Differential expression of genes encoding tight junction proteins in colorectal cancer: Frequent

dysregulation of claudin-1, -8 and -12. *Int. J. Colorectal Dis.* **2007**, *22*, 651–659. [CrossRef] [PubMed]
96. Kinugasa, T.; Huo, Q.; Higashi, D.; Shibaguchi, H.; Kuroki, M.; Tanaka, T.; Futami, K.; Yamashita, Y.; Hachimine, K.; Maekawa, S.; et al. Selective up-regulation of claudin-1 and claudin-2 in colorectal cancer. *Anticancer Res.* **2007**, *27*, 3729–3734. [CrossRef]
97. de Oliveira, S.S.; de Oliveira, I.M.; De Souza, W.; Morgado-Diaz, J.A. Claudins upregulation in human colorectal cancer. *FEBS Lett.* **2005**, *579*, 6179–6185. [CrossRef]
98. Nakagawa, S.; Miyoshi, N.; Ishii, H.; Mimori, K.; Tanaka, F.; Sekimoto, M.; Doki, Y.; Mori, M. Expression of cldn1 in colorectal cancer: A novel marker for prognosis. *Int. J. Oncol.* **2011**, *39*, 791–796.
99. Huo, Q.; Kinugasa, T.; Wang, L.; Huang, J.; Zhao, J.; Shibaguchi, H.; Kuroki, M.; Tanaka, T.; Yamashita, Y.; Nabeshima, K.; et al. Claudin-1 protein is a major factor involved in the tumorigenesis of colorectal cancer. *Anticancer Res.* **2009**, *29*, 851–857.
100. Wang, K.; Xu, C.; Li, W.; Ding, L. Emerging clinical significance of claudin-7 in colorectal cancer: A review. *Cancer Manag. Res.* **2018**, *10*, 3741–3752. [CrossRef]
101. Caruso, M.; Fung, K.Y.; Moore, J.; Brierley, G.V.; Cosgrove, L.J.; Thomas, M.; Cheetham, G.; Brook, E.; Fraser, L.M.; Tin, T.; et al. Claudin-1 expression is elevated in colorectal cancer precursor lesions harboring the braf v600e mutation. *Transl. Oncol.* **2014**, *7*, 456–463. [CrossRef]
102. Schatoff, E.M.; Leach, B.I.; Dow, L.E. Wnt signaling and colorectal cancer. *Curr. Colorectal Cancer Rep.* **2017**, *13*, 101–110. [CrossRef] [PubMed]
103. Katoh, M. Multilayered prevention and treatment of chronic inflammation, organ fibrosis and cancer associated with canonical wnt/betacatenin signaling activation (review). *Int. J. Mol. Med.* **2018**, 713–725. [CrossRef]
104. Krishnamurthy, N.; Kurzrock, R. Targeting the wnt/beta-catenin pathway in cancer: Update on effectors and inhibitors. *Cancer Treat. Rev.* **2018**, *62*, 50–60. [CrossRef] [PubMed]
105. Miwa, N.; Furuse, M.; Tsukita, S.; Niikawa, N.; Nakamura, Y.; Furukawa, Y. Involvement of claudin-1 in the beta-catenin/tcf signaling pathway and its frequent upregulation in human colorectal cancers. *Oncol. Res.* **2001**, *12*, 469–476. [CrossRef]
106. Bhat, A.A.; Sharma, A.; Pope, J.; Krishnan, M.; Washington, M.K.; Singh, A.B.; Dhawan, P. Caudal homeobox protein cdx-2 cooperates with wnt pathway to regulate claudin-1 expression in colon cancer cells. *PLoS ONE* **2012**, *7*, e37174. [CrossRef]
107. Cherradi, S.; Ayrolles-Torro, A.; Vezzo-Vié, N.; Gueguinou, N.; Denis, V.; Combes, E.; Boissière, F.; Busson, M.; Canterel-Thouennon, L.; Mollevi, C.; et al. Antibody targeting of claudin-1 as a potential colorectal cancer therapy. *J. Exp. Clin. Cancer Res.* **2017**, *36*, 89. [CrossRef]
108. Jung, H.; Jun, K.H.; Jung, J.H.; Chin, H.M.; Park, W.B. The expression of claudin-1, claudin-2, claudin-3, and claudin-4 in gastric cancer tissue. *J. Surg. Res.* **2011**, *167*, e185–e191. [CrossRef]
109. Huang, J.; Li, J.; Qu, Y.; Zhang, J.; Zhang, L.; Chen, X.; Liu, B.; Zhu, Z. The expression of claudin 1 correlates with beta-catenin and is a prognostic factor of poor outcome in gastric cancer. *Int. J. Oncol.* **2014**, *44*, 1293–1301. [CrossRef]
110. Huang, J.; Zhang, L.; He, C.; Qu, Y.; Li, J.; Zhang, J.; Du, T.; Chen, X.; Yu, Y.; Liu, B.; et al. Claudin-1 enhances tumor proliferation and metastasis by regulating cell anoikis in gastric cancer. *Oncotarget* **2015**, *6*, 1652–1665. [CrossRef]
111. Tokuhara, Y.; Morinishi, T.; Matsunaga, T.; Ohsaki, H.; Kushida, Y.; Haba, R.; Hirakawa, E. Claudin-1, but not claudin-4, exhibits differential expression patterns between well- to moderately-differentiated and poorly-differentiated gastric adenocarcinoma. *Oncol. Lett.* **2015**, *10*, 93–98. [CrossRef]
112. Parkin, D.M.; Bray, F.; Ferlay, J.; Pisani, P. Global cancer statistics, 2002. *CA Cancer J. Clin.* **2005**, *55*, 74–108. [CrossRef] [PubMed]
113. Ouban, A.; Hamdan, H.; Hakam, A.; Ahmed, A.A. Claudin-1 expression in squamous cell carcinomas of different organs: Comparative study of cancerous tissues and normal controls. *Int. J. Surg. Pathol.* **2012**, *20*, 132–138. [CrossRef] [PubMed]
114. Ghouri, Y.A.; Mian, I.; Rowe, J.H. Review of hepatocellular carcinoma: Epidemiology, etiology, and carcinogenesis. *J. Carcinog.* **2017**, *16*, 1–18. [PubMed]
115. Yoon, C.-H.; Kim, M.-J.; Park, M.-J.; Park, I.-C.; Hwang, S.-G.; An, S.; Choi, Y.-H.; Yoon, G.; Lee, S.-J. Claudin-1 acts through c-abl-protein kinase cdelta (pkcdelta) signaling and has a causal role in the acquisition of invasive capacity in human liver cells. *J. Biol. Chem.* **2010** *285*, 226–233. [CrossRef] [PubMed]

116. Chen, Y.-J.; You, M.-L.; Chong, Q.-Y.; Pandey, V.; Zhuang, Q.-S.; Liu, D.-X.; Ma, L.; Zhu, T.; Lobie, P.E. Autocrine human growth hormone promotes invasive and cancer stem cell-like behavior of hepatocellular carcinoma cells by stat3 dependent inhibition of claudin-1 expression. *Int. J. Mol. Sci.* **2017**, *18*, 1274. [CrossRef] [PubMed]
117. Jung, J.H.; Jung, C.K.; Choi, H.J.; Jun, K.H.; Yoo, J.; Kang, S.J.; Lee, K.Y. Diagnostic utility of expression of claudins in non-small cell lung cancer: Different expression profiles in squamous cell carcinomas and adenocarcinomas. *Pathol. Res. Pract.* **2009**, *205*, 409–416. [CrossRef]
118. Soini, Y. Claudins in lung diseases. *Respir. Res.* **2011**, *12*, 70. [CrossRef]
119. Sun, B.-S.; Yao, Y.-Q.; Pei, B.-X.; Zhang, Z.-F.; Wang, C.-L. Claudin-1 correlates with poor prognosis in lung adenocarcinoma. *Thorac. Cancer* **2016**, *7*, 556–563. [CrossRef]
120. Landskron, G.; De la Fuente, M.; Thuwajit, P.; Thuwajit, C.; Hermoso, M.A. Chronic inflammation and cytokines in the tumor microenvironment. *J. Immunol. Res.* **2014**, *2014*, 149185. [CrossRef]
121. Shiozaki, A.; Bai, X.H.; Shen-Tu, G.; Moodley, S.; Takeshita, H.; Fung, S.Y.; Wang, Y.; Keshavjee, S.; Liu, M. Claudin 1 mediates tnfalpha-induced gene expression and cell migration in human lung carcinoma cells. *PLoS ONE* **2012**, *7*, e38049. [CrossRef]
122. Bhat, A.A.; Ahmad, R.; Uppada, S.B.; Singh, A.B.; Dhawan, P. Claudin-1 promotes tnf-α-induced epithelial-mesenchymal transition and migration in colorectal adenocarcinoma cells. *Exp. Cell Res.* **2016**, *349*, 119–127. [CrossRef] [PubMed]
123. Jemal, A.; Siegel, R.; Xu, J.; Ward, E. Cancer statistics, 2010. *CA Cancer J. Clin.* **2010**, *60*, 277–300. [CrossRef] [PubMed]
124. Tan, C.; Cruet-Hennequart, S.; Troussard, A.; Fazli, L.; Costello, P.; Sutton, K.; Wheeler, J.; Gleave, M.; Sanghera, J.; Dedhar, S. Regulation of tumor angiogenesis by integrin-linked kinase (ilk). *Cancer Cell* **2004**, *5*, 79–90. [CrossRef]
125. Reid, B.M.; Permuth, J.B.; Sellers, T.A. Epidemiology of ovarian cancer: A review. *Cancer Biol. Med.* **2017**, *14*, 9–32. [PubMed]
126. Liu, T.; Cheng, W.; Lai, D.; Huang, Y.; Guo, L. Characterization of primary ovarian cancer cells in different culture systems. *Oncol. Rep.* **2010**, *23*, 1277–1284. [PubMed]
127. Zhu, Y.; Sundfeldt, K. Tight junction formation in epithelial ovarian adenocarcinoma. *Acta Obs. Gynecol. Scand.* **2007**, *86*, 1011–1019. [CrossRef]
128. Tandon, P.; Dadhich, A.; Saluja, H.; Bawane, S.; Sachdeva, S. The prevalence of squamous cell carcinoma in different sites of oral cavity at our rural health care centre in loni, maharashtra—A retrospective 10-year study. *Contemp. Oncol.* **2017**, *21*, 178–183. [CrossRef]
129. Habelhah, H.; Okada, F.; Kobayashi, M.; Nakai, K.; Choi, S.; Hamada, J.; Moriuchi, T.; Kaya, M.; Yoshida, K.; Fujinaga, K.; et al. Increased e1af expression in mouse fibrosarcoma promotes metastasis through induction of mt1-mmp expression. *Oncogene* **1999**, *18*, 1771–1776. [CrossRef]
130. Sappayatosok, K.; Phattarataratip, E. Overexpression of claudin-1 is associated with advanced clinical stage and invasive pathologic characteristics of oral squamous cell carcinoma. *Head Neck Pathol.* **2015**, *9*, 173–180. [CrossRef]
131. Lourenco, S.V.; Coutinho-Camillo, C.M.; Buim, M.E.; Pereira, C.M.; Carvalho, A.L.; Kowalski, L.P.; Soares, F.A. Oral squamous cell carcinoma: Status of tight junction claudins in the different histopathological patterns and relationship with clinical parameters. A tissue-microarray-based study of 136 cases. *J. Clin. Pathol.* **2010**, *63*, 609–614. [CrossRef]
132. De Vicente, J.C.; Fernández-Valle, Á.; Vivanco-Allende, B.; Santamarta, T.R.; Lequerica-Fernández, P.; Hernández-Vallejo, G.; Allonca-Campa, E. The prognostic role of claudins -1 and -4 in oral squamous cell carcinoma. *Anticancer Res.* **2015**, *35*, 2949–2959. [PubMed]
133. Jerant, A.F.; Johnson, J.T.; Sheridan, C.D.; Caffrey, T.J. Early detection and treatment of skin cancer. *Am. Fam. Physician* **2000**, *62*, 357–368, 375–376, 381–382.
134. Weeraratna, A.T.; Becker, D.; Carr, K.M.; Duray, P.H.; Rosenblatt, K.P.; Yang, S.; Chen, Y.; Bittner, M.; Strausberg, R.L.; Riggins, G.J.; et al. Generation and analysis of melanoma sage libraries: Sage advice on the melanoma transcriptome. *Oncogene* **2004**, *23*, 2264–2274. [CrossRef] [PubMed]
135. Izraely, S.; Sagi-Assif, O.; Klein, A.; Meshel, T.; Ben-Menachem, S.; Zaritsky, A.; Ehrlich, M.; Prieto, V.G.; Bar-Eli, M.; Pirker, C.; et al. The metastatic microenvironment: Claudin-1 suppresses the malignant phenotype of melanoma brain metastasis. *Int. J. Cancer* **2015**, *136*, 1296–1307. [CrossRef]

136. Wong, M.C.S.; Goggins, W.B.; Wang, H.H.X.; Fung, F.D.H.; Leung, C.; Wong, S.Y.S.; Ng, C.F.; Sung, J.J.Y. Global incidence and mortality for prostate cancer: Analysis of temporal patterns and trends in 36 countries. *Eur. Urol.* **2016**, *70*, 862–874. [CrossRef]
137. Sugawara, T.; Iwamoto, N.; Akashi, M.; Kojima, T.; Hisatsune, J.; Sugai, M.; Furuse, M. Tight junction dysfunction in the stratum granulosum leads to aberrant stratum corneum barrier function in claudin-1-deficient mice. *J. Dermatol. Sci.* **2013**, *70*, 12–18. [CrossRef]
138. Inai, T.; Kobayashi, J.; Shibata, Y. Claudin-1 contributes to the epithelial barrier function in mdck cells. *Eur. J. Cell Biol.* **1999**, *78*, 849–855. [CrossRef]
139. McCarthy, K.M.; Francis, S.A.; McCormack, J.M.; Lai, J.; Rogers, R.A.; Skare, I.B.; Lynch, R.D.; Schneeberger, E.E. Inducible expression of claudin-1-myc but not occludin-vsv-g results in aberrant tight junction strand formation in mdck cells. *J. Cell Sci.* **2000**, *113*, 3387.
140. Grosse, B.; Cassio, D.; Yousef, N.; Bernardo, C.; Jacquemin, E.; Gonzales, E. Claudin-1 involved in neonatal ichthyosis sclerosing cholangitis syndrome regulates hepatic paracellular permeability. *Hepatology* **2012**, *55*, 1249–1259. [CrossRef]
141. Poritz, L.S.; Harris, L.R., 3rd; Kelly, A.A.; Koltun, W.A. Increase in the tight junction protein claudin-1 in intestinal inflammation. *Dig. Dis. Sci.* **2011**, *56*, 2802–2809. [CrossRef]
142. Sladojevic, N.; Stamatovic, S.M.; Johnson, A.M.; Choi, J.; Hu, A.; Dithmer, S.; Blasig, I.E.; Keep, R.F.; Andjelkovic, A.V. Claudin-1-dependent destabilization of the blood–brain barrier in chronic stroke. *J. Neurosci.* **2019**, *39*, 743. [CrossRef] [PubMed]
143. Hashimoto, Y.; Fukasawa, M.; Kuniyasu, H.; Yagi, K.; Kondoh, M. Claudin-targeted drug development using anti-claudin monoclonal antibodies to treat hepatitis and cancer. *Ann. N. Y. Acad. Sci.* **2017**, *1397*, 5–16. [CrossRef] [PubMed]
144. Hashimoto, Y.; Tada, M.; Iida, M.; Nagase, S.; Hata, T.; Watari, A.; Okada, Y.; Doi, T.; Fukasawa, M.; Yagi, K.; et al. Generation and characterization of a human–mouse chimeric antibody against the extracellular domain of claudin-1 for cancer therapy using a mouse model. *Biochem. Biophys. Res. Commun.* **2016**, *477*, 91–95. [CrossRef] [PubMed]
145. Nordin, B.E.; Robertson, A.; Seamark, R.F.; Bridges, A.; Philcox, J.C.; Need, A.G.; Horowitz, M.; Morris, H.A.; Deam, S. The relation between calcium absorption, serum dehydroepiandrosterone, and vertebral mineral density in postmenopausal women. *J. Clin. Endocrinol. Metab.* **1985**, *60*, 651–657. [CrossRef]
146. Fukasawa, M.; Nagase, S.; Shirasago, Y.; Iida, M.; Yamashita, M.; Endo, K.; Yagi, K.; Suzuki, T.; Wakita, T.; Hanada, K.; et al. Monoclonal antibodies against extracellular domains of claudin-1 block hepatitis c virus infection in a mouse model. *J. Virol.* **2015**, *89*, 4866–4879. [CrossRef] [PubMed]
147. Hollandsworth, H.M.; Lwin, T.M.; Amirfakhri, S.; Filemoni, F.; Batra, S.K.; Hoffman, R.M.; Dhawan, P.; Bouvet, M. Anti-claudin-1 conjugated to a near-infrared fluorophore targets colon cancer in pdox mouse models. *J. Surg. Res.* **2019**, *242*, 145–150. [CrossRef]
148. Kominsky, S.L. Claudins: Emerging targets for cancer therapy. *Expert Rev. Mol. Med.* **2006**, *8*, 1–11. [CrossRef]
149. Saeki, R.; Kondoh, M.; Kakutani, H.; Tsunoda, S.-i.; Mochizuki, Y.; Hamakubo, T.; Tsutsumi, Y.; Horiguchi, Y.; Yagi, K. A novel tumor-targeted therapy using a claudin-4-targeting molecule. *Mol. Pharmacol.* **2009**, *76*, 918. [CrossRef]
150. Hashimoto, Y.; Kawahigashi, Y.; Hata, T.; Li, X.; Watari, A.; Tada, M.; Ishii-Watabe, A.; Okada, Y.; Doi, T.; Fukasawa, M.; et al. Efficacy and safety evaluation of claudin-4-targeted antitumor therapy using a human and mouse cross-reactive monoclonal antibody. *Pharmacol. Res. Perspect.* **2016**, *4*, e00266. [CrossRef]
151. Torres, J.B.; Knight, J.C.; Mosley, M.J.; Kersemans, V.; Koustoulidou, S.; Allen, D.; Kinchesh, P.; Smart, S.; Cornelissen, B. Imaging of claudin-4 in pancreatic ductal adenocarcinoma using a radiolabelled anti-claudin-4 monoclonal antibody. *Mol. Imaging Biol.* **2018**, *20*, 292–299. [CrossRef]
152. Hashimoto, Y.; Tachibana, K.; Krug, S.M.; Kunisawa, J.; Fromm, M.; Kondoh, M. Potential for tight junction protein-directed drug development using claudin binders and angubindin-1. *Int. J. Mol. Sci.* **2019**, *20*, 4016. [CrossRef] [PubMed]
153. Simon, D.B.; Lu, Y.; Choate, K.A.; Velazquez, H.; Al-Sabban, E.; Praga, M.; Casari, G.; Bettinelli, A.; Colussi, G.; Rodriguez-Soriano, J.; et al. Paracellin-1, a renal tight junction protein required for paracellular Mg2+ resorption. *Science* **1999**, *285*, 103–106. [CrossRef] [PubMed]
154. Sanjad, S.A.; Hariri, A.; Habbal, Z.M.; Lifton, R.P. A novel pcln-1 gene mutation in familial hypomagnesemia with hypercalciuria and atypical phenotype. *Pediatr. Nephrol.* **2007**, *22*, 503–508. [CrossRef] [PubMed]

155. Hirano, T.; Kobayashi, N.; Itoh, T.; Takasuga, A.; Nakamaru, T.; Hirotsune, S.; Sugimoto, Y. Null mutation of pcln-1/claudin-16 results in bovine chronic interstitial nephritis. *Genome Res.* **2000**, *10*, 659–663. [CrossRef]
156. Hadj-Rabia, S.; Baala, L.; Vabres, P.; Hamel-Teillac, D.; Jacquemin, E.; Fabre, M.; Lyonnet, S.; de Prost, Y.; Munnich, A.; Hadchouel, M.; et al. Claudin-1 gene mutations in neonatal sclerosing cholangitis associated with ichthyosis: A tight junction disease. *Gastroenterology* **2004**, *127*, 1386–1390. [CrossRef]
157. Sineni, C.J.; Yildirim-Baylan, M.; Guo, S.; Camarena, V.; Wang, G.; Tokgoz-Yilmaz, S.; Duman, D.; Bademci, G.; Tekin, M. A truncating cldn9 variant is associated with autosomal recessive nonsyndromic hearing loss. *Hum. Genet.* **2019**, *138*, 1071–1075. [CrossRef]
158. Hadj-Rabia, S.; Brideau, G.; Al-Sarraj, Y.; Maroun, R.C.; Figueres, M.-L.; Leclerc-Mercier, S.; Olinger, E.; Baron, S.; Chaussain, C.; Nochy, D.; et al. Multiplex epithelium dysfunction due to cldn10 mutation: The helix syndrome. *Genet. Med.* **2018**, *20*, 190–201. [CrossRef]
159. Wilcox, E.R.; Burton, Q.L.; Naz, S.; Riazuddin, S.; Smith, T.N.; Ploplis, B.; Belyantseva, I.; Ben-Yosef, T.; Liburd, N.A.; Morell, R.J.; et al. Mutations in the gene encoding tight junction claudin-14 cause autosomal recessive deafness dfnb29. *Cell* **2001**, *104*, 165–172. [CrossRef]
160. Hampson, G.; Konrad, M.A.; Scoble, J. Familial hypomagnesaemia with hypercalciuria and nephrocalcinosis (fhhnc): Compound heterozygous mutation in the claudin 16 (cldn16) gene. *BMC Nephrol.* **2008**, *9*, 12. [CrossRef]
161. Konrad, M.; Schaller, A.; Seelow, D.; Pandey, A.V.; Waldegger, S.; Lesslauer, A.; Vitzthum, H.; Suzuki, Y.; Luk, J.M.; Becker, C.; et al. Mutations in the tight-junction gene claudin 19 (cldn19) are associated with renal magnesium wasting, renal failure, and severe ocular involvement. *Am. J. Hum. Genet.* **2006**, *79*, 949–957. [CrossRef]
162. Seker, M.; Fernandez-Rodriguez, C.; Martinez-Cruz, L.A.; Müller, D. Mouse models of human claudin-associated disorders: Benefits and limitations. *Int. J. Mol. Sci.* **2019**, *20*, 5504. [CrossRef] [PubMed]

11

Structure and Junctional Complexes of Endothelial, Epithelial and Glial Brain Barriers

Mariana Castro Dias *, Josephine A. Mapunda, Mykhailo Vladymyrov and Britta Engelhardt *

Theodor Kocher Institute, University of Bern, 3012 Bern, Switzerland; josephine.mapunda@tki.unibe.ch (J.A.M.); mykhailo.vladymyrov@tki.unibe.ch (M.V.)

* Correspondence: mariana.dias@tki.unibe.ch (M.C.D.); bengel@tki.unibe.ch (B.E.)

Abstract: The homeostasis of the central nervous system (CNS) is ensured by the endothelial, epithelial, mesothelial and glial brain barriers, which strictly control the passage of molecules, solutes and immune cells. While the endothelial blood-brain barrier (BBB) and the epithelial blood-cerebrospinal fluid barrier (BCSFB) have been extensively investigated, less is known about the epithelial and mesothelial arachnoid barrier and the glia limitans. Here, we summarize current knowledge of the cellular composition of the brain barriers with a specific focus on describing the molecular constituents of their junctional complexes. We propose that the brain barriers maintain CNS immune privilege by dividing the CNS into compartments that differ with regard to their role in immune surveillance of the CNS. We close by providing a brief overview on experimental tools allowing for reliable in vivo visualization of the brain barriers and their junctional complexes and thus the respective CNS compartments.

Keywords: brain barriers; blood-brain barrier; neurovascular unit; blood-cerebrospinal fluid barrier; arachnoid barrier; glia limitans; tight junctions; adherens junctions

---

1. Introduction

The brain barriers established by the endothelial blood-brain barrier (BBB), the epithelial blood-cerebrospinal fluid barrier (BCSFB), the meningeal brain barriers and the blood spinal cord barrier are essential for maintaining central nervous system (CNS) homeostasis [1]. While the structural and junctional components of the BBB and of the BCSFB of the choroid plexus (ChP) have been vastly explored and described, much less is known about the cells and junctional complexes establishing the meningeal barriers and the glia limitans.

Development of the CNS vasculature begins with the process of vasculogenesis, where angioblasts, which originate from the mesoderm, move to the head region and form the primary perineural vascular plexus around the developing brain. The proliferating neuroectodermal tissue is invaded by vascular sprouts from the primary perineural plexus establishing the brain vasculature by a process called angiogenesis. The unique barrier properties of the brain endothelial cells are not intrinsic and are rather induced by a process referred to as barriergenesis by the continuous crosstalk with the developing neuroectodermal tissue [2–5].

The ChP develops from different locations along the dorsal axis of the neural tube, with the hindbrain ChP of the fourth ventricle being the first to be formed, once the neural tube is closed [6]. While the ChP stroma develops from mesenchymal cells, the ChP epithelium derives from neuroepithelial cells. Despite the different origin along the rostral-caudal axis of the developing nervous system, the mature ChP epithelia are morphologically similar (reviewed in Reference [6]).

The meningeal barriers comprise three layers: the dura mater, the arachnoid mater and the pia mater, and cover the brain and the spinal cord from its earliest stages of development, establishing a protective covering of the CNS in the adult. Our current understanding of the development of the meninges has not much advanced beyond identifying the origins of pial, arachnoid and dural fibroblasts in the frontal brain from neural crest, and in the midbrain, hindbrain and spinal cord from mesoderm [7]. Despite their differences in developmental origin, the cells of the pia, arachnoid and dura mater have been described to establish functionally similar barriers. However, the respective barrier function may be established by different molecular components of their junctional complexes [8–11].

Taken together, junctional complexes established between cells of the brain barriers might not only differ between the different barriers but also within one respective barrier along the rostral-caudal axis. Here, we summarize the current knowledge of the brain barriers, with a particular focus on the molecular components of their junctional complexes. Additionally, we suggest that visualization of brain barriers junctional complexes provides useful landmarks for in vivo imaging of immune surveillance of the CNS.

2. Blood-Brain Barrier and Its Central Players: Cellular and Acellular Components of the Neurovascular Unit (NVU)

The brain vasculature is a big and complex network composed of arteries and arterioles, capillaries, venules and veins, which allows for the vital distribution of nutrients and oxygen to the CNS at the level of the microvasculature, namely arterioles, capillaries and post-capillary venules [12]. Brain microvascular endothelial cells, referred to as the blood-brain barrier (BBB), possess unique features restricting the free passage of ions, molecules and cells from the blood into the CNS parenchyma, while facilitating the transport of toxic compounds out of the CNS [13,14]. Development and maintenance of BBB characteristics in CNS microvascular endothelium relies on the continuous crosstalk between cellular and acellular elements of the CNS. In the adult, the entity of BBB endothelium, pericytes, astrocytes, and the basal membranes are thus collectively referred to as the neurovascular unit (NVU) (represented in Figure 1) [15]. Maintenance of a functional NVU is a prerequisite for BBB function and requires understanding of the specific contributions of these cells to barrier function.

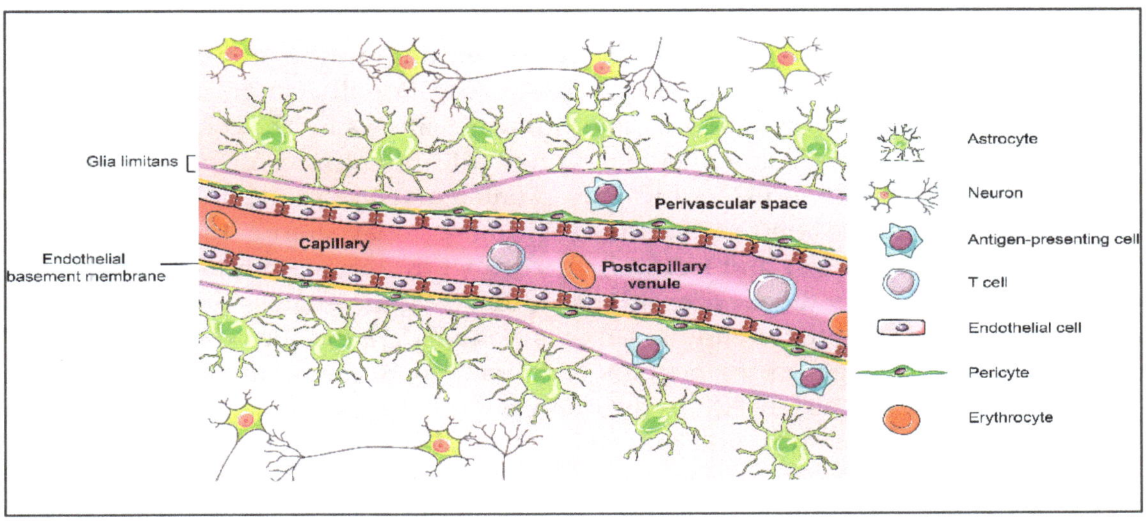

Figure 1. Schematic representation of the components of the neurovascular unit at the level of brain capillaries and post-capillary venules. Drawings of the individual cell types were adapted from Servier Medical Art (http://smart.servier.com/), licensed under a Creative Common Attribution 3.0 Generic License.

2.1. Endothelial Cells

Most studies of the BBB focus on the capillary endothelial cells, due to their unique properties [16,17]. Besides lacking fenestrae, these endothelial cells are brought together by the presence of continuous and complex tight junctions that prohibit free diffusion of molecules across the BBB via the paracellular route [18–20]. Additionally, they have very low pinocytosis rates when compared to the peripheral endothelium [21,22]. Of notice, the major facilitator superfamily domain-containing protein 2 (Mfsd2a), a membrane transport protein, is specifically expressed in the CNS endothelium and acts as a key regulator suppressing vesicular activity at the BBB [23,24]. To ensure the passage of nutrients from the blood into the CNS and to remove potential toxic agents from the brain, the BBB endothelial cells express specific transporters and efflux pumps [25–27]. In particular, the efflux pump MDR1/P-glycoprotein (Pgp) translocates potentially harmful lipophilic or endogenous molecules from the CNS to the blood [28]. The glucose transporter GLUT-1 (*SLC2A1*) is also highly enriched in the BBB endothelium, allowing for glucose delivery to the CNS [29]. The barrier characteristics referred to above are also present at the level of the endothelial cells forming the BBB in the post-capillary venules. While the capillaries represent a barrier for solutes and ions, leukocyte trafficking is regulated at the level of the post-capillary venules where the endothelial cells express specific adhesion molecules [1,30].

Additionally, the luminal side of the brain endothelium is covered by the glycocalyx, a carbohydrate-rich mesh of anionic polymers, which acts as a first physical barrier between the blood and the vessel wall [31]. In peripheral microvessels, the glycocalyx regulates vascular permeability to water and to albumin, and glycocalyx breakdown results in plasma leakage and enhanced leukocyte recruitment, in vivo [32,33]. Recent advances in in vivo imaging have allowed to confirm barrier forming role of the glycocalyx in peripheral vessels [34] and at the level of the BBB [35].

However, there are areas of the brain, the circumventricular organs, where the capillaries lack BBB properties and rather have fenestrae allowing blood components to freely pass towards specialized neurons serving neuroendocrine and neurosensory functions [36]. In addition, the ChP vasculature is characterized by fenestrae allowing free access of blood components into the ChP stroma [37].

2.2. Mural Cells: Smooth Muscle Cells and Pericytes

Mural cells comprise smooth muscle cells and pericytes. While pericytes, embedded in the endothelial basement membrane, incompletely cover the capillaries, smooth muscle cells seem to completely surround arterioles and to a lower degree venules and, together with the endothelial basement membrane, compose the tunica media [38]. Additionally, smooth muscle cells express contractile proteins such as alpha-smooth muscle actin, which allows for the control of vessel diameter and blood flow [38–41]. Recent studies have assigned additional roles of pericytes in regulating blood flow at the level of capillaries [42], which has initiated a debate on the role of arteriolar smooth muscle cell or capillary pericyte contractility in regulating regional cerebral blood flow. At the same time, pericytes are morphologically and functionally distinct from smooth muscle cells [43], which is further supported by the recent discoveries on their different gene expression profile in CNS blood vessels [44]. Pericytes interact with the endothelial cells via specific adhesion points that can either form peg-and-socket junctions, due to the presence of N-cadherin [45], or can occur as adhesion plaques, gap junctions and even tight junctions [46–48]. In the brain, endothelial coverage by pericytes is extremely high, with an endothelial cell/pericyte ratio between 1:1 to 3:1 [49]. Additionally, pericytes are involved in wound healing, angiogenesis, deposition of extracellular matrix and, during embryogenesis, have a strong role in BBB formation by suppressing vesicular activity in brain endothelial cells [43,50,51]. In the adult stage, loss of pericytes impairs BBB properties of endothelial cells by affecting their gene expression profile leading to increased vesicular trafficking and loss of polarization of astrocytic endfeet [51]. Moreover, pericytes regulate expression of cellular adhesion molecules in the BBB endothelium [50–52].

2.3. Basement Membrane

Pericytes and endothelial cells secrete and are surrounded by the extracellular matrix proteins that comprise the endothelial basement membrane. It is a crosslinked network composed of type IV collagen, $\alpha4$ and $\alpha5$ laminins, nidogen, heparan sulphate proteoglycans and some glycoproteins [53]. A unique feature of CNS microvessels is the presence of a second basement membrane, referred to as parenchymal basement membrane, which is secreted by astrocytes [53,54]. With a composition of $\alpha 1$ and $\alpha 2$ laminins at the brain surface and with $\alpha 2$ laminins at the perivascular spaces, the parenchymal basement membranes are molecularly distinct from the endothelial basement membrane [55]. At the capillary level it merges with the endothelial basement membrane, while at the level of post-capillary venules there is a small perivascular space between those two basement membranes. The three-dimensional configuration and the components of these basement membranes host specific signaling cues for cellular growth and survival during the development and further maintenance of the BBB, while giving physical support to the vasculature [56] (reviewed in Reference [57]).

2.4. Astrocytes

Astrocytes also assume a tight relationship with the endothelium and the mural cells. These glial cells are widely present throughout the CNS and perform a supportive role that regulates the neuronal microenvironment by controlling neurotransmitters and electrolytes balance, synapse formation and clearance of axonal debris [16,58,59]. Furthermore, they have extensive cellular protrusions that cover blood vessels and neuronal synapses, promoting a direct link between neuronal circuits and the vasculature [16]. The astrocytic endfeet are in direct contact with the microvasculature of the CNS, covering a large part of the abluminal aspect of CNS microvessels [39]. Astrocytic processes express a range of proteins, including α-dystroglycan and dystrophin, which link the astrocyte endfoot to the parenchymal basement membrane, and the water channel aquaporin 4, which plays a crucial role regulating water homeostasis in the CNS [60,61]. Astrocytes have been described as an important intermediate in promoting endothelial cell barrier properties, as well as secreting the components of the parenchymal basement membrane, as mentioned above [16,53,62]. Additionally, the astrocytic endfeet and the parenchymal basement membrane form a second barrier of the CNS called glia limitans [1]. Several in vitro studies showed that with a co-culture of astrocytes and endothelial cells, the junctional complexes between adjacent endothelial cells upregulate its main proteins, contributing to an increase in barrier integrity [63,64]. Moreover, secretion of sonic hedgehog, retinoic acid and angiopoietin-1 by astrocytes supports the barrier properties of brain endothelial cells [65,66]. On the other hand, the interaction of astrocytes with endothelial cells is also very important, with endothelial-secreted factors such as LIF1 promoting astrocytic differentiation [67].

3. Molecular Constituents of the BBB

The BBB acts as a physical and metabolic barrier due to the unique biochemical make-up of the BBB endothelial cells. Here we will only focus on describing the molecules establishing the paracellular diffusion barrier of the BBB, which are proteins composing tight junctions, adherens junctions, transmembrane proteins localized at cell-to-cell contacts outside the organized tight junctions and adherens junctions, and intracellular junctional scaffolding proteins.

3.1. BBB Tight Junctions (TJs)

TJs localized between adjacent endothelial cells are core elements actively involved in the establishment of a paracellular barrier, which limits free diffusion of ions and molecules at cell-cell junctions, adopting a "gate" function. TJs were discovered by transmission electron microscopy studies and described as focal points were the membranes of adjacent cells come together, obliterating the intercellular cleft [68]. Freeze-fracture electron microscopy showed that the TJ strands of the BBB endothelium resemble rather those of epithelial cells by producing predominantly protoplasmic (P-face) associated continuous and complex particle strands [69–74]. Although the exact mechanism underlying P-face and exoplasmic (E-face) association of TJ particles remain to be shown, occludin and claudins are localized to the TJ particle strands and thus it is hypothesized that a high fraction of P-face associated TJs correlates with increased avidity of the transmembrane TJ proteins to the actin cytoskeleton elements [73].

Besides blocking paracellular diffusion, TJs are proposed to act as a "fence" by creating an intramembrane diffusion barrier. This not only stops free passage of proteins and lipids across the lipid bilayer [75], but also prohibits an intermixing of the components of the plasma membrane, creating distinct apical and basolateral membrane sites [76]. Additionally, they distribute selected membrane components to the cell surface and allow for accumulation of internal scaffolding proteins, essential to establish a link to the cytoskeleton [77]. Altogether, the "gate and fence" function that TJs assume is essential to assure the regulation of a physiological barrier. Most of the gathered knowledge about TJs is derived from epithelial cells and readily assumed to hold true for endothelial cells, despite lacking formal proof [78]. In terms of molecular composition, TJs are composed of transmembrane proteins, cytoplasmic plaque proteins, signaling proteins and adapters that bind these complexes to the actin cytoskeleton. Transmembrane proteins are suggested to be the elements playing the barrier and fence role, since they are constituted by transmembrane, cytoplasmic and extracellular domains. Three groups of transmembrane proteins are found to be localized in TJs: the claudins, the immunoglobulin superfamily junctional adhesion molecules (JAMs) and the tight junction-associated MARVEL proteins (TAMP) [78,79].

Claudins are the major components of TJs and to date, 27 members of this family have been identified in mammals [80]. Their general structure consists of an intracellular NH_2 domain, four transmembrane domains, two different extracellular loops (ECL1 and ECL2) and one intracellular COOH terminus [79]. In their C-terminus, most of the claudins possess a PDZ-binding motif used to interact with the TJ-associated scaffolding proteins [81]. ECL1 is involved in regulating the paracellular tightness and ion permeability, while ECL2 mediates cis/trans claudin-claudin interactions [82,83]. In vitro data showed that in contrast to other TJ transmembrane proteins, claudins are sufficient to form TJs strands in fibroblasts and that paracellular permeability is increased upon disruption of claudins, which suggests that this family of proteins is essential for barrier formation [84–87]. It has been suggested that specific barrier functions demand expression of different combinations of claudins, which is believed to influence the charge and size-selective properties of the barrier [88]. Claudins such as claudins-1, -3, -5, -11, -14 and -19 act as TJ sealing membrane proteins, while claudins-2, -10, -15 and -17 are involved in ion pore formation between two adjacent cells [89]. Claudin-deficient mice have been extremely helpful to understand the specific function of individual claudins in specific tissue barriers. For example, claudin-1 deficiency leads to epidermal permeability barrier defects in the skin, while lack of claudin-11 causes defects in CNS myelin sheaths [90,91]. Regarding the BBB endothelium, claudin-1, claudin-3, claudin-5, claudin-11 and claudin-12 were described to be expressed in the BBB endothelial cells [92–95]. Claudin-5 is specifically expressed in endothelial cells with very high expression in brain endothelium [50]. Lack of claudin-5 leads to perinatal death in mice, due to exacerbated BBB permeability to tracers smaller than 800 Da [92,96]. Supporting the important role of claudin-5 in BBB TJs, inducible knock-down of claudin-5 in adult mice was found to lead to seizures and behavioral changes, due to loss of BBB properties [97]. Despite this dramatic failure of BBB TJ functions, mice lacking claudin-5 still present morphologically intact and P-face associated TJs in the BBB, which suggests that claudins other than claudin-5 maintain BBB TJs in these mice [92]. Claudin-3 has been described as an additional component of BBB TJs, mediating TJ maturation during development by acting as a downstream target of Wnt/β-catenin signaling [50,98,99]. Additional evidence for a role of claudin-3 in regulating BBB integrity was derived from studies in an animal model for multiple sclerosis or in human glioblastoma multiforme, were claudin-3 immunostaining was found to be specifically lost in inflamed CNS microvessels [98]. These observations supported a role for claudin-3 in maintaining junctional BBB integrity. Making use of a novel claudin-3 deficient mouse and advanced transcriptomic profiling of brain endothelial cells we and others recently made the unexpected observation that claudin-3 mRNA cannot be detected in brain endothelial cells and thus claudin-3 is not expressed in brain endothelial cells and does not contribute to BBB TJs [100–102]. Disturbingly, we found most of the claudin-3 detecting antibodies to produce comparable junctional immunostainings of brain endothelial cells in wild-type and claudin-3-deficient mice suggesting that available claudin-3 antibodies cross-react with another junctional molecule and do not allow to the reliable detection of expression and junctional localization of claudin-3 protein [102]. In addition, positive immunostaining for claudin-1 and claudin-11 has been reported in mouse brain endothelium [103] [93,95]. However, we and others showed that claudin-1 mRNA is not detectable in brain endothelial cells [44,104,105]. In fact, endothelial specific inducible ectopic expression of claudin-1 in BBB endothelium reduces BBB leakiness in an animal model of multiple sclerosis and ameliorates clinical signs of the disease further supporting absence of claudin-1 at the BBB [104]. Similarly, claudin-11 mRNA was not detected in brain endothelial cells in single-cell RNA sequencing datasets from mouse brain microvessels, but its expression was readily detected in oligodendrocytes [44,105].

Finally, claudin-12 is an atypical member of the claudin family because it does not have a PDZ-binding motif [69]. Prominent junctional immunostaining for claudin-12 [92] suggested its localization to BBB TJs. However, by establishing a claudin-12 deficient mouse we could recently show that expression of claudin-12 in microvascular brain endothelial cells is low to absent and is not required for barrier function of the BBB [106]. Analysis of the localization of claudin-12 protein in brain endothelial cells was again hampered by the lack of antibodies specifically detecting claudin-12 in wild-type but not in claudin-12 deficient mice.

Taken together, reports describing junctional localization of claudin-1, claudin-3, and claudin-12 proteins in brain endothelial cells may be inaccurate due to not well-defined cross-reactivities and thus false-positive staining of claudin antibodies. Cross-reactivity of the anti-claudin antibodies may be due to the high degree of conservation and a common amino acid sequence within some members of the claudin family, which could potentiate a cross-reactivity of anti-claudin antibodies [83]. Furthermore, recent studies described detection of claudin-11 at the protein level in BBB endothelial cells in mice especially in the spinal cord and man [93,94]. As single-cell RNAseq analysis of brain endothelial cells failed to detect claudin-11 in these cells [44,105] and so, it is mandatory to explore if claudin 11 mRNA levels are too low for detection in the single-cell RNAseq screens or if detection of claudin-11 protein in these studies is again due to the cross-reaction of claudin-11 antibodies with other junctional entities.

Thus, at present, claudin-5 is the only claudin confirmed to be expressed at the mRNA and protein level in brain endothelial cells, to be localized to BBB TJs and to play a major role in BBB TJs integrity. Advanced transcriptome profiling studies have identified additional claudins, e.g., claudin-25/CLDND1 to be expressed in brain microvascular endothelial cells, with their exact function to be determined [44,105].

The TAMP family comprises the proteins occludin, tricellulin and MARVELD3 [107–109], which have four transmembrane domains and a MARVEL (MAL-related proteins for vesicle trafficking and membrane link) domain. To date, occludin and tricellulin were found to be expressed at the TJs of the BBB [110,111]. Occludin, the first transmembrane protein discovered to be localized exclusively in TJs, is highly expressed in endothelial cells of the CNS and was suggested to play a role in barrier integrity [107,110–113]. However, occludin-deficient epithelial cells develop morphologically intact TJs and consequently maintain barrier function [114]. On the other hand, phosphorylation of occludin was shown to regulate endothelial permeability and to be necessary for cortical actin organization in various endothelial cell models [115–117]. Additionally, occludin-deficient mice do not show impaired TJs, but are characterized by chronic inflammation, poor TJ integrity and by calcifications in the brain [118]. These results suggest that occludin may be involved in calcium movement across the BBB and in the regulation of TJ stability and barrier function, rather than TJ assembly. Another member of the TAMP family is tricellulin, which, as the name indicates, is specifically localized to tricellular junctions, a point in which three adjacent cells meet (reviewed in Reference [119]). Albeit being widely described in epithelial monolayers, recently expression of tricellulin was described in endothelial cells forming the BBB and the blood–retinal barrier [111,120], underscoring the epithelial nature of these endothelial barriers. The precise function of tricellulin in endothelial cells forming the BBB and blood–retinal barrier remains however to be described.

The third group of proteins found in BBB TJs are the members of the JAM family, JAM-A, JAM-B and JAM-C. These proteins are immunoglobulin (Ig)-like molecules composed of a single transmembrane domain, two extracellular Ig domains and a PDZ-binding motif in their C-terminus that allows for an interaction with cytoplasmic proteins linking the JAMs to the actin cytoskeleton [121,122]. JAMs are highly enriched in the TJs of epithelial and endothelial cells [123–127] and are the tight junction-associated transmembrane proteins that regulate cell polarity by interacting with Par3 [128,129]. JAM-A immunostaining was found at the BBB and as JAM-A upon transfection into CHO cells establishes a barrier, JAM-A has been suggested to contribute to BBB integrity [126]. It was additionally described to regulate the migration of monocytes across BBB cell-to-cell contacts [130]. Recently, we have described JAM-B to be localized to BBB TJs but discovered that JAM-B-deficient mice have an intact BBB, which suggests that JAM-B is not required for proper BBB function [131]. JAM-B has been found to bind $\alpha 4\beta 1$-integrin and interestingly mediates $\alpha 4\beta 1$-integrin mediated migration of $CD8^{+}$ T cells into the CNS but is not involved in the $\alpha 4\beta 1$-integrin-mediated migration of $CD4^{+}$ T cells across the BBB [131–134]. The third member of the family, JAM-C was also found to be expressed in brain endothelial cells. Albeit C57BL/6 mice deficient for JAM-C develop a hydrocephalus, this is not due to impairment of BBB function, suggesting that JAM-C is not required for BBB junctional integrity [125,135].

The claudins, occludin and the JAMs expressed in TJs connect to the cytoskeleton via the interaction with intracellular scaffolding proteins, which form the TJ plaque. The membrane-associated guanylate kinase (MAGUK) proteins represent the major subgroup of scaffolding proteins at the TJs [136,137]. These proteins are structurally similar, since they share one or more PDZ domains, an SH3 domain and a catalytically inactive guanylate kinase (GUK) domain and are overall involved in the establishment of cell-cell adhesion, cell polarity and cell survival [138]. Zona occludens (ZO) proteins were the first proteins of the MAGUK family to be identified, with ZO-1 and ZO-2 localizing to endothelial TJs, which is essential for the formation of TJ strands [139,140]. Lack of ZO-1 and ZO-2 in mice is embryonically lethal [140,141]. ZO-1 can be found in the BBB TJs, interacting with claudins, occludin, JAMs and ZO-2, and promoting a link between these proteins and the actin cytoskeleton, by binding to F-actin [142]. It is also a main regulator of tension at vascular endothelial (VE)-cadherin-based adherens junction complexes, mainly binding to α-catenin and recruiting important mediators for junctional assembly and stability [69,143]. Additionally, cingulin, afadin (AF-6) and 7H6 antigen are also involved in the coupling of the junctional complexes to the cytoskeleton, supporting TJ structure and stability [144,145]. Cingulin connects ZO-2, AF-6 and JAMs to F-actin, while 7H6 acts towards maturation and maintenance of TJs [146–148].

3.2. *Adherens Junctions (AJs) of the BBB*

Adherens junctions (Ajs) have a role distinct from TJs. Prior to TJ formation, cell-cell contacts are established by AJs, which is a prerequisite for TJ maturation and maintenance. Thus, AJs are generally required for TJ formation, and a continuous crosstalk between AJs and TJs is necessary for the organization and preservation of the junctional complexes (reviewed in Reference [149]). VE-cadherin is the main protein of the endothelial AJs and is involved in blood vessel assembly and endothelial stabilization and survival [150,151]. Its intracellular domains engage p120 catenin and β-catenin [152,153]. By binding to β-catenin, α-catenin acts as a bridge between the cadherins and the actin cytoskeleton, since it binds to vinculin, α-actinin, ZO-1 or formin-1 [154,155]. Interestingly, VE-cadherin promotes expression of claudin-5 in endothelial cells derived from murine embryonic stem cells [156], supporting the notion that mature AJs are prerequisite for the establishment of TJs. Nectin is an additional component of the AJs of the BBB and connects to the actin cytoskeleton via AF-6. The nectin-AF-6 complex contributes to the formation of AJs and stabilization of TJs [157,158].

Additional molecules are found to be localized at the endothelial cell-cell junctions, outside of the organized AJs and TJs. Platelet endothelial cell adhesion molecule-1 (PECAM-1) is a transmembrane protein that belongs to the Ig superfamily. It is highly concentrated and restricted to the endothelial junctions but is found outside the organized TJs and AJs complexes [159,160]. However, it plays a very important role in angiogenesis with a mechanosensory function and is involved in the regulation of vascular integrity as PECAM-1-deficient mice show impaired BBB junctional integrity [161–163]. CD99 is an additional protein localized in the endothelial cell junctional complexes, outside of TJs or AJs [164,165]. This protein is a highly O-glycosylated type I transmembrane protein not belonging to any protein family. CD99 mediates leukocyte trafficking across the BBB, with any additional role in BBB junctional integrity remaining to be shown [166]. A schematic representation of the molecular composition of the junctional complexes of the BBB endothelium as known today is illustrated in Figure 2.

Although a large number of molecules localized at the cell-to-cell contacts of the BBB within TJ, AJs and beyond has thus been described, their dynamic interplay and regulation during CNS homeostasis and how their expression and localization changes during neurological disorders when BBB function is impaired remains to be investigated in more detail. In addition, species differences in the expression of molecular constituents of the BBB junctional complexes have been described [94,167,168], an aspect that is important when aiming to translate observations from animal models to the human BBB.

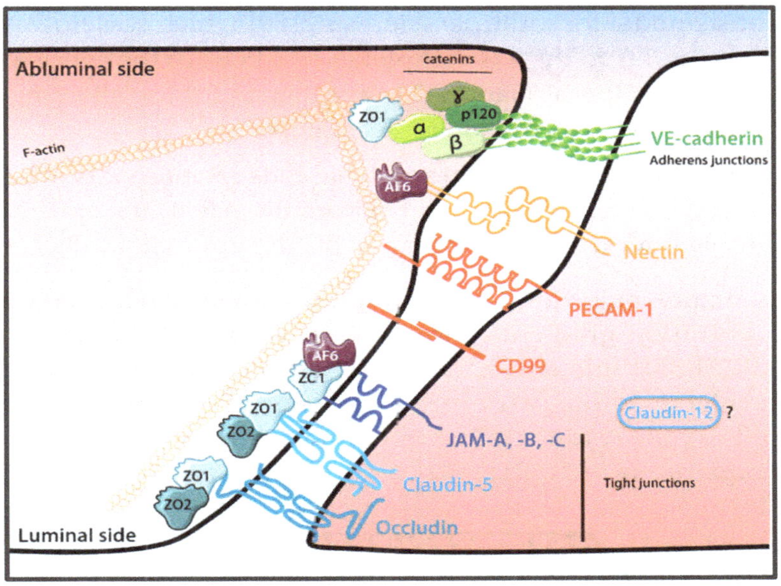

Figure 2. Schematic representation of the junctional complexes of the BBB. The proteins that compose the tight and adherens junctions are connected to the cytoskeleton via intracellular scaffolding proteins, ZO-1, ZO-2 and AF-6. Despite being expressed at low levels by the BBB endothelial cells [106], the subcellular location and function of claudin-12 remains to be defined. The forms of the individual proteins were adapted from Servier Medical Art (http://smart.servier.com/), licensed under a Creative Common Attribution 3.0 Generic License.

4. The Epithelial Blood-Cerebrospinal Fluid Barrier of the Choroid Plexus

The development and maintenance of the CNS is assured by the cerebrospinal fluid (CSF), found in the spinal and brain subarachnoid space, cisterns, sulci and in the cerebral ventricles [169,170]. Importantly, the CSF is produced by the ChP, a highly vascularized secretory tissue that extends into all four cerebral ventricles. The ChP folds out from the ventricular walls and is composed of a highly vascularized stroma that is surrounded by a monolayer of epithelial cells [171]. Because the dense network of capillaries that irrigate the ChP is fenestrated and thus permissive to the passage of blood derived molecules, the epithelial cells surrounding the ChP stroma build a BCSFB [18,172]. Accordingly, the molecules that diffuse from the blood across the ChP vascular wall reach the ChP stroma but are hindered by the epithelial BCSFB to reach the CSF [37].

In accordance to the endothelial BBB, TJs and AJs are present in the BCSFB epithelium, ensuring not only the barrier function but also the apicobasal polarity of the BCSFB [173]. Claudin-1, claudin-2, claudin-3 and claudin-11, together with occludin and ZO-1, were identified to be expressed in the TJs of the BCSFB [100,174–176]. JAM-A, but not JAM-B, is found in the epithelium of the BCSFB, while JAM-C is only present in the BCSFB of the ChP of the fourth ventricle [125,131,149]. In the epithelial monolayer, TJs are apically localized and distinguishable from the AJs, which are in the basolateral compartment [149]. Claudin-1-deficient mice die post-natal due to failure of epidermal barrier function [90], suggesting that lack of claudin-1 at the BCSFB is not essential for BCSFB maturation during embryogenesis. However, the precise role of claudin-1 at the BCSFB needs further investigations. Claudin-2 deficient mice show reduced reabsorption of Na^+ in the proximal tubule of the kidney suggesting a role for claudin-2 in paracellular transport of small cations across the BCSFB [177]. Interestingly, claudin-11 is suggested to induce parallel running TJs strands observed at the BCSFB, but also in myelin sheaths of oligodendrocytes and Sertoli cells, which all express claudin-11 [37,91]. According to the claudin-11 expression pattern, claudin-11 deficient mice display defects in spermatogenesis [91] and show behavioral defects [178]. If any of the CNS related defects is due to lack of claudin-11 at the BCSFB has not been addressed to date [90,91,177,179]. In addition, claudin-3 may be involved in maintenance of BCSFB TJ integrity, since its absence was found to impair BCSFB integrity in a mouse model for multiple sclerosis [180] which was however not confirmed in a second study [102]. Thus, similar to the TJs complexes of the BBB, the precise understanding of the interplay between TJ sealing claudins as claudin-1, claudin-3 and those allowing for transport of water and cations, as claudin-2 in BCSFB TJs, remain to be explored.

Like any other epithelial barrier, the BCSFB epithelial cells establish AJs, where the transmembrane E-cadherin binds to the catenin complex, which anchors the AJs to the epithelial actin cytoskeleton allowing to control adhesive interactions between the BCSFB epithelial cells [149,181].

A schematic representation of the junctional complexes at the ChP BCSFB can be found in Figure 3. In order to set up concentration gradients and assure proper brain function, cellular transporters are also present in the BCSFB. While efflux transporters from the ABC family oversee the return of lipophilic solutes to the blood, solute carrier (SLC) transporters are responsible for the delivery of ions and amino acids to the CSF. Additionally, water flow from the blood to the CSF is in addition to claudin-2 mediated by aquaporin 1 (AQP1) [182].

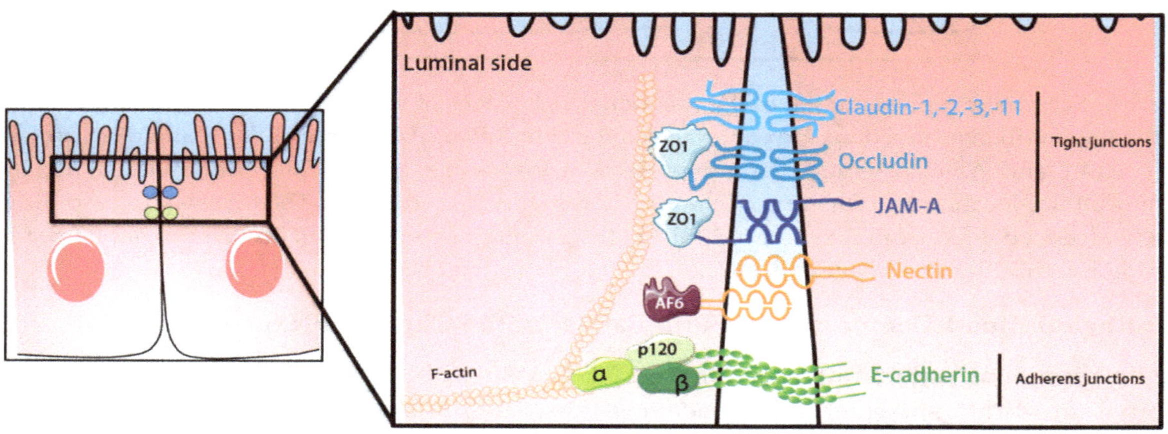

Figure 3. Schematic representation of the junctional complexes of the BCSFB at the choroid plexus. In similarity to the BBB, the proteins that compose the tight and adherens junctions of the choroid plexus BCSFB are connected to the cytoskeleton via intracellular scaffolding proteins and are localized in the apical part of the choroid plexus epithelial cells. The shapes of the proteins were adapted from Servier Medical Art (http://smart.servier.com/), licensed under a Creative Common Attribution 3.0 Generic License.

5. The Meningeal Brain Barriers

The meninges constitute three layers, the dura mater, the arachnoid mater and pia mater, which sheath the brain and spinal cord (represented in Figure 4) [183,184]. For decades, the meninges were known for their role in protecting the brain and spinal cord. In recent years, studies have shown that the meninges also play a vital role in the development of the skull and the brain [9,185,186]. Additionally, studies in rats have shown the presence of stomata on the leptomeningeal coverings of the blood vessels in the sub arachnoid space suggesting that the meningeal layers might be a potential route for the transportation of humoral immune factors from the CSF to perivascular or vascular mural spaces [187].

In humans the dura mater is characterized by three different layers, specifically the periosteal, the meningeal and the border cell layer, with the latter forming the border to the arachnoid barrier cells. The dural border cell layer is composed of flattened fibroblasts with scattered intercellular junctions and pronounced extracellular spaces lacking extracellular collagen [10,183,184,188]. It has been shown that the dura and dural border cell layer are immunoreactive to AQP1 [8]. Although junctions have been described between dura mater fibroblasts, their molecular composition is unknown. The dura mater lacks a BBB [189], thus the arachnoid barrier establishes a BCSFB between the dura mater and the subarachnoid space.

The arachnoid mater is divided into two layers—the arachnoid barrier cell layer and the arachnoid trabeculae, extending into the subarachnoid space. The arachnoid barrier cell layer is characterized by closely joined cells connected by an extensive and continuous system of TJs [190] [191,192]. On the other hand, the arachnoid trabecular cells are characterized by the presence of sporadic fibroblasts [1,10,184]. The molecular composition of arachnoid barrier TJs is not yet well explored and may depend on the origin of these cells from the neural crest or mesoderm, which may lead to the development of epithelial or mesothelial junctional complexes, respectively. Recent studies have shown positive immunostaining for claudin-11 in arachnoid barrier cells in the brains of both humans and rodents [8,189], suggesting that the arachnoid BCSFB establishes TJs similar to those at the ChP BCSFB. Arachnoid barrier cells also express AQP1, the water channels expressed by ChP BCSFB epithelial cells. In addition, existence of AJs between arachnoid barrier fibroblasts has been supported by detection of junctional localization of E-cadherin between the arachnoid cells of the brain in mice [55,193,194]. Presence of junctional proteins in the arachnoidal cells has been further supported as culture of human arachnoid granulations, which are described as projections of the arachnoid membrane into the dural sinuses, have demonstrated the presence of cytoskeletal and junctional proteins [194]. In addition, gap junctions and desmosomes ensure arachnoid barrier function. The presence of connexin 43 composed gap junctions provides electrical and metabolic coupling of arachnoid barrier cells relevant for the regulation of CSF passage across this barrier [194,195]. Desmosomes between arachnoid barrier epithelial or mesothelial cells establish mechanical stability of this barrier by anchoring these junctional structures to the intermediate filament network of the cells [194,196–198]. A potential functional impact on the different developmental origin of arachnoid barrier cells from the neural crest or mesoderm and thus the possible maturation of epithelial versus mesothelial junctional complexes has not yet been investigated. Thus, a detailed analysis of region-specific differences in the molecular composition of arachnoid barrier junctional complexes and their role in regulating barrier formation remains to be performed.

Below the arachnoid barrier there is a subarachnoid space filled with cerebrospinal fluid. Towards the brain and spinal cord tissue the subarachnoid space is lined by the cells forming the pia mater. The pia mater is composed of evenly flattened cells that line all the surface of the brain and spinal cord. The cells of the pia mater lack TJs and do not form a barrier for solutes. At the same time, they are joined by desmosomes and gap junctions [199]. In accordance to the arachnoid barrier, the precise junctional compositions and the regional differences in their composition due to the different developmental origin of the pia mater cells remains to be investigated.

The pia mater is separated from the brain and spinal cord parenchyma by glia limitans formed by the parenchymal basement membrane and astrocyte endfeet [1,184]. In the healthy CNS there is no TJs between astrocyte endfeet, which are rather linked by gap junctions and not well-defined intercellular junctions [190]. Bradbury and colleagues speculated already in 1975 that this suggests that the astrocytic glia limitans establishes a border between the CSF and the brain parenchyma [190]. This notion is further supported by the observation that during neuroinflammation, when BBB integrity is impaired, reactive astrocytes can form TJs possibly prohibiting the parenchymal entry of humoral and cellular factors from the blood stream [200,201]. TJs in reactive astrocytes are composed of JAM-A, claudin-1 and claudin-4. While both claudins have been shown to play a vital role in glial scar formation, JAM-A rather seems to regulate the migration of immune cells across the glia limitans [201,202].

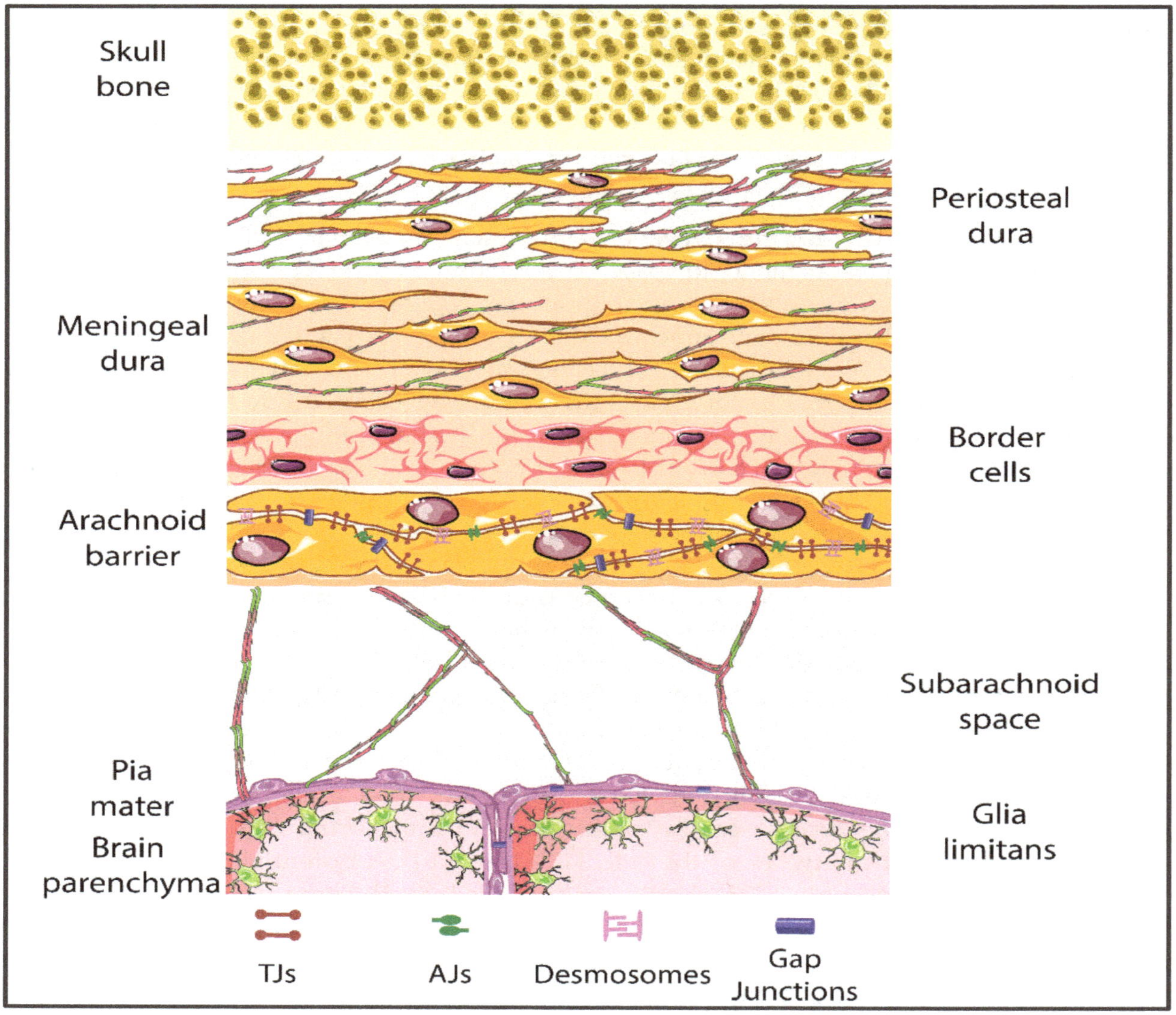

Figure 4. Schematic representation of the meningeal layers. The different layers and cellular composition of the meningeal layers are displayed. Tight junctions are highlighted between the arachnoid barrier cells as parallel lines, with adherens junctions, desmosomes and gap junctions also being represented. The shapes of the cell types were adapted from Servier Medical Art (http://smart.servier.com/), licensed under a Creative Common Attribution 3.0 Generic License.

6. Visualization of the Junctional Complexes of the Brain Barriers

Recent advances in high end in vivo imaging techniques including two-photon intravital microscopy (2P-IVM) have allowed for unprecedented observations of immune cell entry into the CNS during immune surveillance and neuroinflammation. We have proposed that the brain barriers play a vital role in separating compartments in the CNS that provide different access to the immune cells. Advanced imaging has also allowed for precise localization of cellular junctional molecules [203,204] in vitro and in vivo. Exploring the visualization of molecularly distinct junctional complexes in epithelial, endothelial, mesothelial and glial barriers thus provides a novel option for visualization of these barriers by fluorescent imaging using immunofluorescence staining or transgenic reporter mouse models. Furthermore, visualization of junctional complexes in these barriers allows for the study of cellular pathways of immune cell trafficking across these barriers using in vivo live cell imaging [205]. Here, we provide a brief overview of the main tools allowing for reliable visualization of CNS barrier components and thus CNS compartments by confocal and 2P-IVM imaging in healthy and neuroinflammatory conditions. In that context, several published studies have employed endothelial and epithelial junctional reporter mice models to study immune cell trafficking into the CNS.

There are few major approaches allowing for visualization of junctional complexes in brain barriers. First, transgenic mouse models, where a fluorescent reporter is expressed under the promoter of the junctional protein expressed in the respective cell, can be employed. In this case, the soluble fluorescent protein will be visible in the whole volume of the cells expressing the corresponding junctional protein. For example, claudin-5, which has been identified as a major constituent of the TJs of CNS endothelial cells, was used to create the Tg (Cldn5-GFP) Cbet/U reporter mouse line [44], where green fluorescence protein (GFP) is expressed under the control of the claudin-5 promoter. The strong GFP signal in the CNS endothelial cells makes this mouse model very useful for direct visualization of CNS blood vessels by 2P-IVM (Figure 5).

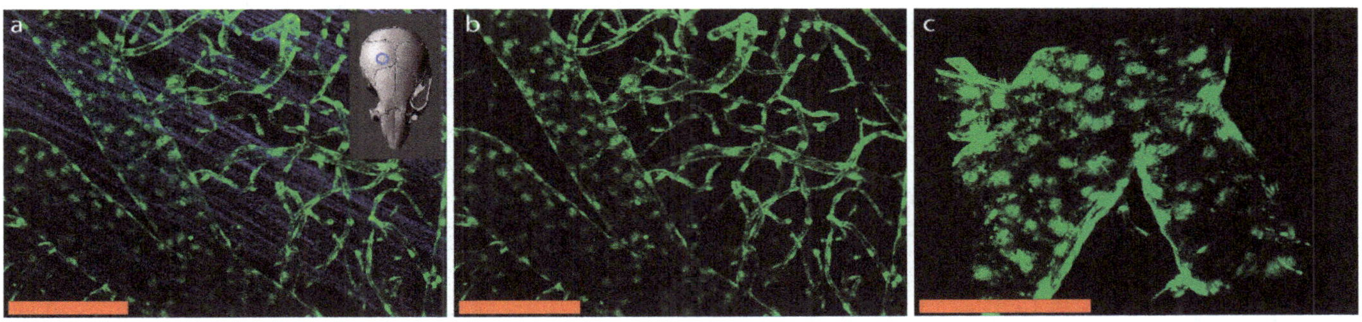

Figure 5. Representative images of the cranial and spinal cord window 2P-IVM imaging of the claudin-5-GFP reporter mouse. (**a**) The cranial window was placed over the right hemisphere of the mouse brain as depicted on the insert. Second harmonic generation in blue derives from collagen fibers in the dura mater. The strong GFP signal visible throughout the brain endothelial cells allows for imaging of the vessels along the entire vascular tree. (**b**) Cranial window region from (a) after removal of the dura mater. (**c**) Cervical spinal cord window. High magnification image showing the dorsal vein and branching veins. Scale bars = 100μm.

Alternatively, knock-in mouse models can be employed where the fluorescent protein is fused to the respective junctional molecule. This approach enables visualization of the subcellular localization of the respective junctional molecule in the cells via the fluorescent tag. For example, the VE-cadherin-GFP knock-in mouse [206], expressing a C-terminal GFP fusion protein of VE-cadherin in the endogenous VE-cadherin locus, enables live imaging of vascular AJs. This model provides a GFP signal, the strength of which is dependent on endogenous VE-cadherin expression. It is suitable for 2P-IVM of vascular AJs in blood vessels of the brain and spinal cord of healthy mice, but also in mice suffering from experimental autoimmune encephalomyelitis (aEAE), an animal model for multiple sclerosis

(Figure 6). In the dura mater, in addition to the blood vessels lined by VE-cadherin expressing vascular endothelial cells, the lymphatic vessels composed of lymphatic endothelial cells, expressing lower levels of VE-cadherin, can be visualized using the VE-cadherin-GFP knock-in reporter mouse model. VE-cadherin-GFP+ blood vessels can be distinguished from VE-cadherin GFP+ lymphatic vessels by the absence of signal of a fluorescent tracer such as orange TRITC Dextran (500 kDa) injected into the blood stream prior to imaging (Figure 6, red arrows). The VE-cadherin-GFP knock-in mouse is thus highly suitable for visualization of the endothelial AJs in the brain and spinal cord along the entire vascular tree. Additionally, it allows for imaging and distinguishing the dural blood and lymphatic vessels on the surface of the brain.

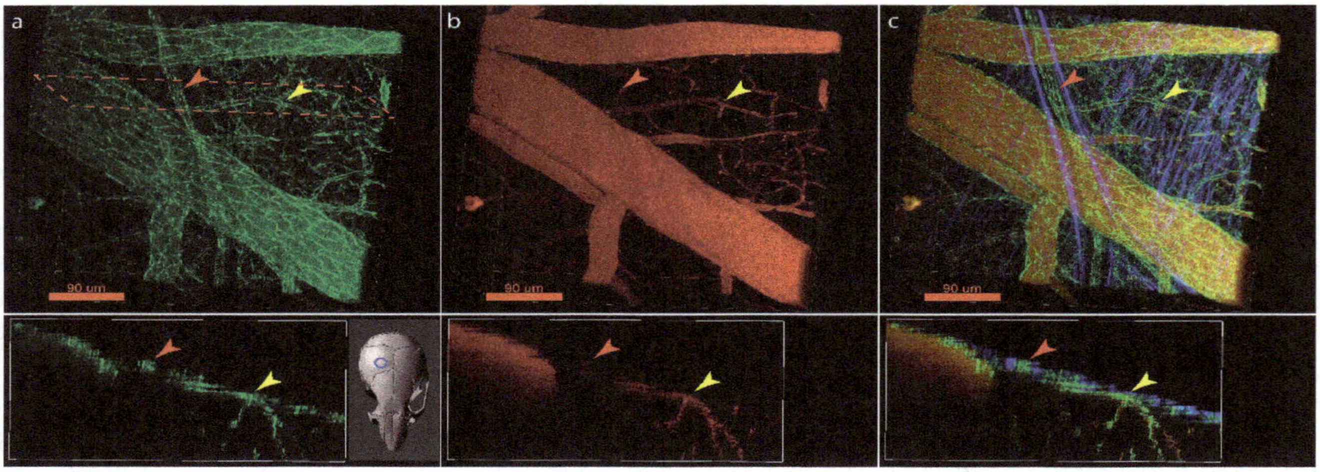

Figure 6. Visualization of CNS and meningeal endothelial AJs using the VE-Cadherin-GFP reporter mouse. (**a**) Cranial window allowing to visualize meningeal, subarachnoid, subpial and cortical vascular VE-cadherin-GFP+ AJs. (**b**) Cranial window preparation from (a) highlighting blood vessels after i.v. injection of TRITC Dextran. (**c**) Overlay of a and b and the second harmonic generation of the collagen fibers in the dura allow to distinguish VE-cadherin-GFP+ TRITC+ blood vessels and VE-cadherin-GFP+ TRITC^{neg} lymphatic vessels in the dura mater. Examples of blood and lymphatic vessels of similar caliber size are highlighted with a yellow and red arrowhead, respectively. The cranial window was placed over the right hemisphere of the mouse brain as depicted in the insert below (a). Top row—3D stack, bottom—XZ maximum intensity projection of 20 μm along Y at the cross section highlighted at the top. Scale bars = 90μm.

Similarly, the E-Cadherin-CFP knock-in mouse expressing the fluorescent protein monomer Cyan (mCFP) fused to the C-terminus of E-cadherin enables visualization of the AJs of epithelial cells [207]. This mouse model may be potentially suitable to visualize AJs between neural crest derived arachnoid barrier cells in the frontal brain, as a recent report demonstrated specific expression of E-cadherin in AJs between arachnoid barrier cells [55]. Once in vivo imaging techniques for the ChP are established it may also be used for visualization of AJs complexes of the ChP BCSFB in vivo.

In vivo imaging of BBB TJ complexes in the mouse has already been successfully reported [205]. To this end a reporter mouse expressing claudin-5 fused to eGFP under the control of the Tie-2 promoter was established [205]. This mouse model has allowed for visualization of claudin-5 in BBB TJs in vivo and direct observation of the cellular pathway of T-cell migration across the BBB owing to the strong expression of eGFP-CLN-5. Potential shortcomings of these models are that claudin-5-GFP is expressed under the control of the Tie-2 promoter, activity of which is regulated by inflammatory mediators. In addition, Tie-2 driven expression is not restricted to endothelium but allows for expression of the transgene in myeloid cells. This may impact on the molecular composition of the BBB TJ and thus its integrity and may be significantly different from physiological conditions.

Another shortcoming of the fusion protein models is the potential impact of the relatively big fluorescent proteins on the appropriate localization of junctional molecules and their interaction with intracellular scaffolding proteins. This issue can be addressed by employing self-complementing split fluorescent proteins [208,209], where only a small part of the florescent protein is used to tag the complex of interest, for example the GFP_{11} fragment of only 16 amino acids formed of a single β-strand in case of the $GFP_{1-10D7/11M3\ OPT}$ split protein [210]. The GFP1-10 fragment overexpressed in the corresponding cell remains non-fluorescent until complementation with the GFP_{11} fused to the target protein.

Additional fluorescent reporter mouse models can be obtained with the versatile aid of the Cre/loxP system. For example, by crossing the Ai14 line [211], where the fluorescent protein is expressed upon deletion of the loxP-flanked stop cassette in the Rosa26 locus, with lines expressing Cre under a brain barrier specific promoter [212], will allow to produce reporter mouse models with the respective fluorescent reporter.

Visualization of the BCSFB can be achieved by employing the FOXJ1-GFP reporter mouse or FOXJ1 Cre models [213], specific for the ciliated epithelial cells, or the transthyretin reporter [214], specific for the ChP epithelial cells.

To distinguish capillary and post-capillary segments of the CNS vascular tree, differential visualization of smooth muscle cells and pericytes can be used. For example, the expression of Cre and eGFP under the promoter of smooth muscle myosin heavy chain was developed to visualize the vascular and non-vascular smooth muscle cells [215]. Visualization of pericytes can be achieved with the Pdgfrb-Cre mouse models [216] or the NG2 reporter mouse lines [212]. In the latter cases, one has however to take into account that expression of the reporters may not be entirely exclusive to the brain barrier cells. NG2 is additionally expressed in oligodendrocyte precursor and smooth muscle cells, and Pdgfrb is also found to be expressed in oligodendrocytes, fibroblasts and smooth muscle cells.

Finally, in peripheral vascular beds visualization of junctional components has been achieved by intravenous injection of fluorescently tagged non-function blocking antibodies targeting, e.g., PECAM-1 to label endothelial cell junctions in vivo [217]. This approach may fail to detect junctional components in the complex and tight BBB TJs in the absence of neuroinflammation, as the antibodies may fail to reach their target. However, during neuroinflammation, when BBB TJ integrity is impaired this may provide a rapid approach for in vivo imaging of BBB junctions.

At the surface of the brain or spinal cord assignment of a vascular structure to a specific layer within the meninges or a compartment in the CNS can be facilitated by making use of the second harmonic generation (SHG) signal [218] obtained from collagen type I secreted in large amounts mainly by the fibroblasts in the dura mater. In 2P-IVM, using a short-pulse laser with an excitation wavelength in the range of 800–1300 nm, the excitation light can be separated from the emitted SHG signal, the wavelength of which is half of the excitation wavelength. SHG allows for localization of CNS compartments rich in collagens, especially the dura mater and, to a certain degree, the arachnoid barrier and pial layers (Figure 6). Furthermore, in some cases these compartments can be distinguished by the collagen fibers morphology [219] and can serve as a reliable landmark in the context of visualization of brain barriers by 2P-IVM.

Various small molecule tracers, such as TRITC and FITC Dextran, nanoparticles can be employed to visualize the luminal compartment of blood or lymphatic vessels. In neuroinflammation, imaging junctional complexes combined with the injection of vascular fluorescent tracers may allow to image BBB junctional leakiness. By injecting tracers into the ventricles or into the cisterna magna, distribution of the fluorescent tracers has been successfully applied to visualize paravascular CSF drainage pathways along subarachnoid vessels [220] allowing to determine CSF flow and the distribution of humoral factors in the CNS. Combining these methodologies with reporter mice for the brain barriers will allow advances in our understanding of CNS immune surveillance.

7. Conclusions

Despite decades of research, there are still a lot of unresolved questions concerning the precise molecular composition and interaction of TJs and AJs of the brain barriers, which have hampered the development of therapeutic strategies targeting the brain barriers in neuroinflammatory disorders. Regarding the BBB, it has been recently confirmed by several reports that claudin-1 and claudin-3 are absent from BBB endothelial cells and thus, claudin-5 is the only critical TJ claudin proven to date to contribute for the mouse BBB function [102]. However, when BBB endothelial cells lack caudin-5, TJs remain morphologically intact [92], which indicates that another claudin must maintain TJ morphology as visualized by transmission electron microscopy. Since we have recently reported that also claudin-12 is not required for proper BBB TJ function [106], the protein that maintains morphologically intact BBB TJs in the absence of claudin-5 remains to be identified. This approach is currently hampered by the lack of suitable antibodies detecting claudins with high and reliably specificity.

Structural and cellular junctions of meningeal barriers are less studied as compared to the BBB and the BCSFB. Meningeal barriers from the front and posterior region of the nervous system have been described to develop from different origins and therefore will have a different molecular make-up of their respective junctional complexes which may lead to regional differences in meningeal barrier functions.

A first set of available tools for in vivo visualization of the junctional components of the different brain barriers allows to study the dynamic interactions of immune cells with the brain barriers during health and neuroinflammation. Identifying the junctional components of the endothelial, epithelial, meningeal and glial brain barriers will set the stage for developing further mouse models, which will allow for in vivo imaging of the brain barriers. With these tools, one can thus explore their function in maintaining CNS immune surveillance and the impact of brain barrier disruption in neuroinflammation in vivo.

Author Contributions: M.C.D.: writing and editing; J.A.M.: writing and editing; M.V.: writing and editing; B.E.: concept, writing, editing, supervision, funding acquisition.

Acknowledgments: We thank Urban Deutsch for the valuable scientific input.

Abbreviations

aEAE	Active EAE
AF-6	Afadin
AJs	Adherens junctions
AQP1	Aquaporin 1
BBB	Blood-brain barrier
BCSFB	Blood-cerebrospinal-fluid barrier
CLDN	Claudin
CLDND1	Claudin domain containing 1
CNS	Central nervous system
ChP	Choroid plexus
CSF	Cerebrospinal fluid
ECL	Extracellular loop
E-cadherin	Epithelial cadherin
E-face	Exoplasmic face
FITC	Fluorescein
GFP	Green fluorescent protein
JAMs	Junctional adhesion molecules
MAGUK	Membrane-associated guanylate kinase
Mfsd2a	Major facilitator superfamily domain-containing protein 2

mCFP	monomer cyan fluorescent protein
N-cadherin	Neural cadherin
NG2	Neuronal-glial antigen 2
NVU	Neurovascular unit
Pdgfrb	Platelet derived growth factor receptor beta
PECAM-1	Platelet endothelial cell adhesion molecule 1
P-face	Protoplasmic face
Pgp	P-glycoprotein
SHG	Second harmonic generation
SLC	Solute carrier
TAMP	Tight junction-associated MARVEL proteins
Tg	Transgenic
TJs	Tight junctions
TRITC	Tetramethylrhodamine
VCAM-1	Vascular cell adhesion molecule 1
VE-cadherin	Vascular endothelial cadherin
ZO	Zona occludens
2P-IVM	Two-photon intravital microscopy

References

1. Engelhardt, B.; Vajkoczy, P.; Weller, R.O. The movers and shapers in immune privilege of the cns. *Nat. Immunol.* **2017**, *18*, 123–131. [CrossRef]
2. Engelhardt, B. Development of the blood-brain barrier. *Cell Tissue Res.* **2003**, *314*, 119–129. [CrossRef] [PubMed]
3. Engelhardt, B.; Sorokin, L. The blood-brain and the blood-cerebrospinal fluid barriers: Function and dysfunction. *Semin. Immunopathol.* **2009**, *31*, 497–511. [CrossRef] [PubMed]
4. Greenberg, D.A.; Jin, K. From angiogenesis to neuropathology. *Nature* **2005**, *438*, 954–959. [CrossRef] [PubMed]
5. Lee, H.S.; Han, J.; Bai, H.J.; Kim, K.W. Brain angiogenesis in developmental and pathological processes: Regulation, molecular and cellular communication at the neurovascular interface. *FEBS J.* **2009**, *276*, 4622–4635. [CrossRef]
6. Lun, M.P.; Monuki, E.S.; Lehtinen, M.K. Development and functions of the choroid plexus-cerebrospinal fluid system. *Nat. Rev. Neurosci.* **2015**, *16*, 445–457. [CrossRef]
7. DeSisto, J.; O'Rourke, R.; Bonney, S.; Jones, H.E.; Guimiot, F.; Jones, K.L.; Siegenthaler, J. A Cellular Atlas of the Developing Meninges Reveals Meningeal Fibroblast Diversity and Function. *Dev. Cell* **2019**, 53. Available online: https://papers.ssrn.com/sol3/papers.cfm?abstract_id=3396490 (accessed on 30 May 2019). [CrossRef]
8. Weller, R.O.; Sharp, M.M.; Christodoulides, M.; Carare, R.O.; Mollgard, K. The meninges as barriers and facilitators for the movement of fluid, cells and pathogens related to the rodent and human cns. *Acta Neuropathol.* **2018**, *135*, 363–385. [CrossRef]
9. Richtsmeier, J.T.; Flaherty, K. Hand in glove: Brain and skull in development and dysmorphogenesis. *Acta Neuropathol.* **2013**, *125*, 469–489. [CrossRef]
10. Mack, J.; Squier, W.; Eastman, J.T. Anatomy and development of the meninges: Implications for subdural collections and csf circulation. *Pediatr. Radiol.* **2009**, *39*, 200–210. [CrossRef]
11. Angelov, D.N.; Vasilev, V.A. Morphogenesis of rat cranial meninges. A light—And electron-microscopic study. *Cell Tissue Res.* **1989**, *257*, 207–216. [CrossRef] [PubMed]
12. Daneman, R.; Prat, A. The blood-brain barrier. *Cold Spring Harb. Perspect. Biol.* **2015**, *7*, a020412. [CrossRef] [PubMed]
13. Zlokovic, B.V. The blood-brain barrier in health and chronic neurodegenerative disorders. *Neuron* **2008**, *57*, 178–201. [CrossRef] [PubMed]

14. Liebner, S.; Dijkhuizen, R.M.; Reiss, Y.; Plate, K.H.; Agalliu, D.; Constantin, G. Functional morphology of the blood-brain barrier in health and disease. *Acta Neuropathol.* **2018**, *135*, 311–336. [CrossRef] [PubMed]
15. Muoio, V.; Persson, P.B.; Sendeski, M.M. The neurovascular Unit-Concept review. *Acta Physiol.* **2014**, *210*, 790–798. [CrossRef] [PubMed]
16. Abbott, N.J.; Ronnback, L.; Hansson, E. Astrocyte-Endothelial interactions at the Blood-Brain barrier. *Nat. Rev. Neurosci.* **2006**, *7*, 41–53. [CrossRef] [PubMed]
17. Ge, S.; Song, L.; Pachter, J.S. Where is the blood-brain barrier... Really? *J. Neurosci. Res.* **2005**, *79*, 421–427. [CrossRef]
18. Brightman, M.W.; Reese, T.S. Junctions between intimately apposed cell membranes in the vertebrate brain. *J. Cell Biol.* **1969**, *40*, 648–677. [CrossRef]
19. Wolburg, H.; Lippoldt, A. Tight junctions of the blood-brain barrier: Development, composition and regulation. *Vasc. Pharmacol.* **2002**, *38*, 323–337. [CrossRef]
20. Huber, J.D.; Egleton, R.D.; Davis, T.P. Molecular physiology and pathophysiology of tight junctions in the blood-brain barrier. *Trends Neurosci.* **2001**, *24*, 719–725. [CrossRef]
21. Reese, T.S.; Karnovsky, M.J. Fine structural localization of a blood-brain barrier to exogenous peroxidase. *J. Cell Biol.* **1967**, *34*, 207–217. [CrossRef] [PubMed]
22. Coomber, B.L.; Stewart, P.A. Morphometric analysis of cns microvascular endothelium. *Microvasc. Res.* **1985**, *30*, 99–115. [CrossRef]
23. Ben-Zvi, A.; Lacoste, B.; Kur, E.; Andreone, B.J.; Mayshar, Y.; Yan, H.; Gu, C. Mfsd2a is critical for the formation and function of the blood-brain barrier. *Nature* **2014**, *509*, 507–511. [CrossRef] [PubMed]
24. Andreone, B.J.; Chow, B.W.; Tata, A.; Lacoste, B.; Ben-Zvi, A.; Bullock, K.; Deik, A.A.; Ginty, D.D.; Clish, C.B.; Gu, C. Blood-brain barrier permeability is regulated by lipid transport-dependent suppression of caveolae-mediated transcytosis. *Neuron* **2017**, *94*, 581–594. [CrossRef]
25. Betz, A.L.; Goldstein, G.W. Polarity of the blood-brain barrier: Neutral amino acid transport into isolated brain capillaries. *Science* **1978**, *202*, 225–227. [CrossRef]
26. Betz, A.L.; Firth, J.A.; Goldstein, G.W. Polarity of the blood-brain barrier: Distribution of enzymes between the luminal and antiluminal membranes of brain capillary endothelial cells. *Brain Res.* **1980**, *192*, 17–28. [CrossRef]
27. Daneman, R. The blood-brain barrier in health and disease. *Ann. Neurol.* **2012**, *72*, 648–672. [CrossRef]
28. Ha, S.N.; Hochman, J.; Sheridan, R.P. Mini review on molecular modeling of p-glycoprotein (pgp). *Curr. Top. Med. Chem.* **2007**, *7*, 1525–1529. [CrossRef]
29. Cornford, E.M.; Hyman, S.; Swartz, B.E. The human brain glut1 glucose transporter: Ultrastructural localization to the blood-brain barrier endothelia. *J. Cereb. Blood Flow Metab. Off. J. Int. Soc. Cereb. Blood Flow Metab.* **1994**, *14*, 106–112. [CrossRef]
30. Aird, W.C. Phenotypic heterogeneity of the endothelium: I. Structure, function, and mechanisms. *Circ. Res.* **2007**, *100*, 158–173. [CrossRef]
31. Nieuwdorp, M.; Meuwese, M.C.; Vink, H.; Hoekstra, J.B.; Kastelein, J.J.; Stroes, E.S. The endothelial glycocalyx: A potential barrier between health and vascular disease. *Curr. Opin. Lipidol.* **2005**, *16*, 507–511. [CrossRef] [PubMed]
32. Salmon, A.H.; Ferguson, J.K.; Burford, J.L.; Gevorgyan, H.; Nakano, D.; Harper, S.J.; Bates, D.O.; Peti-Peterdi, J. Loss of the endothelial glycocalyx links albuminuria and vascular dysfunction. *J. Am. Soc. Nephrol. JASN* **2012**, *23*, 1339–1350. [CrossRef] [PubMed]
33. Lukasz, A.; Hillgruber, C.; Oberleithner, H.; Kusche-Vihrog, K.; Pavenstadt, H.; Rovas, A.; Hesse, B.; Goerge, T.; Kumpers, P. Endothelial glycocalyx breakdown is mediated by angiopoietin-2. *Cardiovasc. Res.* **2017**, *113*, 671–680. [CrossRef] [PubMed]
34. Betteridge, K.B.; Arkill, K.P.; Neal, C.R.; Harper, S.J.; Foster, R.R.; Satchell, S.C.; Bates, D.O.; Salmon, A.H.J. Sialic acids regulate microvessel permeability, revealed by novel in vivo studies of endothelial glycocalyx structure and function. *J. Physiol.* **2017**, *595*, 5015–5035. [CrossRef]
35. Kutuzov, N.; Flyvbjerg, H.; Lauritzen, M. Contributions of the glycocalyx, endothelium, and extravascular compartment to the blood-brain barrier. *Proc. Natl. Acad. Sci. USA* **2018**, *115*, E9429–E9438. [CrossRef]
36. Duvernoy, H.M.; Risold, P.Y. The circumventricular organs: An atlas of comparative anatomy and vascularization. *Brain Res. Rev.* **2007**, *56*, 119–147. [CrossRef]

37. Wolburg, H.; Paulus, W. Choroid plexus: Biology and pathology. *Acta Neuropathol.* **2010**, *119*, 75–88. [CrossRef]
38. Hartmann, D.A.; Underly, R.G.; Grant, R.I.; Watson, A.N.; Lindner, V.; Shih, A.Y. Pericyte structure and distribution in the cerebral cortex revealed by high-resolution imaging of transgenic mice. *Neurophotonics* **2015**, *2*, 041402. [CrossRef]
39. Mathiisen, T.M.; Lehre, K.P.; Danbolt, N.C.; Ottersen, O.P. The perivascular astroglial sheath provides a complete covering of the brain microvessels: An electron microscopic 3d reconstruction. *Glia* **2010**, *58*, 1094–1103. [CrossRef]
40. Khennouf, L.; Gesslein, B.; Brazhe, A.; Octeau, J.C.; Kutuzov, N.; Khakh, B.S.; Lauritzen, M. Active role of capillary pericytes during stimulation-induced activity and spreading depolarization. *Brain J. Neurol.* **2018**, *141*, 2032–2046. [CrossRef]
41. Hill, R.A.; Tong, L.; Yuan, P.; Murikinati, S.; Gupta, S.; Grutzendler, J. Regional blood flow in the normal and ischemic brain is controlled by arteriolar smooth muscle cell contractility and not by capillary pericytes. *Neuron* **2015**, *87*, 95–110. [CrossRef] [PubMed]
42. Hall, C.N.; Reynell, C.; Gesslein, B.; Hamilton, N.B.; Mishra, A.; Sutherland, B.A.; O'Farrell, F.M.; Buchan, A.M.; Lauritzen, M.; Attwell, D. Capillary pericytes regulate cerebral blood flow in health and disease. *Nature* **2014**, *508*, 55–60. [CrossRef] [PubMed]
43. Armulik, A.; Genove, G.; Betsholtz, C. Pericytes: Developmental, physiological, and pathological perspectives, problems, and promises. *Dev. Cell* **2011**, *21*, 193–215. [CrossRef] [PubMed]
44. Vanlandewijck, M.; He, L.; Mae, M.A.; Andrae, J.; Ando, K.; Del Gaudio, F.; Nahar, K.; Lebouvier, T.; Lavina, B.; Gouveia, L.; et al. A molecular atlas of cell types and zonation in the brain vasculature. *Nature* **2018**, *554*, 475–480. [CrossRef]
45. Gerhardt, H.; Wolburg, H.; Redies, C. N-Cadherin mediates pericytic-endothelial interaction during brain angiogenesis in the chicken. *Dev. Dyn. Off. Publ. Am. Assoc. Anat.* **2000**, *218*, 472–479. [CrossRef]
46. Courtoy, P.J.; Boyles, J. Fibronectin in the microvasculature: Localization in the pericyte-endothelial interstitium. *J. Ultrastruct. Res.* **1983**, *83*, 258–273. [CrossRef]
47. Cuevas, P.; Gutierrez-Diaz, J.A.; Reimers, D.; Dujovny, M.; Diaz, F.G.; Ausman, J.I. Pericyte endothelial gap junctions in human cerebral capillaries. *Anat. Embryol.* **1984**, *170*, 155–159. [CrossRef]
48. Larson, D.M.; Carson, M.P.; Haudenschild, C.C. Junctional transfer of small molecules in cultured bovine brain microvascular endothelial cells and pericytes. *Microvasc. Res.* **1987**, *34*, 184–199. [CrossRef]
49. Shepro, D.; Morel, N.M. Pericyte physiology. *FASEB J. Off. Publ. Fed. Am. Soc. Exp. Biol.* **1993**, *7*, 1031–1038. [CrossRef]
50. Daneman, R.; Zhou, L.; Agalliu, D.; Cahoy, J.D.; Kaushal, A.; Barres, B.A. The mouse blood-brain barrier transcriptome: A new resource for understanding the development and function of brain endothelial cells. *PLoS ONE* **2010**, *5*, e13741. [CrossRef]
51. Armulik, A.; Genove, G.; Mae, M.; Nisancioglu, M.H.; Wallgard, E.; Niaudet, C.; He, L.; Norlin, J.; Lindblom, P.; Strittmatter, K.; et al. Pericytes regulate the blood-brain barrier. *Nature* **2010**, *468*, 557–561. [CrossRef] [PubMed]
52. Villasenor, R.; Ozmen, L.; Messaddeq, N.; Gruninger, F.; Loetscher, H.; Keller, A.; Betsholtz, C.; Freskgard, P.O.; Collin, L. Trafficking of endogenous immunoglobulins by endothelial cells at the blood-brain barrier. *Sci. Rep.* **2016**, *6*, 25658. [CrossRef] [PubMed]
53. Sorokin, L. The impact of the extracellular matrix on inflammation. *Nat. Rev. Immunol.* **2010**, *10*, 712–723. [CrossRef]
54. Hallmann, R.; Horn, N.; Selg, M.; Wendler, O.; Pausch, F.; Sorokin, L.M. Expression and function of laminins in the embryonic and mature vasculature. *Physiol. Rev.* **2005**, *85*, 979–1000. [CrossRef] [PubMed]
55. Hannocks, M.J.; Pizzo, M.E.; Huppert, J.; Deshpande, T.; Abbott, N.J.; Thorne, R.G.; Sorokin, L. Molecular characterization of perivascular drainage pathways in the murine brain. *J. Cereb. Blood Flow Metab. Off. J. Int. Soc. Cereb. Blood Flow Metab.* **2018**, *38*, 669–686. [CrossRef] [PubMed]
56. Yurchenco, P.D.; Patton, B.L. Developmental and pathogenic mechanisms of basement membrane assembly. *Curr. Pharm. Des.* **2009**, *15*, 1277–1294. [CrossRef] [PubMed]
57. Thomsen, M.S.; Routhe, L.J.; Moos, T. The vascular basement membrane in the healthy and pathological brain. *J. Cereb. Blood Flow Metab. Off. J. Int. Soc. Cereb. Blood Flow Metab.* **2017**, *37*, 3300–3317. [CrossRef]

58. Zhang, Y.; Barres, B.A. Astrocyte heterogeneity: An underappreciated topic in neurobiology. *Curr. Opin. Neurobiol.* **2010**, *20*, 588–594. [CrossRef]
59. Allaman, I.; Belanger, M.; Magistretti, P.J. Astrocyte-neuron metabolic relationships: For better and for worse. *Trends Neurosci.* **2011**, *34*, 76–87. [CrossRef]
60. Noell, S.; Wolburg-Buchholz, K.; Mack, A.F.; Beedle, A.M.; Satz, J.S.; Campbell, K.P.; Wolburg, H.; Fallier-Becker, P. Evidence for a role of dystroglycan regulating the membrane architecture of astroglial endfeet. *Eur. J. Neurosci.* **2011**, *33*, 2179–2186. [CrossRef]
61. Wolburg, H.; Wolburg-Buchholz, K.; Fallier-Becker, P.; Noell, S.; Mack, A.F. Structure and functions of aquaporin-4-based orthogonal arrays of particles. *Int. Rev. Cell Mol. Biol.* **2011**, *287*, 1–41. [PubMed]
62. Janzer, R.C.; Raff, M.C. Astrocytes induce blood-brain barrier properties in endothelial cells. *Nature* **1987**, *325*, 253–257. [CrossRef] [PubMed]
63. Siddharthan, V.; Kim, Y.V.; Liu, S.; Kim, K.S. Human astrocytes/astrocyte-conditioned medium and shear stress enhance the barrier properties of human brain microvascular endothelial cells. *Brain Res.* **2007**, *1147*, 39–50. [CrossRef] [PubMed]
64. Colgan, O.C.; Collins, N.T.; Ferguson, G.; Murphy, R.P.; Birney, Y.A.; Cahill, P.A.; Cummins, P.M. Influence of basolateral condition on the regulation of brain microvascular endothelial tight junction properties and barrier function. *Brain Res.* **2008**, *1193*, 84–92. [CrossRef] [PubMed]
65. Alvarez, J.I.; Dodelet-Devillers, A.; Kebir, H.; Ifergan, I.; Fabre, P.J.; Terouz, S.; Sabbagh, M.; Wosik, K.; Bourbonniere, L.; Bernard, M.; et al. The hedgehog pathway promotes blood-brain barrier integrity and cns immune quiescence. *Science* **2011**, *334*, 1727–1731. [CrossRef]
66. Gurnik, S.; Devraj, K.; Macas, J.; Yamaji, M.; Starke, J.; Scholz, A.; Sommer, K.; Di Tacchio, M.; Vutukuri, R.; Beck, H.; et al. Angiopoietin-2-induced blood-brain barrier compromise and increased stroke size are rescued by ve-ptp-dependent restoration of tie2 signaling. *Acta Neuropathol.* **2016**, *131*, 753–773. [CrossRef]
67. Engelhardt, B.; Liebner, S. Novel insights into the development and maintenance of the blood-brain barrier. *Cell Tissue Res.* **2014**, *355*, 687–699. [CrossRef]
68. Farquhar, M.G.; Palade, G.E. Junctional complexes in various epithelia. *J. Cell Biol.* **1963**, *17*, 375–412. [CrossRef]
69. Itoh, M.; Furuse, M.; Morita, K.; Kubota, K.; Saitou, M.; Tsukita, S. Direct binding of three tight junction-associated maguks, zo-1, zo-2, and zo-3, with the cooh termini of claudins. *J. Cell Biol.* **1999**, *147*, 1351–1363. [CrossRef]
70. Simionescu, M.; Simionescu, N.; Palade, G.E. Segmental differentiations of cell junctions in the vascular endothelium. Arteries and veins. *J. Cell Biol.* **1976**, *68*, 705–723. [CrossRef]
71. Muhleisen, H.; Wolburg, H.; Betz, E. Freeze-fracture analysis of endothelial cell membranes in rabbit carotid arteries subjected to short-term atherogenic stimuli. *Virchows Arch. B Cell Pathol. Incl. Mol. Pathol.* **1989**, *56*, 413–417. [CrossRef] [PubMed]
72. Nagy, Z.; Peters, H.; Huttner, I. Fracture faces of cell junctions in cerebral endothelium during normal and hyperosmotic conditions. *Lab. Investig. J. Tech. Methods Pathol.* **1984**, *50*, 313–322.
73. Liebner, S.; Kniesel, U.; Kalbacher, H.; Wolburg, H. Correlation of tight junction morphology with the expression of tight junction proteins in blood-brain barrier endothelial cells. *Eur. J. Cell Biol.* **2000**, *79*, 707–717. [CrossRef] [PubMed]
74. Lippoldt, A.; Kniesel, U.; Liebner, S.; Kalbacher, H.; Kirsch, T.; Wolburg, H.; Haller, H. Structural alterations of tight junctions are associated with loss of polarity in stroke-prone spontaneously hypertensive rat blood-brain barrier endothelial cells. *Brain Res.* **2000**, *885*, 251–261. [CrossRef]
75. Diamond, J.M. Twenty-first bowditch lecture. The epithelial junction: Bridge, gate, and fence. *Physiologist* **1977**, *20*, 10–18.
76. van Meer, G.; Gumbiner, B.; Simons, K. The tight junction does not allow lipid molecules to diffuse from one epithelial cell to the next. *Nature* **1986**, *322*, 639–641. [CrossRef]
77. Fanning, A.S.; Anderson, J.M. Protein modules as organizers of membrane structure. *Curr. Opin. Cell Biol.* **1999**, *11*, 432–439. [CrossRef]
78. Bauer, H.C.; Krizbai, I.A.; Bauer, H.; Traweger, A. "You shall not pass"—Tight junctions of the blood brain barrier. *Front. Neurosci.* **2014**, *8*, 392. [CrossRef]
79. Gunzel, D.; Yu, A.S. Claudins and the modulation of tight junction permeability. *Physiol. Rev.* **2013**, *93*, 525–569. [CrossRef]

80. Mineta, K.; Yamamoto, Y.; Yamazaki, Y.; Tanaka, H.; Tada, Y.; Saito, K.; Tamura, A.; Igarashi, M.; Endo, T.; Takeuchi, K.; et al. Predicted expansion of the claudin multigene family. *FEBS Lett.* **2011**, *585*, 606–612. [CrossRef]
81. Stiffler, M.A.; Chen, J.R.; Grantcharova, V.P.; Lei, Y.; Fuchs, D.; Allen, J.E.; Zaslavskaia, L.A.; MacBeath, G. Pdz domain binding selectivity is optimized across the mouse proteome. *Science* **2007**, *317*, 364–369. [CrossRef] [PubMed]
82. Colegio, O.R.; Van Itallie, C.; Rahner, C.; Anderson, J.M. Claudin extracellular domains determine paracellular charge selectivity and resistance but not tight junction fibril architecture. *Am. J. Physiol. Cell Physiol.* **2003**, *284*, C1346–C1354. [CrossRef] [PubMed]
83. Krause, G.; Winkler, L.; Piehl, C.; Blasig, I.; Piontek, J.; Muller, S.L. Structure and function of extracellular claudin domains. *Ann. N. Y. Acad. Sci.* **2009**, *1165*, 34–43. [CrossRef]
84. Furuse, M.; Fujita, K.; Hiiragi, T.; Fujimoto, K.; Tsukita, S. Claudin-1 and -2: Novel integral membrane proteins localizing at tight junctions with no sequence similarity to occludin. *J. Cell Biol.* **1998**, *141*, 1539–1550. [CrossRef] [PubMed]
85. Furuse, M.; Furuse, K.; Sasaki, H.; Tsukita, S. Conversion of zonulae occludentes from tight to leaky strand type by introducing claudin-2 into madin-darby canine kidney i cells. *J. Cell Biol.* **2001**, *153*, 263–272. [CrossRef]
86. Van Itallie, C.; Rahner, C.; Anderson, J.M. Regulated expression of claudin-4 decreases paracellular conductance through a selective decrease in sodium permeability. *J. Clin. Investig.* **2001**, *107*, 1319–1327. [CrossRef]
87. Amasheh, S.; Schmidt, T.; Mahn, M.; Florian, P.; Mankertz, J.; Tavalali, S.; Gitter, A.H.; Schulzke, J.D.; Fromm, M. Contribution of claudin-5 to barrier properties in tight junctions of epithelial cells. *Cell Tissue Res.* **2005**, *321*, 89–96. [CrossRef]
88. Anderson, J.M.; Van Itallie, C.M. Physiology and function of the tight junction. *Cold Spring Harb. Perspect. Biol.* **2009**, *1*, a002584. [CrossRef]
89. Amasheh, S.; Meiri, N.; Gitter, A.H.; Schoneberg, T.; Mankertz, J.; Schulzke, J.D.; Fromm, M. Claudin-2 expression induces cation-selective channels in tight junctions of epithelial cells. *J. Cell Sci.* **2002**, *115*, 4969–4976. [CrossRef]
90. Furuse, M.; Hata, M.; Furuse, K.; Yoshida, Y.; Haratake, A.; Sugitani, Y.; Noda, T.; Kubo, A.; Tsukita, S. Claudin-based tight junctions are crucial for the mammalian epidermal barrier: A lesson from claudin-1-deficient mice. *J. Cell Biol.* **2002**, *156*, 1099–1111. [CrossRef]
91. Gow, A.; Southwood, C.M.; Li, J.S.; Pariali, M.; Riordan, G.P.; Brodie, S.E.; Danias, J.; Bronstein, J.M.; Kachar, B.; Lazzarini, R.A. Cns myelin and sertoli cell tight junction strands are absent in osp/claudin-11 null mice. *Cell* **1999**, *99*, 649–659. [CrossRef]
92. Nitta, T.; Hata, M.; Gotoh, S.; Seo, Y.; Sasaki, H.; Hashimoto, N.; Furuse, M.; Tsukita, S. Size-selective loosening of the blood-brain barrier in claudin-5-deficient mice. *J. Cell Biol.* **2003**, *161*, 653–660. [CrossRef] [PubMed]
93. Uchida, Y.; Sumiya, T.; Tachikawa, M.; Yamakawa, T.; Murata, S.; Yagi, Y.; Sato, K.; Stephan, A.; Ito, K.; Ohtsuki, S.; et al. Involvement of claudin-11 in disruption of blood-brain, -spinal cord, and -arachnoid barriers in multiple sclerosis. *Mol. Neurobiol.* **2019**, *56*, 2039–2056. [CrossRef] [PubMed]
94. Berndt, P.; Winkler, L.; Cording, J.; Breitkreuz-Korff, O.; Rex, A.; Dithmer, S.; Rausch, V.; Blasig, R.; Richter, M.; Sporbert, A.; et al. Tight junction proteins at the blood-brain barrier: Far more than claudin-5. *Cell. Mol. Life Sci. CMLS* **2019**, *76*, 1987–2002. [CrossRef]
95. Sladojevic, N.; Stamatovic, S.M.; Johnson, A.M.; Choi, J.; Hu, A.; Dithmer, S.; Blasig, I.E.; Keep, R.F.; Andjelkovic, A.V. Claudin-1-dependent destabilization of the blood-brain barrier in chronic stroke. *J. Neurosci. Off. J. Soc. Neurosci.* **2019**, *39*, 743–757. [CrossRef]
96. Morita, K.; Sasaki, H.; Furuse, M.; Tsukita, S. Endothelial claudin: Claudin-5/tmvcf constitutes tight junction strands in endothelial cells. *J. Cell Biol.* **1999**, *147*, 185–194. [CrossRef]
97. Greene, C.; Kealy, J.; Humphries, M.M.; Gong, Y.; Hou, J.; Hudson, N.; Cassidy, L.M.; Martiniano, R.; Shashi, V.; Hooper, S.R.; et al. Dose-dependent expression of claudin-5 is a modifying factor in schizophrenia. *Mol. Psychiatry* **2018**, *23*, 2156–2166. [CrossRef]

98. Wolburg, H.; Wolburg-Buchholz, K.; Kraus, J.; Rascher-Eggstein, G.; Liebner, S.; Hamm, S.; Duffner, F.; Grote, E.H.; Risau, W.; Engelhardt, B. Localization of claudin-3 in tight junctions of the blood-brain barrier is selectively lost during experimental autoimmune encephalomyelitis and human glioblastoma multiforme. *Acta Neuropathol.* **2003**, *105*, 586–592.
99. Liebner, S.; Corada, M.; Bangsow, T.; Babbage, J.; Taddei, A.; Czupalla, C.J.; Reis, M.; Felici, A.; Wolburg, H.; Fruttiger, M.; et al. Wnt/beta-catenin signaling controls development of the blood-brain barrier. *J. Cell Biol.* **2008**, *183*, 409–417. [CrossRef]
100. Steinemann, A.; Galm, I.; Chip, S.; Nitsch, C.; Maly, I.P. Claudin-1, -2 and -3 are selectively expressed in the epithelia of the choroid plexus of the mouse from early development and into adulthood while claudin-5 is restricted to endothelial cells. *Front. Neuroanat.* **2016**, *10*, 16. [CrossRef]
101. Kominsky, S.L.; Tyler, B.; Sosnowski, J.; Brady, K.; Doucet, M.; Nell, D.; Smedley, J.G.; McClane, B.; Brem, H.; Sukumar, S. Clostridium perfringens enterotoxin as a novel-targeted therapeutic for brain metastasis. *Cancer Res.* **2007**, *67*, 7977–7982. [CrossRef] [PubMed]
102. Castro Dias, M.; Coisne, C.; Lazarevic, I.; Baden, P.; Hata, M.; Iwamoto, N.; Francisco, D.M.F.; Vanlandewijck, M.; He, L.; Baier, F.A.; et al. Claudin-3-deficient c57bl/6j mice display intact brain barriers. *Sci. Rep.* **2019**, *9*, 203. [CrossRef] [PubMed]
103. Liebner, S.; Fischmann, A.; Rascher, G.; Duffner, F.; Grote, E.H.; Kalbacher, H.; Wolburg, H. Claudin-1 and claudin-5 expression and tight junction morphology are altered in blood vessels of human glioblastoma multiforme. *Acta Neuropathol.* **2000**, *100*, 323–331. [CrossRef] [PubMed]
104. Pfeiffer, F.; Schafer, J.; Lyck, R.; Makrides, V.; Brunner, S.; Schaeren-Wiemers, N.; Deutsch, U.; Engelhardt, B. Claudin-1 induced sealing of blood-brain barrier tight junctions ameliorates chronic experimental autoimmune encephalomyelitis. *Acta Neuropathol.* **2011**, *122*, 601–614. [CrossRef] [PubMed]
105. Zhang, Y.; Chen, K.; Sloan, S.A.; Bennett, M.L.; Scholze, A.R.; O'Keeffe, S.; Phatnani, H.P.; Guarnieri, P.; Caneda, C.; Ruderisch, N.; et al. An rna-sequencing transcriptome and splicing database of glia, neurons, and vascular cells of the cerebral cortex. *J. Neurosci. Off. J. Soc. Neurosci.* **2014**, *34*, 11929–11947. [CrossRef] [PubMed]
106. Castro Dias, M.; Coisne, C.; Baden, P.; Enzmann, G.; Garrett, L.; Becker, L.; Holter, S.M.; Hrabe de Angelis, M.; Deutsch, U.; Engelhardt, B. Claudin-12 is not required for blood-brain barrier tight junction function. *Fluids Barriers CNS* **2019**, *16*, 30. [CrossRef]
107. Furuse, M.; Hirase, T.; Itoh, M.; Nagafuchi, A.; Yonemura, S.; Tsukita, S.; Tsukita, S. Occludin: A novel integral membrane protein localizing at tight junctions. *J. Cell Biol.* **1993**, *123*, 1777–1788. [CrossRef]
108. Ikenouchi, J.; Furuse, M.; Furuse, K.; Sasaki, H.; Tsukita, S.; Tsukita, S. Tricellulin constitutes a novel barrier at tricellular contacts of epithelial cells. *J. Cell Biol.* **2005**, *171*, 939–945. [CrossRef]
109. Steed, E.; Rodrigues, N.T.; Balda, M.S.; Matter, K. Identification of marveld3 as a tight junction-associated transmembrane protein of the occludin family. *BMC Cell Biol.* **2009**, *10*, 95. [CrossRef]
110. Hirase, T.; Staddon, J.M.; Saitou, M.; Ando-Akatsuka, Y.; Itoh, M.; Furuse, M.; Fujimoto, K.; Tsukita, S.; Rubin, L.L. Occludin as a possible determinant of tight junction permeability in endothelial cells. *J. Cell Sci.* **1997**, *110*, 1603–1613.
111. Iwamoto, N.; Higashi, T.; Furuse, M. Localization of angulin-1/lsr and tricellulin at tricellular contacts of brain and retinal endothelial cells in vivo. *Cell Struct. Funct.* **2014**, *39*, 13015. [CrossRef] [PubMed]
112. Balda, M.S.; Whitney, J.A.; Flores, C.; Gonzalez, S.; Cereijido, M.; Matter, K. Functional dissociation of paracellular permeability and transepithelial electrical resistance and disruption of the apical-basolateral intramembrane diffusion barrier by expression of a mutant tight junction membrane protein. *J. Cell Biol.* **1996**, *134*, 1031–1049. [CrossRef] [PubMed]
113. McCarthy, K.M.; Skare, I.B.; Stankewich, M.C.; Furuse, M.; Tsukita, S.; Rogers, R.A.; Lynch, R.D.; Schneeberger, E.E. Occludin is a functional component of the tight junction. *J. Cell Sci.* **1996**, *109*, 2287–2298. [PubMed]
114. Saitou, M.; Fujimoto, K.; Doi, Y.; Itoh, M.; Fujimoto, T.; Furuse, M.; Takano, H.; Noda, T.; Tsukita, S. Occludin-deficient embryonic stem cells can differentiate into polarized epithelial cells bearing tight junctions. *J. Cell Biol.* **1998**, *141*, 397–408. [CrossRef]
115. Kuwabara, H.; Kokai, Y.; Kojima, T.; Takakuwa, R.; Mori, M.; Sawada, N. Occludin regulates actin cytoskeleton in endothelial cells. *Cell Struct. Funct.* **2001**, *26*, 109–116. [CrossRef] [PubMed]

116. Murakami, T.; Felinski, E.A.; Antonetti, D.A. Occludin phosphorylation and ubiquitination regulate tight junction trafficking and vascular endothelial growth factor-induced permeability. *J. Biol. Chem.* **2009**, *284*, 21036–21046. [CrossRef]
117. Wachtel, M.; Frei, K.; Ehler, E.; Fontana, A.; Winterhalter, K.; Gloor, S.M. Occludin proteolysis and increased permeability in endothelial cells through tyrosine phosphatase inhibition. *J. Cell Sci.* **1999**, *112*, 4347–4356.
118. Saitou, M.; Furuse, M.; Sasaki, H.; Schulzke, J.D.; Fromm, M.; Takano, H.; Noda, T.; Tsukita, S. Complex phenotype of mice lacking occludin, a component of tight junction strands. *Mol. Biol. Cell* **2000**, *11*, 4131–4142. [CrossRef]
119. Higashi, T.; Miller, A.L. Tricellular junctions: How to build junctions at the trickiest points of epithelial cells. *Mol. Biol. Cell* **2017**, *28*, 2023–2034. [CrossRef]
120. Mariano, C.; Palmela, I.; Pereira, P.; Fernandes, A.; Falcao, A.S.; Cardoso, F.L.; Vaz, A.R.; Campos, A.R.; Goncalves-Ferreira, A.; Kim, K.S.; et al. Tricellulin expression in brain endothelial and neural cells. *Cell Tissue Res.* **2013**, *351*, 397–407. [CrossRef]
121. Martin-Padura, I.; Lostaglio, S.; Schneemann, M.; Williams, L.; Romano, M.; Fruscella, P.; Panzeri, C.; Stoppacciaro, A.; Ruco, L.; Villa, A.; et al. Junctional adhesion molecule, a novel member of the immunoglobulin superfamily that distributes at intercellular junctions and modulates monocyte transmigration. *J. Cell Biol.* **1998**, *142*, 117–127. [CrossRef] [PubMed]
122. Garrido-Urbani, S.; Bradfield, P.F.; Imhof, B.A. Tight junction dynamics: The role of junctional adhesion molecules (jams). *Cell Tissue Res.* **2014**, *355*, 701–715. [CrossRef] [PubMed]
123. Aurrand-Lions, M.; Duncan, L.; Ballestrem, C.; Imhof, B.A. Jam-2, a novel immunoglobulin superfamily molecule, expressed by endothelial and lymphatic cells. *J. Biol. Chem.* **2001**, *276*, 2733–2741. [CrossRef] [PubMed]
124. Liu, Y.; Nusrat, A.; Schnell, F.J.; Reaves, T.A.; Walsh, S.; Pochet, M.; Parkos, C.A. Human junction adhesion molecule regulates tight junction resealing in epithelia. *J. Cell Sci.* **2000**, *113*, 2363–2374. [PubMed]
125. Wyss, L.; Schafer, J.; Liebner, S.; Mittelbronn, M.; Deutsch, U.; Enzmann, G.; Adams, R.H.; Aurrand-Lions, M.; Plate, K.H.; Imhof, B.A.; et al. Junctional adhesion molecule (jam)-c deficient c57bl/6 mice develop a severe hydrocephalus. *PLoS ONE* **2012**, *7*, e45619. [CrossRef] [PubMed]
126. Padden, M.; Leech, S.; Craig, B.; Kirk, J.; Brankin, B.; McQuaid, S. Differences in expression of junctional adhesion molecule-a and beta-catenin in multiple sclerosis brain tissue: Increasing evidence for the role of tight junction pathology. *Acta Neuropathol.* **2007**, *113*, 177–186. [CrossRef]
127. Arrate, M.P.; Rodriguez, J.M.; Tran, T.M.; Brock, T.A.; Cunningham, S.A. Cloning of human junctional adhesion molecule 3 (jam3) and its identification as the jam2 counter-receptor. *J. Biol. Chem.* **2001**, *276*, 45826–45832. [CrossRef]
128. Ebnet, K.; Aurrand-Lions, M.; Kuhn, A.; Kiefer, F.; Butz, S.; Zander, K.; Meyer zu Brickwedde, M.K.; Suzuki, A.; Imhof, B.A.; Vestweber, D. The junctional adhesion molecule (jam) family members jam-2 and jam-3 associate with the cell polarity protein par-3: A possible role for jams in endothelial cell polarity. *J. Cell Sci.* **2003**, *116*, 3879–3891. [CrossRef]
129. Ebnet, K.; Suzuki, A.; Horikoshi, Y.; Hirose, T.; Meyer Zu Brickwedde, M.K.; Ohno, S.; Vestweber, D. The cell polarity protein asip/par-3 directly associates with junctional adhesion molecule (jam). *EMBO J.* **2001**, *20*, 3738–3748. [CrossRef]
130. Williams, D.W.; Calderon, T.M.; Lopez, L.; Carvallo-Torres, L.; Gaskill, P.J.; Eugenin, E.A.; Morgello, S.; Berman, J.W. Mechanisms of hiv entry into the cns: Increased sensitivity of hiv infected cd14+cd16+ monocytes to ccl2 and key roles of ccr2, jam-a, and alcam in diapedesis. *PLoS ONE* **2013**, *8*, e69270. [CrossRef]
131. Tietz, S.; Perinat, T.; Greene, G.; Enzmann, G.; Deutsch, U.; Adams, R.; Imhof, B.; Aurrand-Lions, M.; Engelhardt, B. Lack of junctional adhesion molecule (jam)-b ameliorates experimental autoimmune encephalomyelitis. *Brain Behav. Immun.* **2018**, *73*, 3–20. [CrossRef] [PubMed]
132. Cunningham, S.A.; Rodriguez, J.M.; Arrate, M.P.; Tran, T.M.; Brock, T.A. Jam2 interacts with alpha$_4$beta$_1$. Facilitation by jam3. *J. Biol. Chem.* **2002**, *277*, 27589–27592. [CrossRef] [PubMed]
133. Ludwig, R.J.; Hardt, K.; Hatting, M.; Bistrian, R.; Diehl, S.; Radeke, H.H.; Podda, M.; Schon, M.P.; Kaufmann, R.; Henschler, R.; et al. Junctional adhesion molecule (jam)-b supports lymphocyte rolling and adhesion through interaction with alpha4beta1 integrin. *Immunology* **2009**, *128*, 196–205. [CrossRef] [PubMed]

134. Martin-Blondel, G.; Pignolet, B.; Tietz, S.; Yshii, L.; Gebauer, C.; Perinat, T.; Van Weddingen, I.; Blatti, C.; Engelhardt, B.; Liblau, R. Migration of encephalitogenic cd8 t cells into the central nervous system is dependent on the alpha4beta1-integrin. *Eur. J. Immunol.* **2015**, *45*, 3302–3312. [CrossRef] [PubMed]
135. Mochida, G.H.; Ganesh, V.S.; Felie, J.M.; Gleason, D.; Hill, R.S.; Clapham, K.R.; Rakiec, D.; Tan, W.H.; Akawi, N.; Al-Saffar, M.; et al. A homozygous mutation in the tight-junction protein jam3 causes hemorrhagic destruction of the brain, subependymal calcification, and congenital cataracts. *Am. J. Hum. Genet.* **2010**, *87*, 882–889. [CrossRef] [PubMed]
136. Funke, L.; Dakoji, S.; Bredt, D.S. Membrane-associated guanylate kinases regulate adhesion and plasticity at cell junctions. *Annu. Rev. Biochem.* **2005**, *74*, 219–245. [CrossRef]
137. Dimitratos, S.D.; Woods, D.F.; Stathakis, D.G.; Bryant, P.J. Signaling pathways are focused at specialized regions of the plasma membrane by scaffolding proteins of the maguk family. *Bioessays News Rev. Mol. Cell. Dev. Biol.* **1999**, *21*, 912–921. [CrossRef]
138. Gonzalez-Mariscal, L.; Betanzos, A.; Avila-Flores, A. Maguk proteins: Structure and role in the tight junction. *Semin. Cell Dev. Biol.* **2000**, *11*, 315–324. [CrossRef]
139. Umeda, K.; Ikenouchi, J.; Katahira-Tayama, S.; Furuse, K.; Sasaki, H.; Nakayama, M.; Matsui, T.; Tsukita, S.; Furuse, M.; Tsukita, S. Zo-1 and zo-2 independently determine where claudins are polymerized in tight-junction strand formation. *Cell* **2006**, *126*, 741–754. [CrossRef]
140. Katsuno, T.; Umeda, K.; Matsui, T.; Hata, M.; Tamura, A.; Itoh, M.; Takeuchi, K.; Fujimori, T.; Nabeshima, Y.; Noda, T.; et al. Deficiency of zonula occludens-1 causes embryonic lethal phenotype associated with defected yolk sac angiogenesis and apoptosis of embryonic cells. *Mol. Biol. Cell* **2008**, *19*, 2465–2475. [CrossRef]
141. Xu, J.; Kausalya, P.J.; Phua, D.C.; Ali, S.M.; Hossain, Z.; Hunziker, W. Early embryonic lethality of mice lacking zo-2, but not zo-3, reveals critical and nonredundant roles for individual zonula occludens proteins in mammalian development. *Mol. Cell. Biol.* **2008**, *28*, 1669–1678. [CrossRef] [PubMed]
142. Van Itallie, C.M.; Anderson, J.M. Architecture of tight junctions and principles of molecular composition. *Semin. Cell Dev. Biol.* **2014**, *36*, 157–165. [CrossRef] [PubMed]
143. Tornavaca, O.; Chia, M.; Dufton, N.; Almagro, L.O.; Conway, D.E.; Randi, A.M.; Schwartz, M.A.; Matter, K.; Balda, M.S. Zo-1 controls endothelial adherens junctions, cell-cell tension, angiogenesis, and barrier formation. *J. Cell Biol.* **2015**, *208*, 821–838. [CrossRef] [PubMed]
144. Hawkins, B.T.; Davis, T.P. The blood-brain barrier/neurovascular unit in health and disease. *Pharmacol. Rev.* **2005**, *57*, 173–185. [CrossRef]
145. Davies, D.C. Blood-brain barrier breakdown in septic encephalopathy and brain tumours. *J. Anat.* **2002**, *200*, 639–646. [CrossRef]
146. Cordenonsi, M.; D'Atri, F.; Hammar, E.; Parry, D.A.; Kendrick-Jones, J.; Shore, D.; Citi, S. Cingulin contains globular and coiled-coil domains and interacts with zo-1, zo-2, zo-3, and myosin. *J. Cell Biol.* **1999**, *147*, 1569–1582. [CrossRef]
147. Balda, M.S.; Matter, K. The tight junction protein zo-1 and an interacting transcription factor regulate erbb-2 expression. *EMBO J.* **2000**, *19*, 2024–2033. [CrossRef]
148. Satoh, H.; Zhong, Y.; Isomura, H.; Saitoh, M.; Enomoto, K.; Sawada, N.; Mori, M. Localization of 7h6 tight junction-associated antigen along the cell border of vascular endothelial cells correlates with paracellular barrier function against ions, large molecules, and cancer cells. *Exp. Cell Res.* **1996**, *222*, 269–274. [CrossRef]
149. Tietz, S.; Engelhardt, B. Brain barriers: Crosstalk between complex tight junctions and adherens junctions. *J. Cell Biol.* **2015**, *209*, 493–506. [CrossRef]
150. Carmeliet, P.; Lampugnani, M.G.; Moons, L.; Breviario, F.; Compernolle, V.; Bono, F.; Balconi, G.; Spagnuolo, R.; Oosthuyse, B.; Dewerchin, M.; et al. Targeted deficiency or cytosolic truncation of the ve-cadherin gene in mice impairs vegf-mediated endothelial survival and angiogenesis. *Cell* **1999**, *98*, 147–157. [CrossRef]
151. Crosby, C.V.; Fleming, P.A.; Argraves, W.S.; Corada, M.; Zanetta, L.; Dejana, E.; Drake, C.J. Ve-cadherin is not required for the formation of nascent blood vessels but acts to prevent their disassembly. *Blood* **2005**, *105*, 2771–2776. [CrossRef] [PubMed]
152. Yap, A.S.; Niessen, C.M.; Gumbiner, B.M. The juxtamembrane region of the cadherin cytoplasmic tail supports lateral clustering, adhesive strengthening, and interaction with $p120^{ctn}$. *J. Cell Biol.* **1998**, *141*, 779–789. [CrossRef] [PubMed]

153. Shibamoto, S.; Hayakawa, M.; Takeuchi, K.; Hori, T.; Miyazawa, K.; Kitamura, N.; Johnson, K.R.; Wheelock, M.J.; Matsuyoshi, N.; Takeichi, M.; et al. Association of p120, a tyrosine kinase substrate, with e-cadherin/catenin complexes. *J. Cell Biol.* **1995**, *128*, 949–957. [CrossRef] [PubMed]
154. Rimm, D.L.; Koslov, E.R.; Kebriaei, P.; Cianci, C.D.; Morrow, J.S. Alpha 1(e)-catenin is an actin-binding and -bundling protein mediating the attachment of f-actin to the membrane adhesion complex. *Proc. Natl. Acad. Sci. USA* **1995**, *92*, 8813–8817. [CrossRef] [PubMed]
155. Kobielak, A.; Fuchs, E. Alpha-catenin: At the junction of intercellular adhesion and actin dynamics. *Nat. Rev. Mol. Cell Biol.* **2004**, *5*, 614–625. [CrossRef]
156. Taddei, A.; Giampietro, C.; Conti, A.; Orsenigo, F.; Breviario, F.; Pirazzoli, V.; Potente, M.; Daly, C.; Dimmeler, S.; Dejana, E. Endothelial adherens junctions control tight junctions by ve-cadherin-mediated upregulation of claudin-5. *Nat. Cell Biol.* **2008**, *10*, 923–934. [CrossRef]
157. Indra, I.; Hong, S.; Troyanovsky, R.; Kormos, B.; Troyanovsky, S. The adherens junction: A mosaic of cadherin and nectin clusters bundled by actin filaments. *J. Investig. Dermatol.* **2013**, *133*, 2546–2554. [CrossRef]
158. Takai, Y.; Irie, K.; Shimizu, K.; Sakisaka, T.; Ikeda, W. Nectins and nectin-like molecules: Roles in cell adhesion, migration, and polarization. *Cancer Sci.* **2003**, *94*, 655–667. [CrossRef]
159. Newman, P.J. The biology of pecam-1. *J. Clin. Investig.* **1997**, *99*, 3–8. [CrossRef]
160. Lyck, R.; Ruderisch, N.; Moll, A.G.; Steiner, O.; Cohen, C.D.; Engelhardt, B.; Makrides, V.; Verrey, F. Culture-induced changes in blood-brain barrier transcriptome: Implications for amino-acid transporters in vivo. *J. Cereb. Blood Flow Metab. Off. J. Int. Soc. Cereb. Blood Flow Metab.* **2009**, *29*, 1491–1502. [CrossRef]
161. Graesser, D.; Solowiej, A.; Bruckner, M.; Osterweil, E.; Juedes, A.; Davis, S.; Ruddle, N.H.; Engelhardt, B.; Madri, J.A. Altered vascular permeability and early onset of experimental autoimmune encephalomyelitis in pecam-1-deficient mice. *J. Clin. Investig.* **2002**, *109*, 383–392. [CrossRef]
162. Wimmer, I.; Tietz, S.; Nishihara, H.; Deutsch, U.; Sallusto, F.; Gosselet, F.; Lyck, R.; Muller, W.A.; Lassmann, H.; Engelhardt, B. Pecam-1 stabilizes blood-brain barrier integrity and favors paracellular t-cell diapedesis across the blood-brain barrier during neuroinflammation. *Front. Immunol.* **2019**, *10*, 711. [CrossRef] [PubMed]
163. Privratsky, J.R.; Newman, P.J. Pecam-1: Regulator of endothelial junctional integrity. *Cell Tissue Res.* **2014**, *355*, 607–619. [CrossRef] [PubMed]
164. Schenkel, A.R.; Mamdouh, Z.; Chen, X.; Liebman, R.M.; Muller, W.A. Cd99 plays a major role in the migration of monocytes through endothelial junctions. *Nat. Immunol.* **2002**, *3*, 143–150. [CrossRef] [PubMed]
165. Bixel, G.; Kloep, S.; Butz, S.; Petri, B.; Engelhardt, B.; Vestweber, D. Mouse cd99 participates in t-cell recruitment into inflamed skin. *Blood* **2004**, *104*, 3205–3213. [CrossRef]
166. Winger, R.C.; Harp, C.T.; Chiang, M.Y.; Sullivan, D.P.; Watson, R.L.; Weber, E.W.; Podojil, J.R.; Miller, S.D.; Muller, W.A. Cutting edge: CD99 is a novel therapeutic target for control of t cell-mediated central nervous system autoimmune disease. *J. Immunol.* **2016**, *196*, 1443–1448. [CrossRef]
167. Hoshi, Y.; Uchida, Y.; Tachikawa, M.; Inoue, T.; Ohtsuki, S.; Terasaki, T. Quantitative atlas of blood-brain barrier transporters, receptors, and tight junction proteins in rats and common marmoset. *J. Pharm. Sci.* **2013**, *102*, 3343–3355. [CrossRef]
168. O'Brown, N.M.; Pfau, S.J.; Gu, C. Bridging barriers: A comparative look at the blood-brain barrier across organisms. *Genes Dev.* **2018**, *32*, 466–478. [CrossRef]
169. Lehtinen, M.K.; Bjornsson, C.S.; Dymecki, S.M.; Gilbertson, R.J.; Holtzman, D.M.; Monuki, E.S. The choroid plexus and cerebrospinal fluid: Emerging roles in development, disease, and therapy. *J. Neurosci. Off. J. Soc. Neurosci.* **2013**, *33*, 17553–17559. [CrossRef]
170. Redzic, Z.B.; Preston, J.E.; Duncan, J.A.; Chodobski, A.; Szmydynger-Chodobska, J. The choroid plexus-cerebrospinal fluid system: From development to aging. *Curr. Top. Dev. Biol.* **2005**, *71*, 1–52.
171. Kaur, C.; Rathnasamy, G.; Ling, E.A. The choroid plexus in healthy and diseased brain. *J. Neuropathol. Exp. Neurol.* **2016**, *75*, 198–213. [CrossRef] [PubMed]
172. Hurley, J.V.; Anderson, R.M.; Sexton, P.T. The fate of plasma protein which escapes from blood vessels of the choroid plexus of the rat—An electron microscope study. *J. Pathol.* **1981**, *134*, 57–70. [CrossRef] [PubMed]
173. Liddelow, S.A.; Temple, S.; Mollgard, K.; Gehwolf, R.; Wagner, A.; Bauer, H.; Bauer, H.C.; Phoenix, T.N.; Dziegielewska, K.M.; Saunders, N.R. Molecular characterisation of transport mechanisms at the developing mouse blood-csf interface: A transcriptome approach. *PLoS ONE* **2012**, *7*, e33554. [CrossRef]

174. Wolburg, H.; Wolburg-Buchholz, K.; Liebner, S.; Engelhardt, B. Claudin-1, claudin-2 and claudin-11 are present in tight junctions of choroid plexus epithelium of the mouse. *Neurosci. Lett.* **2001**, *307*, 77–80. [CrossRef]
175. Aijaz, S.; Balda, M.S.; Matter, K. Tight junctions: Molecular architecture and function. *Int. Rev. Cytol.* **2006**, *248*, 261–298. [PubMed]
176. Kratzer, I.; Vasiljevic, A.; Rey, C.; Fevre-Montange, M.; Saunders, N.; Strazielle, N.; Ghersi-Egea, J.F. Complexity and developmental changes in the expression pattern of claudins at the blood-csf barrier. *Histochem. Cell Biol.* **2012**, *138*, 861–879. [CrossRef]
177. Muto, S.; Hata, M.; Taniguchi, J.; Tsuruoka, S.; Moriwaki, K.; Saitou, M.; Furuse, K.; Sasaki, H.; Fujimura, A.; Imai, M.; et al. Claudin-2-deficient mice are defective in the leaky and cation-selective paracellular permeability properties of renal proximal tubules. *Proc. Natl. Acad. Sci. USA* **2010**, *107*, 8011–8016. [CrossRef]
178. Maheras, K.J.; Peppi, M.; Ghoddoussi, F.; Galloway, M.P.; Perrine, S.A.; Gow, A. Absence of claudin 11 in cns myelin perturbs behavior and neurotransmitter levels in mice. *Sci. Rep.* **2018**, *8*, 3798. [CrossRef]
179. Gow, A.; Davies, C.; Southwood, C.M.; Frolenkov, G.; Chrustowski, M.; Ng, L.; Yamauchi, D.; Marcus, D.C.; Kachar, B. Deafness in claudin 11-null mice reveals the critical contribution of basal cell tight junctions to stria vascularis function. *J. Neurosci. Off. J. Soc. Neurosci.* **2004**, *24*, 7051–7062. [CrossRef]
180. Kooij, G.; Kopplin, K.; Blasig, R.; Stuiver, M.; Koning, N.; Goverse, G.; van der Pol, S.M.; van Het Hof, B.; Gollasch, M.; Drexhage, J.A.; et al. Disturbed function of the blood-cerebrospinal fluid barrier aggravates neuro-inflammation. *Acta Neuropathol.* **2014**, *128*, 267–277. [CrossRef]
181. Vestweber, D. Cadherins in tissue architecture and disease. *J. Mol. Med.* **2015**, *93*, 5–11. [CrossRef] [PubMed]
182. Liddelow, S.A. Development of the choroid plexus and blood-csf barrier. *Front. Neurosci.* **2015**, *9*, 32. [CrossRef] [PubMed]
183. Decimo, I.; Fumagalli, G.; Berton, V.; Krampera, M.; Bifari, F. Meninges: From protective membrane to stem cell niche. *Am. J. Stem. Cells* **2012**, *1*, 92–105. [PubMed]
184. Haines, D.E.; Mihailoff, G.A. Fundamental neuroscience for basic and clinical application. In *Science Direct*; Elsevier Publishing Company: Amsterdam, The Netherlands, 2018; pp. 107–121.
185. Bifari, F.; Berton, V.; Pino, A.; Kusalo, M.; Malpeli, G.; Di Chio, M.; Bersan, E.; Amato, E.; Scarpa, A.; Krampera, M.; et al. Meninges harbor cells expressing neural precursor markers during development and adulthood. *Front. Cell. Neurosci.* **2015**, *9*, 383. [CrossRef]
186. Bjornsson, C.S.; Apostolopoulou, M.; Tian, Y.; Temple, S. It takes a village: Constructing the neurogenic niche. *Dev. Cell* **2015**, *32*, 435–446. [CrossRef]
187. Pizzo, M.E.; Wolak, D.J.; Kumar, N.N.; Brunette, E.; Brunnquell, C.L.; Hannocks, M.J.; Abbott, N.J.; Meyerand, M.E.; Sorokin, L.; Stanimirovic, D.B.; et al. Intrathecal antibody distribution in the rat brain: Surface diffusion, perivascular transport and osmotic enhancement of delivery. *J. Physiol.* **2018**, *596*, 445–475. [CrossRef]
188. Gagan, J.R.; Tholpady, S.S.; Ogle, R.C. Cellular dynamics and tissue interactions of the dura mater during head development. *Birth Defects Res. Part C Embryo Today Rev.* **2007**, *81*, 297–304. [CrossRef]
189. Saunders, N.R.; Habgood, M.D.; Mollgard, K.; Dziegielewska, K.M. The biological significance of brain barrier mechanisms: Help or hindrance in drug delivery to the central nervous system? *F1000Research* **2016**, *10*, 5. [CrossRef]
190. Nabeshima, S.; Reese, T.S.; Landis, D.M.; Brightman, M.W. Junctions in the meninges and marginal glia. *J. Comp. Neurol.* **1975**, *164*, 127–169. [CrossRef]
191. Coles, J.A.; Myburgh, E.; Brewer, J.M.; McMenamin, P.G. Where are we? The anatomy of the murine cortical meninges revisited for intravital imaging, immunology, and clearance of waste from the brain. *Prog. Neurobiol.* **2017**, *156*, 107–148. [CrossRef]
192. Yasuda, K.; Cline, C.; Vogel, P.; Onciu, M.; Fatima, S.; Sorrentino, B.P.; Thirumaran, R.K.; Ekins, S.; Urade, Y.; Fujimori, K.; et al. Drug transporters on arachnoid barrier cells contribute to the blood-cerebrospinal fluid barrier. *Drug Metab. Dispos. Biol. Fate Chem.* **2013**, *41*, 923–931. [CrossRef] [PubMed]
193. Figarella-Branger, D.; Pellissier, J.F.; Bouillot, P.; Bianco, N.; Mayan, M.; Grisoli, F.; Rougon, G. Expression of neural cell-adhesion molecule isoforms and epithelial cadherin adhesion molecules in 47 human meningiomas: Correlation with clinical and morphological data. *Mod. Pathol. Off. J. U. S. Can. Acad. Pathol.* **1994**, *7*, 752–761.

194. Holman, D.W.; Grzybowski, D.M.; Mehta, B.C.; Katz, S.E.; Lubow, M. Characterization of cytoskeletal and junctional proteins expressed by cells cultured from human arachnoid granulation tissue. *Cereb. Fluid Res.* **2005**, *2*, 9. [CrossRef] [PubMed]
195. Grafstein, B.; Liu, S.; Cotrina, M.L.; Goldman, S.A.; Nedergaard, M. Meningeal cells can communicate with astrocytes by calcium signaling. *Ann. Neurol.* **2000**, *47*, 18–25. [CrossRef]
196. Kartenbeck, J.; Schwechheimer, K.; Moll, R.; Franke, W.W. Attachment of vimentin filaments to desmosomal plaques in human meningiomal cells and arachnoidal tissue. *J. Cell Biol.* **1984**, *98*, 1072–1081. [CrossRef]
197. Murphy, M.; Chen, J.N.; George, D.L. Establishment and characterization of a human leptomeningeal cell line. *J. Neurosci. Res.* **1991**, *30*, 475–483. [CrossRef]
198. Hasegawa, M.; Yamashima, T.; Kida, S.; Yamashita, J. Membranous ultrastructure of human arachnoid cells. *J. Neuropathol. Exp. Neurol.* **1997**, *56*, 1217–1227. [CrossRef]
199. Rudolf Nieuwenhuys, J.V.; van Huijzen, C. *The Human Central Nervous System: A Synopsis and Atlas*; Springer Berlin Heidelberg: Berlin, Germany, 2007; Chapter 4.
200. Kebir, H.; Kreymborg, K.; Ifergan, I.; Dodelet-Devillers, A.; Cayrol, R.; Bernard, M.; Giuliani, F.; Arbour, N.; Becher, B.; Prat, A. Human T_H17 lymphocytes promote blood-brain barrier disruption and central nervous system inflammation. *Nat. Med.* **2007**, *13*, 1173–1175. [CrossRef]
201. Horng, S.; Therattil, A.; Moyon, S.; Gordon, A.; Kim, K.; Argaw, A.T.; Hara, Y.; Mariani, J.N.; Sawai, S.; Flodby, P.; et al. Astrocytic tight junctions control inflammatory cns lesion pathogenesis. *J. Clin. Investig.* **2017**, *127*, 3136–3151. [CrossRef]
202. Stamatovic, S.M.; Sladojevic, N.; Keep, R.F.; Andjelkovic, A.V. Relocalization of junctional adhesion molecule a during inflammatory stimulation of brain endothelial cells. *Mol. Cell. Biol.* **2012**, *32*, 3414–3427. [CrossRef]
203. Schlingmann, B.; Overgaard, C.E.; Molina, S.A.; Lynn, K.S.; Mitchell, L.A.; Dorsainvil White, S.; Mattheyses, A.L.; Guidot, D.M.; Capaldo, C.T.; Koval, M. Regulation of claudin/zonula occludens-1 complexes by hetero-claudin interactions. *Nat. Commun.* **2016**, *7*, 12276. [CrossRef] [PubMed]
204. Kaufmann, R.; Piontek, J.; Grull, F.; Kirchgessner, M.; Rossa, J.; Wolburg, H.; Blasig, I.E.; Cremer, C. Visualization and quantitative analysis of reconstituted tight junctions using localization microscopy. *PLoS ONE* **2012**, *7*, e31128. [CrossRef] [PubMed]
205. Paul, D.; Baena, V.; Ge, S.; Jiang, X.; Jellison, E.R.; Kiprono, T.; Agalliu, D.; Pachter, J.S. Appearance of claudin-5$^+$ leukocytes in the central nervous system during neuroinflammation: A novel role for endothelial-derived extracellular vesicles. *J. Neuroinflamm.* **2016**, *13*, 292. [CrossRef] [PubMed]
206. Winderlich, M.; Keller, L.; Cagna, G.; Broermann, A.; Kamenyeva, O.; Kiefer, F.; Deutsch, U.; Nottebaum, A.F.; Vestweber, D. Ve-ptp controls blood vessel development by balancing tie-2 activity. *J. Cell Biol.* **2009**, *185*, 657–671. [CrossRef] [PubMed]
207. Snippert, H.J.; van der Flier, L.G.; Sato, T.; van Es, J.H.; van den Born, M.; Kroon-Veenboer, C.; Barker, N.; Klein, A.M.; van Rheenen, J.; Simons, B.D.; et al. Intestinal crypt homeostasis results from neutral competition between symmetrically dividing lgr5 stem cells. *Cell* **2010**, *143*, 134–144. [CrossRef]
208. Feng, S.; Sekine, S.; Pessino, V.; Li, H.; Leonetti, M.D.; Huang, B. Improved split fluorescent proteins for endogenous protein labeling. *Nat. Commun.* **2017**, *8*, 370. [CrossRef]
209. Kamiyama, D.; Sekine, S.; Barsi-Rhyne, B.; Hu, J.; Chen, B.; Gilbert, L.A.; Ishikawa, H.; Leonetti, M.D.; Marshall, W.F.; Weissman, J.S.; et al. Versatile protein tagging in cells with split fluorescent protein. *Nat. Commun.* **2016**, *7*, 11046. [CrossRef]
210. Cabantous, S.; Terwilliger, T.C.; Waldo, G.S. Protein tagging and detection with engineered self-assembling fragments of green fluorescent protein. *Nat. Biotechnol.* **2005**, *23*, 102–107. [CrossRef]
211. Madisen, L.; Zwingman, T.A.; Sunkin, S.M.; Oh, S.W.; Zariwala, H.A.; Gu, H.; Ng, L.L.; Palmiter, R.D.; Hawrylycz, M.J.; Jones, A.R.; et al. A robust and high-throughput cre reporting and characterization system for the whole mouse brain. *Nat. Neurosci.* **2010**, *13*, 133–140. [CrossRef]
212. Hartmann, D.A.; Underly, R.G.; Watson, A.N.; Shih, A.Y. A murine toolbox for imaging the neurovascular unit. *Microcirculation* **2015**, *22*, 168–182. [CrossRef]
213. Ostrowski, L.E.; Hutchins, J.R.; Zakel, K.; O'Neal, W.K. Targeting expression of a transgene to the airway surface epithelium using a ciliated cell-specific promoter. *Mol. Ther. J. Am. Soc. Gene Ther.* **2003**, *8*, 637–645. [CrossRef]

214. Johnson, B.A.; Coutts, M.; Vo, H.M.; Hao, X.; Fatima, N.; Rivera, M.J.; Sims, R.J.; Neel, M.J.; Kang, Y.J.; Monuki, E.S. Accurate, strong, and stable reporting of choroid plexus epithelial cells in transgenic mice using a human transthyretin bac. *Fluids Barriers CNS* **2018**, *15*, 22. [CrossRef] [PubMed]
215. Xin, H.B.; Deng, K.Y.; Rishniw, M.; Ji, G.; Kotlikoff, M.I. Smooth muscle expression of cre recombinase and egfp in transgenic mice. *Physiol. Genom.* **2002**, *10*, 211–215. [CrossRef]
216. Cuttler, A.S.; LeClair, R.J.; Stohn, J.P.; Wang, Q.; Sorenson, C.M.; Liaw, L.; Lindner, V. Characterization of pdgfrb-cre transgenic mice reveals reduction of rosa26 reporter activity in remodeling arteries. *Genesis* **2011**, *49*, 673–680. [CrossRef] [PubMed]
217. Girbl, T.; Lenn, T.; Perez, L.; Rolas, L.; Barkaway, A.; Thiriot, A.; Del Fresno, C.; Lynam, E.; Hub, E.; Thelen, M.; et al. Distinct compartmentalization of the chemokines cxcl1 and cxcl2 and the atypical receptor ackr1 determine discrete stages of neutrophil diapedesis. *Immunity* **2018**, *49*, 1062–1076. [CrossRef] [PubMed]
218. Zipfel, W.R.; Williams, R.M.; Christie, R.; Nikitin, A.Y.; Hyman, B.T.; Webb, W.W. Live tissue intrinsic emission microscopy using multiphoton-excited native fluorescence and second harmonic generation. *Proc. Natl. Acad. Sci. USA* **2003**, *100*, 7075–7080. [CrossRef] [PubMed]
219. Nam, M.H.; Baek, M.; Lim, J.; Lee, S.; Yoon, J.; Kim, J.; Lee, M.S.; Soh, K.S. Discovery of a novel fibrous tissue in the spinal pia mater by polarized light microscopy. *Connect. Tissue Res.* **2014**, *55*, 147–155. [CrossRef]
220. Iliff, J.J.; Wang, M.; Liao, Y.; Plogg, B.A.; Peng, W.; Gundersen, G.A.; Benveniste, H.; Vates, G.E.; Deane, R.; Goldman, S.A.; et al. A paravascular pathway facilitates csf flow through the brain parenchyma and the clearance of interstitial solutes, including amyloid beta. *Sci. Transl. Med.* **2012**, *4*, 147ra111. [CrossRef]

12

Contributions of Myosin Light Chain Kinase to Regulation of Epithelial Paracellular Permeability and Mucosal Homeostasis

Wei-Qi He [1,*], Jing Wang [1], Jian-Ying Sheng [1], Juan-Min Zha [1], W. Vallen Graham [2,3] and Jerrold R. Turner [4,*]

1 Jiangsu Key Laboratory of Neuropsychiatric Diseases and Cambridge-Suda (CAM-SU) Genomic Resource Center, Medical College of Soochow University, Suzhou 215123, China; 20184250017@stu.suda.edu.cn (J.W.); 20184250012@stu.suda.edu.cn (J.-Y.S.); zhajuanmin@suda.edu.cn (J.-M.Z.)

2 Thelium Therapeutics, New York, NY 10014, USA; vallen@theliumtx.com

3 Laboratory of Chemical Biology & Signal Transduction, The Rockefeller University, New York, NY 10065, USA

4 Laboratory of Mucosal Barrier Pathobiology, Department of Pathology, Brigham and Women's Hospital and Harvard Medical School, Boston, MA 02115, USA

* Correspondence: whe@suda.edu.cn (W.-Q.H.); jrturner@bwh.harvard.edu (J.R.T.)

Abstract: Intestinal barrier function is required for the maintenance of mucosal homeostasis. Barrier dysfunction is thought to promote progression of both intestinal and systemic diseases. In many cases, this barrier loss reflects increased permeability of the paracellular tight junction as a consequence of myosin light chain kinase (MLCK) activation and myosin II regulatory light chain (MLC) phosphorylation. Although some details about MLCK activation remain to be defined, it is clear that this triggers perijunctional actomyosin ring (PAMR) contraction that leads to molecular reorganization of tight junction structure and composition, including occludin endocytosis. In disease states, this process can be triggered by pro-inflammatory cytokines including tumor necrosis factor-α (TNF), interleukin-1β (IL-1β), and several related molecules. Of these, TNF has been studied in the greatest detail and is known to activate long MLCK transcription, expression, enzymatic activity, and recruitment to the PAMR. Unfortunately, toxicities associated with inhibition of MLCK expression or enzymatic activity make these unsuitable as therapeutic targets. Recent work has, however, identified a small molecule that prevents MLCK1 recruitment to the PAMR without inhibiting enzymatic function. This small molecule, termed Divertin, restores barrier function after TNF-induced barrier loss and prevents disease progression in experimental chronic inflammatory bowel disease.

Keywords: tight junction; barrier function; inflammatory bowel disease; drug development; mucosal immunology; cytokines; ZO-1; occludin; claudin; actomyosin

---

1. Structure of Epithelial Intercellular Junctions

Intestinal mucosal surfaces are covered by a single layer of columnar epithelial cells required for absorptive and defensive functions. This requires cellular polarization in order to ensure appropriately-oriented membrane specializations, protein and lipid trafficking, and vectorial transport of ions and larger solutes [1–3]. For example, the dense forests of microvilli that increase apical surface area and facilitate nutrition absorption [4] are not needed on the basolateral surface, which interacts with adjacent epithelial cells, immune cells, and the basement membrane.

In order to maintain polarized function and regulate the stimuli to which apical and basolateral surfaces are exposed, intestinal epithelial cells must establish a physical barrier that prevents free diffusion across the paracellular shunt pathway [5–14]. This barrier is formed by a series of junctions that provide different types of intercellular connections. This apical junctional complex cannot be resolved by light microscopy but can be seen as the terminal bar, a dense spot or bar where apical and lateral membranes meet [5,15–17]. Transmission electron microscopy allows visualization of distinct regions such as, from apical to basal, tight junctions (zonulae occludens), adherens junctions (zonulae adherens), and desmosomes [5] (Figure 1A,B). Tight junctions are sites of close apposition of adjacent cell membranes, termed kiss sites, where the adjacent plasma membrane outer leaflets appear to fuse into a single layer. This corresponds to the site at which paracellular flux of macromolecular probes is blocked. Although some microfilaments are associated with tight junctions, they are most concentrated at adherens junctions [18–20], which do not form barriers to paracellular flux but are critical to intercellular adhesion and maintenance of the apical junctional complex. Finally, desmosomes can be recognized as electron dense membrane structures associated with intermediate, i.e., cytokeratin, filaments [21–24].

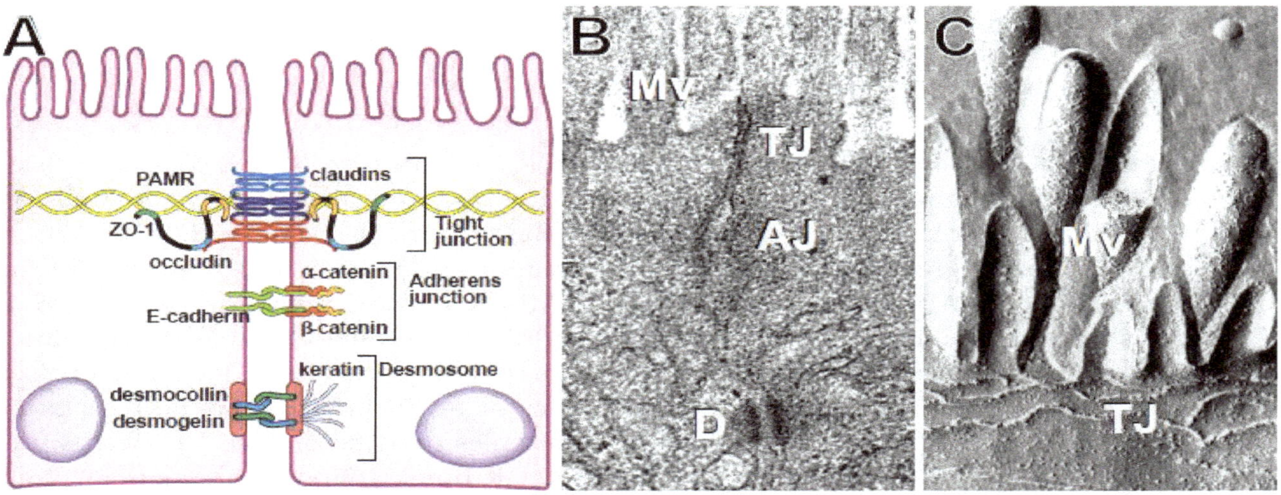

Figure 1. The structure of epithelial intercellular junctions. (**A**) Schematic showing interactions between the perijunctional actomyosin ring (PAMR), zonula occludens-1 (ZO-1), occludin, claudins at the tight junction; E-cadherin, α-catenin, and β-catenin at the adherens junction; and desmogelin, desmocollin, and intermediate filaments at the desmosome. (**B**) Transmission electron micrograph showing the tight junction (TJ), adherens junction (AJ), desmosome (D), and microvilli (Mv). From Turner. Nat Rev Immunol 2009. (**C**) Freeze-fracture electron micrograph of intramembranous tight junction strands. From Shen et al. *Annu. Rev. Physiol.* 2011.

Tight junction structure is far more interesting when viewed by freeze-fracture electron microscopy [25–27]. This reveals an anastomosing, mesh-like network of intramembranous strands (Figure 1C). Closer examination shows that the strands are composed of individual particles, causing some observers to compare the appearance to a string of pearls. The particles are thought to represent tight junction protein complexes that include polymers of claudin family proteins [7,28–30]. Consistent with this, alterations in the ensemble of claudin proteins expressed can modify the architecture of the strand network [31]. Although lipids must also be associated with tight junction structures, these are less well-characterized. It is, however, known that tight junctions are cholesterol- and sphingolipid-rich microdomains and that cholesterol depletion reduces both strand network complexity and paracellular barrier function [32–34].

2. The Paracellular "Shunt" Pathway

The intestinal mucosa confines potentially injurious contents within the lumen. The paracellular barrier, however, cannot be absolute; it must be selectively permeable to water, ions, small nutrients, and selected macromolecules in order to facilitate passive transport that is essential for nutrition and metabolism. Permeability of tight junction flux pathways must, therefore, be precisely regulated. For example, tight junction permeability is increased during nutrient absorption. This is triggered by Na^+–nutrient cotransport, which increases paracellular permeability by activating myosin light chain kinase (MLCK) to cause perijunctional actomyosin ring (PAMR) remodeling [35–39] (Figure 2). In the context of nutrient absorption, these permeability increases are limited to small, nutrient-sized molecules [35,40]. This couples with the transepithelial gradients established by active, transcellular transport, i.e., Na^+ and nutrient release into the basal extracellular milieu, to drive passive paracellular fluid absorption [37,41,42]. The absorbed fluid, from the unstirred layer, which contains high concentrations of nutrient monomers as a consequence of brush border hydrolase, e.g., disaccharidase and peptidase, activity [43,44]. Fluid absorption therefore carries nutrients, against their concentration gradient, by the mechanisms of solvent drag [42,44,45]. Increased tight junction permeability amplifies this process and allows total transepithelial nutrient absorption to exceed the maximum capacity of transcellular transport pathways [37,38,41,45–48]. A similar process allows claudin-2-mediated paracellular Na^+ transport to complement transcellular Na^+ transport and enhance the efficiency of Na^+ reabsorption in the renal proximal tubule [49].

In contrast to Na^+–nutrient cotransport [35,40], MLCK activation by inflammatory stimuli, e.g., tumor necrosis factor α (TNF), increases paracellular permeability to larger macromolecules, up to ~125 Å in diameter, thereby activating the low capacity leak pathway [50–55] (Figure 2). The differences between these two forms of MLCK-dependent barrier regulation are incompletely understood, but it is notable that occludin endocytosis occurs in response to TNF but not Na^+–nutrient cotransport (Figure 2).

Some claudin proteins, e.g., claudin-2, form actively-gated paracellular channels that define the pore pathway [52,53,56] In contrast to the leak pathway, the high capacity pore pathway channels are exquisitely size- and charge-selective, with a cutoff of ~8 Å diameter [57,58]. This limits the pore pathway to small ions and water and is too small to accommodate even small nutrients, e.g., glucose and amino acids. The pore pathway is, however, essential for nutrient transport as it allows Na^+ ions within the lamina propria, i.e., beneath the epithelial cells, to leak back into the gut lumen [59,60]. This provides the lumenal Na^+ that is required for Na^+–nutrient cotransport, the dominant route of intestinal nutrient absorption. Thus, mice lacking the two principal claudins that form paracellular cation channels within the intestinal epithelium die in the first few weeks of life as a result of nutrient malabsorption [59]. The remainder of this review will focus on the leak pathway. Claudin channels and the pore pathway are discussed elsewhere [61–64].

Na^+–nutrient cotransport at the apical brush border activates MLCK. Nutrients and Na^+ exit across the basolateral membrane via diffusive exchangers and the Na^+/K^+-ATPase, respectively. Although not indicated here, activation of other transporters, e.g., apical NHE3-mediated Na^+ absorption, further increases basolateral Na^+ [65–69]. Together, these events increase lamina propria osmolarity [70] to draw fluid across the tight junction. The modest, size-selective, increases in leak pathway permeability elicited by MLCK allow small nutrient-sized molecules to be carried with this fluid via solvent drag [44]. The end result is amplification of transcellular nutrient absorption by paracellular water and nutrient absorption. MLCK is also activated by tumor necrosis factor (TNF), but, in this case, myosin light chain phosphorylation triggers caveolar endocytosis of occludin that causes much greater increases in macromolecular permeability. Although not shown here, TNF also inhibits apical NHE3-mediated Na^+ absorption, in part explaining why fluid secretion accompanies TNF-induced increases in paracellular permeability [71].

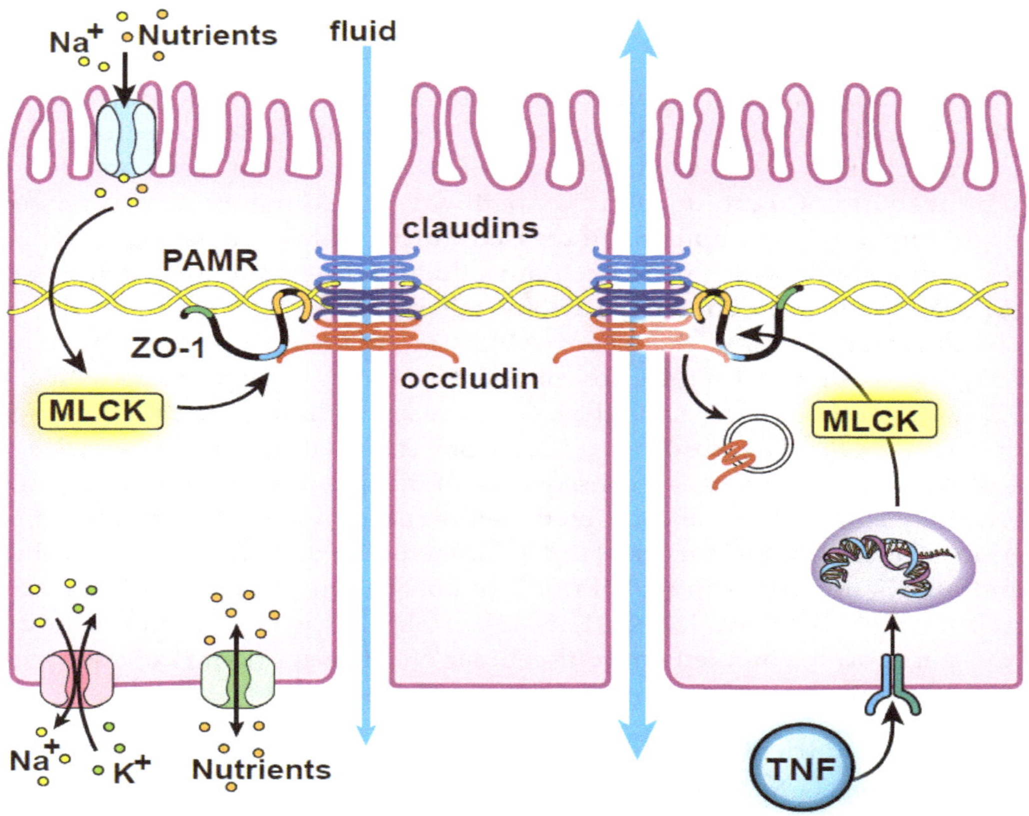

Figure 2. Roles of myosin light chain kinase (MLCK) in physiological and pathophysiological tight junction regulation. PAMR: perijunctional actomyosin ring; TNF: tumor necrosis factor.

3. MLCK, ZO-1, and Occludin Regulate the Leak Pathway

3.1. MLCK Regulates Leak Pathway Permeability

Morphological analyses of rodent mucosae identified the appearance of intrajunctional dilatations and PAMR condensation as the morphological correlates of increased paracellular permeability induced by Na^+–nutrient cotransport [37–39]. In vitro models were then used to show that MLCK-mediated phosphorylation of myosin II regulatory light chain (MLC) was required for paracellular permeability increases that follow activation of Na^+–nutrient cotransport [35], enteropathogenic E. coli infection [72], or TNF stimulation [50]. MLCK inhibition also prevented Na^+–nutrient cotransport-induced paracellular permeability increases in rodent mucosae [35] and was associated with Na^+–nutrient cotransport-induced permeability increases in human intestine [36]. Finally, transgenic expression of constitutive-active MLCK increased paracellular permeability in vitro [73,74] and in vivo [75]. MLCK is, therefore, a key signaling node in physiological and pathophysiological regulation of epithelial tight junctions.

3.2. MLCK Regulates Tight Junction Protein Interactions and Structure

The importance of MLCK to tight junction regulation in vivo was initially demonstrated in the context of acute, systemic T cell activation [51]. Administration of T cell-activating anti-CD3 antibodies to mice and humans induces a cytokine storm that includes massive systemic release of TNF [76–78]. This results in an acute, self-limited, TNF-dependent diarrhea [79,80]. Ultrastructural examination revealed PAMR condensation similar to that induced by Na^+–nutrient cotransport; intrajunctional dilatations were not detected and are likely a consequence of the massive paracellular water absorption driven by Na^+–nutrient cotransport [51]. Further study confirmed increased intestinal epithelial MLC phosphorylation following T cell activation. Notably, the peak of epithelial MLC phosphorylation coincided with maximal intestinal fluid accumulation, and both MLC phosphorylation and fluid accumulation resolved with similar time courses [51]. Analysis of tight junction protein distributions by immunofluorescence microscopy demonstrated subtle changes in ZO-1 localization that included reduced ZO-1 staining and increased waviness of bicellular tight junction profiles within intestinal epithelia of anti-CD3-treated mice [51]. Genetic MLCK activation in vitro also induced undulations within ZO-1-labeled tight junctions [74]. Enzymatic MLCK inhibition reversed these changes, both in vivo and in vitro [51,74]. It may be that these morphological alterations reflect changes in ZO-1 and ZO-2 phase separation [81].

In vitro ZO-1 reorganization induced by MLCK activation was accompanied by similar changes in occludin and F-actin profiles, and these proteins continued to be closely associated [74]. In contrast, claudins 1 and 2 appeared to remain at tight junctions at sites of ZO-1, occludin, and F-actin invaginations [74]. Consistent with this, a previous ultrastructural study of rodent mucosae demonstrated ZO-1 displacement from junctional fibrils, which we now understand to be comprised of claudin polymers, in the context of Na^+–nutrient cotransport-induced increases in paracellular permeability [82]. Thus, structural changes induced by MLCK-dependent MLC phosphorylation include reorganization of tight junction protein complexes.

Recognition that tight junction protein complexes undergo continuous remodeling, even at steady-state, i.e., in the absence of exogenous stimuli [83], led to the hypothesis that MLCK-mediated leak pathway regulation might be a consequence of altered remodeling dynamics. In vitro MLCK inhibition had no effect on anchoring and exchange of claudin-1, occludin, or F-actin and epithelial tight junctions [84]. In contrast, ZO-1 exchange was markedly reduced following MLCK inhibition [84]. Remarkably, enzymatic MLCK inhibition also reduced ZO-1 exchange in vivo [84]. This increase in ZO-1 anchoring at the tight junction following MLCK inhibition was mapped to the actin binding region (ABR) of ZO-1 [84]. Mutation of ZO-1 to delete the ABR caused a modest increase in basal mobile fraction but rendered ZO-1 insensitive to MLCK inhibition. Moreover, the free ABR was able to act as a dominant negative inhibitor of MLCK-mediated barrier regulation [84]. These data suggest that one mechanism by which MLCK regulates tight junction structure and function involves interactions mediated by the ZO-1 ABR.

3.3. MLCK Activation Triggers Tight-Junction Protein Endocytosis

In addition to subtle ZO-1 reorganization, TNF induced endocytosis of the tight junction protein occludin in vitro and in vivo [51,54,85] (Figure 2). Although TNF also triggered endocytosis of other tight junction proteins in vitro, only occludin was internalized in vivo. This was, initially, difficult to understand, as intestinal barrier function in occludin knockout mice has been reported to be similar to that of wildtype mice [86,87]. MLCK-dependent occludin endocytosis was, however, the first change that accompanied actin depolymerization-induced barrier loss in vitro [88]. This internalization occurred via caveolae, and inhibition of caveolar endocytosis prevented actin depolymerization-induced barrier loss in vitro [88]. MLCK inhibition blocked the caveolin-1 dependent endocytosis of occludin.

Further investigation showed that occludin endocytosis triggered by TNF and related cytokines was mediated by caveolae, both in vitro and in vivo [85,89,90]. Moreover, inhibition of caveolar endocytosis blocked this inflammation-induced barrier loss. Most strikingly, neither occludin endocytosis nor barrier loss occurred after TNF treatment of caveolin-1 knockout mice [85]. This does not, however, demonstrate that occludin endocytosis is essential for TNF-induced barrier loss; it only shows that occludin is a reliable marker of the caveolar endocytosis that drives such barrier loss. To determine the specific contribution(s) of occludin to TNF-induced barrier loss, responses of transgenic mice expressing enhanced green fluorescent protein (EGFP)-occludin within intestinal epithelial cells were analyzed. TNF did not induce diarrhea, i.e., net fluid secretion, in these occludin overexpressing mice [85]. Barrier loss was also markedly attenuated in EGFP-occludin transgenic mice [85]. Thus, removal of occludin from the tight junction is required for TNF-induced increases in tight junction permeability.

3.4. Interactions Mediated by the Occludin OCEL Domain Regulate Leak Pathway Barrier Function

Despite the reported absence of barrier dysfunction in occludin knockout mice, several studies have shown that occludin knockdown in vitro increases paracellular permeability to macromolecules [89,91]. Similarly, acute occludin downregulation by miR-122a transfection in vivo increased paracellular permeability to 10 kDa dextran [92]. It may, therefore, be that compensation by other members of the tight junction associated Marvel protein (TAMP) family [93–98] supports normal macromolecular barrier function in occludin knockout mice.

Detailed analyses of Caco-2 intestinal epithelial cell monolayers demonstrated that occludin knockdown specifically increased paracellular permeability by a pathway with a theoretical diameter of 125 Å, presumably the leak pathway [89]. Similar studies of occludin-knockdown MDCK cell monolayers demonstrated increased permeability to molecules with diameters greater than ~4 Å, although no upper limit was defined [91]. These MDCK cells were protected from cytokine-induced increases in 3 kDa dextran flux [99]. Conversely, occludin overexpression augmented such cytokine-induced macromolecular flux [99]. However, the studies found that cytokine treatment paradoxically increased transepithelial electrical resistance (TER) in MDCK monolayers and that these TER increases were either attenuated or exaggerated by occludin knockdown or overexpression, respectively [99]. Subsequent studies of Caco-2 monolayers demonstrated the opposite, that occludin knockdown prevented TNF-induced TER loss [89]. TNF was also unable to increase paracellular macromolecular permeability of occludin knockdown Caco-2 monolayers. Thus, macromolecular permeability of TNF-treated, occludin-sufficient Caco-2 monolayers was similar to that of occludin-deficient monolayers regardless of TNF treatment [89]. The impact of occludin on barrier function required the C-terminal coiled-coil occludin/ELL domain (OCEL) domain, as barrier function of occludin-deficient Caco-2 monolayers was enhanced by EGFP-occludin, but not EGFP-occludin$^{\Delta OCEL}$, expression [89]. EGFP-occludin expression also restored TNF-sensitivity to occludin-knockdown Caco-2 monolayers, but monolayers expressing EGFP-occludin$^{\Delta OCEL}$ remained insensitive to TNF [89]. These data indicate that the same barrier defects are induced by genetic occludin knockdown or TNF-induced occludin removal from the tight junction [89]. Thus, while not essential for tight junction assembly, occludin is a critical regulator of the macromolecular, leak pathway barrier.

3.5. MLCK-Induced Occludin Endocytosis Requires ZO-1 Interactions with the Occludin OCEL Domain

Despite a close functional relationship, molecular sites of interactions between occludin and microfilaments have not been defined. It is, therefore, not clear how MLCK-mediated actomyosin contraction induces occludin endocytosis. One possibility is that ZO-1 acts as an intermediate that links occludin to F-actin. Although this has not been explored in detail, the free OCEL domain, which mediates occludin binding to ZO-1, does act as a dominant negative to prevent TNF-induced occludin endocytosis [89]. Moreover, occludin K433, located within the ZO-1 binding site, is critical to this dominant negative OCEL activity [89]. Thus, although further study is needed, ZO-1 may link occludin endocytosis to MLCK-dependent actomyosin contraction. This may also be related to tight junction-dependent mechanosensation, which involves ZO-1, ZO-2, the transcription factor DbpA/ZONAB, and their interactions with the cytoskeleton [100,101].

4. Regulation of MLCK Expression and Localization

4.1. Regulation of MLCK Transcription

The critical role of MLCK in acute, TNF-induced barrier loss was initially demonstrated in vitro using pharmacological inhibitors [50,54,55] and in vivo using a combination of pharmacological inhibitors and mice lacking epithelial long MLCK [51]. Further study demonstrated that, beyond MLCK enzymatic activity, transcriptional upregulation of long MLCK expression was essential to TNF-induced barrier loss [54]. TNF activated long MLCK transcription via the high-affinity TNF receptor TNFR2, thereby explaining how extremely low, nanogram concentrations of TNF were sufficient to trigger MLCK upregulation and barrier dysfunction in vitro [102]. Moreover, these data demonstrated that requirement for interferon-γ (IFN-γ) pretreatment before some cell monolayers were able to respond to TNF reflected IFN-γ-induced TNFR2 transcription [54,102]. Consistent with this, in vivo studies demonstrated that TNFR2 signaling was required for epithelial long MLCK upregulation during disease progression in T cell transfer-induced, experimental, chronic inflammatory bowel disease [103].

In vitro studies from two groups conflict as to whether TNF-induced MLCK upregulation depended on NFκB or AP-1 signaling [54,104]. One group found that the NFκB inhibitors curcumin, triptolide, and pyrrolidine dithiocarbamate prevented TNF-induced MLCK upregulation and barrier loss [104,105]. A second group found that a series of NFκB inhibitors, including MG132, capsaicin, curcumin, and triptolide were unable to prevent TNF-induced barrier loss [54]; MG132 and triptolide actually enhanced TNF-induced barrier loss. This group found that sulfasalazine was able to block TNF-induced barrier loss but did so in a dose dependent manner; barrier loss was prevented at 0.5 mM sulfasalazine but exaggerated at 2 mM sulfasalazine. Biochemical studies of NFκB RelA translocation to the nucleus and luciferase expression from an NFκB-responsive promoter showed that MG132 and 2 mM sulfasalazine, but not 0.5 mM sulfasalazine, inhibited TNF-induced NFκB activation. In contrast, 0.5 mM sulfasalazine, but not 2 mM sulfasalazine or MG132, blocked TNF-induced MLCK upregulation and MLC phosphorylation [54]. These data clearly separated NFκB from MLCK upregulation and suggested that another signaling pathway activated by TNF was responsible for barrier loss. Although both of these groups used Caco-2 cells in their studies, one possible explanation could be that the first group used a relatively undifferentiated Caco-2 clone while the second group used the differentiated, enterocyte-like Caco-2_{BBe} subclone [106].

Both groups went on to clone the human long MLCK promoter, which contained functional binding sites for both NFκB and AP-1 [107,108]. However, results again differed between the groups. The first group used a human genome database search to identify the 2 kb upstream of the human long MLCK transcriptional start site and eight NFκB binding sites within that region [104,107]. They found that two of these were active, with one repressing and the other increasing transcription ~2-fold [104]. In contrast, the second group used 5′ Rapid Amplification of cDNA Ends (5′-RACE) to identify two different long MLCK transcriptional start sites [108]. The upstream 4 kb sequence contained three functional AP-1 sites and two functional NFκB sites, all of which could regulate long MLCK transcription [108]. However, the intact 4 kb long MLCK promoter was not responsive to TNF prior to 14 days after confluence, suggesting that Caco-2 differentiation, which occurs progressively in the 2 weeks following growth to confluence, was required for full promoter activation [108]. From 3 to 14 days after confluence, activity of the MLCK promoter in response to TNF increased ~7-fold and transcription from an AP-1-responsive promoter increased ~2-fold, but transcription from an NFκB -responsive promoter decreased ~2-fold. Taken as a whole, these data suggest a unified model that explains the data from both groups. In this model, undifferentiated intestinal epithelial cells modestly upregulate long MLCK via NFκB signaling, while TNF induces more extensive MLCK upregulation via AP-1 signaling within differentiated cells. Consistent with this, the first group has gone on to show that AP-1-activating elements, such as mitogen activated protein (MAP) kinases, are involved in cytokine-induced MLCK upregulation [109,110].

4.2. MLCK Expression in Chronic Intestinal Disease

Quantitative morphometry of biopsies from human patients showed that MLCK expression and activity were increased in ulcerative colitis and Crohn's disease [111]. Consistent with a role for TNF signaling, the magnitude of MLCK activation correlated directly with the degree of inflammatory activity in these biopsies.

Further analyses of the specific contributions of MLCK activation to disease were performed using transgenic mice expressing constitutively-active MLCK within the intestinal epithelium or global long MLCK knockout mice. Mice expressing constitutively-active MLCK displayed increased intestinal paracellular leak pathway permeability and increased MLC phosphorylation relative to non-transgenic littermates [75]. Both of these could be corrected by enzymatic MLCK inhibition [75]. Although these barrier defects were insufficient to induce spontaneous disease, compensatory mucosal immune activation including increased lamina propria T cell numbers, mild Th1 polarization, and heightened acute responses to infectious pathogens were detected [75,112]. However, when studied using the T cell transfer model of chronic inflammatory bowel disease, the transgenic mice developed disease more rapidly than non-transgenic littermates [75]. Moreover, overall disease severity was greater in transgenic mice and survival was markedly reduced. Thus, MLCK upregulation can impact progression of experimental inflammatory bowel disease.

Conversely, global long MLCK knockout mice were markedly protected from T cell transfer-induced colitis [103]. Although reduced disease could, potentially, reflect deletion of long MLCK in cells other than intestinal epithelium, the protection afforded by long MLCK knockout could be overcome by intestinal epithelial-specific constitutively-active MLCK expression [75]. Thus, MLCK inhibition is a potential target in chronic inflammatory bowel disease. It should, however, be noted that MLCK has many functions beyond tight junction regulation. For example, MLCK activation is critical to epithelial migration and wound repair [113–115]. Consistent with this, long MLCK knockout mice fared worse than their wild type counterparts when subjected to dextran sulfate sodium (DSS)-induced epithelial damage [103]. Similarly, in vivo knockout of nonmuscle myosin II markedly disrupts intestinal homeostasis, as demonstrated by goblet cell and barrier loss; as expected, these severely compromised mice are hypersensitive to DSS-induced injury [116]. Finally, MLCK loss may also have other consequences, either directly or as a secondary effect following disruption of cytoskeletal organization, protein trafficking, and signal transduction [117]. Thus, while intestinal epithelial MLCK inhibition and tight junction barrier preservation may be helpful in immune-mediated disease, they may also be associated with toxicities when epithelial damage predominates.

4.3. Enzymatic MLCK Inhibition is Not Feasible as a Therapeutic Intervention

Three distinct MLCK genes are present in mammals. These are *MYLK*, the smooth muscle/non-muscle MLCK located on human chromosome 3; skeletal muscle *MYLK2* on human chromosome 20; and cardiac *MYLK3* on human chromosome 16. *MYLK* encodes a long MLCK expressed in intestinal epithelium as well as short MLCK and telokin [118,119]. Telokin has regulatory functions but cannot phosphorylate MLC [120–122]. In contrast, both short MLCK, an ~110-130 kDa protein expressed in smooth muscle, and long MLCK, an ~215 kDa protein expressed in non-muscle cells, including epithelium, share identical catalytic and calmodulin-binding regulatory domains (Figure 3A). Thus, neither chemical nor genetic inhibitors can distinguish between short and long MLCK.

Enzymatic MLCK inhibition could not be used therapeutically since, as noted above, long and short MLCK have identical catalytic domains. In mice, short MLCK knockout leads to death in the early perinatal period [123]. Development of inducible knockout mice with loxP sites flanking the *MYLK* sequence encoding the catalytic domain overcame this limitation [124]. However, tissue-specific knockout in smooth muscle resulted in hypotension, bladder dysfunction, and severe intestinal dysmotility, and death [124]. Thus, while targeted inhibition of MLCK-mediated tight junction barrier regulation might have therapeutic benefit, loss of MLCK enzymatic activity in smooth muscle [124] and non-muscle [103,113] cells would have unacceptable toxicities.

4.4. TNF Induces IgCAM3-Mediated Long MLCK1 Recruitment to the Perijunctional Actomyosin Ring

The difference between long and short MLCK is the presence of a long stretch of N-terminal sequence in long MLCK [125–127]. The exons that encode that part of the protein undergo extensive alternative splicing, resulting in a number of different long MLCK splice variants [127]. Only isoforms MLCK1 and MLCK2, which differ by a single 207 nucleotide exon, are expressed in intestinal epithelia [125]. The differential functions of these and other long MLCK splice variants are not well-defined. However, it has been reported that the 69 amino acids encoded by the 207 nucleotide exon that distinguishes MLCK1 from MLCK2 include a src kinase target site whose phosphorylation can regulate MLCK activity [126].

Tissue analysis using an antibody specific for long MLCK1 demonstrated that, relative to total epithelial long MLCK, long MLCK1 was concentrated within the perijunctional actomyosin ring [125] (Figure 3B). Moreover, targeted knockdown of long MLCK1 within Caco-2 monolayers reduced paracellular permeability [125]. Further study showed that, in addition to increasing long MLCK expression, TNF specifically induced long MLCK1 recruitment to the perijunctional actomyosin ring [128]. This selective effect on MLCK1 was only post-translational, as intestinal epithelial MLCK1 and MLCK2 transcripts were comparably increased by TNF (unpublished data, Graham and Turner).

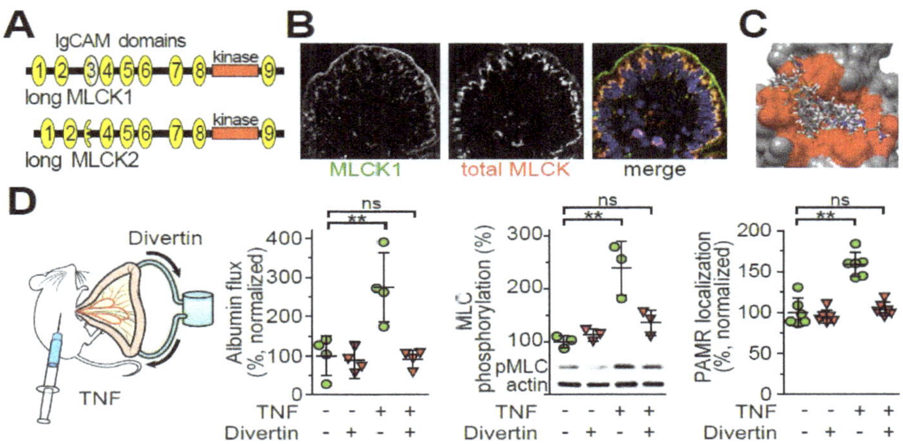

Figure 3. Specific targeting of long MLCK isoform 1 (MLCK1) prevents TNF-induced barrier loss in vivo. (**A**) Protein domain structure of long MLCK isoforms 1 and 2. Immunoglobulin-cell adhesion molecule (IgCAM) domains are numbered from the amino terminus. (**B**) Long MLCK1 (green), total MLCK (red), and nuclei (blue) in normal human jejunum. MLCK1 is preferentially-localized to the perijunctional actomyosin ring. (**C**) Virtually screened compounds docked to a binding pocket within the unique IgCAM3 domain of MLCK1. (**D**) Mice were injected with vehicle or TNF, and jejunal loops were perfused with either saline-containing vehicle or Divertin. TNF-induced increases in albumin flux (from bloodstream into the gut lumen) were blocked by Divertin. Divertin also blocked TNF-induced myosin II regulatory light chain (MLC) phosphorylation and MLCK1 recruitment to the PAMR. Notably, Divertin does not inhibit MLCK enzymatic activity. ** $p < 0.01$ by ANOVA with Dunn's multiple comparison test. From Graham et al. *Nat. Med.* 2019.

Structural analysis indicated that the 69 amino acids unique to long MLCK1 completed the immunoglobulin-cell adhesion molecule (IgCAM) domain 3, one of nine IgCAM domains within long MLCK1 [128]. Because this is the only difference between MLCK1 and MLCK2, it stands to reason that the key features responsible for TNF-induced MLCK 1 recruitment to the perijunctional actomyosin ring reside within IgCAM3. Consistent with this, expression of the IgCAM3 domain alone attenuated TNF-induced MLCK1 recruitment to the perijunctional actomyosin ring and barrier regulation, suggesting that it disrupted interactions between MLCK1 and other intracellular molecules (unpublished data, He and Turner).

4.5. *TNF Induces IgCAM3-Mediated Long MLCK1 Recruitment to the Perijunctional Actomyosin Ring*

The observation that an enzymatically inactive region, i.e., IgCAM3, was responsible for long MLCK1 recruitment to the perijunctional actomyosin ring presented an opportunity for blocking MLCK-dependent tight junction regulation in disease. After solving the IgCAM3 crystal structure [128,129], a hydrophobic drug-binding pocket within IgCAM3, but not the other long MLCK1 IgCAM domains, was identified. Importantly, this drug binding pocket was conserved between human and mouse long MLCK1 [128]. An in silico screen using molecular docking software identified a small number of candidate molecules with putative binding to the targeted hydrophobic pocket (Figure 3C). In vitro screening demonstrated that one of these was able to increase intestinal epithelial barrier function without inhibiting MLCK enzymatic activity in a cell-free assay, disrupting smooth muscle contraction, or interfering with epithelial wound repair, i.e., migration. This molecule, termed Divertin, was, however, able to displace long MLCK1 from the perijunctional actomyosin ring, reverse TNF-induced barrier loss, and reduce MLC phosphorylation in Caco-2 monolayers [128].

Consistent with its targeted effect on MLCK1-dependent tight junction regulation, mice could receive daily Divertin administration for 30 days without any apparent toxicity. Divertin was, however, able to prevent MLCK1 recruitment to the perijunctional actomyosin ring, MLC phosphorylation, occludin internalization, and leak pathway barrier loss in response to acute TNF challenge in mice (Figure 3D). Divertin also prevented perijunctional MLCK1 recruitment, MLC phosphorylation, and occludin internalization in human intestinal biopsies treated with TNF in vitro. Thus, Divertin was able to reverse and prevent acute, TNF-induced MLC phosphorylation and barrier loss in vitro and in vivo [128] (Figure 3D).

The efficacy of Divertin and restoring intestinal barrier function in chronic disease was initially assessed using IL-10 knockout mice, which develop in intestinal barrier defect early in the course of disease [128,130]. Although Divertin did not affect intestinal permeability in wild type mice, it was able to restore the intestinal barrier in IL-10 knockout mice [128]. Moreover, Divertin was markedly effective in limiting disease during T cell transfer colitis regardless of whether it was administered just prior to or after clinical disease presentation [128] (Figure 4). Remarkably, all measures assessed showed that Divertin was superior or equivalent to anti-TNF in limiting T cell transfer colitis severity [128] (Figure 4). Some data did, however, indicate that the beneficial effects of Divertin might be additive to, or even synergistic with, those of anti-TNF. Thus, preventing MLCK1 recruitment to the perijunctional actomyosin ring may be a non-toxic approach to limiting or preventing immune-mediated colitis either alone or in combination with immunomodulatory agents, e.g., anti-TNF [128] (Figure 4).

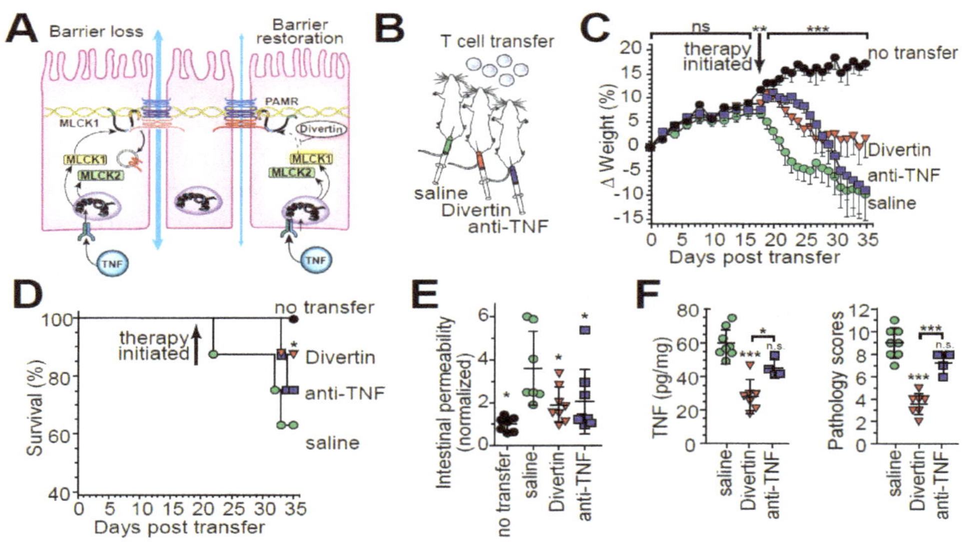

Figure 4. Inhibition of MLCK1 recruitment to the perijunctional actomyosin ring attenuates immune-mediated colitis. (**A**) Proposed mechanism of Divertin action. (**B**) Immunodeficient mice received naïve CD4+ effector T cells. Therapy with saline (vehicle), divertin, anti-TNF, or combined Divertin and anti-TNF was initiated after definitive features of disease developed (day 19). (**C**) Divertin limited weight loss after T cell transfer and was superior to anti-TNF antibody treatment. ** $p < 0.01$ by two-tailed t test for no transfer vs. all other mice at day 18. *** $p < 0.001$ by ANOVA with Tukey's multiple comparison test over the interval from 19–35 days. (**D**) Divertin enhanced survival after T cell transfer. * $p < 0.05$, versus saline-treated mice, by Gehan–Breslow–Wilcoxon test. (**E**) Divertin limited intestinal barrier loss. * $p < 0.05$, versus saline-treated mice, by ANOVA with Newman–Keuls multiple comparison test. (**F**) Divertin was superior to anti-TNF antibody treatment in limiting mucosal cytokine production and histopathology after T cell transfer. * $p < 0.05$; *** $p < 0.001$ by ANOVA with Bonferroni correction. From Graham et al. *Nat. Med.* 2019.

5. Perspective and Future Directions

The past 25 years have seen tremendous growth in our understanding of tight junction structure, cell biology, and pathobiology. Most recently, this has resulted in the develop of a proof-of-concept molecule that may provide a foundation for creation of actual therapeutic agents. Despite the enthusiasm this has engendered, there remains much to be learned. Specific topics include elucidation of mechanisms that regulate MLCK1 recruitment to the perijunctional actomyosin ring, further definition of pore pathway function in health and disease, and characterization of the structural and functional properties of defined and to-be-discovered tight junction components.

Author Contributions: Manuscript conceptualization, W.-Q.H., and J.R.T.; initial draft, W.-Q.H., J.W., J.-Y.S., and J.R.T.; figure preparation, W.-Q.H., J.-M.Z., W.V.G., and J.R.T.; manuscript revision, W.-Q.H., J.-M.Z., W.V.G., and J.R.T.; All authors have read and agreed to the published version of the manuscript.

References

1. Blasky, A.J.; Mangan, A.; Prekeris, R. Polarized protein transport and lumen formation during epithelial tissue morphogenesis. *Annu. Rev. Cell Dev. Biol.* **2015**, *31*, 575–591. [CrossRef] [PubMed]
2. Venkatasubramanian, J.; Ao, M.; Rao, M.C. Ion transport in the small intestine. *Curr. Opin. Gastroenterol.* **2010**, *26*, 123–128. [CrossRef] [PubMed]
3. Cereijido, M.; Contreras, R.G.; Shoshani, L.; Flores-Benitez, D.; Larre, I. Tight junction and polarity interaction in the transporting epithelial phenotype. *Biochim. Biophys. Acta* **2008**, *1778*, 770–793. [CrossRef] [PubMed]
4. Crawley, S.W.; Mooseker, M.S.; Tyska, M.J. Shaping the intestinal brush border. *J. Cell Biol.* **2014**, *207*, 441–451. [CrossRef] [PubMed]
5. Farquhar, M.; Palade, G. Junctional complexes in various epithelia. *J. Cell Biol.* **1963**, *17*, 375–412. [CrossRef] [PubMed]
6. Cereijido, M.; Robbins, E.S.; Dolan, W.J.; Rotunno, C.A.; Sabatini, D.D. Polarized monolayers formed by epithelial cells on a permeable and translucent support. *J. Cell Biol.* **1978**, *77*, 853–880. [CrossRef]
7. Martinez-Palomo, A.; Meza, I.; Beaty, G.; Cereijido, M. Experimental modulation of occluding junctions in a cultured transporting epithelium. *J. Cell Biol.* **1980**, *87*, 736–745. [CrossRef]
8. Otani, T.; Nguyen, T.P.; Tokuda, S.; Sugihara, K.; Sugawara, T.; Furuse, K.; Miura, T.; Ebnet, K.; Furuse, M. Claudins and JAM-A coordinately regulate tight junction formation and epithelial polarity. *J. Cell Biol.* **2019**, *218*, 3372–3396. [CrossRef]
9. Nalle, S.C.; Zuo, L.; Ong, M.; Singh, G.; Worthylake, A.M.; Choi, W.; Manresa, M.C.; Southworth, A.P.; Edelblum, K.L.; Baker, G.J.; et al. Graft-versus-host disease propagation depends on increased intestinal epithelial tight junction permeability. *J. Clin. Investig.* **2019**, *129*, 902–914. [CrossRef]
10. Kuo, W.T.; Shen, L.; Zuo, L.; Shashikanth, N.; Ong, M.; Wu, L.; Zha, J.; Edelblum, K.L.; Wang, Y.; Wang, Y.; et al. Inflammation-induced Occludin Downregulation Limits Epithelial Apoptosis by Suppressing Caspase-3 Expression. *Gastroenterology* **2019**, *157*, 1323–1337. [CrossRef]
11. Kohno, T.; Konno, T.; Kojima, T. Role of Tricellular Tight Junction Protein Lipolysis-Stimulated Lipoprotein Receptor (LSR) in Cancer Cells. *Int. J. Mol. Sci.* **2019**, *20*, 3555. [CrossRef] [PubMed]
12. Camilleri, M. Leaky gut: Mechanisms, measurement and clinical implications in humans. *Gut* **2019**, *68*, 1516–1526. [CrossRef] [PubMed]
13. Brazil, J.C.; Quiros, M.; Nusrat, A.; Parkos, C.A. Innate immune cell-epithelial crosstalk during wound repair. *J. Clin. Investig.* **2019**, *129*, 2983–2993. [CrossRef] [PubMed]
14. Powell, D.W. Barrier function of epithelia. *Am. J. Physiol.* **1981**, *241*, G275–G288. [CrossRef] [PubMed]
15. Kottra, G.; Fromter, E. Functional properties of the paracellular pathway in some leaky epithelia. *J. Exp. Biol.* **1983**, *106*, 217–229. [PubMed]
16. Overton, J.; Shoup, J. Fine Structure of Cell Surface Specializations in the Maturing Duodenal Mucosa of the Chick. *J. Cell Biol.* **1964**, *21*, 75–85. [CrossRef]
17. Tsubouchi, S.; Leblond, C.P. Migration and turnover of entero-endocrine and caveolated cells in the epithelium of the descending colon, as shown by radioautography after continuous infusion of 3H-thymidine into mice. *Am. J. Anat.* **1979**, *156*, 431–451. [CrossRef]

18. Harris, T.J.; Tepass, U. Adherens junctions: From molecules to morphogenesis. *Nat. Rev. Mol. Cell Biol.* **2010**, *11*, 502–514. [CrossRef]
19. Pinheiro, D.; Bellaiche, Y. Mechanical Force-Driven Adherens Junction Remodeling and Epithelial Dynamics. *Dev. Cell* **2018**, *47*, 3–19. [CrossRef]
20. Tsukita, S.; Tsukita, S.; Nagafuchi, A.; Yonemura, S. Molecular linkage between cadherins and actin filaments in cell-cell adherens junctions. *Curr. Opin. Cell Biol.* **1992**, *4*, 834–839. [CrossRef]
21. Brooke, M.A.; Nitoiu, D.; Kelsell, D.P. Cell-cell connectivity: Desmosomes and disease. *J. Pathol.* **2012**, *226*, 158–171. [CrossRef] [PubMed]
22. Capaldo, C.T.; Farkas, A.E.; Nusrat, A. Epithelial adhesive junctions. *F1000prime Rep.* **2014**, *6*, 1. [CrossRef] [PubMed]
23. Nava, P.; Laukoetter, M.G.; Hopkins, A.M.; Laur, O.; Gerner-Smidt, K.; Green, K.J.; Parkos, C.A.; Nusrat, A. Desmoglein-2: A novel regulator of apoptosis in the intestinal epithelium. *Mol. Biol. Cell* **2007**, *18*, 4565–4578. [CrossRef] [PubMed]
24. Zen, K.; Babbin, B.A.; Liu, Y.; Whelan, J.B.; Nusrat, A.; Parkos, C.A. JAM-C is a component of desmosomes and a ligand for CD11b/CD18-mediated neutrophil transepithelial migration. *Mol. Biol. Cell* **2004**, *15*, 3926–3937. [CrossRef] [PubMed]
25. Chalcroft, J.P.; Bullivant, S. An interpretation of liver cell membrane and junction structure based on observation of freeze-fracture replicas of both sides of the fracture. *J. Cell Biol.* **1970**, *47*, 49–60. [CrossRef] [PubMed]
26. Claude, P.; Goodenough, D.A. Fracture faces of zonulae occludentes from "tight" and "leaky" epithelia. *J. Cell Biol.* **1973**, *58*, 390–400. [CrossRef]
27. Wade, J.B.; Karnovsky, M.J. The structure of the zonula occludens. A single fibril model based on freeze-fracture. *J. Cell Biol.* **1974**, *60*, 168–180. [CrossRef]
28. Furuse, M.; Fujita, K.; Hiiragi, T.; Fujimoto, K.; Tsukita, S. Claudin-1 and -2: Novel integral membrane proteins localizing at tight junctions with no sequence similarity to occludin. *J. Cell Biol.* **1998**, *141*, 1539–1550. [CrossRef]
29. Furuse, M.; Sasaki, H.; Fujimoto, K.; Tsukita, S. A single gene product, claudin-1 or -2, reconstitutes tight junction strands and recruits occludin in fibroblasts. *J. Cell Biol.* **1998**, *143*, 391–401. [CrossRef]
30. Furuse, M.; Sasaki, H.; Tsukita, S. Manner of interaction of heterogeneous claudin species within and between tight junction strands. *J. Cell Biol.* **1999**, *147*, 891–903. [CrossRef]
31. Yamazaki, Y.; Tokumasu, R.; Kimura, H.; Tsukita, S. Role of claudin species-specific dynamics in reconstitution and remodeling of the zonula occludens. *Mol. Biol. Cell* **2011**, *22*, 1495–1504. [CrossRef] [PubMed]
32. Francis, S.A.; Kelly, J.M.; McCormack, J.; Rogers, R.A.; Lai, J.; Schneeberger, E.E.; Lynch, R.D. Rapid reduction of MDCK cell cholesterol by methyl-beta-cyclodextrin alters steady state transepithelial electrical resistance. *Eur. J. Cell Biol.* **1999**, *78*, 473–484. [CrossRef]
33. Nusrat, A.; Parkos, C.A.; Verkade, P.; Foley, C.S.; Liang, T.W.; Innis-Whitehouse, W.; Eastburn, K.K.; Madara, J.L. Tight junctions are membrane microdomains. *J. Cell Sci.* **2000**, *113*, 1771–1781.
34. Nusrat, A.; von Eichel-Streiber, C.; Turner, J.R.; Verkade, P.; Madara, J.L.; Parkos, C.A. Clostridium difficile toxins disrupt epithelial barrier function by altering membrane microdomain localization of tight junction proteins. *Infect. Immun.* **2001**, *69*, 1329–1336. [CrossRef]
35. Turner, J.R.; Rill, B.K.; Carlson, S.L.; Carnes, D.; Kerner, R.; Mrsny, R.J.; Madara, J.L. Physiological regulation of epithelial tight junctions is associated with myosin light-chain phosphorylation. *Am. J. Physiol.* **1997**, *273*, C1378–C1385. [CrossRef]
36. Berglund, J.J.; Riegler, M.; Zolotarevsky, Y.; Wenzl, E.; Turner, J.R. Regulation of human jejunal transmucosal resistance and MLC phosphorylation by Na(+)-glucose cotransport. *Am. J. Physiol.- Gastrointest. Liver Physiol.* **2001**, *281*, G1487–G1493. [CrossRef]
37. Madara, J.L.; Pappenheimer, J.R. Structural basis for physiological regulation of paracellular pathways in intestinal epithelia. *J. Membr. Biol.* **1987**, *100*, 149–164. [CrossRef]
38. Atisook, K.; Carlson, S.; Madara, J.L. Effects of phlorizin and sodium on glucose-elicited alterations of cell junctions in intestinal epithelia. *Am. J. Physiol.* **1990**, *258*, C77–C85. [CrossRef]
39. Atisook, K.; Madara, J.L. An oligopeptide permeates intestinal tight junctions at glucose-elicited dilatations. *Gastroenterology* **1991**, *100*, 719–724. [CrossRef]
40. Turner, J.R.; Cohen, D.E.; Mrsny, R.J.; Madara, J.L. Noninvasive in vivo analysis of human small intestinal paracellular absorption: Regulation by Na+-glucose cotransport. *Dig. Dis. Sci.* **2000**, *45*, 2122–2126.

[CrossRef]
41. Pappenheimer, J.R. Physiological regulation of transepithelial impedance in the intestinal mucosa of rats and hamsters. *J. Membr. Biol.* **1987**, *100*, 137–148. [CrossRef] [PubMed]
42. Pappenheimer, J.R.; Reiss, K.Z. Contribution of solvent drag through intercellular junctions to absorption of nutrients by the small intestine of the rat. *J. Membr. Biol.* **1987**, *100*, 123–136. [CrossRef]
43. Pappenheimer, J.R. Role of pre-epithelial "unstirred" layers in absorption of nutrients from the human jejunum. *J. Membr. Biol.* **2001**, *179*, 185–204. [CrossRef] [PubMed]
44. Pappenheimer, J.R. On the coupling of membrane digestion with intestinal absorption of sugars and amino acids. *Am. J. Physiol.* **1993**, *265*, G409–G417. [CrossRef] [PubMed]
45. Pappenheimer, J.R.; Dahl, C.E.; Karnovsky, M.L.; Maggio, J.E. Intestinal absorption and excretion of octapeptides composed of D amino acids. *Proc. Natl. Acad. Sci. USA* **1994**, *91*, 1942–1945. [CrossRef]
46. Pappenheimer, J.R. Physiological regulation of epithelial junctions in intestinal epithelia. *Acta Physiol. Scand. Suppl.* **1988**, *571*, 43–51.
47. Pappenheimer, J.R. Paracellular intestinal absorption of glucose, creatinine, and mannitol in normal animals: Relation to body size. *Am. J. Physiol.* **1990**, *259*, G290–G299. [CrossRef]
48. Meddings, J.B.; Westergaard, H. Intestinal glucose transport using perfused rat jejunum in vivo: Model analysis and derivation of corrected kinetic constants. *Clin. Sci.* **1989**, *76*, 403–413. [CrossRef]
49. Pei, L.; Solis, G.; Nguyen, M.T.; Kamat, N.; Magenheimer, L.; Zhuo, M.; Li, J.; Curry, J.; McDonough, A.A.; Fields, T.A.; et al. Paracellular epithelial sodium transport maximizes energy efficiency in the kidney. *J. Clin. Investig.* **2016**, *126*, 2509–2518. [CrossRef]
50. Zolotarevsky, Y.; Hecht, G.; Koutsouris, A.; Gonzalez, D.E.; Quan, C.; Tom, J.; Mrsny, R.J.; Turner, J.R. A membrane-permeant peptide that inhibits MLC kinase restores barrier function in in vitro models of intestinal disease. *Gastroenterology* **2002**, *123*, 163–172. [CrossRef]
51. Clayburgh, D.R.; Barrett, T.A.; Tang, Y.; Meddings, J.B.; Van Eldik, L.J.; Watterson, D.M.; Clarke, L.L.; Mrsny, R.J.; Turner, J.R. Epithelial myosin light chain kinase-dependent barrier dysfunction mediates T cell activation-induced diarrhea in vivo. *J. Clin. Investig.* **2005**, *115*, 2702–2715. [CrossRef] [PubMed]
52. Turner, J.R. Intestinal mucosal barrier function in health and disease. *Nat. Rev. Immunol.* **2009**, *9*, 799–809. [CrossRef]
53. Anderson, J.M.; Van Itallie, C.M. Physiology and function of the tight junction. *Cold Spring Harb. Perspect Biol.* **2009**, *1*, a002584. [CrossRef] [PubMed]
54. Wang, F.; Graham, W.V.; Wang, Y.; Witkowski, E.D.; Schwarz, B.T.; Turner, J.R. Interferon-gamma and tumor necrosis factor-alpha synergize to induce intestinal epithelial barrier dysfunction by up-regulating myosin light chain kinase expression. *Am. J. Pathol.* **2005**, *166*, 409–419. [CrossRef]
55. Ma, T.Y.; Boivin, M.A.; Ye, D.; Pedram, A.; Said, H.M. Mechanism of TNF-{alpha} modulation of Caco-2 intestinal epithelial tight junction barrier: Role of myosin light-chain kinase protein expression. *Am. J. Physiol. Gastrointest. Liver Physiol.* **2005**, *288*, G422–G430. [CrossRef] [PubMed]
56. Weber, C.R.; Liang, G.H.; Wang, Y.; Das, S.; Shen, L.; Yu, A.S.; Nelson, D.J.; Turner, J.R. Claudin-2-dependent paracellular channels are dynamically gated. *eLife* **2015**, *4*, e09906. [CrossRef]
57. Yu, A.S.; Cheng, M.H.; Angelow, S.; Gunzel, D.; Kanzawa, S.A.; Schneeberger, E.E.; Fromm, M.; Coalson, R.D. Molecular basis for cation selectivity in claudin-2-based paracellular pores: Identification of an electrostatic interaction site. *J. Gen. Physiol.* **2009**, *133*, 111–127. [CrossRef]
58. Li, J.; Zhuo, M.; Pei, L.; Rajagopal, M.; Yu, A.S. Comprehensive cysteine-scanning mutagenesis reveals Claudin-2 pore-lining residues with different intrapore locations. *J. Biol. Chem.* **2014**, *289*, 6475–6484. [CrossRef]
59. Wada, M.; Tamura, A.; Takahashi, N.; Tsukita, S. Loss of claudins 2 and 15 from mice causes defects in paracellular Na+ flow and nutrient transport in gut and leads to death from malnutrition. *Gastroenterology* **2013**, *144*, 369–380. [CrossRef]
60. Tamura, A.; Hayashi, H.; Imasato, M.; Yamazaki, Y.; Hagiwara, A.; Wada, M.; Noda, T.; Watanabe, M.; Suzuki, Y.; Tsukita, S. Loss of claudin-15, but not claudin-2, causes Na+ deficiency and glucose malabsorption in mouse small intestine. *Gastroenterology* **2011**, *140*, 913–923. [CrossRef]
61. Suzuki, H.; Tani, K.; Fujiyoshi, Y. Crystal structures of claudins: Insights into their intermolecular interactions. *Ann. N. Y. Acad. Sci.* **2017**, *1397*, 25–34. [CrossRef] [PubMed]
62. Rosenthal, R.; Gunzel, D.; Theune, D.; Czichos, C.; Schulzke, J.D.; Fromm, M. Water channels and barriers formed by claudins. *Ann. N. Y. Acad. Sci.* **2017**, *1397*, 100–109. [CrossRef] [PubMed]

63. Garcia-Hernandez, V.; Quiros, M.; Nusrat, A. Intestinal epithelial claudins: Expression and regulation in homeostasis and inflammation. *Ann. N. Y. Acad. Sci.* **2017**, *1397*, 66–79. [CrossRef]
64. Muto, S. Physiological roles of claudins in kidney tubule paracellular transport. *Am. J. Physiol. Renal Physiol.* **2016**. [CrossRef] [PubMed]
65. Turner, J.R.; Black, E.D. NHE3-dependent cytoplasmic alkalinization is triggered by Na(+)-glucose cotransport in intestinal epithelia. *Am. J. Physiol. Cell Physiol.* **2001**, *281*, C1533–C1541. [CrossRef]
66. Zhao, H.; Shiue, H.; Palkon, S.; Wang, Y.; Cullinan, P.; Burkhardt, J.K.; Musch, M.W.; Chang, E.B.; Turner, J.R. Ezrin regulates NHE3 translocation and activation after Na+-glucose cotransport. *Proc. Natl. Acad. Sci. USA* **2004**, *101*, 9485–9490. [CrossRef]
67. Shiue, H.; Musch, M.W.; Wang, Y.; Chang, E.B.; Turner, J.R. Akt2 phosphorylates ezrin to trigger NHE3 translocation and activation. *J. Biol. Chem.* **2005**, *280*, 1688–1695. [CrossRef]
68. Hu, Z.; Wang, Y.; Graham, W.V.; Su, L.; Musch, M.W.; Turner, J.R. MAPKAPK-2 is a critical signaling intermediate in NHE3 activation following Na^+-glucose cotransport. *J. Biol. Chem.* **2006**, *281*, 24247–24253. [CrossRef]
69. Lin, R.; Murtazina, R.; Cha, B.; Chakraborty, M.; Sarker, R.; Chen, T.E.; Lin, Z.; Hogema, B.M.; de Jonge, H.R.; Seidler, U.; et al. D-glucose acts via sodium/glucose cotransporter 1 to increase NHE3 in mouse jejunal brush border by a Na+/H+ exchange regulatory factor 2-dependent process. *Gastroenterology* **2011**, *140*, 560–571. [CrossRef]
70. Bohlen, H.G. Na+-induced intestinal interstitial hyperosmolality and vascular responses during absorptive hyperemia. *Am. J. Physiol.* **1982**, *242*, H785–H789. [CrossRef]
71. Clayburgh, D.R.; Musch, M.W.; Leitges, M.; Fu, Y.X.; Turner, J.R. Coordinated epithelial NHE3 inhibition and barrier dysfunction are required for TNF-mediated diarrhea in vivo. *J. Clin. Investig.* **2006**, *116*, 2682–2694. [CrossRef] [PubMed]
72. Yuhan, R.; Koutsouris, A.; Savkovic, S.D.; Hecht, G. Enteropathogenic Escherichia coli-induced myosin light chain phosphorylation alters intestinal epithelial permeability. *Gastroenterology* **1997**, *113*, 1873–1882. [CrossRef]
73. Hecht, G.; Pestic, L.; Nikcevic, G.; Koutsouris, A.; Tripuraneni, J.; Lorimer, D.D.; Nowak, G.; Guerriero, V., Jr.; Elson, E.L.; Lanerolle, P.D. Expression of the catalytic domain of myosin light chain kinase increases paracellular permeability. *Am. J. Physiol.* **1996**, *271*, C1678–C1684. [CrossRef] [PubMed]
74. Shen, L.; Black, E.D.; Witkowski, E.D.; Lencer, W.I.; Guerriero, V.; Schneeberger, E.E.; Turner, J.R. Myosin light chain phosphorylation regulates barrier function by remodeling tight junction structure. *J. Cell Sci.* **2006**, *119*, 2095–2106. [CrossRef] [PubMed]
75. Su, L.; Shen, L.; Clayburgh, D.R.; Nalle, S.C.; Sullivan, E.A.; Meddings, J.B.; Abraham, C.; Turner, J.R. Targeted epithelial tight junction dysfunction causes immune activation and contributes to development of experimental colitis. *Gastroenterology* **2009**, *136*, 551–563. [CrossRef]
76. Charpentier, B.; Hiesse, C.; Lantz, O.; Ferran, C.; Stephens, S.; O'Shaugnessy, D.; Bodmer, M.; Benoit, G.; Bach, J.F.; Chatenoud, L. Evidence that antihuman tumor necrosis factor monoclonal antibody prevents OKT3-induced acute syndrome. *Transplantation* **1992**, *54*, 997–1002. [CrossRef]
77. Ferran, C.; Dy, M.; Merite, S.; Sheehan, K.; Schreiber, R.; Leboulenger, F.; Landais, P.; Bluestone, J.; Bach, J.F.; Chatenoud, L. Reduction of morbidity and cytokine release in anti-CD3 MoAb-treated mice by corticosteroids. *Transplantation* **1990**, *50*, 642–648. [CrossRef]
78. Ferran, C.; Sheehan, K.; Dy, M.; Schreiber, R.; Merite, S.; Landais, P.; Noel, L.H.; Grau, G.; Bluestone, J.; Bach, J.F.; et al. Cytokine-related syndrome following injection of anti-CD3 monoclonal antibody: Further evidence for transient in vivo T cell activation. *Eur. J. Immunol.* **1990**, *20*, 509–515. [CrossRef]
79. Musch, M.W.; Clarke, L.L.; Mamah, D.; Gawenis, L.R.; Zhang, Z.; Ellsworth, W.; Shalowitz, D.; Mittal, N.; Efthimiou, P.; Alnadjim, Z.; et al. T cell activation causes diarrhea by increasing intestinal permeability and inhibiting epithelial Na+/K+-ATPase. *J. Clin. Investig.* **2002**, *110*, 1739–1747. [CrossRef]
80. Tang, Y.; Clayburgh, D.R.; Mittal, N.; Goretsky, T.; Dirisina, R.; Zhang, Z.; Kron, M.; Ivancic, D.; Katzman, R.B.; Grimm, G.; et al. Epithelial NF-kappaB enhances transmucosal fluid movement by altering tight junction protein composition after T cell activation. *Am. J. Pathol.* **2010**, *176*, 158–167. [CrossRef]
81. Beutel, O.; Maraspini, R.; Pombo-Garcia, K.; Martin-Lemaitre, C.; Honigmann, A. Phase Separation of Zonula Occludens Proteins Drives Formation of Tight Junctions. *Cell* **2019**, *179*, 923–936.e911. [CrossRef]
82. Madara, J.L.; Carlson, S.; Anderson, J.M. ZO-1 maintains its spatial distribution but dissociates from junctional fibrils during tight junction regulation. *Am. J. Physiol.* **1993**, *264*, C1096–C1101. [CrossRef] [PubMed]

83. Shen, L.; Weber, C.R.; Turner, J.R. The tight junction protein complex undergoes rapid and continuous molecular remodeling at steady state. *J. Cell Biol.* **2008**, *181*, 683–695. [CrossRef] [PubMed]
84. Yu, D.; Marchiando, A.M.; Weber, C.R.; Raleigh, D.R.; Wang, Y.; Shen, L.; Turner, J.R. MLCK-dependent exchange and actin binding region-dependent anchoring of ZO-1 regulate tight junction barrier function. *Proc. Natl. Acad. Sci. USA* **2010**, *107*, 8237–8241. [CrossRef] [PubMed]
85. Marchiando, A.M.; Shen, L.; Graham, W.V.; Weber, C.R.; Schwarz, B.T.; Austin, J.R., 2nd; Raleigh, D.R.; Guan, Y.; Watson, A.J.; Montrose, M.H.; et al. Caveolin-1-dependent occludin endocytosis is required for TNF-induced tight junction regulation in vivo. *J. Cell Biol.* **2010**, *189*, 111–126. [CrossRef] [PubMed]
86. Saitou, M.; Furuse, M.; Sasaki, H.; Schulzke, J.D.; Fromm, M.; Takano, H.; Noda, T.; Tsukita, S. Complex phenotype of mice lacking occludin, a component of tight junction strands. *Mol. Biol. Cell* **2000**, *11*, 4131–4142. [CrossRef] [PubMed]
87. Schulzke, J.D.; Gitter, A.H.; Mankertz, J.; Spiegel, S.; Seidler, U.; Amasheh, S.; Saitou, M.; Tsukita, S.; Fromm, M. Epithelial transport and barrier function in occludin-deficient mice. *Biochim. Biophys. Acta* **2005**, *1669*, 34–42. [CrossRef]
88. Shen, L.; Turner, J.R. Actin depolymerization disrupts tight junctions via caveolae-mediated endocytosis. *Mol. Biol. Cell* **2005**, *16*, 3919–3936. [CrossRef]
89. Buschmann, M.M.; Shen, L.; Rajapakse, H.; Raleigh, D.R.; Wang, Y.; Wang, Y.; Lingaraju, A.; Zha, J.; Abbott, E.; McAuley, E.M.; et al. Occludin OCEL-domain interactions are required for maintenance and regulation of the tight junction barrier to macromolecular flux. *Mol. Biol. Cell* **2013**, *24*, 3056–3068. [CrossRef]
90. Schwarz, B.T.; Wang, F.; Shen, L.; Clayburgh, D.R.; Su, L.; Wang, Y.; Fu, Y.X.; Turner, J.R. LIGHT signals directly to intestinal epithelia to cause barrier dysfunction via cytoskeletal and endocytic mechanisms. *Gastroenterology* **2007**, *132*, 2383–2394. [CrossRef]
91. Yu, A.S.; McCarthy, K.M.; Francis, S.A.; McCormack, J.M.; Lai, J.; Rogers, R.A.; Lynch, R.D.; Schneeberger, E.E. Knockdown of occludin expression leads to diverse phenotypic alterations in epithelial cells. *Am. J. Physiol. Cell Physiol.* **2005**, *288*, C1231–C1241. [CrossRef] [PubMed]
92. Ye, D.; Guo, S.; Al-Sadi, R.; Ma, T.Y. MicroRNA regulation of intestinal epithelial tight junction permeability. *Gastroenterology* **2011**, *141*, 1323–1333. [CrossRef]
93. Cording, J.; Berg, J.; Kading, N.; Bellmann, C.; Tscheik, C.; Westphal, J.K.; Milatz, S.; Gunzel, D.; Wolburg, H.; Piontek, J.; et al. In tight junctions, claudins regulate the interactions between occludin, tricellulin and marvelD3, which, inversely, modulate claudin oligomerization. *J. Cell Sci.* **2013**, *126*, 554–564. [CrossRef] [PubMed]
94. Raleigh, D.R.; Marchiando, A.M.; Zhang, Y.; Shen, L.; Sasaki, H.; Wang, Y.; Long, M.; Turner, J.R. Tight junction-associated MARVEL proteins marveld3, tricellulin, and occludin have distinct but overlapping functions. *Mol. Biol. Cell* **2010**, *21*, 1200–1213. [CrossRef] [PubMed]
95. Steed, E.; Elbediwy, A.; Vacca, B.; Dupasquier, S.; Hemkemeyer, S.A.; Suddason, T.; Costa, A.C.; Beaudry, J.B.; Zihni, C.; Gallagher, E.; et al. MarvelD3 couples tight junctions to the MEKK1-JNK pathway to regulate cell behavior and survival. *J. Cell Biol.* **2014**, *204*, 821–838. [CrossRef] [PubMed]
96. Steed, E.; Rodrigues, N.T.; Balda, M.S.; Matter, K. Identification of MarvelD3 as a tight junction-associated transmembrane protein of the occludin family. *BMC Cell Biol.* **2009**, *10*, 95. [CrossRef] [PubMed]
97. Krug, S.M.; Amasheh, S.; Richter, J.F.; Milatz, S.; Gunzel, D.; Westphal, J.K.; Huber, O.; Schulzke, J.D.; Fromm, M. Tricellulin forms a barrier to macromolecules in tricellular tight junctions without affecting ion permeability. *Mol. Biol. Cell* **2009**, *20*, 3713–3724. [CrossRef] [PubMed]
98. Krug, S.M.; Bojarski, C.; Fromm, A.; Lee, I.M.; Dames, P.; Richter, J.F.; Turner, J.R.; Fromm, M.; Schulzke, J.D. Tricellulin is regulated via interleukin-13-receptor alpha2, affects macromolecule uptake, and is decreased in ulcerative colitis. *Mucosal. Immunol.* **2018**, *11*, 345–356. [CrossRef]
99. Van Itallie, C.M.; Fanning, A.S.; Holmes, J.; Anderson, J.M. Occludin is required for cytokine-induced regulation of tight junction barriers. *J. Cell Sci.* **2010**, *123*, 2844–2852. [CrossRef]
100. Schwayer, C.; Shamipour, S.; Pranjic-Ferscha, K.; Schauer, A.; Balda, M.; Tada, M.; Matter, K.; Heisenberg, C.P. Mechanosensation of Tight Junctions Depends on ZO-1 Phase Separation and Flow. *Cell* **2019**, *179*, 937–952 e918. [CrossRef]
101. Spadaro, D.; Le, S.; Laroche, T.; Mean, I.; Jond, L.; Yan, J.; Citi, S. Tension-Dependent Stretching Activates ZO-1 to Control the Junctional Localization of Its Interactors. *Curr. Biol.* **2017**, *27*, 3783–3795 e3788. [CrossRef] [PubMed]

102. Wang, F.; Schwarz, B.T.; Graham, W.V.; Wang, Y.; Su, L.; Clayburgh, D.R.; Abraham, C.; Turner, J.R. IFN-gamma-induced TNFR2 expression is required for TNF-dependent intestinal epithelial barrier dysfunction. *Gastroenterology* **2006**, *131*, 1153–1163. [CrossRef] [PubMed]
103. Su, L.; Nalle, S.C.; Shen, L.; Turner, E.S.; Singh, G.; Breskin, L.A.; Khramtsova, E.A.; Khramtsova, G.; Tsai, P.Y.; Fu, Y.X.; et al. TNFR2 activates MLCK-dependent tight junction dysregulation to cause apoptosis-mediated barrier loss and experimental colitis. *Gastroenterology* **2013**, *145*, 407–415. [CrossRef] [PubMed]
104. Ye, D.; Ma, I.; Ma, T.Y. Molecular mechanism of tumor necrosis factor-alpha modulation of intestinal epithelial tight junction barrier. *Am. J. Physiol. Gastrointest. Liver Physiol.* **2006**, *290*, G496–G504. [CrossRef] [PubMed]
105. Ma, T.Y.; Iwamoto, G.K.; Hoa, N.T.; Akotia, V.; Pedram, A.; Boivin, M.A.; Said, H.M. TNF-alpha-induced increase in intestinal epithelial tight junction permeability requires NF-kappa B activation. *Am. J. Physiol. Gastrointest. Liver Physiol.* **2004**, *286*, G367–G376. [CrossRef] [PubMed]
106. Peterson, M.D.; Mooseker, M.S. Characterization of the enterocyte-like brush border cytoskeleton of the C2BBe clones of the human intestinal cell line, Caco-2. *J. Cell Sci.* **1992**, *102*, 581–600. [PubMed]
107. Ye, D.; Ma, T.Y. Cellular and molecular mechanisms that mediate basal and tumour necrosis factor-alpha-induced regulation of myosin light chain kinase gene activity. *J. Cell Mol. Med.* **2008**, *12*, 1331–1346. [CrossRef] [PubMed]
108. Graham, W.V.; Wang, F.; Clayburgh, D.R.; Cheng, J.X.; Yoon, B.; Wang, Y.; Lin, A.; Turner, J.R. Tumor necrosis factor-induced long myosin light chain kinase transcription is regulated by differentiation-dependent signaling events. Characterization of the human long myosin light chain kinase promoter. *J. Biol. Chem.* **2006**, *281*, 26205–26215. [CrossRef]
109. Al-Sadi, R.; Guo, S.; Ye, D.; Dokladny, K.; Alhmoud, T.; Ereifej, L.; Said, H.M.; Ma, T.Y. Mechanism of IL-1beta modulation of intestinal epithelial barrier involves p38 kinase and activating transcription factor-2 activation. *J. Immunol.* **2013**, *190*, 6596–6606. [CrossRef]
110. Al-Sadi, R.; Guo, S.; Ye, D.; Ma, T.Y. TNF-alpha Modulation of Intestinal Epithelial Tight Junction Barrier Is Regulated by ERK1/2 Activation of Elk-1. *Am. J. Pathol.* **2013**. [CrossRef]
111. Blair, S.A.; Kane, S.V.; Clayburgh, D.R.; Turner, J.R. Epithelial myosin light chain kinase expression and activity are upregulated in inflammatory bowel disease. *Lab. Investig.* **2006**, *86*, 191–201. [CrossRef] [PubMed]
112. Edelblum, K.L.; Sharon, G.; Singh, G.; Odenwald, M.A.; Sailer, A.; Cao, S.; Ravens, S.; Thomsen, I.; El Bissati, K.; McLeod, R.; et al. The microbiome activates CD4 T-cell-mediated immunity to compensate for increased intestinal permeability. *Cell Mol. Gastroenterol. Hepatol.* **2017**, *4*, 285–297. [CrossRef] [PubMed]
113. Russo, J.M.; Florian, P.; Shen, L.; Graham, W.V.; Tretiakova, M.S.; Gitter, A.H.; Mrsny, R.J.; Turner, J.R. Distinct temporal-spatial roles for rho kinase and myosin light chain kinase in epithelial purse-string wound closure. *Gastroenterology* **2005**, *128*, 987–1001. [CrossRef] [PubMed]
114. Tamada, M.; Perez, T.D.; Nelson, W.J.; Sheetz, M.P. Two distinct modes of myosin assembly and dynamics during epithelial wound closure. *J. Cell Biol.* **2007**, *176*, 27–33. [CrossRef]
115. Bement, W.M.; Forscher, P.; Mooseker, M.S. A novel cytoskeletal structure involved in purse string wound closure and cell polarity maintenance. *J. Cell Biol.* **1993**, *121*, 565–578. [CrossRef]
116. Lechuga, S.; Ivanov, A.I. Disruption of the epithelial barrier during intestinal inflammation: Quest for new molecules and mechanisms. *Biochim. Biophys. Acta* **2017**, *1864*, 1183–1194. [CrossRef]
117. Isobe, K.; Raghuram, V.; Krishnan, L.; Chou, C.L.; Yang, C.R.; Knepper, M.A. CRISPR-Cas9/Phosphoproteomics Identifies Multiple Non-canonical Targets of Myosin Light Chain Kinase. *Am. J. Physiol. Renal Physiol.* **2020**. [CrossRef]
118. Kamm, K.E.; Stull, J.T. Dedicated myosin light chain kinases with diverse cellular functions. *J. Biol. Chem.* **2001**, *276*, 4527–4530. [CrossRef]
119. Garcia, J.G.; Lazar, V.; Gilbert-McClain, L.I.; Gallagher, P.J.; Verin, A.D. Myosin light chain kinase in endothelium: Molecular cloning and regulation. *Am. J. Respir. Cell Mol. Biol.* **1997**, *16*, 489–494. [CrossRef]
120. Khromov, A.S.; Wang, H.; Choudhury, N.; McDuffie, M.; Herring, B.P.; Nakamoto, R.; Owens, G.K.; Somlyo, A.P.; Somlyo, A.V. Smooth muscle of telokin-deficient mice exhibits increased sensitivity to Ca2+ and decreased cGMP-induced relaxation. *Proc. Natl. Acad. Sci. USA* **2006**, *103*, 2440–2445. [CrossRef]
121. Choudhury, N.; Khromov, A.S.; Somlyo, A.P.; Somlyo, A.V. Telokin mediates Ca2+-desensitization through activation of myosin phosphatase in phasic and tonic smooth muscle. *J. Muscle Res. Cell Motil.* **2004**, *25*, 657–665. [CrossRef] [PubMed]

122. Nieznanski, K.; Sobieszek, A. Telokin (kinase-related protein) modulates the oligomeric state of smooth-muscle myosin light-chain kinase and its interaction with myosin filaments. *Biochem. J.* **1997**, *322*, 65–71. [CrossRef] [PubMed]
123. Somlyo, A.V.; Wang, H.; Choudhury, N.; Khromov, A.S.; Majesky, M.; Owens, G.K.; Somlyo, A.P. Myosin light chain kinase knockout. *J. Muscle Res. Cell Motil.* **2004**, *25*, 241–242. [CrossRef] [PubMed]
124. He, W.Q.; Peng, Y.J.; Zhang, W.C.; Lv, N.; Tang, J.; Chen, C.; Zhang, C.H.; Gao, S.; Chen, H.Q.; Zhi, G.; et al. Myosin light chain kinase is central to smooth muscle contraction and required for gastrointestinal motility in mice. *Gastroenterology* **2008**, *135*, 610–620. [CrossRef] [PubMed]
125. Clayburgh, D.R.; Rosen, S.; Witkowski, E.D.; Wang, F.; Blair, S.; Dudek, S.; Garcia, J.G.; Alverdy, J.C.; Turner, J.R. A differentiation-dependent splice variant of myosin light chain kinase, MLCK1, regulates epithelial tight junction permeability. *J. Biol. Chem.* **2004**, *279*, 55506–55513. [CrossRef] [PubMed]
126. Birukov, K.G.; Csortos, C.; Marzilli, L.; Dudek, S.; Ma, S.F.; Bresnick, A.R.; Verin, A.D.; Cotter, R.J.; Garcia, J.G. Differential regulation of alternatively spliced endothelial cell myosin light chain kinase isoforms by p60(Src). *J. Biol. Chem.* **2001**, *276*, 8567–8573. [CrossRef] [PubMed]
127. Lazar, V.; Garcia, J.G. A single human myosin light chain kinase gene (MLCK.; MYLK). *Genomics* **1999**, *57*, 256–267. [CrossRef]
128. Graham, W.V.; He, W.; Marchiando, A.M.; Zha, J.; Singh, G.; Li, H.S.; Biswas, A.; Ong, M.; Jiang, Z.H.; Choi, W.; et al. Intracellular MLCK1 diversion reverses barrier loss to restore mucosal homeostasis. *Nat. Med.* **2019**, *25*, 690–700. [CrossRef]
129. Graham, W.V.; Magis, A.T.; Bailey, K.M.; Turner, J.R.; Ostrov, D.A. Crystallization and preliminary X-ray analysis of the human long myosin light-chain kinase 1-specific domain IgCAM3. *Acta Crystallogr. Sect. F Struct Biol. Cryst Commun.* **2011**, *67*, 221–223. [CrossRef]
130. Madsen, K.L.; Malfair, D.; Gray, D.; Doyle, J.S.; Jewell, L.D.; Fedorak, R.N. Interleukin-10 gene-deficient mice develop a primary intestinal permeability defect in response to enteric microflora. *Inflamm. Bowel Dis.* **1999**, *5*, 262–270. [CrossRef]

13

Role of Tricellular Tight Junction Protein Lipolysis-Stimulated Lipoprotein Receptor (LSR) in Cancer Cells

Takayuki Kohno *, Takumi Konno and Takashi Kojima

Department of Cell Science, Research Institute for Frontier Medicine, Sapporo Medical University, Sapporo 060-8556, Japan

* Correspondence: kohno@sapmed.ac.jp.

Abstract: Maintaining a robust epithelial barrier requires the accumulation of tight junction proteins, LSR/angulin-1 and tricellulin, at the tricellular contacts. Alterations in the localization of these proteins temporarily cause epithelial barrier dysfunction, which is closely associated with not only physiological differentiation but also cancer progression and metastasis. In normal human endometrial tissues, the endometrial cells undergo repeated proliferation and differentiation under physiological conditions. Recent observations have revealed that the localization and expression of LSR/angulin-1 and tricellulin are altered in a menstrual cycle-dependent manner. Moreover, it has been shown that endometrial cancer progression affects these alterations. This review highlights the differences in the localization and expression of tight junction proteins in normal endometrial cells and endometrial cancers and how they cause functional changes in cells.

Keywords: tricellular tight junctions; endometrial cancer; epithelial barrier dysfunction

---

1. Introduction

The endometrium is a regenerative tissue in which the cells undergo proliferation and differentiation depending on the levels of estrogen, progesterone, or various cytokines. The organization of cell-cell junctions, such as tight junctions, adherence junctions, gap junctions, and desmosomes, has important implications for the homeostatic regulation of many tissues, including the endometrium [1]. Cell-cell junctions are formed not only in bicellular regions but also at tricellular contacts [2]. Several reviews have mentioned that occludin (OCLN) and claudins (CLDNs) have been established as bicellular tight junction proteins involved in the formation and maintenance of epithelial barriers [3–5]. A recent study revealed that their expression and localization are affected by the menstrual cycle [6]. According to the report, CLDN-1, -3, -4, and -7 localized in the subapical region during the proliferative phase of the endometrium, while they were broadly distributed to the lateral region during the secretory phase (Figure 1). Furthermore, it has been shown that robust epithelial barrier formation requires localization of these tight junction proteins at the subapical region by analyzing primary cultured normal human endometrial cells. Recent studies have revealed that the localization of tricellular tight junction proteins, tricellulin and LSR/angulin-1, to tricellular contacts is required for epithelial barrier maturation based on the proper localization of OCLN and CLDNs [7]. A recent study demonstrated that tricellulin localized in the subapical region during the endometrial secretory phase, whereas LSR was broadly distributed to the lateral region [8]. In contrast, during the proliferative phase of endometrium formation, both proteins localized in the subapical region. Furthermore, analysis using primary cultured normal human endometrial cells revealed that localization of LSR to the tricellular contacts is required for the formation of mature epithelial polarity with sufficient barrier function. These findings suggested that LSR and tricellulin are closely related to the functional regulation of periodic morphological changes in the endometrial tissue. In the normal human endometrium, a part of the mechanism that regulates the localization and expression of tricellular tight junction proteins has been elucidated below.

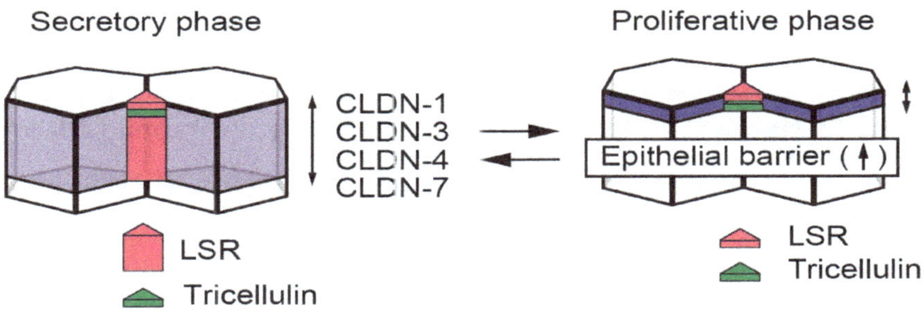

Figure 1. The localization of tight junction proteins is affected by menstrual cycle. In secretory phase of human endometrium, CLDN-1, -3, -4, and -7 are widely distributed to the lateral region. Tricellulin localized in tricellular contacts of the subapical region, whereas LSR is widely distributed to the lateral region. In proliferative phase, CLDNs localized in the subapical tight junction region. Tricellulin and LSR localized in the subapical tricellular contacts.

2. Tricellular Tight Junction Proteins and Cancer

Many oncogenic processes are known to be involved in genetic instability based on failure of DNA mismatch repair pathways [9]. It is an established fact that the abnormal cell growth, dedifferentiation, and EMT are induced by the activation of oncogenes, such as Ras, and/or the inactivation of tumor suppressor genes, such as PTEN and p53 [10]. These adverse events, like a cancer metastasis, are certainly accompanied with reconstitution of cell-cell junctions [11]. While most of the differentiated epithelial cells have established tight junctions, disruption of tight junctions abolishes cell polarity and promotes dedifferentiation [3,12]. Furthermore, a decrease in epithelial barrier function is implicated in cancer cell invasion and metastasis [13]. Epithelial barrier homeostasis is disrupted by decreased expression of tight junction proteins as well as by their overexpression [14]. It still remains largely unknown how expression of tight junction proteins is regulated during the oncogenic process. Interestingly, decreased expression of tricellulin, which regulates epithelial barrier maturation, has been reported to be associated with tumor progression. For instance, in human tonsillar squamous cell carcinoma, decreased expression of tricellulin and CLDN-7 and increased expression of CLDN-1 have been identified [15]. In hepatocellular carcinoma cells, decreased expression of tricellulin has been observed as compared to that in normal hepatocytes [16]. In addition, lower prognosis of intrahepatic cholangiocarcinoma (iCCC) has been shown to correlate with decreased expression of tricellulin [17]. In pancreatic cancer, the decreased expression of tricellulin exhibits a correlation with decreased differentiation [18]. In gastric carcinoma, Snail-induced EMT negatively regulates the expression of tricellulin [19].

Increasing number of studies have reported the relationship between malignant transformation and expression of LSR, which is another tricellular tight junction protein. It has been reported that the expression of LSR is higher in invasive ductal carcinomas compared to that in invasive lobular carcinomas [20]. In addition, LSR is considered as a candidate prognostic biomarker in colon cancer patients [21]. Recent observations have revealed that the expression levels of LSR, tricellulin, and CLDN-1 were higher in head and neck squamous cell carcinoma tissues compared with those in normal palatine tonsils [22]. In addition, by analyzing the immunohistochemical staining using paraffin sections of head and neck squamous cell carcinoma tissue, it has been shown that the expression levels of both LSR and CLDN-1 are increased in cancerous tissues, especially in invasive tissues, compared to those in adjacent dysplasia tissues. Increased expression of CLDN-1 has been observed in advanced head and neck cancer [23]. CLDN-1 has also been shown to be significantly expressed in hypopharyngeal squamous cell carcinoma tissues, suggesting that CLDN-1 is associated with tumor differentiation and lymph node metastasis [24]. As described above, various cancerous malignancies are associated with changes in the expression and localization of not only bicellular tight junction proteins but also tricellular tight junction proteins. These findings suggested that tricellular tight junction proteins may interact closely with bicellular junctions during malignant transformation in response to reduction of the barrier function.

3. Expression and Localization of the Tricellular Tight Junction Proteins, LSR and Tricellulin, during Endometriosis and Endometrial Carcinoma

During endometriosis, decreased expression levels of CLDN-3, -4, and -7 have been observed [25], and in endometrial cancer, increased expression levels of CLDN-3 and -4 have been reported [26,27]. Since changes in the expression levels of bicellular tight junction proteins were observed during the pathogenesis of the endometrial cancer, it is reasonable to consider that these processes were also accompanied by changes in expression levels of tricellular tight junction proteins. Recently, by analyzing the immunohistochemical staining using paraffin sections of endometriotic and endometrial cancer tissue, it has been found that during endometriosis tricellulin was localized in the subapical region similar to normal human endometrial tissue, while LSR was localized in the subapical region of tricellular contacts in addition to the lateral region [8]. In endometrial carcinoma G1, where the formation of gland-like structure is retained, the expression levels of tricellulin and LSR were distributed unevenly from the subapical to the lateral region of cell-cell junctions. In G2 and G3 endometrial carcinoma, their expression levels were decreased. Taken together, these findings revealed that the grade of malignancy correlated with the decreased expression levels of tricellulin and LSR in addition to changes in the localizations of these proteins (Figure 2). Among cultured cells derived from endometrial cancer, we were able to confirm the expression levels of both tricellulin and LSR in Sawano, HHUA, and JHMUE-1 cells, all of which exhibit an epithelial phenotype, whereas little or no expression was observed in JHMUE-2, which exhibits a fibroblast-like morphology. Since the expression levels of tricellulin and LSR contribute to the maintenance of the morphology of epithelial cells, we hypothesized that depletion of these proteins enhances cell motility. The endometrial cancer cell line, Sawano, endogenously expresses tricellulin and LSR. In Sawano cells with LSR knockdown, the epithelial barrier function was reduced, and thereby, cell motility, cell invasion, and proliferation were enhanced compared to those in the parental control. Thus, the localization of LSR at tricellular contacts is necessary for maintaining the robustness of the epithelial barrier function. The relationship between the exclusion of LSR from tricellular contacts and cancer progression has been discussed below, with a focus on endometrial cancer.

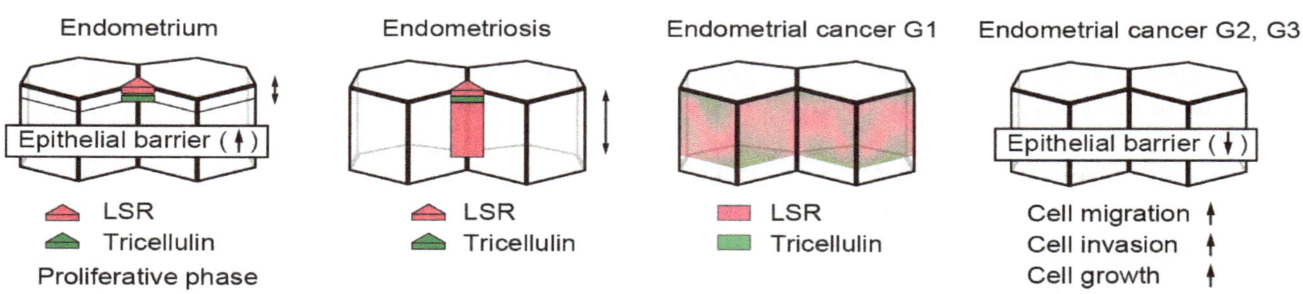

Figure 2. Expression and localization of LSR and tricellulin during endometriosis and in endometrial cancers. LSR and tricellulin localized in tricellular contacts in endometrium. During endometriosis, tricellulin is localized in the subapical region of tricellular contacts and LSR is localized in not only the subapical tricellular contacts but also in the lateral tricellular contacts. In endometrial cancer G1, tricellulin and LSR were distributed unevenly from the subapical to the lateral region of bicellular junctions. In endometrial cancer G2 and G3, the expression levels of tricellulin and LSR were downregulated, resulting in decrease of epithelial barrier and increase of cell migration, cell invasion, and cell growth.

4. Obesity and Endometrial Cancer

Diagnoses of endometrial cancer have increased worldwide in recent years [28]. Obesity is a major risk factor for endometrial cancer [29]. Bioinformatics analysis using cBioProtal and DAVID bioinformatics resources confirmed that expression of genes related to glucose metabolism and lipid metabolism is increased in endometrial cancer [30]. Increase in estrogen, decrease in adiponectin, and increase in inflammatory cytokines are all known as typical cancer-inducing factors [31]. Leptin has also been reported to be involved in endometrial cell proliferation [32]. Previous studies have reported that an increase in circulating adiponectin and leptin-adiponectin ratio may be potential risk factors for breast cancer, colorectal cancer, pancreatic cancer, and endometrial cancer [33,34]. Leptin is produced not only from an adipose tissue, but also from follicles and placenta, and its production is associated with menstrual cycle and pregnancy [35,36]. Leptin is involved in facilitating endometrial cancer progression and metastasis of pancreatic cancer via the activation of JAK2/STAT3 pathway [37,38]. Adiponectin suppresses the progression and development of cancer by antagonizing this pathway [39]. It has been found that in endometrial cancer cells, leptin suppressed the expression of LSR, while adiponectin increased its expression [8]. Moreover, studies using inhibitors suggested that the stimulation with leptin or adiponectin induced an alteration of LSR expression via the PI3K and JAK2/3 pathways.

It has been speculated that there is an interface between the regulatory pathways of the epithelial barrier formation and signaling pathways via the adipocytokine receptor. The knockdown of LSR enhanced cell motility and invasion in Sawano cells. This finding correlated with the cellular response associated with leptin-dependent downregulation of LSR (Figure 3). Interestingly, even in normal human endometrial cells, leptin suppressed LSR expression, while adiponectin increased its expression. It is assumed that obesity is involved in the malignant transformation of endometrial cancer besides attenuating the robust tight junctions of normal endometrium. LSR has been identified as a lipid receptor involved in lipid clearance [40]. In mice, suppression of LSR expression in the liver causes systemic hyperlipidemia, resulting in obesity and weight gain [41]. The differences in the function and role of LSR as a lipoprotein receptor and the involvement of LSR in obesity-dependent epithelial barrier attenuation should be clarified in future studies.

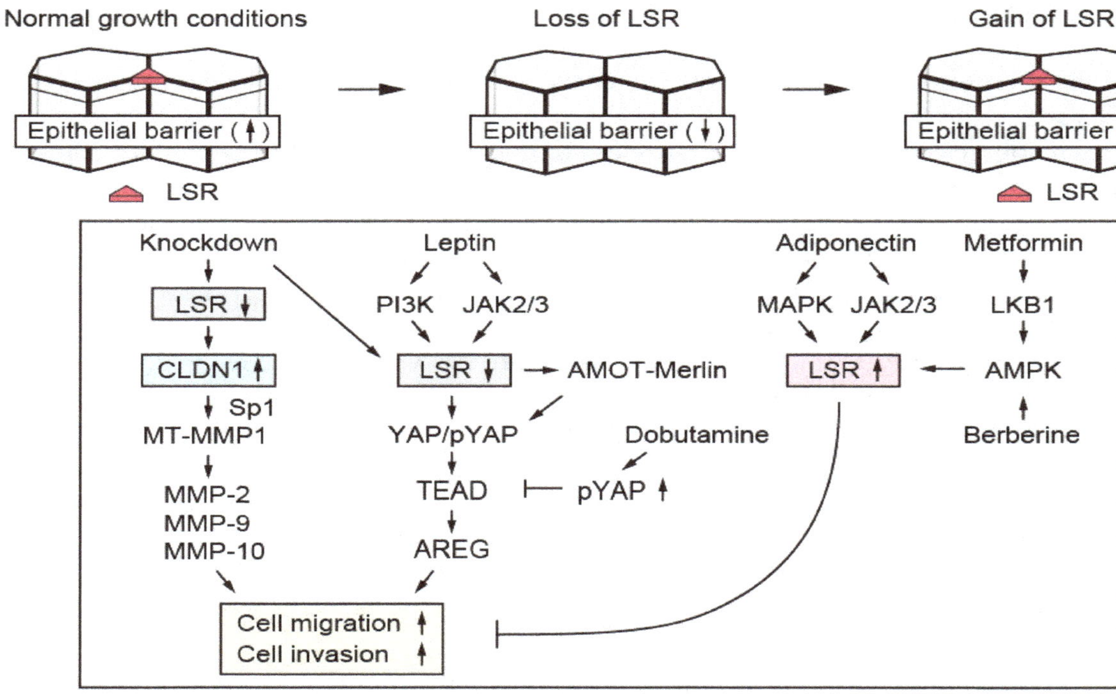

Figure 3. Changes in cellular functions by repression and re-expression of LSR. Under normal growth conditions, LSR localized in tricellular contacts in primary cultured normal human endometrial cells and Sawano cells. The knockdown of LSR enhanced cell motility and cell growth accompanying with decrease in barrier function. Leptin suppressed LSR expression; in contrast, adiponectin induced an increase in its expression. AMPK activator metformin and berberine also induced an increase in LSR expression at the subapical region of tricellular contacts, resulting in the rescue of the LSR-knockdown phenotypes.

5. Glucose Metabolism and Endometrial Cancer

Obesity has been reported to be an independent risk factor for the development of diabetes [42]. Epidemiological studies have shown that metformin, a therapeutic agent for type 2 diabetes, reduces the incidence of endometrial cancer [43]. In addition, berberine, which is a herbal medicine component, has been reported to be not only effective in type 2 diabetes, but also in suppression of growth of cancer [44]. We found that metformin and berberine both increased LSR expression in endometrial cancer cells. The upregulation of LSR expression by these drugs contributed to the suppression of motility and invasion of endometrial cancer cells enhanced by leptin administration. Metformin and berberine also increased LSR expression in primary cultured normal human endometrial cells (Figure 3). Therefore, these drugs may be used to treat diseases based on epithelial barrier disruption. In fact, these drugs, which are categorized as AMPK activators, are currently being considered as potential therapeutic agents for endometrial cancer [43–45].

AMPK is an energy sensor that regulates the levels of intracellular ATP and centrally regulates metabolism [46,47]. Initially, depletion of intracellular ATP was reported to temporarily and reversibly disrupt tight junctions [48]. However, recent studies have indicated that AMPK, rather than affecting the intracellular ATP levels, may directly regulate tight junction proteins [49]. In the report, the authors revealed that AMPK regulates the relocalization of ZO-1 after Ca switch, independently of the intracellular ATP levels. Furthermore, AMPK has been reported to promote stabilization of tight junctions and to enhance barrier function via phosphorylation of the scaffold protein, GIV, which regulates cell polarity [50]. Metformin acts as a therapeutic agent for diabetes via LKB1-mediated phosphorylation of AMPK, which is accompanied by mitochondrial OxPhos suppression [51], suggesting that, in epithelial cells, metformin stabilizes tight junctions via the activation of AMPK.

Interestingly, it has been previously reported that the progression of endometrial cancer correlates with the decrease in AMPK expression [52]. It is necessary to elucidate the signal transduction pathways involved in AMPK-regulated glucose metabolism and the regulation of epithelial barrier function.

6. Mechanisms of Enhancement of Cell Invasion Caused by Decreased LSR Expression

Using immunohistochemical analysis of paraffinized sections of endometrial cancer tissues, we observed a positive expression of LSR and negative expression of CLDN-1 in the gland-like structure region. In contrast, in the invasive front area, LSR expression decreased and CLDN-1 expression increased. Following knockdown of LSR in endometrial cancer Sawano cells, CLDN-1 expression increased, while there was no significant change in the expression levels of CLDN-3, -4, -7, and OCLN. Before LSR knockdown, although CLDN-1 localized in the subapical region, it was widely distributed not only to the subapical region but also to the lateral region after LSR knockdown. These findings suggested that there was a negative relationship between the expression levels of LSR and CLDN-1 (Figure 3). In intestinal epithelial cells, it has been reported that regulation of CLDN-1 expression requires Sp1 binding to the CLDN-1 promoter region [53]. It has been reported that CLDN-1, -4, and -19 harbor Sp1 binding sites in the promoter region [54–56]. We confirmed that Sp1-dependent transcriptional regulation was involved in the enhancement of CLDN-1 expression associated with LSR repression [57].

It has been reported that cell invasion is enhanced via the cleavage of laminin-5 gamma 2 chains by activation of MT-MMP1 and MMP2 in CLDN-1-overexpressing OSC cells [58]. In addition, in SW480 cells, overexpressing CLDN-1, cell invasion is enhanced through the activation of MMP2 and MMP9 [59]. The initial process of cell invasion requires reconstitution of extracellular matrix components, along with the attenuation of cell junctions [60]. Twenty four MMP family members have been identified so far [61]. It has been found that knockdown of LSR increased the expression levels of MT-MMP1, MMP2, MMP9, and MMP10 in Sawano cells [57]. MT-MMP1 has been reported to be a initiating factor that regulates the MMP cascade following the activation of proMMP2 [62]. Interestingly, double knockdown of LSR and CLDN-1 suppressed the increase in cell invasion by LSR knockdown [57]. Little is known about the precise molecular mechanisms underlying the activation of MMPs accompanying the expression of CLDN-1 in endometrial cancer tissues. The suppression of LSR downregulation may regulate the malignant transformation of endometrial cancer.

7. Hippo Pathway and Endometrial Cancer

Relaxation of cell-cell junctions and abnormality of epithelial polarity suppress contact inhibition in epithelial cells, resulting in the induction of abnormal proliferation. The Hippo pathway comprehensively regulates these mechanisms [63]. When the Hippo pathway is turned on, LATS1/2 is phosphorylated via MST1/2. Phosphorylated LATS1/2 phosphorylates YAP, and phosphorylated YAP is degraded via 14-3-3. On the other hand, when the Hippo pathway is blocked, the phosphorylation of YAP is suppressed. The non-phosphorylated form of YAP translocates from the cytoplasm to the nucleus as a transcription cofactor and induces the expression of target genes, such as AREG and DKK1, depending on the expression of the transcription factor TEAD.

We found that YAP is localized in the cytoplasm of the endometrial tissue and in the nucleus in G1, G2, and G3 endometrioid carcinoma, as revealed by immunohistochemical staining using paraffinized sections of endometriotic and endometrial cancer tissues. As mentioned above, cell motility and invasion enhanced by knockdown of LSR were decreased by double knockdown of LSR and YAP. These findings suggested that the decrease in epithelial barrier function caused by the suppression of LSR expression is involved in the regulation of cell motility and invasion via YAP (Figure 3). The β-adrenergic receptor agonist, dobutamine, decreases nuclear YAP levels and increases the amount of cytosolic phosphorylated YAP in human osteoblastoma U2OS cells [64]. Dobutamine has also been reported to suppress the enhancement in the expression of YAP in gastric carcinoma, resulting in the suppression of cell motility and invasion [65]. In addition, in LSR-knocked down Sawano cells, dobutamine administration suppressed the enhancement in cell motility and invasion via the increase of phosphorylated YAP. The precise molecular mechanisms underlying the phosphorylation of Hippo kinases, such as MST1/2 and LATS1/2, via LSR-mediated epithelial barrier modulation still need to be elucidated.

Under glucose starvation conditions, AMPK is phosphorylated by LKB1 [66]. Phosphorylated AMPK has been reported to suppress nuclear translocation of YAP via the phosphorylation of LATS1/2 and/or direct phosphorylation of YAP [47]. In Sawano cells, under glucose-starving conditions, YAP is localized in the proximity of the cell-cell junctions [67]. In addition, both AMPK and YAP were phosphorylated. Moreover, both cell invasion and cell motility enhanced by LSR knockdown were rescued by glucose starvation. It has been speculated that these mechanisms are probably similar to the effect of treatment of AMPK activator, metformin or berberine, as mentioned above. Glucose starvation also increased LSR expression. Further studies are needed to elucidate the signaling pathways by which glucose starvation regulates epithelial barrier functions in endometrial cancer.

Using DNA microarray and qPCR analysis, it has been found that the expression levels of the transcription factors TEAD and AREG, increased in LSR-knockdown Sawano cells [67]. Moreover, immunohistochemical analysis using paraffinized sections of endometriotic and endometrial cancer tissue showed that AREG was expressed in the cytoplasm and that the expression increased with the progression of cancer stage. In Sawano cells, increasing cell motility and invasion by LSR knockdown was suppressed by knockdown of AREG. These effects were also observed after TEAD knockdown. In parental Sawano cells, knockdown of AREG did not affect cell motility and invasion. Therefore, it is concluded that TEAD-dependent AREG expression via the Hippo pathway is involved in the enhancement of cell motility accompanied by the suppression of LSR expression.

8. Crosstalk between the Hippo Pathway and Tight Junctions

Merlin/NF2 is known as one of the tumor suppressor factors that regulate the Hippo pathway [68]. Merlin localizes to adherens junctions by interacting with E-cadherin, PAR3, and catenin [69]. Merlin also interacts with YAP and AMOT, a scaffold protein of Mst1/2 and LATS1/2 at tight junctions and contributes to the regulation of EMT [70]. It has been suggested that changes in the cell adhesion between adjacent cells, that is, modulation of tight junctions and adherens junctions, regulate the phosphorylation of YAP via the Hippo pathway, leading to the disruption of contact inhibition and normal growth. However, the precise molecular mechanisms have yet to be elucidated.

By immunohistochemical analysis, we found that AMOT localized in the subapical region and the lateral region of endometriosis tissues [67]. In endometrioid adenocarcinoma, positive expression of AMOT was observed in the gland-like structure region. Compared with that in endometrial carcinoma G1, decreased expression of AMOT was observed in G2 and G3. The Motin family consists of AMOT (angiomotin), AMOTL1 (angiomotin-like 1), and AMOTL2 (angiomotin-like 2) [71]. In addition, two isoforms of AMOT, AMOT-p130 and AMOT-p80, have been identified. AMOT-p80 has been identified as an oncogene in hemangioendothelioma, head and neck squamous cell carcinoma, and prostate cancer [72–74]. AMOT-p130 has been reported to exhibit oncogenic functions as well as tumor suppressive functions [71]. AMOTL1 has been shown to act as an oncogene in breast cancer [75] and cervical cancer [76], and AMOTL2 has been reported to act as an oncogene in breast cancer [77] and suppressor glioblastoma carcinogenesis [78]. In endometrial cancer, decreased expression of AMOT was observed during cancer progression [67]. Molecular mechanisms related to AMOT in endometrial cancer would be clarified in the near future.

Using immunostaining analysis, we found that, in Sawano cells, endogenous Merlin localized in the vicinity of the cell-cell junctions identically to the other cells [68,79]. Under these conditions, AMOT is localized in tight junctions. It is known that AMOT interacts with Patj, Pals2, and Mupp1 at the tight junctions and that Merlin binds to the coiled-coil region of AMOT [80]. The Rac GTPase-activating protein, Rich1, binds through this region of AMOT. In mature tight junctions because Merlin binds to AMOT, Rich1 cannot interact with AMOT and localizes to the cytosol, resulting in the inactivation of Rac. When Merlin is dissociated from AMOT, which is localized at tight junctions, Rich1 binds to the coiled-coil region of AMOT, thereby activating Rac and enhancing cell proliferation and cell motility. In Sawano cells, LSR knockdown decreased the expression levels of both AMOT and Merlin, and AMOT and LSR double knockdown further reduced the expression of Merlin [67]. The report revealed that the increased invasion and motility of Sawano cells by LSR knockdown was suppressed by AMOT knockdown. In the parental Sawano cells, AMOT knockdown increased cell invasion and motility, which were, in turn, suppressed by YAP knockdown. These findings suggested that YAP as well as Rac are involved in the malignant transformation of endometrial cancer cells. Identification of the crosstalk between AMOT/Merlin pathway and Rac/Rich1 pathway is considered to contribute to the elucidation of the malignant transformation mechanisms of endometrial cancer.

9. Changes in LSR Localization Are Associated with Changes in Cell Size Following Changes in Cell Density

An inverse correlation has been reported between cell density and motility of epithelial cells [81,82]. At high cell densities, apparent cell size decreases with increase in cell thickness accompanying decreasing cell motility. Conversely, at low cell densities, cells spread thinly and cell motility increases. In Sawano cells, LSR and tricellulin localized in tricellular contacts under high cell density conditions, while these proteins migrated to the bicellular region under low cell density conditions despite the presence of tricellular contacts [83]. Under these conditions, TER measurements indicated that an epithelial barrier was present even at low cell densities. The localizations of the bicellular tight junction proteins, OCLN and CLDN-4, were not affected by changes in cell density. Furthermore, knockdown of LSR under high cell density conditions induced thin spreading of cells and enhanced cell motility (unpublished observation). These findings suggested that the localization of LSR at tricellular contacts is necessary for the maintenance of static epithelial cell sheets.

An increase in cell density affects the intracellular tension mediated by actomyosin [84]. MRLC2 is localized at bicellular regions in the low density culture of Sawano cells [83]. Phosphorylated form of MRLC2, which represents activated myosin, is also localized in these regions. Under high cell density conditions, MRLC2 accumulated in vesicles or aggregated as particles near the apical cell surface membrane without localization to the apical bicellular region. It has been reported that activated MRLC2 is dephosphorylated by MYPT1, which is a component of the myosin phosphatase complex [85]. MYPT1 is localized in bicellular regions under high cell density conditions; however, under low cell density conditions, this protein was largely delocalized. These findings suggested that there is a high intracellular tension at bicellular regions of low density-grown cells as compared to that of high density-grown cells. When phosphorylated MRLC2 is localized in bicellular regions and MYPT1 is delocalized from there, LSR is localized in bicellular tight junctions. On the other hand, when MYPT1 is localized in bicellular regions, LSR is localized in apical tricellular contacts (Figure 4). Taken together, the localization of LSR altered in a cell density- and/or cellular tension-dependent manner. It is necessary to elucidate how LSR recognizes cell size and intracellular tension.

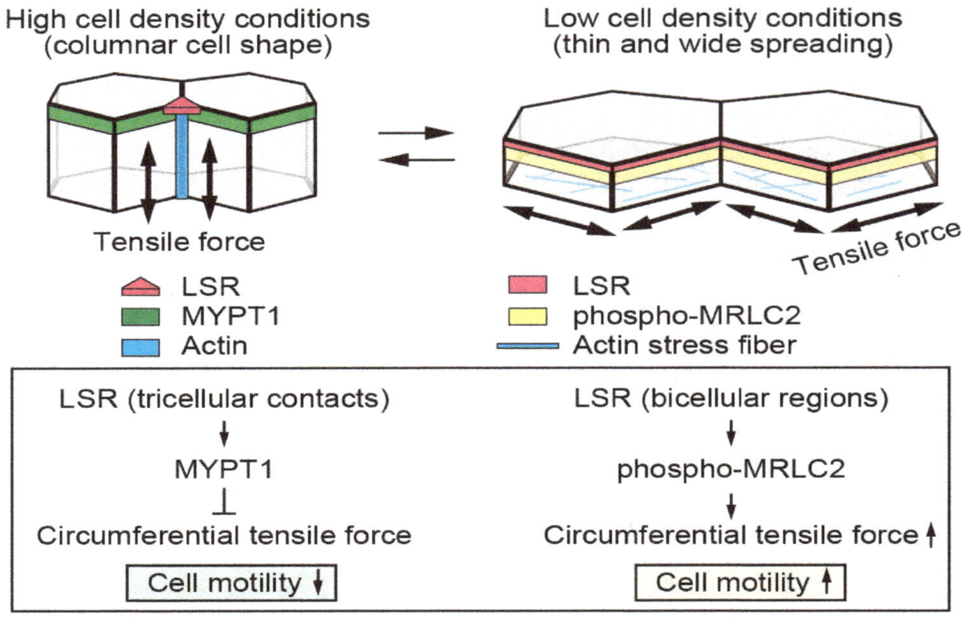

Figure 4. Decreased cellular tension causes LSR to localize at tricellular contacts. At high cell density, where cells were grown with columnar shape, the cellular tension decreased in the circumferential direction because MYPT1 is localized in these regions. Under these conditions, LSR is localized in tricellular contacts. In contrast, higher tension existed in cell-cell junctions at lower cell density because the phosphorylated form of MRLC2 is localized in these regions. In addition, actin stress fibers formed adjacent to the basal membrane. Under these conditions, LSR localized in bicellular junctions. It is noteworthy that thin and wide spreading cells increase cell motility.

10. Decreased Cellular Tension Causes LSR to Localize at Tricellular Contacts

Cellular tension is formed at cell-cell junctions and cell-substrate interface [86]. When the cells occupy a wide spread area, the number of focal contacts, where paxillin and integrins bind to the extracellular matrix, increases at basal membrane, facilitating polymerization of actin cytoskeleton [87]. Myosin, a cross-linked protein of actin fibers, generates cellular tension by contracting actin fibers [88]. Sawano cells cultured at low density conditions were spread thin and wide, and significant stress fiber formation was induced [83]. This suggested that high cellular tension is present under these conditions. Cellular tension is reportedly reduced by ROCK inhibitor, Y27632, muscle and non-muscle myosin II inhibitor, blebbistatin, or MLCK inhibitor, ML-7 [89]. When these reagents were added under low density conditions of Sawano cells where LSR localized in bicellular junctions, LSR localization decreased at bicellular junctions and increased at tricellular contacts. In our preliminary experiments, the focal adhesion kinase FAK localized in bicellular regions under low density of Sawano cells. FAK has been reported to be involved in the control of intestinal barrier functions [90] and blood-testis barrier functions [91]. In Sawano cells, FAK was dislocated from bicellular regions with increasing cell densities; however, the underlying mechanisms are still unclear. It is thought that the relationship between the regulatory mechanism of cellular tension involving FAK and the regulatory mechanism of epithelial barrier functions involving LSR will become clear in future studies.

11. Crosstalk between Intracellular Tension and Cell Junctions

In Sawano cells, LSR is reversibly translocated in a cellular tension- and/or cell size-dependent manner, whereas the localizations of OCLN and CLDN-4 were not affected by these conditions. Interestingly, accumulation of F-actin was observed not only along the lateral region of tricellular contacts but also the circumferential subapical region, leading to increase in cell thickness with the high cell density culture of Sawano cells [83]. Moreover, the actin polymerization inhibitor, Cytochalasin

D, excluded LSR from tricellular contacts, resulting in partial stratification of monolayered Sawano cell sheets. In long-term high-density culture, Sawano cells spontaneously and partially stratified in absence of Cytochalasin D. Under these conditions, the accumulated F-actin disappeared from lateral regions of tricellular contacts near the stratified area. In addition, LSR was translocated from tricellular contacts to the cell surface besides intracellular vesicles. Simultaneously, the epithelial barrier function decreased (unpublished observations). These findings suggested that the accumulation of F-actin at tricellular contacts is implicated in LSR localization to tricellular contacts. It is still not clear how the accumulated actin at tricellular regions regulates cellular tension. It is necessary to analyze the precise mechanism by which LSR localization is regulated by tension formed by actomyosin at tricellular contacts.

There have been almost no reports that LSR interacts directly with the actomyosin cytoskeleton at tricellular contacts. In contrast, it has been revealed that tricellulin promotes localization of both actin and myosin at tricellular contacts via interaction with Cdc42GEF protein Tuba (DNMBP, ARHGEF36) [92]. Cdc42 is one of the key proteins involved in the formation and maturation of epithelial polarity and contributes to the enhancement of cellular tension via MRCK(Cdc42BPA) [93]. MYPT1 is inactivated via phosphorylation by MRCK or ROCK [94,95]. In addition, the accumulation of tricellulin at tricellular contacts is controlled by LSR [7,96]. However, little is known about the downregulation of cellular tension during epithelial cell maturation by these proteins. Reportedly, the non-phosphorylated form of LSR in which mutations have been introduced to serine residues localizes to bicellular junctions [97]. However, the significance of LSR localization at bicellular junctions and its underlying molecular mechanism have not been elucidated yet. It is thought that changes in LSR localization along with changes in cellular tension are associated with acquisition of endometrial cancer motility. Further studies are required in order to understand the precise molecular mechanisms underlying regulation of LSR localization based on the changes in actomyosin activity and cellular tension.

Adherens junctions organize prior to tight junctions during the intercellular closure [98]. Adherens junctions interact with the actin cytoskeleton via the nectin-afadin complex and the cadherin-α/β-catenin complex [4]. Thereby, the actin cytoskeleton forms circumferential actomyosin bundles, contributing to cell polarity and cell thickness [99]. Shroom3 and Lulu1/2 are involved in the regulation of circumferential actomyosin bundles via ROCK [100,101]. Tight junctions are organized by the polarity complex consisting of PAR3, PAR6, and aPKC, which interact with circumferential actomyosin bundles via ZO1/2/3 [102]. For polarized epithelial cells, the cell thickness increases with long-term culture. The process of cell thickening is involved in the rearrangement of the actin cytoskeleton and the microtubule network [103]. Drebrin has been identified as a factor linking the actin cytoskeleton to the microtubule network through the interaction with complex consisting of myosin IIB, spectrin βII, and EB3 [104]. The report has revealed that, in Drebrin knockout cells, the cell thickness was reduced without compromising the cell polarity and the barrier function. Drebrin also localizes to gap junctions by interacting with connexin-43 [105]. Further studies are required to investigate whether Drebrin regulates LSR localization.

12. Hippo Pathway and Cytoskeletal Dynamics

Junctional complexes comprised of tight junctions and adherens junctions are important for the maintenance of apico-basal polarity in planar epithelial cells [4]. Many of these complexes interact with various adaptor proteins via PDZ domains and are linked to actomyosin networks [106]. These interactions allow epithelial cells to maintain apico-basal polarity [107]. Insufficient planar cell polarity is formed when only a few cells exist in a population per unit area [4]. In contrast, planar cell mobility decreases as the number of cells increases in the population until contact inhibition occurs, thereby resulting in apico-basal polarity formation [108]. It is already known that the Hippo pathway participates in these processes. In polarized epithelial cells, the Hippo pathway is involved with the maturation of tight junctions and adherens junctions, as well as the planar cell polarity pathway, mechanotransduction pathway, and growth factor signaling [82]. YAP (yes-associated protein) is a regulator of cell size [109]. This report revealed that a signal from widely spread cells induced the activation of transcription factors following the nuclear localization of YAP, whereas a signal from narrowly spread cells induced the inhibition of transcription factors following the degradation of YAP outside the nucleus [110]. Moreover, ARHGAP18, a Rho GTPase-activating protein, regulates the cortical actin network through the YAP signaling cascade [111]. A recent study has indicated that in endometrial carcinoma, the nuclear localization of YAP is involved in increased malignancy [112]. MYPT1 is known to activate Merlin [68,79]. The localization of MYPT1 was sensitive to changes in cell density [83]. Therefore, in endometrial cancer cells, the activity of Merlin might be altered by changes in cell density. It is necessary to identify the molecules involved in the crosstalk between the mechanism that regulates the cell size based on YAP expression and the mechanism that regulates the tight junction integrity based on the cellular tension.

13. Conclusions

During endometrial cancer progression, a decreased expression of LSR and increased expression of CLDN-1 have been observed. In primary cultured normal human endometrial cells, leptin reduced the expression of LSR. Obesity is one of the risk factors for endometrial cancer. These findings demonstrated that disruption of epithelial barrier integrity due to translocation of LSR was related to the mechanism of malignant transformation of endometrial cancer. These mechanisms were closely related to the Hippo pathway and also involved the reconstitution of extracellular matrix components. Furthermore, changes in cellular tension were associated with changes in LSR localization. The Hippo pathway has also been reported to be involved in the regulation of cellular tension. The cellular tension influences the translocation of LSR from bicellular junctions to tricellular contacts. On the contrary, the mechanism that causes the disruption of the robust epithelial barrier is poorly understood in endometrial cancer cells. In order to elucidate functional changes in the epithelial barrier during the malignant transformation of normal endometrial cells, it is required to accurately analyze the molecular mechanisms that regulate the localization of tricellular tight junction proteins.

Acknowledgments: This work was supported by the Ministry of Education, Culture, Sports, Science, and Technology of Japan.

Abbreviations

CLDNs	claudins
EMT	epithelial-mesenchymal transition
HCC	hepatocellular carcinoma
HNSCC	head and neck squamous cell carcinoma
iCCC	intrahepatic cholangiocarcinoma
LSR	lipolysis-stimulated lipoprotein receptor
MLCK	myosin light chain kinase
MYPT1	myosin phosphatase target subunit 1
MRLC2	myosin regulatory light chain 2
OCLN	occludin
OSC	oral squamous cell carcinoma
TER	transepithelial electrical resistance

References

1. Garcia, M.A.; Nelson, W.J.; Chavez, N. Cell-cell junctions organize structural and signaling networks. *Cold Spring Harb. Perspect. Biol.* **2018**, *10*, a029181. [CrossRef] [PubMed]
2. Furuse, M.; Izumi, Y.; Oda, Y.; Higashi, T.; Iwamoto, N. Molecular organization of tricellular tight junctions. *Tissue Barriers* **2014**, *2*, e28960. [CrossRef] [PubMed]
3. Tsukita, S.; Yamazaki, Y.; Katsuno, T.; Tamura, A.; Tsukita, S. Tight junction-based epithelial microenvironment

and cell proliferation. *Oncogene* **2008**, *27*, 6930–6938. [CrossRef] [PubMed]

4. Zihni, C.; Mills, C.; Matter, K.; Balda, M.S. Tight junctions: From simple barriers to multifunctional molecular gates. *Nat. Rev. Mol. Cell Biol.* **2016**, *17*, 564–580. [CrossRef] [PubMed]
5. Förster, C. Tight junctions and the modulation of barrier function in disease. *Histochem. Cell Biol.* **2008**, *130*, 55–70. [CrossRef] [PubMed]
6. Someya, M.; Kojima, T.; Ogawa, M.; Ninomiya, T.; Nomura, K.; Takasawa, A.; Murata, M.; Tanaka, S.; Saito, T.; Sawada, N. Regulation of tight junctions by sex hormones in normal human endometrial epithelial cells and uterus cancer cell line Sawano. *Cell Tissue Res.* **2013**, *354*, 481–494. [CrossRef] [PubMed]
7. Furuse, M.; Oda, Y.; Higashi, T.; Iwamoto, N.; Masuda, S. Lipolysis-stimulated lipoprotein receptor: A novel membrane protein of tricellular tight junctions. *Ann. N. Y. Acad. Sci.* **2012**, *1257*, 54–58. [CrossRef]
8. Shimada, H.; Satohisa, S.; Kohno, T.; Takahashi, S.; Hatakeyama, T.; Konno, T.; Tsujiwaki, M.; Saito, T.; Kojima, T. The roles of tricellular tight junction protein lipolysis-stimulated lipoprotein receptor in malignancy of human endometrial cancer cells. *Oncotarget* **2016**, *7*, 27735–27752. [CrossRef]
9. Bilyk, O.; Coatham, M.; Jewer, M.; Postovit, L.M. Epithelial-to-Mesenchymal Transition in the female reproductive tract: From normal functioning to disease pathology. *Front. Oncol.* **2017**, *7*, 1–21. [CrossRef]
10. Sherr, C.J. Principles of tumor suppression. *Cell* **2004**, *116*, 235–246. [CrossRef]
11. Heerboth, S.; Housman, G.; Leary, M.; Longacre, M.; Byler, S.; Lapinska, K.; Willbanks, A.; Sarkar, S. EMT and tumor metastasis. *Clin. Transl. Med.* **2015**, *4*. [CrossRef]
12. Martin, T.A.; Jiang, W.G. Loss of tight junction barrier function and its role in cancer metastasis. *Biochim. Biophys. Acta Biomembr.* **2009**, *1788*, 872–891. [CrossRef]
13. Martin, T.A. The role of tight junctions in cancer metastasis. *Semin. Cell Dev. Biol.* **2014**, *36*, 224–231. [CrossRef]
14. Leech, A.O.; Cruz, R.G.B.; Hill, A.D.K.; Hopkins, A.M. Paradigms lost—An emerging role for over-expression of tight junction adhesion proteins in cancer pathogenesis. *Ann. Transl. Med.* **2015**, *3*, 1–15.
15. Kondoh, A.; Takano, K.I.; Kojima, T.; Ohkuni, T.; Kamekura, R.; Ogasawara, N.; Go, M.; Sawada, N.; Himi, T. Altered expression of claudin-1, claudin-7, and tricellulin regardless of human papilloma virus infection in human tonsillar squamous cell carcinoma. *Acta Otolaryngol.* **2011**, *131*, 861–868. [CrossRef]
16. Patonai, A.; Erdélyi-Belle, B.; Korompay, A.; Somorácz, A.; Straub, B.K.; Schirmacher, P.; Kovalszky, I.; Lotz, G.; Kiss, A.; Schaff, Z. Claudins and tricellulin in fibrolamellar hepatocellular carcinoma. *Virchows Arch.* **2011**, *458*, 679–688. [CrossRef]
17. Somorácz, A.; Korompay, A.; Törzsök, P.; Patonai, A.; Erdélyi-Belle, B.; Lotz, G.; Schaff, Z.; Kiss, A. Tricellulin expression and its prognostic significance in primary liver carcinomas. *Pathol. Oncol. Res.* **2014**, *20*, 755–764. [CrossRef]
18. Korompay, A.; Borka, K.; Lotz, G.; Somorácz, A.; Törzsök, P.; Erdélyi-Belle, B.; Kenessey, I.; Baranyai, Z.; Zsoldos, F.; Kupcsulik, P.; et al. Tricellulin expression in normal and neoplastic human pancreas. *Histopathology* **2012**, *60*, E76–E86. [CrossRef]
19. Masuda, R.; Semba, S.; Mizuuchi, E.; Yanagihara, K.; Yokozaki, H. Negative regulation of the tight junction protein tricellulin by snail-induced epithelial-mesenchymal transition in gastric carcinoma cells. *Pathobiology* **2010**, *77*, 106–113. [CrossRef]
20. Reaves, D.K.; Fagan-Solis, K.D.; Dunphy, K.; Oliver, S.D.; Scott, D.W.; Fleming, J.M. The role of lipolysis stimulated lipoprotein receptor in breast cancer and directing breast cancer cell behavior. *PLoS ONE* **2014**, *9*, e91747. [CrossRef]
21. García, J.M.; Peña, C.; García, V.; Domínguez, G.; Muñoz, C.; Silva, J.; Millán, I.; Diaz, R.; Lorenzo, Y.; Rodriguez, R.; et al. Prognostic value of LISCH7 mRNA in plasma and tumor of colon cancer patients. *Clin. Cancer Res.* **2007**, *13*, 6351–6358. [CrossRef]
22. Takano, K.; Kakuki, T.; Obata, K.; Nomura, K.; Miyata, R.; Kondo, A.; Kurose, M.; Kakiuchi, A.; Kaneko, Y.; Kohno, T.; et al. The behavior and role of lipolysis-stimulated lipoprotein receptor, a component of tricellular tight junctions, in head and neck squamous cell carcinomas. *Anticancer Res.* **2016**, *36*, 5895–5904. [CrossRef]
23. Sappayatosok, K.; Phattarataratip, E. Overexpression of claudin-1 is associated with advanced clinical stage and invasive pathologic characteristics of oral squamous cell carcinoma. *Head Neck Pathol.* **2015**, *9*, 173–180. [CrossRef]
24. Li, W.; Dong, Q.; Li, L.; Zhang, Z.; Cai, X.; Pan, X. Prognostic significance of claudin-1 and cyclin B1 protein expression in patients with hypopharyngeal squamous cell carcinoma. *Oncol. Lett.* **2016**, *11*, 2995–3002. [CrossRef]

25. Grund, S.; Grümmer, R. Direct cell-cell interactions in the endometrium and in endometrial pathophysiology. *Int. J. Mol. Sci.* **2018**, *19*, 2227. [CrossRef]
26. Pan, X.Y.; Wang, B.; Che, Y.C.; Weng, Z.P.; Dai, H.Y.; Peng, W. Expression of claudin-3 and claudin-4 in normal, hyperplastic, and malignant endometrial tissue. *Int. J. Gynecol. Cancer* **2007**, *17*, 233–241. [CrossRef]
27. Konecny, G.E.; Agarwal, R.; Keeney, G.A.; Winterhoff, B.; Jones, M.B.; Mariani, A.; Riehle, D.; Neuper, C.; Dowdy, S.C.; Wang, H.J.; et al. Claudin-3 and claudin-4 expression in serous papillary, clear-cell, and endometrioid endometrial cancer. *Gynecol. Oncol.* **2008**, *109*, 263–269. [CrossRef]
28. Lortet-Tieulent, J.; Ferlay, J.; Bray, F.; Jemal, A. International patterns and trends in endometrial cancer incidence, 1978–2013. *J. Natl. Cancer Inst.* **2018**, *110*, 354–361. [CrossRef]
29. Hopkins, B.D.; Goncalves, M.D.; Cantley, L.C. Obesity and cancer mechanisms: Cancer metabolism. *J. Clin. Oncol.* **2016**, *34*, 4277–4283. [CrossRef]
30. Byrne, F.L.; Poon, I.K.H.; Modesitt, S.C.; Tomsig, J.L.; Chow, J.D.Y.; Healy, M.E.; Baker, W.D.; Atkins, K.A.; Lancaster, J.M.; Marchion, D.C.; et al. Metabolic vulnerabilities in endometrial cancer. *Cancer Res.* **2014**, *74*, 5832–5845. [CrossRef]
31. Gelsomino, L.; Naimo, G.D.; Catalano, S.; Mauro, L.; Andò, S. The emerging role of adiponectin in female malignancies. *Int. J. Mol. Sci.* **2019**, *20*, 2127. [CrossRef]
32. Stern, J.H.; Rutkowski, J.M.; Scherer, P.E. Adiponectin, leptin, and fatty acids in the maintenance of metabolic homeostasis through adipose tissue crosstalk. *Cell Metab.* **2016**, *23*, 770–784. [CrossRef]
33. Ashizawa, N.; Yahata, T.; Quan, J.; Adachi, S.; Yoshihara, K.; Tanaka, K. Serum leptin-adiponectin ratio and endometrial cancer risk in postmenopausal female subjects. *Gynecol. Oncol.* **2010**, *119*, 65–69. [CrossRef]
34. Gong, T.T.; Wu, Q.J.; Wang, Y.L.; Ma, X.X. Circulating adiponectin, leptin and adiponectin-leptin ratio and endometrial cancer risk: Evidence from a meta-analysis of epidemiologic studies. *Int. J. Cancer* **2015**, *137*, 1967–1978. [CrossRef]
35. Dupont, J.Ë.; Reverchon, M.; Cloix, L.; Froment, P.; Ramé, C. Involvement of adipokines, AMPK, PI3K and the PPAR signaling pathways in ovarian follicle development and cancer. *Int. J. Dev. Biol.* **2012**, *56*, 959–967. [CrossRef]
36. Pérez-Pérez, A.; Toro, A.; Vilariño-García, T.; Maymó, J.; Guadix, P.; Dueñas, J.L.; Fernández-Sánchez, M.; Varone, C.; Sánchez-Margalet, V. Leptin action in normal and pathological pregnancies. *J. Cell. Mol. Med.* **2018**, *22*, 716–727. [CrossRef]
37. Fan, Y.; Gan, Y.; Shen, Y.; Cai, X.; Song, Y.; Zhao, F.; Yao, M.; Gu, J.; Tu, H. Leptin signaling enhances cell invasion and promotes the metastasis of human pancreatic cancer via increasing MMP-13 production. *Oncotarget* **2015**, *6*, 16120–16134. [CrossRef]
38. Ahn, J.H.; Choi, Y.S.; Choi, J.H. Leptin promotes human endometriotic cell migration and invasion by up-regulating MMP-2 through the JAK2/STAT3 signaling pathway. *Mol. Hum. Reprod.* **2015**, *21*, 792–802. [CrossRef]
39. Handy, J.A.; Fu, P.P.; Kumar, P.; Mells, J.E.; Sharma, S.; Saxena, N.K.; Anania, F.A. Adiponectin inhibits leptin signalling via multiple mechanisms to exert protective effects against hepatic fibrosis. *Biochem. J.* **2011**, *440*, 385–395. [CrossRef]
40. Stenger, C.; Hanse, M.; Pratte, D.; Mbala, M.L.; Akbar, S.; Koziel, V.; Escanyé, M.C.; Kriem, B.; Malaplate-Armand, C.; Olivier, J.L.; et al. Up-regulation of hepatic lipolysis stimulated lipoprotein receptor by leptin: A potential lever for controlling lipid clearance during the postprandial phase. *FASEB J.* **2010**, *24*, 4218–4228. [CrossRef]
41. Yen, F.T.; Roitel, O.; Bonnard, L.; Notet, V.; Pratte, D.; Stenger, C.; Magueur, E.; Bihain, B.E. Lipolysis stimulated lipoprotein receptor: A novel molecular link between hyperlipidemia, weight gain, and atherosclerosis in mice. *J. Biol. Chem.* **2008**, *283*, 25650–25659. [CrossRef]
42. Parida, S.; Siddharth, S.; Sharma, D. Adiponectin, obesity, and cancer: Clash of the bigwigs in health and disease. *Int. J. Mol. Sci.* **2019**, *20*, 2519. [CrossRef]
43. Lee, T.Y.; Martinez-Outschoorn, U.E.; Schilder, R.J.; Kim, C.H.; Richard, S.D.; Rosenblum, N.G.; Johnson, J.M. Metformin as a therapeutic target in endometrial cancers. *Front. Oncol.* **2018**, *8*, 341. [CrossRef]
44. Wang, H.; Zhu, C.; Ying, Y.; Luo, L.; Huang, D.; Luo, Z. Metformin and berberine, two versatile drugs in treatment of common metabolic diseases. *Oncotarget* **2018**, *9*, 10135–10146. [CrossRef]
45. Umene, K.; Banno, K.; Kisu, I.; Yanokura, M.; Nogami, Y.; Tsuji, K.; Masuda, K.; Ueki, A.; Kobayashi, Y.; Yamagami, W.; et al. New candidate therapeutic agents for endometrial cancer: Potential for clinical practice (Review). *Oncol. Rep.* **2013**, *29*, 855–860. [CrossRef]

46. Mo, J.S.; Meng, Z.; Kim, Y.C.; Park, H.W.; Hansen, C.G.; Kim, S.; Lim, D.S.; Guan, K.L. Cellular energy stress induces AMPK-mediated regulation of YAP and the Hippo pathway. *Nat. Cell Biol.* **2015**, *17*, 500–510. [CrossRef]
47. Wang, W.; Xiao, Z.D.; Li, X.; Aziz, K.E.; Gan, B.; Johnson, R.L.; Chen, J. AMPK modulates Hippo pathway activity to regulate energy homeostasis. *Nat. Cell Biol.* **2015**, *17*, 490–499. [CrossRef]
48. Canfield, P.E.; Geerdes, A.M.; Molitoris, B.A. Effect of reversible ATP depletion on tight-junction integrity in LLC-PK1 cells. *Am. J. Physiol. Physiol.* **2017**, *261*, F1038–F1045. [CrossRef]
49. Zhang, L.; Li, J.; Young, L.H.; Caplan, M.J. AMP-activated protein kinase regulates the assembly of epithelial tight junctions. *Proc. Natl. Acad. Sci. USA* **2006**, *103*, 17272–17277. [CrossRef]
50. Aznar, N.; Patel, A.; Rohena, C.C.; Dunkel, Y.; Joosen, L.P.; Taupin, V.; Kufareva, I.; Farquhar, M.G.; Ghosh, P. AMP-activated protein kinase fortifies epithelial tight junctions during energetic stress via its effector GIV/Girdin. *eLife* **2016**, *5*, 1–33. [CrossRef]
51. Shackelford, D.B.; Shaw, R.J. The LKB1-AMPK pathway: Metabolism and growth control in tumour suppression. *Nat. Rev. Cancer* **2009**, *9*, 563–575. [CrossRef]
52. Zou, J.; Hong, L.; Luo, C.; Li, Z.; Zhu, Y.; Huang, T.; Zhang, Y.; Yuan, H.; Hu, Y.; Wen, T.; et al. Metformin inhibits estrogen-dependent endometrial cancer cell growth by activating the AMPK–FOXO1 signal pathway. *Cancer Sci.* **2016**, *107*, 1806–1817. [CrossRef]
53. Wang, H.B.; Wang, P.Y.; Wang, X.; Wan, Y.L.; Liu, Y.C. Butyrate enhances intestinal epithelial barrier function via up-regulation of tight junction protein claudin-1 transcription. *Dig. Dis. Sci.* **2012**, *57*, 3126–3135. [CrossRef]
54. Dufresne, J.; Cyr, D.G. Activation of an SP binding site is crucial for the expression of claudin 1 in rat epididymal principal cells. *Biol. Reprod.* **2007**, *76*, 825–832. [CrossRef]
55. Honda, H.; Pazin, M.J.; Ji, H.; Wernyj, R.P.; Morin, P.J. Crucial roles of Sp1 and epigenetic modifications in the regulation of the cldn4 promoter in ovarian cancer cells. *J. Biol. Chem.* **2006**, *281*, 21433–21444. [CrossRef]
56. Luk, J.M.; Tong, M.K.; Mok, B.W.; Tam, P.C.; Yeung, W.S.B.; Lee, K.F. Sp1 site is crucial for the mouse claudin-19 gene expression in the kidney cells. *FEBS Lett.* **2004**, *578*, 251–256. [CrossRef]
57. Shimada, H.; Satohisa, S.; Kohno, T.; Konno, T.; Arimoto, C.; Saito, T.; Kojima, T. Downregulation of lipolysis-stimulated lipoprotein receptor promotes cell invasion via claudin-1-mediated matrix metalloproteinases in human endometrial cancer. *Oncol. Lett.* **2017**, *16*, 6776–6782. [CrossRef]
58. Oku, N.; Sasabe, E.; Ueta, E.; Yamamoto, T.; Osaki, T. Tight junction protein claudin-1 enhances the invasive activity of oral squamous cell carcinoma cells by promoting cleavage of laminin-5 γ2 chain via matrix metalloproteinase (MMP)-2 and membrane-type MMP-1. *Cancer Res.* **2006**, *66*, 5251–5257. [CrossRef]
59. Van Itallie, C.M.; Fanning, A.S.; Anderson, J.M.; Boireau, S.; Samuel, M.S.; Pannequin, J.; Ryan, J.L.; Choquet, A.; Chapuis, H.; Rebillard, X.; et al. Claudin-1 regulates cellular transformation and metastatic behavior in colon cancer. *J. Clin. Investig.* **2005**, *115*, 1765–1776.
60. Castro-Castro, A.; Marchesin, V.; Monteiro, P.; Lodillinsky, C.; Rossé, C.; Chavrier, P. Cellular and molecular mechanisms of MT1-MMP-dependent cancer cell invasion. *Annu. Rev. Cell Dev. Biol.* **2016**, *32*, 555–576. [CrossRef]
61. Brown, G.T.; Murray, G.I. Current mechanistic insights into the roles of matrix metalloproteinases in tumour invasion and metastasis. *J. Pathol.* **2015**, *237*, 273–281. [CrossRef]
62. Itoh, Y.; Seiki, M. MT1-MMP: A potent modifier of pericellular microenvironment. *J. Cell. Physiol.* **2006**, *206*, 1–8. [CrossRef]
63. Harvey, K.F.; Zhang, X.; Thomas, D.M. The Hippo pathway and human cancer. *Nat. Rev. Cancer* **2013**, *13*, 246–257. [CrossRef]
64. Bao, Y.; Nakagawa, K.; Yang, Z.; Ikeda, M.; Withanage, K.; Ishigami-Yuasa, M.; Okuno, Y.; Hata, S.; Nishina, H.; Hata, Y. A cell-based assay to screen stimulators of the Hippo pathway reveals the inhibitory effect of dobutamine on the YAP-dependent gene transcription. *J. Biochem.* **2011**, *150*, 199–208. [CrossRef]
65. Zheng, H.X.; Wu, L.N.; Xiao, H.; Du, Q.; Liang, J.F. Inhibitory effects of dobutamine on human gastric adenocarcinoma. *World J. Gastroenterol.* **2014**, *20*, 17092–17099. [CrossRef]
66. Hardie, D.G. AMP-activated protein kinase-an energy sensor that regulates all aspects of cell function. *Genes Dev.* **2011**, *25*, 1895–1908. [CrossRef]
67. Shimada, H.; Abe, S.; Kohno, T.; Satohisa, S.; Konno, T.; Takahashi, S.; Hatakeyama, T.; Arimoto, C.; Kakuki, T.; Kaneko, Y.; et al. Loss of tricellular tight junction protein LSR promotes cell invasion and migration via upregulation of TEAD1/AREG in human endometrial cancer. *Sci. Rep.* **2017**, *7*, 37049. [CrossRef]

68. Cooper, J.; Giancotti, F.G. Molecular insights into NF2/Merlin tumor suppressor function. *FEBS Lett.* **2014**, *588*, 2743–2752. [CrossRef]
69. Gladden, A.B.; Hebert, A.M.; Schneeberger, E.E.; McClatchey, A.I. The NF2 tumor suppressor, Merlin, regulates epidermal development through the establishment of a junctional polarity complex. *Dev. Cell* **2010**, *19*, 727–739. [CrossRef]
70. Zhao, B.; Li, L.; Lu, Q.; Wang, L.H.; Liu, C.Y.; Lei, Q.; Guan, K.L. Angiomotin is a novel Hippo pathway component that inhibits YAP oncoprotein. *Genes Dev.* **2011**, *25*, 51–63. [CrossRef]
71. Huang, T.; Zhou, Y.; Zhang, J.; Cheng, A.S.L.; Yu, J.; To, K.F.; Kang, W. The physiological role of Motin family and its dysregulation in tumorigenesis. *J. Transl. Med.* **2018**, *16*, 98. [CrossRef]
72. Ortiz, A.; Lee, Y.C.; Yu, G.; Liu, H.C.; Lin, S.C.; Bilen, M.A.; Cho, H.; Yu-Lee, L.Y.; Lin, S.H. Angiomotin is a novel component of cadherin-11/β-catenin/p120 complex and is critical for cadherin-11-mediated cell migration. *FASEB J.* **2015**, *29*, 1080–1091. [CrossRef]
73. Hakami, F.; Darda, L.; Stafford, P.; Woll, P.; Lambert, D.W.; Hunter, K.D. The roles of HOXD10 in the development and progression of head and neck squamous cell carcinoma (HNSCC). *Br. J. Cancer* **2014**, *111*, 807–816. [CrossRef]
74. Levchenko, T.; Bratt, A.; Arbiser, J.L.; Holmgren, L. Angiomotin expression promotes hemangioendothelioma invasion. *Oncogene* **2004**, *23*, 1469–1473. [CrossRef]
75. Couderc, C.; Boin, A.; Fuhrmann, L.; Vincent-Salomon, A.; Mandati, V.; Kieffer, Y.; Mechta-Grigoriou, F.; Del Maestro, L.; Chavrier, P.; Vallerand, D.; et al. AMOTL1 promotes breast cancer progression and is antagonized by Merlin. *Neoplasia* **2016**, *18*, 10–24. [CrossRef]
76. Wan, H.Y.; Li, Q.Q.; Zhang, Y.; Tian, W.; Li, Y.N.; Liu, M.; Li, X.; Tang, H. MiR-124 represses vasculogenic mimicry and cell motility by targeting AmotL1 in cervical cancer cells. *Cancer Lett.* **2014**, *355*, 148–158. [CrossRef]
77. Mojallal, M.; Zheng, Y.; Hultin, S.; Audebert, S.; van Harn, T.; Johnsson, P.; Lenander, C.; Fritz, N.; Mieth, C.; Corcoran, M.; et al. AmotL2 disrupts apical–basal cell polarity and promotes tumour invasion. *Nat. Commun.* **2014**, *5*, 4557. [CrossRef]
78. Artinian, N.; Cloninger, C.; Holmes, B.; Benavides-Serrato, A.; Bashir, T.; Gera, J. Phosphorylation of the Hippo pathway component AMOTL2 by the mTORC2 kinase promotes YAP signaling, resulting in enhanced glioblastoma growth and invasiveness. *J. Biol. Chem.* **2015**, *290*, 19387–19401. [CrossRef]
79. Li, W.; Cooper, J.; Karajannis, M.A.; Giancotti, F.G. Merlin: A tumour suppressor with functions at the cell cortex and in the nucleus. *EMBO Rep.* **2012**, *13*, 204–215. [CrossRef]
80. Yi, C.; Troutman, S.; Fera, D.; Stemmer-Rachamimov, A.; Avila, J.L.; Christian, N.; Persson, N.L.; Shimono, A.; Speicher, D.W.; Marmorstein, R.; et al. A tight junction-associated Merlin-Angiomotin complex mediates Merlin's regulation of mitogenic signaling and tumor suppressive functions. *Cancer Cell* **2011**, *19*, 527–540. [CrossRef]
81. Leontieva, O.V.; Demidenko, Z.N.; Blagosklonny, M. V Contact inhibition and high cell density deactivate the mammalian target of rapamycin pathway, thus suppressing the senescence program. *Proc. Natl. Acad. Sci. USA* **2014**, *111*, 8832–8837. [CrossRef]
82. Gumbiner, B.M.; Kim, N.-G. The Hippo-YAP signaling pathway and contact inhibition of growth. *J. Cell Sci.* **2014**, *127*, 709–717. [CrossRef]
83. Kohno, T.; Kikuchi, S.; Ninomiya, T.; Kojima, T. The bicellular tensile force sorts the localization of LSRs in bicellular and tricellular junctions. *Ann. N. Y. Acad. Sci.* **2017**, *1397*, 185–194. [CrossRef]
84. Levayer, R.; Lecuit, T. Biomechanical regulation of contractility: Spatial control and dynamics. *Trends Cell Biol.* **2012**, *22*, 61–81. [CrossRef]
85. Gutzman, J.H.; Sive, H. Epithelial relaxation mediated by the myosin phosphatase regulator Mypt1 is required for brain ventricle lumen expansion and hindbrain morphogenesis. *Development* **2010**, *137*, 795–804. [CrossRef]
86. Mui, K.L.; Chen, C.S.; Assoian, R.K. The mechanical regulation of integrin-cadherin crosstalk organizes cells, signaling and forces. *J. Cell Sci.* **2016**, *129*, 1093–1100. [CrossRef]
87. Parsons, J.T.; Horwitz, A.R.; Schwartz, M.A. Cell adhesion: Integrating cytoskeletal dynamics and cellular tension. *Nat. Rev. Mol. Cell Biol.* **2010**, *11*, 633–643. [CrossRef]
88. Vicente-Manzanares, M.; Ma, X.; Adelstein, R.S.; Horwitz, A.R. Non-muscle myosin II takes centre stage in cell adhesion and migration. *Nat. Rev. Cell Biol.* **2009**, *10*, 778–790. [CrossRef]
89. Amano, M.; Nakayama, M.; Kaibuchi, K. Rho-kinase/ROCK: A key regulator of the cytoskeleton and cell

polarity. *Cytoskeleton* **2010**, *67*, 545–554. [CrossRef]
90. Ma, Y.; Semba, S.; Khan, R.I.; Bochimoto, H.; Watanabe, T.; Fujiya, M.; Kohgo, Y.; Liu, Y.; Taniguchi, T. Focal adhesion kinase regulates intestinal epithelial barrier function via redistribution of tight junction. *Biochim. Biophys. Acta* **2013**, *1832*, 151–159. [CrossRef]
91. Siu, E.R.; Wong, E.W.P.; Mruk, D.D.; Porto, C.S.; Cheng, C.Y. Focal adhesion kinase is a blood-testis barrier regulator. *Proc. Natl. Acad. Sci. USA* **2009**, *106*, 9298–9303. [CrossRef]
92. Oda, Y.; Otani, T.; Ikenouchi, J.; Furuse, M. Tricellulin regulates junctional tension of epithelial cells at tricellular contacts via Cdc42. *J. Cell Sci.* **2014**, *127*, 4201–4212. [CrossRef]
93. Riento, K.; Ridley, A.J. ROCKs: Multifunctional kinases in cell behaviour. *Nat. Rev. Mol. Cell Biol.* **2003**, *4*, 446–456. [CrossRef]
94. Kale, V.P.; Hengst, J.A.; Desai, D.H.; Dick, T.E.; Choe, K.N.; Colledge, A.L.; Takahashi, Y.; Sung, S.; Amin, S.G.; Yun, J.K. A novel selective multikinase inhibitor of ROCK and MRCK effectively blocks cancer cell migration and invasion. *Cancer Lett.* **2014**, *354*, 299–310 [CrossRef]
95. Ito, M.; Nakano, T.; Erdodi, F.; Hartshorne, D.J. Myosin phosphatase: Structure, regulation and function. *Mol. Cell. Biochem.* **2004**, *259*, 197–209. [CrossRef]
96. Masuda, S.; Oda, Y.; Sasaki, H.; Ikenouchi, J.; Higashi, T.; Akashi, M.; Nishi, E.; Furuse, M. LSR defines cell corners for tricellular tight junction formation in epithelial cells. *J. Cell Sci.* **2011**, *124*, 548–555. [CrossRef]
97. Nakatsu, D.; Kano, F.; Taguchi, Y.; Sugawara, T.; Nishizono, T.; Nishikawa, K.; Oda, Y.; Furuse, M.; Murata, M. JNK1/2-dependent phosphorylation of angulin-1/LSR is required for the exclusive localization of angulin-1/LSR and tricellulin at tricellular contacts in EpH4 epithelial sheet. *Genes Cells* **2014**, *19*, 565–581. [CrossRef]
98. Takai, Y.; Nakanishi, H. Nectin and afadin: Novel organizers of intercellular junctions. *J. Cell Sci.* **2003**, *116*, 17–27. [CrossRef]
99. Zhang, J.; Betson, M.; Erasmus, J.; Zeikos, K.; Bailly, M.; Cramer, L.P.; Braga, V.M.M. Actin at cell-cell junctions is composed of two dynamic and functional populations. *J. Cell Sci.* **2005**, *118*, 5549–5562. [CrossRef]
100. Chu, C.W.; Gerstenzang, E.; Ossipova, O.; Sokol, S.Y. Lulu regulates shroom-induced apical constriction during neural tube closure. *PLoS ONE* **2013**, *8*, e81854. [CrossRef]
101. Nishimura, T.; Takeichi, M. Shroom3-mediated recruitment of Rho kinases to the apical cell junctions regulates epithelial and neuroepithelial planar remodeling. *Development* **2008**, *135*, 1493–1502. [CrossRef]
102. Rodriguez-Boulan, E.; Macara, I.G. Organization and execution of the epithelial polarity programme. *Nat. Rev. Mol. Cell Biol.* **2014**, *15*, 225–242. [CrossRef]
103. Toya, M.; Kobayashi, S.; Kawasaki, M.; Shioi, G.; Kaneko, M.; Ishiuchi, T.; Misaki, K.; Meng, W.; Takeichi, M. CAMSAP3 orients the apical-to-basal polarity of microtubule arrays in epithelial cells. *Proc. Natl. Acad. Sci. USA* **2015**, *113*, 1–6. [CrossRef]
104. Bazellières, E.; Massey-Harroche, D.; Barthélémy-Requin, M.; Richard, F.; Arsanto, J.P.; Le Bivic, A. Apico-basal elongation requires a drebrin-E-EB3 complex in columnar human epithelial cells. *J. Cell Sci.* **2012**, *125*, 919–931. [CrossRef]
105. Butkevich, E.; Hülsmann, S.; Wenzel, D.; Shirao, T.; Duden, R.; Majoul, I. Drebrin is a novel connexin-43 binding partner that links gap junctions to the submembrane cytoskeleton. *Curr. Biol.* **2004**, *14*, 650–658. [CrossRef]
106. Guillemot, L.; Paschoud, S.; Pulimeno, P.; Foglia, A.; Citi, S. The cytoplasmic plaque of tight junctions: A scaffolding and signalling center. *Biochim. Biophys. Acta Biomembr.* **2008**, *1778*, 601–613. [CrossRef]
107. Miyake, Y.; Inoue, N.; Nishimura, K.; Kinoshita, N.; Hosoya, H.; Yonemura, S. Actomyosin tension is required for correct recruitment of adherens junction components and zonula occludens formation. *Exp. Cell Res.* **2006**, *312*, 1637–1650. [CrossRef]
108. Iden, S.; Collard, J.G. Crosstalk between small GTPases and polarity proteins in cell polarization. *Nat. Rev. Mol. Cell Biol.* **2008**, *9*, 846–859. [CrossRef]
109. Halder, G.; Dupont, S.; Piccolo, S. Transduction of mechanical and cytoskeletal cues by YAP and TAZ. *Nat. Rev. Mol. Cell Biol.* **2012**, *13*, 591–600. [CrossRef]
110. Dupont, S.; Morsut, L.; Aragona, M.; Enzo, E.; Giulitti, S.; Cordenonsi, M.; Zanconato, F.; Le Digabel, J.; Forcato, M.; Bicciato, S.; et al. Role of YAP/TAZ in mechanotransduction. *Nature* **2011**, *474*, 179–183. [CrossRef]

111. Porazinski, S.; Wang, H.; Asaoka, Y.; Behrndt, M.; Miyamoto, T.; Morita, H.; Hata, S.; Sasaki, T.; Krens, S.F.G.; Osada, Y.; et al. YAP is essential for tissue tension to ensure vertebrate 3D body shape. *Nature* **2015**, *521*, 217–221. [CrossRef]
112. Tsujiura, M.; Mazack, V.; Sudol, M.; Kaspar, H.G.; Nash, J.; Carey, D.J.; Gogoi, R. Yes-Associated Protein (YAP) modulates oncogenic features and radiation sensitivity in endometrial cancer. *PLoS ONE* **2014**, *9*, e100974. [CrossRef]

PERMISSIONS

All chapters in this book were first published by MDPI; hereby published with permission under the Creative Commons Attribution License or equivalent. Every chapter published in this book has been scrutinized by our experts. Their significance has been extensively debated. The topics covered herein carry significant findings which will fuel the growth of the discipline. They may even be implemented as practical applications or may be referred to as a beginning point for another development.

The contributors of this book come from diverse backgrounds, making this book a truly international effort. This book will bring forth new frontiers with its revolutionizing research information and detailed analysis of the nascent developments around the world.

We would like to thank all the contributing authors for lending their expertise to make the book truly unique. They have played a crucial role in the development of this book. Without their invaluable contributions this book wouldn't have been possible. They have made vital efforts to compile up to date information on the varied aspects of this subject to make this book a valuable addition to the collection of many professionals and students.

This book was conceptualized with the vision of imparting up-to-date information and advanced data in this field. To ensure the same, a matchless editorial board was set up. Every individual on the board went through rigorous rounds of assessment to prove their worth. After which they invested a large part of their time researching and compiling the most relevant data for our readers.

The editorial board has been involved in producing this book since its inception. They have spent rigorous hours researching and exploring the diverse topics which have resulted in the successful publishing of this book. They have passed on their knowledge of decades through this book. To expedite this challenging task, the publisher supported the team at every step. A small team of assistant editors was also appointed to further simplify the editing procedure and attain best results for the readers.

Apart from the editorial board, the designing team has also invested a significant amount of their time in understanding the subject and creating the most relevant covers. They scrutinized every image to scout for the most suitable representation of the subject and create an appropriate cover for the book.

The publishing team has been an ardent support to the editorial, designing and production team. Their endless efforts to recruit the best for this project, has resulted in the accomplishment of this book. They are a veteran in the field of academics and their pool of knowledge is as vast as their experience in printing. Their expertise and guidance has proved useful at every step. Their uncompromising quality standards have made this book an exceptional effort. Their encouragement from time to time has been an inspiration for everyone.

The publisher and the editorial board hope that this book will prove to be a valuable piece of knowledge for researchers, students, practitioners and scholars across the globe.

LIST OF CONTRIBUTORS

Shruthi Venugopal, Shaista Anwer and Katalin Szászi
Keenan Research Centre for Biomedical Science of the St. Michael's Hospital and Department of Surgery, University of Toronto, Toronto, ON M5B 1W8, Canada

Hanna Schmidt, Carolin Schilpp, Robin Lochbaum, Kathrin Neuland, Kristin Thompson, Manfred Frick, Paul Dietl and Oliver H. Wittekindt
Institute of General Physiology, Ulm University, Albert-Einstein-Allee 11, 89081 Ulm, Germany

Peter Braubach and Danny Jonigk
Institute of Pathology, Hannover Medical School, Carl-Neuberg-Str. 130625 Hannover, Germany
German Center of Lung Research (DZL), Partnersite BREATH, 306245 Hannover, Germany

Tomohiro Kitano and Shin-ichiro Kitajiri
Department of Otorhinolaryngology, Shinshu University School of Medicine, 3-1-1 Asahi, Matsumoto, Nagano 390-8621, Japan

Shin-ya Nishio and Shin-ichi Usami
Department of Otorhinolaryngology, Shinshu University School of Medicine, 3-1-1 Asahi, Matsumoto, Nagano 390-8621, Japan
Department of Hearing Implant Sciences, Shinshu University School of Medicine, 3-1-1 Asahi, Matsumoto, Nagano 390-8621, Japan

Kazuto Tsukita
Laboratory of Biological Science, Graduate School of Frontier Biosciences, Osaka University, Suita, Osaka 565-0811, Japan
Department of Neurology, Graduate School of Medicine, Kyoto University, Kyoto 606-8507, Japan

Tomoki Yano
Laboratory of Biological Science, Graduate School of Medicine, Osaka University, Suita, Osaka 565-0811, Japan

Atsushi Tamura and Sachiko Tsukita
Laboratory of Biological Science, Graduate School of Frontier Biosciences, Osaka University, Suita, Osaka 565-0811, Japan
Strategic Innovation and Research Center, Teikyo University, Tokyo 173-8605, Japan

Shinsaku Tokuda
Department of Respiratory Medicine, Graduate School of Medicine, Kyoto University, Kyoto 606-8507, Japan
Division of Nephrology and Hypertension, Department of Internal Medicine, University of Kansas Medical Center, Kansas City, KS 66160, USA

Alan S. L. Yu
Division of Nephrology and Hypertension, Department of Internal Medicine, University of Kansas Medical Center, Kansas City, KS 66160, USA

Lorenza González-Mariscal, Helios Gallego-Gutiérrez, Laura González-González and Christian Hernández-Guzmán
Center for Research and Advanced Studies (Cinvestav), Department of Physiology, Biophysics and Neuroscience, Mexico City 07360, Mexico

Annalisa Ziemens, Markus Bleich and Nina Himmerkus
Institute of Physiology, Christian-Albrecht-University Kiel, Hermann-Rodewald-Str. 5, 24118 Kiel, Germany

Svenja R. Sonntag
Institute of Physiology, Christian-Albrecht-University Kiel, Hermann-Rodewald-Str. 5, 24118 Kiel, Germany
Department of Ophthalmology, University of Lübeck, Ratzeburger Allee 160, 23538 Lübeck, Germany

Vera C. Wulfmeyer
Institute of Physiology, Christian-Albrecht-University Kiel, Hermann-Rodewald-Str. 5, 24118 Kiel, Germany
Department of Nephrology and Hypertension, Hannover Medical School, Carl-Neuberg-Str. 1, 30625 Hannover, Germany

Bayram Edemir
Department of Internal Medicine IV, Hematology and Oncology, University Hospital Halle, Ernst-Grube-Str. 40, 06120 Halle (Saale), Germany

Zachary M. Slifer and Anthony T. Blikslager
Department of Clinical Sciences, College of Veterinary Medicine, North Carolina State University, Raleigh, NC 27607, USA

Judith Radloff
Institute of Veterinary Physiology, Department of Veterinary Medicine, Freie Universität Berlin, 14163 Berlin, Germany
Department for Biomedical Sciences, University of Veterinary Medicine, 1210 Vienna, Austria

Valeria Cornelius and Salah Amasheh
Institute of Veterinary Physiology, Department of Veterinary Medicine, Freie Universität Berlin, 14163 Berlin, Germany

Alexander G. Markov
Department of General Physiology, St. Petersburg State University, Universitetskaya nab. 7/9, 199034 St. Petersburg, Russia

Ajaz A. Bhat, Najeeb Syed, Sabah Nisar, Sheema Hashem and Santosh K. Yadav
Division of Translational Medicine, Research Branch, Sidra Medicine, Doha 26999, Qatar

Lubna Therachiyil
Translational Research Institute, Academic Health System, Hamad Medical Corporation, Doha 3050, Qatar
Department of Pharmaceutical Sciences, College of Pharmacy, QU Health, Qatar University, Doha 2713, Qatar

Muzafar A. Macha
Department of Biotechnology, Central University of Kashmir, Ganderbal, Jammu and Kashmir 191201, India
Department of Biochemistry and Molecular Biology, University of Nebraska Medical Center, Omaha, NE 68198, USA

Roopesh Krishnankutty and Shahab Uddin
Translational Research Institute, Academic Health System, Hamad Medical Corporation, Doha 3050, Qatar

Shanmugakonar Muralitharan and Hamda Al-Naemi
Laboratory Animal Research Center, Qatar University, Doha 2713, Qatar

Puneet Bagga and Ravinder Reddy
Center for Magnetic Resonance and Optical Imaging, Department of Radiology, Perelman School of Medicine at the University of Pennsylvania, Philadelphia, PA 19104, USA

Punita Dhawan
Department of Biochemistry and Molecular Biology, University of Nebraska Medical Center, Omaha, NE 68198, USA

Anthony Akobeng
Department of Pediatric Gastroenterology, Sidra Medicine, Doha 26999, Qatar

Michael P. Frenneaux
Academic Health System, Hamad Medical Corporation, Doha 3050, Qatar

Wael El-Rifai
Department of Surgery, University of Miami Miller School of Medicine, Miami, FL 33136, USA

Mohammad Haris
Division of Translational Medicine, Research Branch, Sidra Medicine, Doha 26999, Qatar
Laboratory Animal Research Center, Qatar University, Doha 2713, Qatar

Mariana Castro Dias, Josephine A. Mapunda, Mykhailo Vladymyrov and Britta Engelhardt
Theodor Kocher Institute, University of Bern, 3012 Bern, Switzerland

Wei-Qi He, Jing Wang, Jian-Ying Sheng and Juan-Min Zha
Jiangsu Key Laboratory of Neuropsychiatric Diseases and Cambridge-Suda (CAM-SU) Genomic Resource Center, Medical College of Soochow University, Suzhou 215123, China

W. Vallen Graham
Thelium Therapeutics, New York, NY 10014, USA
Laboratory of Chemical Biology & Signal Transduction, The Rockefeller University, New York, NY 10065, USA

Jerrold R. Turner
Laboratory of Mucosal Barrier Pathobiology, Department of Pathology, Brigham and Women's Hospital and Harvard Medical School, Boston, MA 02115, USA

Takayuki Kohno, Takumi Konno and Takashi Kojima
Department of Cell Science, Research Institute for Frontier Medicine, Sapporo Medical University, Sapporo 060-8556, Japan

Index

www.ingramcontent.com/pod-product-compliance
Lightning Source LLC
LaVergne TN
LVHW081413160926
848178LV00067B/95
* 9 7 8 1 6 4 7 4 0 3 6 5 2 *